MATERIALS SCIENCE RESEARCH
Volume 11

PROCESSING OF CRYSTALLINE CERAMICS

MATERIALS SCIENCE RESEARCH

A Continuation Order Plan is available for this series. A continuation order will bring delivery of each new volume immediately upon publication. Volumes are billed only upon actual shipment. For further information please contact the publisher.

MATERIALS SCIENCE RESEARCH • Volume 11

PROCESSING OF CRYSTALLINE CERAMICS

Edited by

Hayne Palmour III
R. F. Davis
and
T. M. Hare

North Carolina State University
Raleigh, North Carolina

PLENUM PRESS · NEW YORK AND LONDON

Library of Congress Cataloging in Publication Data

University Conference on Ceramic Science, 14th, North Carolina State University, 1977.
Processing of crystalline ceramics.
(Materials science research; 11)
Includes index.
1. Ceramics—Congresses. I. Palmour, Hayne. II. Davis, Robert Foster, 1942-
III. Hare, Thomas M., 1943-   IV. North Carolina. State University, Raleigh.
Title. VI. Series.
TP785.U53 1977   666   78-18441
ISBN 978-1-4684-3380-7   ISBN 978-1-4684-3378-4 (eBook)
DOI 10.1007/978-1-4684-3378-4

Proceedings of the Fourteenth University Conference on
Ceramic Science held in Raleigh, North Carolina, November 7–9, 1977

©1978 Plenum Press, New York
Softcover reprint of the hardcover 1st edition 1978
A Division of Plenum Publishing Corporation
227 West 17th Street, New York, New York 10011

PREFACE

This volume constitutes the Proceedings of the November 7-9, 1977
Conference on PROCESSING OF CRYSTALLINE CERAMICS, held at North
Carolina State University in Raleigh. It was the Fourteenth in a
series of "University Conferences on Ceramic Science" initiated in
1964 and still coordinated by a founding group of four ceramic-
related institutions, of which North Carolina State University is a
charter member, along with the University of California at Berkeley,
Notre Dame University, and the New York State College of Ceramics at
Alfred University. In addition, two other ceramic-oriented schools,
the University of Florida and Case-Western Reserve University, have
also hosted Conferences in the series. These research-oriented
conferences, each uniquely concerned with a timely ceramic theme,
have been well attended by audiences which typically were both international and interdisciplinary in character; their published
Proceedings have been well received and are frequently cited.

This three day conference was concerned with (a) scientific
aspects of all process steps which must be combined and controlled
effectively and sequentially in producing crystalline ceramics (both
oxides and nonoxides), and (b) utilization of these principles in
developing processes for several classes of advanced ceramics critical
to present and future technology.

The importance and relevance of ceramic processing has expanded
considerably during the last two decades; this expansion has been
intensified by (a) the advent or improvement of a wide variety of
modern dielectric, magnetic, electronic, nuclear and structural
ceramics; (b) hitherto unanticipated needs for, and increasing
availability of, high purity ceramic materials in powder, dense poly-
crystal and single crystal forms; (c) evolution of fully automated -
and most recently, computer controlled - ceramic manufacturing processes and (d) increasingly urgent needs for processing both quasi-
traditional and exotic ceramic materials in unusual sizes and
geometries which must be capable of withstanding in-service environmental extremes. As recent examples, one can cite novel materials
and unusual processing techniques being developed for critical ceramic
components in a variety of energy-related systems, including nuclear

fuels and control components, high temperature gas turbine engines, coal gasification and MHD energy extraction systems, and fast-ion conducting electrodes for high temperature batteries designed for vehicular and load-leveling applications.

Since the whole field of ceramics is so broad, and the kinds of processing required at the extremes of that field are so different, it was necessary in planning the Conference (and in envisioning a resultant, rather concise, single volume Proceedings) to limit its scope. For these reasons, the specific theme, PROCESSING OF CRYSTALLINE CERAMICS, was designed (1) to include virtually all forms of polycrystalline ceramics that are processed directly from particulates (and/or other solid or vapor states), and (2) to exclude other, rather different processes employed primarily for (a) growth of single crystals and (b) melt-formed ceramics such as glasses, devitrified glass-ceramics, glazes and enamels. Processing of single crystals and of glasses has been well covered in the literature; each subfield has tended to generate its own theme oriented symposia generally similar to this one.

The 54 individual papers included in these Proceedings embrace all the principal stages of processing of oxide and nonoxide ceramics with emphasis upon their interactions and interrelationships, thus providing a modern synthesis of the processing aspects of many of the novel ceramic materials now employed or under consideration for important technological areas noted above. Thematically, the book is divided into six sections. The first three are principally concerned with particular aspects of the overall process sequence: (1) Fine Particle Science, Technology and Characterization, (2) Solid State Sintering and Grain Growth and (3) Liquid Phase Sintering and Post Firing Technology. The last three sections focus attention on applications of current processing science and technology on representative classes of ceramics for specific types of applications: (4) Dielectric and Magnetic Ceramics, (5) Energy Related Ceramics I: (a) Fast Ion Conductors and (b) MHD, Refractory and Nuclear Ceramics and (6) Energy Related Ceramics II: Non-Oxide Ceramics, with emphasis on structural applications.

The List of Contributors formally acknowledges the considerable cooperation and assistance rendered the Co-Chairmen by (1) a distinguished Advisory Committee, (2) the Conference Staff and (3) the several Session Chairmen, as well as (4) the creative efforts of one hundred distinguished contributing authors representing many of the world's ceramic research centers. We extend our personal thanks to all of them collectively and individually, for their cooperative attitudes, timely responses and many helpful suggestions which have characterized all our working relationships with them.

On behalf of the participants and the ceramic community, we gratefully acknowledge enabling and indispensable financial support

provided for this Conference from Federal sources by the Army
Research Office, the Department of Energy and the Office of Naval
Research, and from University sources by the School of Engineering,
North Carolina State University.

We also acknowledge support of, and participation in, many of
the Conference activities by administrators and members of the
faculty and staff at North Carolina State University. Special
thanks are due to Dr. R. E. Fadum, Dean of Engineering, who welcomed
attendees to the Conference, to Dr. E. G. Droessler, Administrative
Dean for Research, who presided at the Conference Dinner and to
Dr. G. H. Elkan, Professor of Microbiology, who presented a timely,
lively -- and energy-related -- Dinner address entitled "The World
Protein Shortage".

In an undertaking of this magnitude, the prospects of a favorable outcome really rest upon the skills, enthusiasms, and efforts
of a modest group of persons who work with dedication, but largely
behind the scenes. We wish to acknowledge our very special personal
thanks to Dr. Bruce Winston for coordinating our Conference, to
Ann Royster, Ann Ethridge, and Karen Ward for their secretarial
assistance before, during and after the Conference; and to D. B.
Stansel and Jane E. Hodge for assistance in adapting the excellent
physical facilities of the McKimmon Center to the specific needs of
this Conference. We also acknowledge our endebtedness to Mary N.
Yionoulis for publicity, to M. L. Huckabee for serving as coordinator
of projectionists and on occasion, as photographer; to Materials
Engineering students Calvin Carter, Jeff Church, Tim Early, Jim
Hodson, Jerry Moore, Elizabeth Reynolds and Clifford Soble who
served as pages and as projectionists, and to Peggy King and Pam
Wilson for assistance during the Conference. We pay special tribute
to Ann Ethridge for her faithful assistance to the Editors throughout
the undertaking, and for her special skills and experience in typing
and/or revising the edited Proceedings. Finally, it is appropriate
to acknowledge with real affection the patience, tolerance, and
tangible and moral support we have been accorded by our colleagues,
our families and our friends through those extended periods of time
we have had to commit to planning, organizing and editing.

Hayne Palmour III
Robert F. Davis
Thomas M. Hare

Raleigh, N. C.
July, 1978

CONTENTS

SOME CONSIDERATIONS IN THE STATE OF THE ART IN PROCESSING

CRYSTALLINE CERAMICS

M. G. McLaren and W. R. Ott

Rutgers University

New Brunswick, N. J.

INTRODUCTION

During the race for the moon in the 1960's it became readily apparent that the engineering properties of materials were critical to the development of space flight systems. In this period Government agencies, such as the Department of Defense (DOD) and the National Aeronautical and Space Administration (NASA), developed sophisticated systems approaches to achieve the ultimate goal of manned space flight. From these systems came the realization that more reliable materials had to be developed with more carefully controlled physical properties. The three classes of materials; metals, polymers and ceramics were reassessed for possible utilization in space systems. Ceramic materials because of their unique properties thus became a target for the most intense study to date. Though many ceramic processing techniques had already been developed industrially they were reexamined to determine what variations in raw materials and processing could produce a more useful material.

The criticality of ceramics to the National mission provided the incentive for the National Materials Advisory Board to begin a wide sweeping study of ceramics and in particular on "the processing of ceramics." During the early 1960's initial reports were issued by the NMAB[1] indicating the critical areas for requisite research and development on Ceramic Processing. Based on these reports a detailed analysis of the State of the Art on ceramic processing was developed by the NMAB[2] and was published by the National Academy of Sciences in 1968. Dr. Joseph A. Pask chaired this committee and the list of committee members reads like the Who's

Who of the ceramic and material science world. Never before had
so many talented people been brought together to focus their
attention on the characterization of ceramic processing and its
effect on physical properties. As a result of this intense study
the gaps in the technology of ceramic processing were identified.

The results of this study showed the need for characterization
throughout the forming process, commencing with the raw materials,
the microstructure of the "green" compact, the microstructure of
the sintered ceramic, and finally how these structures related to
the engineering properties.

One of the most significant consequences of this study was the
fact that for the first time a compilation and evaluation of
ceramic processing techniques was available for the education of
ceramic engineering students. This book became widely used in most
of the universities teaching ceramic engineering and materials
science. The relationship of characterization, microstructure, and
properties permeated ceramic engineering education as never before.
Though this was a more intangible result of the 1968 ceramic processing evaluation it has had a great long term beneficial effect.

It is not possible for the purpose of this paper to develop a
complete State of the Art in ceramic processing as it currently
exists. The monumental effort by NMAB committee and the number of
man years that it took to develop their report cannot be duplicated
by these authors. The goal of this paper will be to highlight
several advances in ceramic processing and describe the evolution
of the characterization tools which may have influenced these
advances since 1968. Several ceramic materials which are widely
used industrially will be used to illustrate these advances.

CHARACTERIZATION

The ability to characterize starting raw materials, the effect
of the forming process and the effect of the firing conditions on
the final product has been substantially improved in the past few
years. The development of analytical tools which are highly
accurate, precise, and rapid has had a great impact on the advances
in ceramic processing technology.

In the transfer of process technology from research and
development to the production stage, difficulties frequently arise.
These problems stem from uncontrolled and unidentified process
variables. These may be as a result of raw material variations,
process variations or their interactions, and the results can
affect virtually any of the final physical properties. The ability
to control ceramic processes is therefore frequently dependent not

only on engineering skill but on the availability of powerful
characterization tools to determine the nature of the process controlling variables. Consequently, when the reasons for processing
advances are discussed it is critical to analyze the role of
characterization equipment and techniques.

The range of characterization equipment and techniques is very
large but certain major pieces of equipment have played a substantial
role in characterizing ceramic processes. For the purposes of this
discussion consideration will be given to equipment for determining
the following characteristics:

A. Particle Size
B. Chemical Analysis
C. Thermal Analysis
D. Microstructural Analysis
E. Surface Analysis
F. Mineralogical Analysis

In each of these areas of analysis we will consider whether
there are major new capabilities in the field or whether there is
greater resolution, greater speed in the measurement or analysis,
or lower cost.

A. Particle Size Measurement

Davis[3] recently listed fifteen different physical principles
that can be employed to measure particle size. The choice of
technique employed is largely dependent on the nature of the data
needed: (i.e., particle size distribution or average grain size
data). Most, if not all, of these principles were known prior to
1963. The frequently used Coulter Counter was introduced in 1956.
The development of high speed instruments using X-ray counting
techniques, such as the Sedigraph, allow for a more rapid and precise determination when compared to the standard sedimentation
techniques.

B. Chemical Analysis

1. X-ray Spectroscopy[4] - the principles have long been known,
and since the early 1950's, the method has been in wide practical
use. The major difficulty with X-ray spectorscopy has been the
difficulty in quantitatively interpretating the data. The linking
of this unit first with the computer and later with microprocessors
has greatly increased its power.

2. Atomic Absorption - was developed in 1955[5] and has as its

major use the determination of trace constituents.

    3. Gas Chromatography - was in use well before 1960 as a
technique to identify organic compounds. This capability has been
important in dealing with organic additions to ceramics, firing
atmospheres, and air pollution problems arising from firing the
ceramics. The principal difficulty with the instrument has been in
the interpretation of the results. To remedy this, gas chromatography has been used in conjunction with mass spectrometry[6] to
assist in the identification of species. The linking of mass
spectrometers with microprocessors greatly increased the power of
these units. Recently, gas chromatography has been linked with
infra-red spectroscopy and nuclear magnetic resonance, providing an
even greater analytical capability.

C. Thermal Analysis

    Differential thermal analysis was developed in 1890[7] with wide
use of the technique beginning in the 1950's. The 1960's saw the
introduction of elaborate commercial units. Differential Scanning
Calorimetry (DSC) was developed in 1964 by Watson and O'Neill. The
major emphasis in improving this technique has been to develop
methods to quantify the data.

    Thermogravimetric analysis (TG) units resembling those avail-
able today were in use in the early 1900's.[8]

    Dilatometry, while a mature technique, has also seen a growth
in sophisticated instrumentation.

D. Microstructural Analysis

    1. Optical Microscopy - is a mature technique which has been
on the decline in recent years. McCrone[9] expressed his concern
over the continued decline in microscopy courses at Universities.

    2. Transmission Electron Microscopy - reached commercial real-
ization in 1939.[10] By 1960 the first one million volt unit had been
developed in France. The applicability of this technique to ceramic materials was greatly enhanced with the development of commercial ion thinning devices. Recently, Scanning TEM has become a
reality because of the development of a reliable cold field emission
electron gun.

    3. Scanning Electron Microscopy (SEM) - has been a "work
horse" in the development of process technology. The SEM became
commercially available in 1965 and if any technique can be singled

out to have influenced process technology it must be the SEM.
Coupled with either X-ray wavelength spectroscopy or energy dispersion techniques, chemical analysis as well as microstructural
features can be determined.

Other techniques such as Holographic Microscopy[11] have been
developed but have seen only limited utilization.

E. Surface Analysis

1. ESCA - Electron spectroscopy chemical analysis has been
known for 60 years. Major breakthroughs in the late 1950's made it
a useful technique.[12]

2. Auger Electron Spectroscopy (AES) - allows for the study of
surface contamination, and grain boundary segregation.[13] The
technique was described by Auger in 1925. However, until 1968 when
Weber and Peria adapted LEED (low energy electron diffraction) to
AES it saw little use. Commercial instruments have become available in the 1970's.

3. Low Energy Electron Diffraction (LEED) - was already in
widespread use prior to 1968.[13]

While much of the future of process technology will rest with
an understanding of ceramic surfaces these techniques have not yet
made a substantial impact. These units and other adaptations of
existing technology such as the Raman Microprobe[14] developed by
Rosasco and Etz at the National Bureau of Standards will no doubt
be cited in future years for their importance in Ceramic Processing.

Detection Limits

The detectability limits of analytical instruments has been
steadily improving.[14] Instrumental methods are capable of micro-
gram and nanogram limits while selected methods have demonstrated
picogram (10^{-12} g) and femtogram (10^{-15} g) limits. Each method,
of course, has a range of sensitivities that depend on the compound being studied and the procedure used. The limits have been
reviewed by Karasek[14] for a large variety of instruments.

Process technology has grown at a rate directly proportional
to the development of characterization equipment. As we have
developed the skills to analyze the causes of process variation we
have advanced in our ability to process ceramic materials.

Mature techniques such as particle size measurements, X-ray

fluorescence, atomic absorption, gas chromatography and thermal analysis have grown in the last fifteen years. The advances have been in greater resolution, greater speed in the measurement and greater speed in the analysis of the data. These have come about largely from the commercialization of the equipment. These advances have come, of course, at an increase in the expense of the instrumentation.

Developing techniques such as SEM have come in during the period of commercialization of instruments and hence are available as complete units. Their development has been significant in processing improvements.

Surface analysis will undoubtedly lead the way to future ceramic processing improvements. As the instrumentation has wider distribution its impact on ceramic processing will gradually expand.

Thus using these new or modified tools it is possible to examine more closely the effect of ceramic processes on microstructure and this has enabled the ceramic industry to more closely approach the desired properties over the past fifteen years.

PROCESSING

The summary of Solids-Process Characteristics[2] developed by the Pask committee indicated that slip casting, extrusion, isostatic pressing, dry pressing, vibrational compaction, and high energy rate forming with subsequent sintering could be evaluated for what they termed configurations (dimensional control and shape), internal structure, and polycrystalline structure. These forming methods represented the main industrial capability available at that time. It also gave recognition, on a ranking basis, to the capability in 1968 for controlling ceramic materials by these processing techniques. In general, a wide range of engineering properties could be developed for any single oxide composition using the then existing range of forming and processing technology. From a review of their summary chart it is readily apparent that from 1968 until present there have been no quantum jumps in processing technology but rather there have been many "small breakthroughs" that yield improved products through a better understanding of the preparation of powders and their characterization. This applies to materials when processed by techniques already known in 1968. This is not to say that many unique approaches have not been defined, such as chemical vapor deposition, but that in general for the greatest part of the ceramic industry processing techniques have not changed substantially.

Inasmuch as it would be very difficult to follow many single oxide systems or mixed systems as was done in the 1968 NMAB report attention will be focused only on the change in technology of several materials that have wide utilization in industrial ceramics. Aluminum oxide, silicon nitride and silicon carbide will be reviewed and their progressive development over the past twenty years will be discussed.

A. Alumina

In the mid 1950's the major aluminum producing companies[15] realized the opportunities and indeed, the necessity for developing aluminum oxide raw materials which could be utilized in high technology ceramics. For years the aluminum oxides generated for metallurgical use had also served quite successfully in a wide range of refractory and insulator compositions. They had been used mainly in high temperature refractory applications for both sintered and fused cast alumina refractories and special electronic grades of lower soda content had been made for sparkplug insulators. The emerging requirement for suitable substrate materials of aluminum oxide with high density, fine grained, smooth surface characteristics demanded the development of lower soda aluminas with finer crystal sizes. By the 1960's manufacturing techniques were refined and new ones developed whereby more thermally reactive grades of low soda aluminas could be manufactured with a wide range of crystal sizes and alpha alumina contents. Although these were not available in large quantities, pilot plant quantities were generated and an evaluation of these materials was made. The major benefits were that the processes were capable of reducing the soda content to less than one-tenth of one percent, and the mean crystal size could be controlled to 4 microns. The availability of this aluminum oxide material gave a large impetus to the electronic ceramic industry for the production of substrate materials. In the early 1960's commercial quantities of these materials were made available to the industry and many other high grade technical ceramic products were made possible.

For the most sophisticated technical electrical ceramic[15,16] requirements the purity of the low soda content grade of aluminas was not sufficient nor was the mean particle diameter small enough. The surface requirement for the substrate materials was enormously demanding and only by glazing had the electronics industry been able to produce acceptable surfaces for their purposes. Thus an intense investigation began to further reduce the soda content and to control the manufacture of these low soda grades so as to yield very small mean particle diameters and it was in the mid 1960's that these reactive grades of alumina were developed. This excellent research and development enabled the major alumina companies to produce materials of quality sufficient for the

manufacture of surfaces less than 8 microinches as fired and consequently, it was not necessary to glaze the substrate. In order to
meet this requirement the low soda grade aluminas were characteristically: mean crystal sizes of 1.5 microns, surface areas of
3 m^2/g, and a purity of 99.7 percent.[16] As a result of this development the raw materials requirements of the electronics industry
became even more severe, creating demand for alumina raw materials
of the quality previously described but with a controllable particle
size distribution to allow accurate control of the fired characteristics of the material. This led to techniques for the blending
of aluminas such that predictable relationships could be established
between the surface area, particle size distribution, green bulk
density, and the fired density.[15]

By the early 1970's ultra high purity aluminas were developed
which had purities of 99.98% alumina and a mean crystal size of
approximately 0.5 microns and by the mid 1970's the purity was
increased to 99.99% alumina with a mean crystal size of 0.3 microns. These were available in commercial quantities for applications such as translucent envelopes and ultra smooth substrates.
Surface finishes of 1 to 2 microinches as fired were then possible.

To follow this evaluation reference is made to the review by
Williams of the impact of alumina raw materials on the evolution of
substrates.[16]

Because of the wide general use of alumina in the ceramic industry a wide range of alumina raw materials have been developed.
Many other ceramic oxide raw materials might have been chosen for
illustration such as zircon, zirconia, magnesia, or chromic oxide.
However, to follow the development of any of these materials for
use in highly technical ceramics would show the same trend of the
impact of characterization and the characterizing tools on process
control that was shown for the development of the alumina materials.

B. Silicon Carbide and Silicon Nitride

Silicon carbide and silicon nitride both have intrinsically
excellent properties to function in turbine applications. Their
main restrictive property is the brittleness exhibited by most
every ceramic material. Rice[17] and many others have concentrated
much of their research efforts on understanding the failure mechanisms and by process and compositional variations trying to overcome
the design limitations of ceramics in high temperature turbine engines.

Because of the enormous interest in silicon nitride and silicon carbide for turbine engine application, considerable effort has
been expended to develop new processes for manufacturing "reliable"

parts with these materials. A review of these efforts is necessary
in this discussion because it will point out the imagination and
wide range of process variation that have been used to develop
suitable materials.

Silicon carbide has been a very useful refractory material
since the turn of the century. The main problem with its early use
was the fact that it could not be made nonporous. During use at
higher temperatures the operating atmosphere will permeate the refractory, and there is a growth of the refractory composite and a
gradual degradation of the properties.

Many types of bonds[18] have been developed in order to overcome
the problem of porosity. The oldest and most generally used for a
wide range of general purposes was the silicate bond but this limited the refractoriness and did not completely eliminate the porosity. Silicon nitride bonding was achieved and this was an improvement, but the bond was less refractory than the silicon carbide and
still made a permeable refractory. A good density was achieved by
bonding with silicon and forming by hot pressing techniques; however, this was difficult to apply to commercial production. The
advent of the "self-bonded silicon carbide" material by Taylor[18]
permitted the manufacture of dense silicon carbide parts. In general, this material was manufactured by the infusion of silicon
vapor into a mass of silicon carbide crystals which had been pressed
with a carbonaceous interstitial material. The reaction of the
silicon vapor with the carbonaceous material produced very fine
silicon carbide crystal in the interstices and produced a dense
mass.

Chemical vapor deposition techniques were developed in which
the reaction of silane materials with methane under pressure at
elevated temperatures yielded deposition of silicon carbide on a
heated substrate of graphite. The development of both of these
types of processes led to an examination of the use of very dense,
fine grained silicon carbide as a turbine material.

The most significant characteristic of silicon carbide formed
by CVD processes is that it is essentially dense and the mean crystal size of the mass is perhaps several orders of magnitude
smaller than silicon carbide made by other bonding methods. The
great interest on the part of both government agencies[17] and the
automotive industry is obvious and considerable research effort is
being spent on these materials.

An examination of the Department of Defense DOD R&D programs
active from July 1975 through June 1977 was made by Battelle's
Metals and Ceramics Information Center.[19,20] Reviewing this report,
it is obvious that silicon nitride and silicon carbide have

received great emphasis as potential turbine materials that processing techniques for them have improved remarkably. Again it is obvious that a high reliance on sophisticated characterization tools is implicit in the programs.

Rice[17] has described the ceramic turbine materials needs with relation to silicon carbide and silicon nitride. He lists the primary environmental conditions for which the material must have high resistance, such as: low temperature fracture from normal short term stressing, thermal shock, particle impact, creep and high temperature strength, oxidation, and corrosion. With respect to these conditions the current state of the art indicates that CVD processes may hold the greatest potential for producing materials of usable quality for turbine environments. It is concluded that at the present time only short life engines of small size such as used in missiles, torpedoes, etc., operating within the range 2200-2500°F and having lives of 50 hours or less can be considered at all practical. Other processing techniques are suggested but are unevaluated at present.

C.   Other Ceramic Fields

Up to this point in the review considerable discussion has been given to the highly demanding technical requirements of materials for the electronics and turbine manufacture and it seems that recognition of the fact that materials and processes for the great bulk of the ceramics industry must be mentioned as well; industries such as sanitary ware, refractories, glass, structural clay products where enormous tonnages of raw material are used yearly. The leaders[21] of research in these industries have commented that there have not been any major breakthroughs in materials or processing in the past ten years. Rather it has been the constant attention to detail and improved characterization techniques that have permitted higher temperatures, greater thermal shock resistance and improvement in all of the ceramic properties that must be designed into the systems. Again the theme is repeated by each segment of the ceramic industry, that the ability to characterize, hence the ability to control the process through the use of better characterization tools is what is permitting better economics and more useful products.

The work of Phelps[22] and others in the mineral and whitewares industries and the instruments that they now use to characterize materials and processes have brought about much better selection of raw materials and process controls. Phelps' paper which was cited as the best paper in the Whitewares literature in 1976 by the Whitewares Division of the American Ceramic Society on Reformulation Techniques in Whitewares is truly dependent upon his ability to

characterize accurately the mineral content, the chemical composition, the particle size distribution and the surface characteristics of the particles.

The structural clay products, the cement industry and other
segments of the ceramic industry which command a high percentage
of the ceramic industry dollars all repeat the same message;
characterization vs properties vs product acceptability.

Probably the most innovative process that has been developed
in the last few years has been the Pilkington Process[23] for the
production of plate glass. This process represented a significant
and major breakthrough in process technology and it is not probable
that there will be comparable process breakthrough in the ceramic
industry in the near future.

FUNDING FOR CERAMIC RESEARCH

In the 1963 NMAB report by the ad hoc committee on Processing
of Ceramic Materials, specific recommendations were made as to the
funding of research in the area of ceramic processing.

A total of $3,850,000 was recommended with a requisite of 82
man years of effort. The budget contained provisions for $2,100,000
for the area of science and technology limiting ceramic processing,
$1,350,000 for processes requiring empirical development, $400,000
for programs for size scale-up and quantity production, and a one
time equipment purchase of $500,000. It was conceded that the
first category was long range and should be funded over a long period of time - up to five years.

Based upon the National Science Foundation's report[24] the
total funding for research and development in 1963 was almost 24
billion dollars, thus the expenditure for research on ceramic processing was approximately 0.016% of the nations research efforts.
When viewed with respect to the fact that there are really only
three classes of materials for design, i.e., ceramics, metals, and
polymers, it seems to be an almost insignificant amount to spend on
such an important problem.

It is very difficult to determine if indeed even the recommended
level of spending occurred but an estimate of the current total
expenditure may be made.

Currently the Energy Research and Development Administration
(ERDA) is funding about two-thirds of the energy related federally-
supported materials research or about $175 million dollars in fis-
cal 1976 of which $43 million dollars was basic research support.[25]

To estimate the total federal expenditure on materials is very difficult and the Committee on Materials (COMAT) of the Federal Council for Science and Technology found it difficult to assess this question even with tens of people and a one year effort and the final report has not yet been issued. There is difficulty in trying to determine what the total ERDA funding is on ceramic processing because the definition of "ceramics" and "processing" are different things to different people.[26]

The breakdown of $43 million dollars spent on Material Sciences in 1976 indicates 5.92 million dollars spent on ceramics or 13.8% of the material expenditure. The estimated expenditure on Ceramic Processing was 0.9 million dollars or approximately 0.5% of the total ERDA materials research.

The National Science Foundation recognized the importance of the ceramics effort and in 1974 awarded support under the newly formed Ceramics Program. As of July, 1976 which was comprised of essentially fiscal years 1975 and 1976 the total expenditure for Ceramics, glass, hard materials equipment and conferences was approximated 4.2 million dollars.[27] When the expenditure was classified by phenomena it was evident that only 1.1 million dollars or 20.8% were allocated to synthesis, characterization and processing. In a subsequent report by NSF it was shown that processing received 5.9% of the program funding or 0.26 million dollars. Assuming that 60% of money was 1976 allocation then total expenditures of NSF was 0.16 million dollars for ceramic processing. The combined NSF and ERDA expenditure on ceramic processing would then be approximately 1.1 million dollars.

Based on constant dollars, i.e., based on the GNP implicit price deflator (1972 dollars)[24] the recommendation by NMAB committee in 1963 of 3.85 millions dollars, and assuming a constant requirement over the years to the present, this would translate into a 1976 requirement of 5.43 million dollars spent on the science and technology of ceramic processing. Allowing for the inflation during the period 1963 to 1972 it can be safely estimated that 8.0 million dollars should have been appropriated for the same level of support as recommended in 1963.

CONCLUSION

From even these rather brief calculations it is plainly obvious that we are no where near the required support level for effective resolution of the understanding of Ceramic Processing.

In response to the next obvious question of whether the same rate of expenditure is necessary today, referral is made to the study conducted by the NSF's Metallurgy and Materials Section Advisory Panel.[28] This provided a statistical analysis of the assessment of various aspects of ceramics as viewed by scientists and engineers working in the field of ceramics. This report, issued in July of 1977, has revealed the disappointing result that the scientists contacted overwhelmingly felt that the level of the state of ceramic science is far behind that of solid state physics, solid state chemistry, and metallurgy. Obviously, to correct this disparity in technology increased effort by government agencies should be given with the subsequent allocation of appropriate levels of support for research.

When asked about the estimated importance for fruitful fundamental ceramic research on processing the response was a 3.82 out of a possible 4.0. Thus even though other areas of ceramic research had higher ratings of importance it is shown positively that processing has significant requirements for future funding.

Materials hold the key to unlocking the requirements for our future systems in all areas of development including our most essential current problem of energy related systems. The need to develop new processing methods for ceramics, to be able to understand them, characterize them, and finally utilize them, requires commitment. Let us hope that this great Nation understands the necessity for this commitment better than other nations or we might be at a serious economic and military disadvantage.

REFERENCES

1. Report of the Ad Hoc Committee on the Processing of Ceramic Materials, Nat. Acad. Sic., Washington, D. C., 1963.
2. Ceramic Processing, National Materials Advisory Board, Nat. Acad. Sic., Washington, D. C., 1968.
3. R. Davis, Am. Lab., 5 (12) 12-23 (1973).
4. R. W. Gould, Am. Lab., 6 (7) 12-23 (1974).
5. J. Y. Hwang and G. P. Thomas, Am. Lab., 6 (8) 55-63 (1974).
6. Am. Lab., 1 (11) 68-69 (1975).
7. W. W. Wendtland, Am. Lab., 9 (1) 59-65 (1977).
8. W. W. Wendtland, Am. Lab., 9 (6) 25-31 (1977).
9. W. C. McCrone, Am. Lab., 7 (4) 11-16 (1975).
10. T. W. Drummond, Am. Lab., 8 (4) 82-95 (1976).
11. D. Cox, Am. Lab., 7 (4) 17-20 (1976).
12. W. A. Biers, Res. Dev., 11, 18-24 (1975).
13. F. W. Krasek, Res. Dev., 25 (10) 48-58 (1974).
14. F. W. Karasek, Res. Dev., 7, 20-23 (1975).
15. D. R. Watson, "Historical Review of Reynolds Low Soda Ceramic Aluminum Development", Reynolds Aluminum Research Division, 1976.

16. J. C. Williams, Am. Ceram. Soc. Bull., 56 (6) 58-85 (1977).
17. R. W. Rice, "Overview of the Naval Research Laboratory Ceramic Turbine Materials Program", Naval Research Laboratory, 1977.
18. K. M. Taylor, Mat. Meth., 10, 192-95 (1956).
19. D. Niesz and R. R. Wills, DOD, R & D, Ceramic Processing Programs, Metals and Ceramics Information Center, Battelle, 1977.
20. A. M. Diness, private communication, 1977.
21. M. Van Dreser, private communication, 1977.
22. C. W. Phelps, J. Am. Ceram. Soc., 55, 528-29 (1976).
23. L. A. B. Pilkington, Proc. Roy. Soc. Lond. A., 314, 1-25 (1969).
24. National Science Foundation, NSF 77-310.
25. Material Science Programs, FY 1976, United States Energy Research and Development Administration, 1975.
26. D. W. Readey, private communication, 1977.
27. B. A. Wilcox, Ceramics Programs, NSF, 1976.
28. R. S. Gordon, R. Roy, M. J. Sinnot, J. B. Wachtman and B. A. Wilcox, Report on Responses to a Ceramic Science Questionnaire, Div. Mat. Res., NSF, Washington, D. C., 1977.

ACKNOWLEDGEMENT

The authors wish to acknowledge those who so graciously assisted them in their work. Special thanks are given to Bill Prindle, George Economas and Don Graves at the Materials Advisory Board.

They are also grateful to the following gentlemen: Dennis Readey, Roy Rice, Arthur Diness, Winston Duckworth, Victor Tennery, Thomas Scarano, Bob Katz and Dennis Viechnicki, John Hurt, James Gangler, Henry O'Bryan, Neil Ault, James R. Johnson, Dan Burn, Peter Schultz, Dr. Bement, and William Van Hees.

The authors are especially grateful to Ms. Celeste Brandmayr for her work in compiling the data used in this paper.

PART I

FINE PARTICLE SCIENCE,
TECHNOLOGY AND CHARACTERIZATION

PHYSICAL AND CHEMICAL PARAMETERS CONTROLLING THE HOMOGENEITY OF

FINE GRAINED POWDERS AND SINTERING MATERIALS

Max Paulus

Laboratoire d'Etude et de Synthese des Microstructures

CNRS - ESPCI, 10 rue Vauquelin - 75231 Paris Cedex 05

The necessity of very homogeneous, fine grained powder for controlled grain growth and full density sintering is evaluated on the
basis of Kirkendall effect and the local inhibition or acceleration
of crystal growth. Furthermore, physical and chemical interactions
along the different stages of processing on ceramics may limit the
control of composition, homogeneity, porosity and grain size of the
final product. After a short review of the numerous methods which
have been established in order to improve homogeneity of the powders,
the cryogenic method is considered in detail.

INTRODUCTION

It is now well-known that the microstructures of sintered products
depend strongly on the characteristics of the initial powder.[1]
Generally considered are the particle size distribution of the powder,
the shape of the grains, surface oxidation (in case of metals),
absorbed gases, impurities and crystalline defects. These parameters
are important irrespective of the material. However the chemical
and granular homogeneity are the main parameters, in the case of
multicomponent systems, which must be considered in order to achieve
very low post-sintering porosity and maintain very fine grain size
(sometimes less than a micron). We shall see that heterogeneity of
composition or particle size distribution can oppose complete densification and cause discontinuous grain growth. Moreover, certain
ceramics require continuous adjustment of their degree of oxidation,
as a function of temperature. This is achieved by modifying the
partial pressure of oxygen in the treatment atmosphere, thus avoiding
the introduction of a heterogeneity having a redox origin.[2] Finally,
it is often necessary to prevent the expulsion of certain volatile

elements such as lead in PLZT and lithium and zinc in ferrites.
Hence, relatively low temperature sintering is required to avoid any
surface heterogeneity. This is also valid for alloys.

These observations point to the need for knowledge concerning
the preparation of homogeneous submicroscopic powders with perfectly
defined composition in cations and anions, and sintering at low
temperature. However, before passing on to an examination of the
different powder preparation methods, we shall analyze the role
played by chemical and particle size heterogeneity.

CONSEQUENCES OF SINTERING OF A POWDER WITH HETEROGENEOUS COMPOSITION AND PARTICLE SIZE DISTRIBUTION

Homogeneity is obviously desirable to obtain optimum properties
of the material. However, this aspect of the problem is usually
secondary, if the only important point is the final chemical composition, as the latter can always be improved by suitable heat
treatment, provided that grain growth is acceptable. However, heterogeneities of composition, even slight ones, oppose complete densification of the material and can lead to discontinuous grain growth.
In effect, since the diffusion coefficients of the different elements
are not identical, a Kirkendall effect with migration of initial

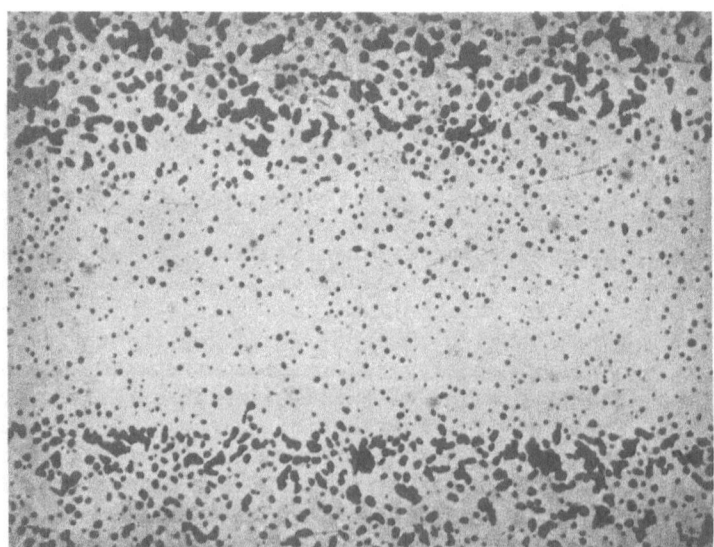

Figure 1. Porosity migration by Kirkendall effect in a manganese-
zinc ferrite containing iron rich heterogeneous zones, due
to settling in the suspension of the mixture.[10]
Magnification x250.

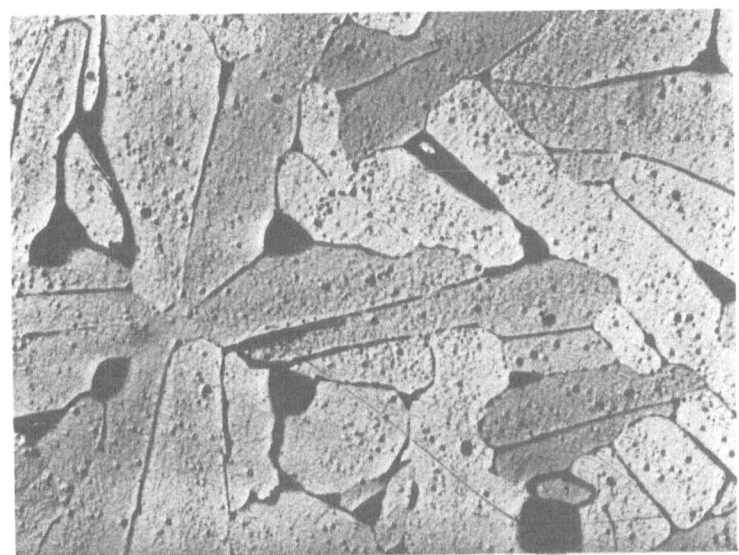

Figure 2. $BaFe_{12}O_{19}$ after one hour at 1400°C in oxygen.   Local excess
of BaO initiates discontinuous grain growth.
Magnification x250.

Porosity and pore coalescence[3,4] is generally observed (Fig. 1) during
homogenization.   Furthermore, the existence of secondary localized
phases causes discontinuous growth, sometimes columnar[5,6] (Fig. 2).
These heterogeneously localized phases may stop normal grain growth
below the solid state dissolution temperature and allow a preferential
migration of the first released grain boundaries.   Finally, powder
with heterogeneous particle size distribution enhances crystal
growth.[7,8]   With the migration rate of a boundary being inversely
proportional to its radius of curvature, it is obvious that big
grains in a matrix of small ones will become larger at the expense
of the matrix.

These processes provide microstructures which are irrecoverable
by heat treatment and require a longer sintering period or higher
temperature.   It should also be noted that in the case of sintering
under pressure, the temperatures are usually relatively low; thus,
the material cannot be readily homogenized.   Also for materials with
a finely dispersed phase, the uniform distribution of the dispersed
phase can never be recovered, since the tendency is towards
coalescence.

Moreover, study of the sintering laws at different stages[9]
reveals the importance of the diffusion rate (in volume, or

intergranular) and of particle diameter. Hence the sintering rate
may be inversely proportional to the cube of the particle diameter.

These considerations highlight the value of obtaining sintering
powders exhibiting significant homogeneity of composition in anions
and cations, and featuring very fine crystal diameter and very low
standard deviation. These powders must also contain a large number
of crystal defects to enhance diffusion and to produce a precise
composition.

These various requirements have called for a major effort for
over ten years, especially in the area of ceramics for electronics,
in order to produce sintering powders with great chemical and
particle size homogeneity and which are fine grained and reactive.

PREPARATION METHODS FOR HOMOGENEOUS POWDERS

The following methods were mainly developed for ceramics, but
are transposable to metals and alloys, either directly or by reduction
of the salt or oxide obtained. They have been tested throughout the
world; in the USA, Japan, Holland, Germany, France and Russia. We
have grouped them in the classification given below.

High-temperature reactions between metallic powders, oxides and
salts. In this classic method, diffusion in the mixture of particles
of different composition causes aggregation of the grains without the
powder reaching perfect homogeneity. It is necessary to grind the
calcined product before sintering, but the crystals remain about a
micron or more in size, which reduces the sintering rate. The powder
particles are neither spherical nor plastic, pourability is poor,
and the filling rates in the dies low. The powder must be converted
into granules of 0.5 to 1 mm, either by humidification and aggregation
on a screen, or by placing in suspension and spray drying at 150°C.
Moreover, the mixers and grinders employed produce contamination
and significant changes in composition. Finally, mixing of particles
with different dimensions and shapes raises reaction order problems
during sintering.[10]

Reactions using molten salts. To accelerate the formation reaction of the compound, one adds to the mixture a compound to provide
a liquid phase which serves as a transfer medium.[11] The process is
improved by employing a double decomposition reaction between a
salt and an oxide, with the cell in the molten or solid state. The
homogeneity of the product is perfect, but its suitability for sintering is not entirely satisfactory.[12]

Thermal decomposition or reduction of solid solutions of coprecipitated salts. These methods were designed to avoid the defects
of the high temperature reaction. However, homogeneity is still

insufficient, although improved by emulsified coprecipitation,[13] but the process is very complex, Moreover, the basic anions and cations employed modify the granular structure after sintering or disturb densification.[14] The composition and concentration of very low additions may be rather difficult to control accurately.

Thermal decomposition or reduction of complex compounds. To obtain perfectly homogeneous, finely divided compounds, a technique is used involving thermal decomposition of complex compounds[15] or metalorganic resins.[16] The homogeneity obtained is good and the grains are very fine after treatment. The sintered density is very high. However, the process is restricted to certain compositions, and it is difficult to impose a precise composition. Finally, the method requires the processing of fairly large quantities of organic matter.

Sol-gel process. This technique first consists of preparing a colloidal suspension of the desired oxides from nitrides or other salts. The suspension is then dispersed in the form of fine droplets at the top of a column of an organic liquid which is immiscible with water, but which acts as a dehydrating agent. This produces a gel which is then sintered, giving rise to high densities, as in the case of thoria,[17] or very high porosity of 60% with very fine pores ranging from 50 to 100Å, as in the case of alumina.[18] The gel process makes it possible to obtain very high density with very fine grain size by prevention of the movement of the cations and anions, but, to our knowledge, the method does not seem to have been used for mixed compounds.

Arc plasma or induction synthesis. Submicron refractory powders from 50 to 500 Å are prepared by arc plasma or induction synthesis, by introducing volatile compounds into a plasma. Carbides, nitrides and borides have thus been obtained.[19] These loose powders can exhibit extremely low density; for example, 1 gram of carbide may occupy a volume of 1 liter.

Thermal decomposition of dehydrated salts. In this process, a liquid solution of metallic elements desired in the final product (sulphates, nitrates, oxalates, etc.) is dehydrated by different methods:
- Spraying in a vessel at 150°C[20] of higher temperature to obtain the calcined product directly.[21] [22]
- Drying and calcination by spraying in a flame.[23] [32]
- Vacuum drying of an emulsion in a lnog-chain hydrocarbon.[13]
- Drying and precipitation with acetone.[24] [25]
- Drying by spraying in a long-chain hydrocarbon raised to a high temperature.[26]

These methods eliminate the defects of coprecipitation (effect
of foreign cations and anions). In practice, however, during dehydration the liquid drops decrease in volume with the relative movement of active materials (Fig. 3), leading to compositional micro-
heterogeneities. Divided and very small droplets of submicron size
might be deemed acceptable,[32] but the contamination by hydrocarbon
would remain.

We have seen above that the desired homogeneity and finely
divided state can only be achieved by using the techniques of thermal
decomposition of metalorganic compounds and gels process which im-
mobilize the metallic elements, but that the composition seems difficult to control. In order to achieve the objective set at the start,
it is therefore necessary to begin with a liquid solution immobilized
by an element which can be eliminated in the solid state. The problem
of mixing does not arise, because the starting material is a liquid
solution. The composition may be very precise and highly varied,
since all the elements placed in solution are retained in the solid
body. It is not necessary to add foreign anions or cations, and no
disturbing effect occurs on the micros tructure, as in the case of
coprecipitation. Since the liquid solution can be sprayed in droplets,
the powder particles obtained after solidification are spherical,
and thus exhibit good pourability and a high filling ration. The
elimination of the immobilizing element leads to a very finely divided,
aerated structure, which is easily deformable. The freeze-drying of
quick-frozen solutions meets these requirements. We give below some
information on the freeze-drying technique and on some results
throughout the world. It is our point of view that, after the
necessary delay for improvements, this technique may be an industrially
promising one.

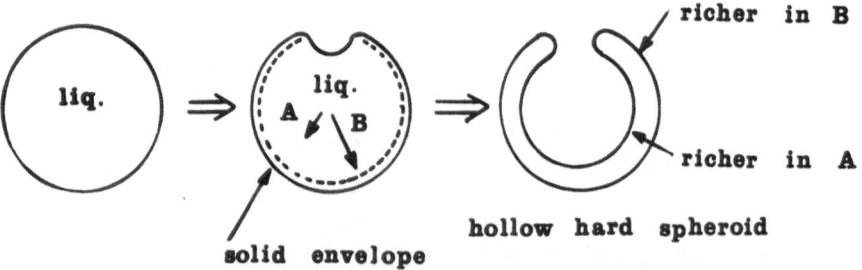

Figure 3.   Formation of heterogeneity by material migration during
            drying of a drop of initially homogeneous liquid.

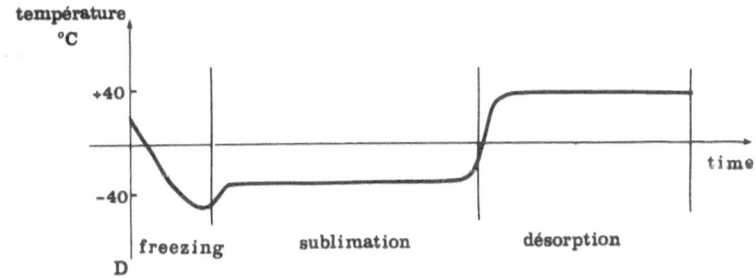

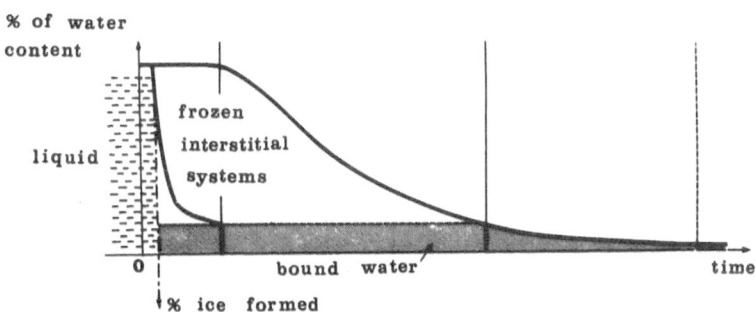

Figure 4.   Thermal cycle as a function of the different states of
            water in the system.

PREPARATION OF FREEZE-DRIED POWDERS: PRINCIPLE OF THE METHOD

    This method consists essentially of rapid freezing without phase
separation of a liquid solution of metallic elements. The solid thus
obtained is then freeze-dried to eliminate the solvent. This sublimation occurs without any variation in bulk volume, as the particles
maintain their shape and dimensions, and remain intact. The solid
product thus obtained is then heat treated for conversion into a
finely divided ceramic or metal product. Several important stages
are involved:

    Dissolution. Water is frequently used as the solvent, involving
sulphates, nitrates, chlorides, formates, oxalates, hydroxides,
etc.[27] The essential condition is to gain sufficient solubility of
the overall cations considered. The salts selected must also be
able to decompose without melting or volatilization during the subsequent thermochemical treatment.

    Quick freezing.   In order to prevent separation of the solution
during freezing, it is necessary to perform freezing very rapidly,
or to provide a eutectic composition, which gives greater latitude
in the cooling rate. In practice, it is preferable to spray fine
droplets ranging from a few hundred microns to a millimeter, which
are cooled by immersion in a liquid bath at a temperature below the

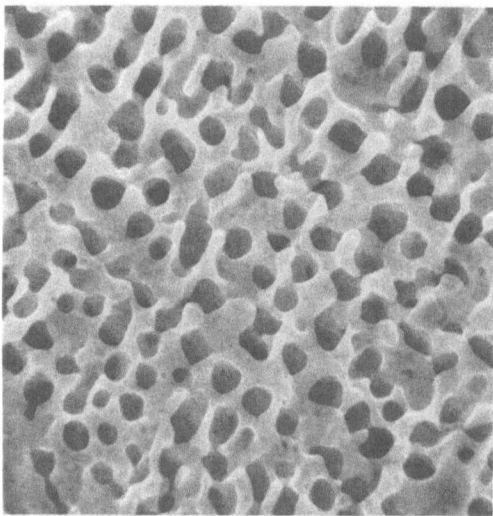

Figure 5.   Solution of Mn and Ni sulphate frozen and freeze-dried.
            The microphotograph reveals the homogeneity of distribution
            and pore dimensions.   Magnification x3000.

cryohydratic temperature.   A convenient laboratory technique consists
of spraying the solution into a liquid nitrogen bath.[28]

   Freeze-drying.   This operation consists essentially of maintaining in a freeze-drying chamber, by means of a very low temperature
condenser (-70°C, for example), a solvent vapor pressure far below
that of the product at any given moment.   As freeze-drying must
always occur in the solid state to prevent any separation, the temperature must be regulated continuously as a function of time, to account
for the state of dehydration of the product.   In effect, when the
solvent is water, it is necessary to consider the existence of very
fine ice crystals (Fig. 4).   As long as these ice crystals are not
sublimed, the product must not exceed the cryohydratic temperature.
It is also necessary to eliminate successively the different waters
of hydration of the solid in question.   Each features a range of
existence as a function of temperature and time, which must be
followed during the freeze-drying process.

   The subsequent stages of calcination, reduction, forming and
sintering are carried out by conventional techniques specific to
each material, but often at very much lower temperatures.

RESULTS AND APPLICATIONS

   The micrograph in Fig. 5 clearly reveals the perfect homogeneity
of distribution of open porosity in freeze-dried products.[29]   This

aeration of the product is an important factor for subsequent effective
thermochemical treatment for conversion of the salt into an oxide or
an alloy, and adjustment of the oxidation degree. The x-ray images
in Fig. 6 obtained on a SEM (Scanning Electron Microscope) reveal
the difference in homogeneity between the product obtained by the
conventional technique of reaction in the solid state, and that
prepared by freeze-drying of a frozen solution.[29] The latter is
perfectly homogeneous irrespective of magnification.

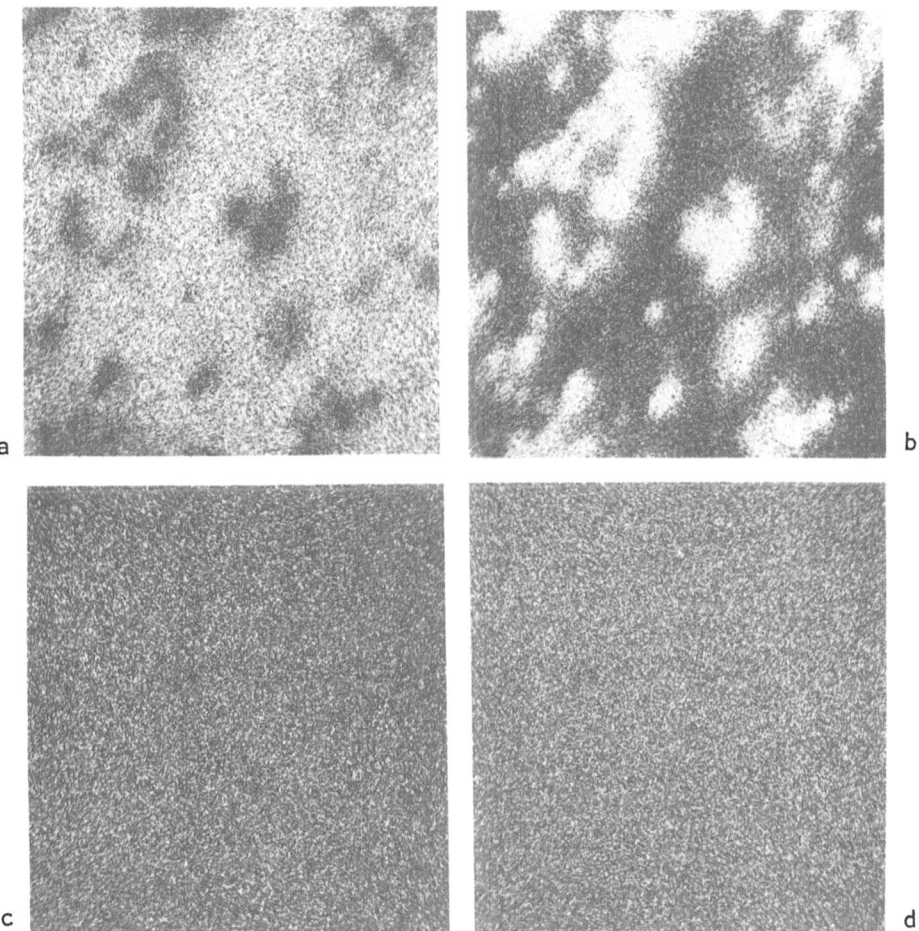

Figure 6.   X-ray pictures enabling a comparison of the distribution
            heterogeneity of Mn and Ni after a mechanical mixing of
            Mn_2O_3 and NiO; a) $Mn_{k\alpha 1}$, b) $Ni_{k\alpha 1}$ with their homogeneity
            of distribution after freeze-drying of a solution of
            $MnSO_4$ and $NiSO_4$, c) $Mn_{k\alpha 1}$ and d) $Ni_{k\alpha 1}$. Magnification
            x2000.

This result implies that the essential principles of the method had been respected:

(a) Freezing-rate greater than 0.1°C/s.[30] In fact the freezing rate must be sufficient in order to avoid any phase separation, but fulfilling this condition is not obvious. The case of Al and Mg sulphate solution is, in this respect, meaningful.[33] An Al sulphate solution frozen in liquid nitrogen is amorphous with a cooling rate between 15°C and 140°C/min and crystallizes only during reheating to room temperature (Fig. 7). But Mg sulphate crystallizes during cooling with a cooling rate of 15°C/min (Fig. 8). As a consequence, the mixed Al-Mg sulphate exhibits during reheating only the thermal effect of recrystallization of Al sulphate thus providing evidence of phase separation (Fig. 9).

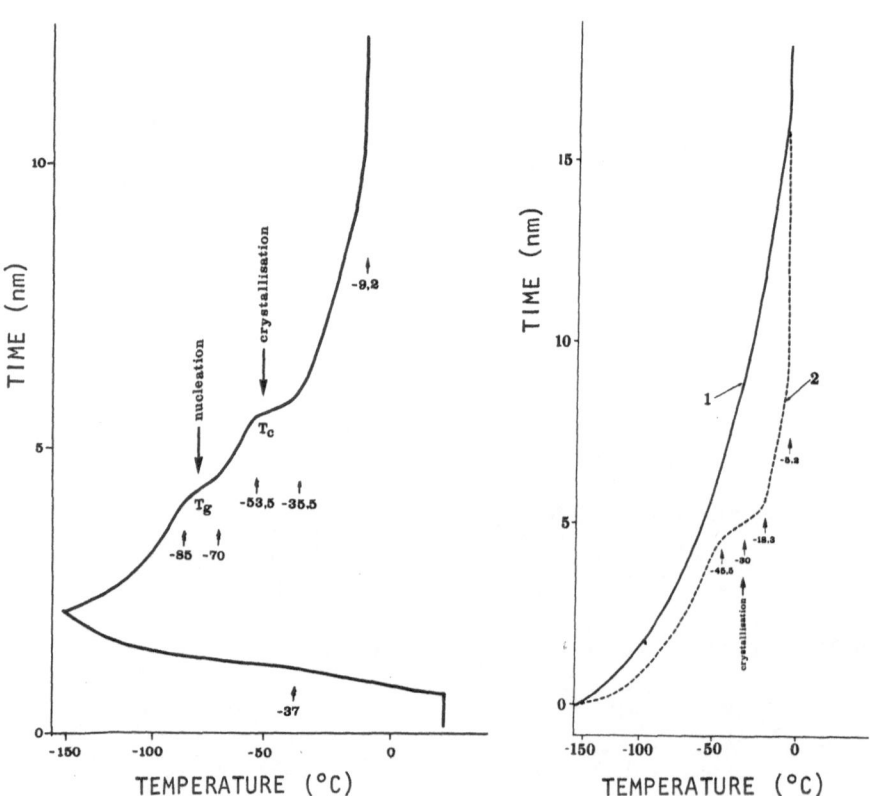

Figure 7.   Thermal analysis of an Al sulphate solution in H_2O. Same results for cooling at 15°C/min and 140°C/min.[33]

Figure 8.   Thermal analysis of a Mg sulphate solution in H_2O[33] (1) after cooling at 15°C/min (2) after cooling at 140°C/min.

(b) The temperature of the frozen product must never be above
its starting melting point. But in some cases the endothermic sublimation, in addition to the cooling system, is not sufficient to
prevent any "pasty melting".[33] Figure 10 shows some transparent
heterogeneities in sintered $MgAl_2O_4$ derived from a freeze-dried
solution of Al and Mg sulphate. These heterogeneities "may be"
attributed to chemical ones, but the origin of the "possible"
heterogeneization is not yet known.

(c) After freeze-drying, the product must not be rehydrated
by deliquescence. Hence dehydration must be intensified sufficiently
to prevent this occurence,[31] as for Mn-Zn ferrite powder formation.

A complete review of the main results obtained from freeze-
drying has been reported elsewhere.[34] But roughly summarized, the
main results are:

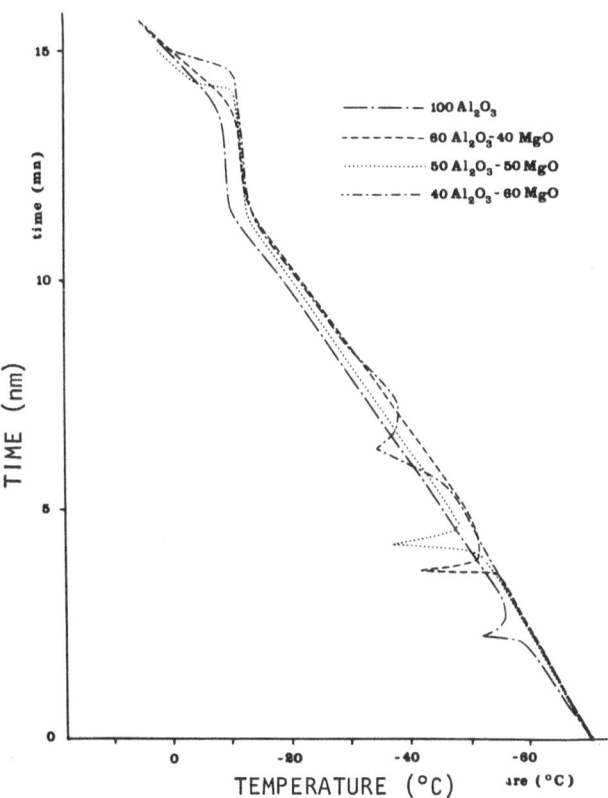

Figure 9. Thermal analysis on reheating of a prefrozen solution
           of Mg-Al sulphate. Thermal effect due to Al sulphate
           recrystallization as a function of the composition.[33]

    (a) The formation of the component occurs at 200-400°C below
the temperature of formation from oxides.

    (b) The mixed salts decomposition temperature falls off
100°C in comparison with the coprecipitated products.

    (c) The sintered porosity may be lowered from 15-10% to
1-0.5%, depending on the material, while the sintering temperature
is lowered by 200-300°C and the crystal diameter decreased from
15-10 μm to 1-0.5 μm.

We have focused our attention on the homogeneity of the starting
material. This is a necessary but not sufficient condition. We have
to maintain this homogeneity during the overall thermochemical treatment of the powder. Even if we avoid the aforementioned (in some
cases, possible) rehydration and melting of the slat before the
formation of the final product, we must still take care to avoid any
sublimation of an element which may locally change the composition.
This phenomena occurs e.g., for Pb in PLZT, or Li and Zn in ferrites.
Furthermore, some oxides such as ferrites and Cr_2O_3 exhibit an
important variation of partial pressure of oxygen (e.g., 10^{-10} atm
at 800°C and 1 atm. at 1400°C for some ferrites). In these conditions
the partial oxygen pressure of the firing atmosphere must always
follow the equilibrium law relative to the material. If these conditions are not fulfilled, we observe a superficial oxidation or a
reduction which may give rise to a cation migration from redox origin.

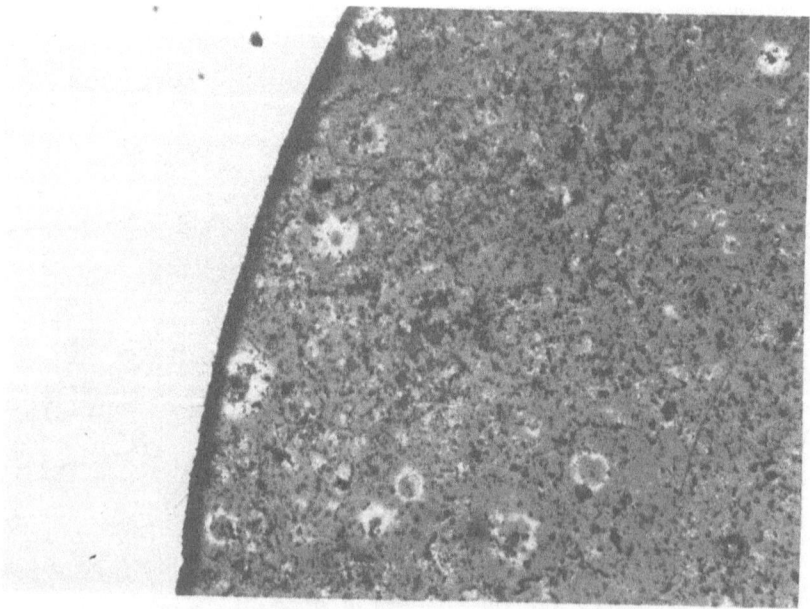

Figure 10. Ring-like zone of transparency in $MgAl_2O_4$.[33]
                   Magnification x300.

These three post powder preparation sources of heterogeneity
(1) superficial sublimation of cations
(2) superficial equilibrium reduction or oxidation
(3) skin - to- core migration of cation from redox origin[2]
    involve:
    (a) diffusion processes for homogenization which may
        give rise to a Kirkendall effect with the corre-
        lated coalescence and the consequences on the
        final density as developed at the beginning of
        this paper, and
    (b) in some cases, formation of a second phase leads
        to discontinuous grain growth by dissolution and
        therefore slows down the sintering process.

SUMMARY

Heterogeneity of composition leads to:
    (a) Homogenization diffusion with Kirkendall effect and
        the correleated coalescence of pores which slows down
        the sintering.
    (b) Presence of inclusions which by dissolution give rise
        to discontinuous growth.

Heterogeneity of grain size enhances grain growth and therefore slows the sintering rate.

Heterogeneity of composition of grain size may rise from non-
equilibrium vapor partial pressure of cation or anion during heat
treatment, but mainly from the starting powder.

Homogeneous and finely divided powder can be achieved only if
the movement of anions and cations are impeded when passing from
the liquid solution to the solid solution of salt or oxide.

Calcination of submicron droplets, thermal decomposition of a
metalorganic compound, the sol-gel process and freeze-drying are the
main techniques which allow the preparation of homogeneous finely
divided powder of ceramics. Although freeze-drying presents some
specific problems, this technique seems to be most convenient to
control the final composition and oxidation degree of the powder.

REFERENCES

1. M. Paulus, In IVe Symposium Europeen de Metallurgie des Poudres Grenoble (France), 13-16th, May 1975. Materiaux et Techniques, 166, 1975.
2. M. Paulus, Ch. Guillaud, Bull. Soc. Chim. Fr., 4, 1175, 1965.
3. M. Paulus, P. Eveno, In 6th Internat. Symposium on the Reactivity of Solids, Schenectady, N.Y., 25-30th August, 1968; Wiley, New York, 585, 1969.
4. P. Eveno, Thesis, Orsay, 1974.
5. C. Lacour, M. Paulus, Phys. Status Solidi, A 27, 441, 1975.
6. C. Lacour, Thesis, Orsay, 1974.
7. C. Lacour, M. Paulus, C. R. Acad. Sci. Paris, Serie C 273, 653, 1971.
8. M. Paulus, 3rd Intern. Conference on Sintering and Related Phenomena, University of Notre-Dame, Indiana, 5-7th June, 1972. (G. C. Kuczynski ed.) Material Research Science, 6, 225.
9. D. L. Johnson, I. B. Cutler, J. Amer. Ceram. Soc., 46, 535, 1963.
10. M. Paulus, International Conference on Ferrites, Kyoto, Japan, July 1970, University of Tokyo Press, 114, 1971.
11. R. G. Wickham, ibid., (Y. Hoshino, S. Iiva and M. Sugimoto ed.), University Park Press, Baltimore, 105, 1971.
12. B. Durand, Thesis, Lyon, 1975.
13. P. J. L. Reijnen, G. P. Th. A. Aarts, R. M. Van de Heuvel, A. L. Stuijts, Joint Meeting Elec. Magn. Ceram., Eindhoven (Holland), April 13th and 14th, 1970.
14. H. M. O'Bryan, P. K. Gallagher, F. R. Monforte, F. Schrey, Amer. Ceram. Soc. Bull., 48, 203, 1969.
15. J. M. Paris, Thesis, Lyon, 1963.
16. M. P. Pechini, Canadian Patent, n°832365, 1968.
17. P. A. Haas, S. D. Clinton, I and EC Prod. Res. Development, 5, 236, 1966.
18. B. E. Yoldas, Amer. Ceram. Soc. Bull., 54, 286, 1975.
19. Exell, R. Roggen, J. Gillot, B. Lux, Int. Conf. on Solid Compounds of Transition Elements, Geneve, 9-13th August, 1973.
20. J. G. M. Delau, Ann. Meeting Amer. Ceram. Soc., Washington, D. C., 7th, May; C. Amer. Ceram. Soc. Bull., 49, 572, 1970.
21. T. Akashi, T. Tsuji, Y. Onoda, 2nd Conf. Sintering Related Phenomena, University of Notre-Dame (U.S.A.), 21-23rd, June. 1965.
22. M. J. Ruthner, H. G. Richter, I. L. Steiner, International Conference on Ferrites, Kyoto, Japan, July 1970, University of Tokyo Press, 75, 1971.
23. J. E. Zneimer, B. Kaplan, K. Lehman, D. A. Lepore, J. Appl. Phys., 35, 1020, 1964.
24. Yu. D. Tre'Yakov, Thesis, Moscow, 1965.
25. Yu. D. Tre'Yakov, I. Ya. Kosinskaya, A. A. Petrova, Inorg. Material 5, 1067, 1969.
26. A. Metzen, Ch. Gorin, Intermag Conference, 14-17th, April, 1975, London.

27. D. Colson, Thesis, Dijon, 1970.
28. Brevet U. S. 3516935 (F. R. Monforte et al.), Bell Telephone
        Laboratories, 23rd, June 1970.
29. C. Lacour, L. Guillaume, A. Tromson, M. Paulus, 4e Symposium
        Européen de Métallurgie des Poudres, Grenoble, 13-15th,
        May, 1975.
30. J. L. Doremieux, A. Giraud, C. Haut, J. C. Le Gal, F. Nicolas,
        P. Dugleux, 4e Symposium Européen de Métallurgie des
        Poudres, Grenoble, 13-15th, May 1975.
31. M. Paulus, C. Lacour, W. Simonet, L. Guillaume, A. Tromson,
        Contrat DGRST, n°74-7-1098, Action concertée "Métallurgie".
32. P. Reijnen, H. Bastium, Powder Metall. Int., Vol. 8, 91, 1976.
33. C. Lacour, "Relationship between thermochemical treatment and
        transparency of $MgAl_2O_4$ prepared from freeze-dried liquid
        solutions", IVth International Round Table Meeting on
        Sintering, Dubrovnik, Yugoslavia, 5 to 10th, September, 1977.
    C. Lacour, M. Paulus, "Lyophilisation Parameters of Ceramic
        Compounds, ibidem.
34. M. Paulus, Ann. Chim., 1, 187 (1976).

DISCUSSION

R. Garvie (CSIRO): Is shrinkage excessive during sintering of
powders prepared by the sol-gel or freeze-drying process? Is it
possible to fabricate large shapes with these powders?
Author: We have no special experience concerning the sol-gel process. But for freeze-drying we get a normal shrinkage due to the
fact that after calcining, the freeze-dried powder remains non-
agglomerated and deformable. So the powder can be pressed to a
relatively good green density, and I don't see any special limitation for large shape fabrication.

L. De Jonghe (Cornell University): I wonder if you could further
speculate on the origin of what seems to be hollow sphere heterogeneities in the local optical properties of the $MgAl_2O_4$ that you
showed.
Author: Solidification of the liquid solution droplets during
cooling is not truly instantaneous due to calefaction occurring in
liquid nitrogen. In fact, for the biggest droplets, solidification from the skin to the core may require a few seconds which may
be sufficient to allow a slight composition gradient by displacement of the solidification boundary. With these conditions a
small shell of the sphere has always the exact composition for
transparency which explains the optical results.

        It must be noticed that in the case of freeze drying, we do
not get hollow spheres as in spray drying.

ULTRAFINE POWDERS OF OXIDE AND NON-OXIDE CERAMIC MATERIALS AND THEIR SINTERABILITY

M. Hoch and K. M. Nair

University of Cincinnati

Cincinnati, Ohio   45221

INTRODUCTION

Ultrafine powders of ceramic materials have been prepared by the decomposition of organometallic complexes,[1,2] thermal decomposition,[3,4] vapor deposition or pyrolysis of inorganic polymers.[5] Processes beginning with preparation of the necessary organometallic complexes from commercially available inorganic salts (using hydrolytic methods which have been fully developed) have already been shown to have numerous advantages over other techniques in the preparation of highly pure and homogenous ceramic powders.[6,7] Thus, we have synthesized our ceramic metal oxides,[8,9,10] as well as our nitrides[11,12] and carbides[13] by methods along these lines. In this article we will summarize some of the general characteristics of these processes for the preparation of ultrafine (<100A) and ultrapure (99.99%) ceramic oxides, nitrides, and carbides, among which are Al_2O_3, Y_2O_3, zyttrite, Al_2O_3-ZrO_2 eutectic, mullite, various spinels, AlN, amorphous Si_3N_4, α-Si_3N_4, β'-sialon, and β'-SiC. We will also report on the sinterability of these materials at or below 1500°C.

Freshly prepared amorphous oxides were used as precursors in the preparation of their respective nitrides and carbides. The structure and crystallinity of these powders were determined by x-ray diffraction analysis, their particle size by transmission and scanning electron microscopy, and their reactivity by powder density.[7,8]

EXPERIMENTAL PROCEDURE

1. Preparation of Alkoxides

To begin with, certain metals or inorganic metal salts were allowed to react with appropriate alcohols, preferably tertiary alcohols.[1,7,14,15] The alkoxides thus formed could be purified and crystallized. In some cases, conversion of the metal to an alkoxide required a catalyst.[1,7] Typical examples of these reactions are:

$$SiCl_4 + 4C_2H_5OH \rightarrow Si(OC_2H_5)_4 + 4HCl \tag{1}$$

$$2Y + 6C_3H_7OH \xrightarrow[\text{reflux}]{HgCl_2} 2Y(OC_3H_7)_3 + 3H_2 \tag{2}$$

After purification, certain commercially available metal alkoxides were also found to be usable.

2. Hydrolysis of Alkoxides

The metal alkoxides were subsequently hydrolyzed through reacting with deionized water or an NH_4OH solution. When preparing mixed hydroxides, the alkoxide solutions were mixed, if necessary, under a nonpolar organic solvent (such as C_6H_6) before reacting with the NH_4OH solution.

The resulting hydroxides were filtered and washed in deionized water until a neutral pH was reached and then washed in isopropyl alcohol, acetone, benzene, or toluene. (In general, materials washed with nonpolar organic liquids have shown higher surface reactivity than the ones washed in H_2O and isopropyl alcohol or H_2O alone.) The washed materials were heated to dryness at 125°C in air. Further heating at 550°C to 600°C for 1 to 2 1/2 hours in air often resulted in the formation of desired amorphous oxides.

3. Reduction-Nitridation Reaction

The major methods for preparation of refractory nitrides are direct nitridation of the metal,[16] decomposition of respective metal amides,[17] reaction between NH_3 and the metal halide,[16,17] and pyrolysis of inorganic polymers.[5] Some of these processes generally result in non-homogeneous products which are mixtures with intermediate products or, sometimes, with metal chelates. We were able to develop a solid-gas reaction process to reduce and nitride amorphous metal oxides in one step, using NH_3 gas as the sole reducing-nitriding agent.[11,12]

Amorphous metal oxides were heated in NH_3-flowing alumina tube furnaces with gas tight ends. Ammonia reacted with the metal oxide and produced either amorphous or crystalline metal nitrides depending on the temperature of the reaction

$$MeO + 2/3\ NH_3 \rightarrow MeN_{2/3} = H_2O \tag{3}$$

where Me is an Al, or Si ion. The amorphous nitride could be crystallized by heating it in an N_2 atmosphere at a higher temperature. Due to the fine particle size of the oxide materials used, gas molecules had a very high probability of reacting with all the materials before decomposition of the NH_3 gas to N_2 and H_2, and before crystallization of the oxide. For all the materials studied, maintenance of the amorphous character of the oxide powder until its reduction-nitridation was a prerequisite for the formation of nitride. The optimum reduction-nitridation temperature was dependent on the free energy of the reaction, the partial pressure of oxygen over the system, and the temperature of crystallization of the oxide material.

4. Reduction-Carburization Reaction

Carburization of the metal or metal oxide by graphite is a common technique in the preparation of metal carbides. This method, however, leads to a carbide containing free carbon or a nonstoichio-metric carbide material. We found we could use CH_4 gas as a single-step reducing-carburizing agent for amorphous metal hydroxide powders[13] with much more desirable results. Since the ultrafine oxide powders have very high surface area and are highly surface reactive, CH_4 gas easily diffuses into the powders and reacts with the oxide, producing carbide materials. The appropriate temperature varies depending on the particular metal hydroxide and the free energy of that reaction as well as the partial pressure of water vapor over the system. Unlike the nitridation process, maintenance of amorphous character does not seem to be a prerequisite for the reduction-carburization reaction. As in the case of nitrides, though, low temperature reduction-carburization produced amorphous carbides and higher temperatures yielded crystalline materials. Amorphous carbides could be converted to crystalline form by heating the H_2 at a higher temperature. The process can be represented by the equation:

$$Si(OH)_4 + CH_4 \rightarrow SiC + 4H_2O \tag{4}$$

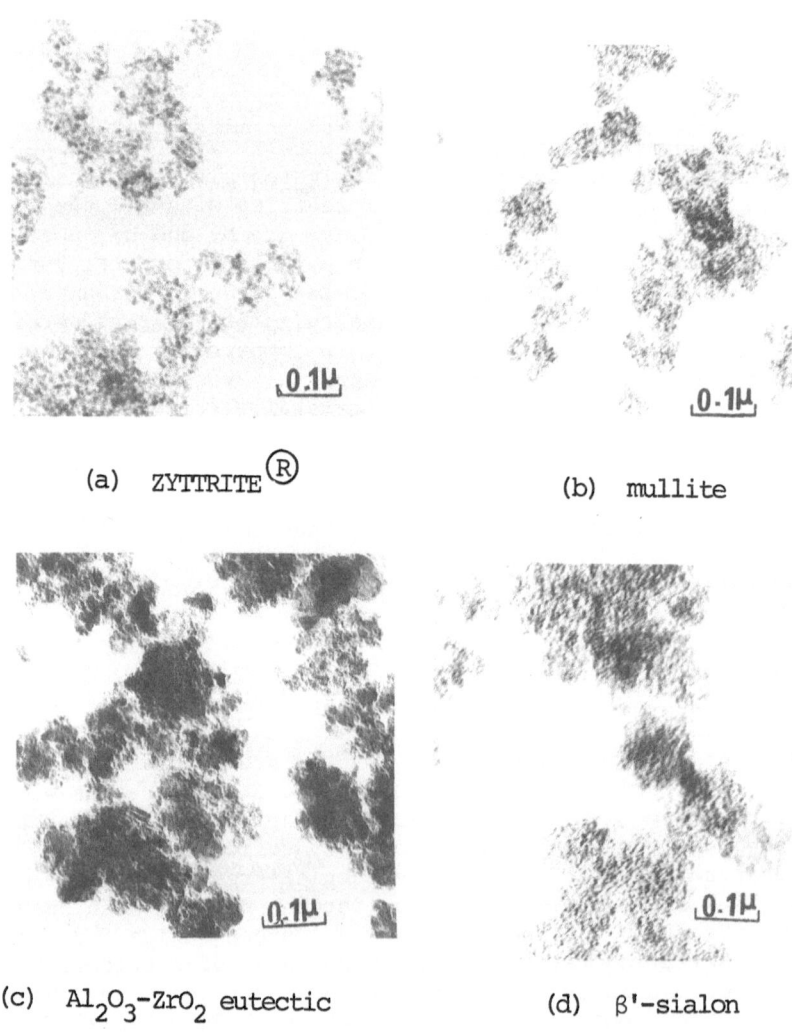

(a) ZYTTRITE®   (b) mullite

(c) Al_2O_3-ZrO_2 eutectic   (d) β'-sialon

Figure 1. Electron micrographs of ultrafine powders.

5. Sinterability of Ultrafine Powders

Sintering of conventional ceramic oxides occurs mainly via lattice or grain boundary diffusion. It has been suggested, on the other hand that ultrafine powders sinter mainly by a mechanism similar to plastic flow.[9] Since ultrafine powders are highly surface reactive, one would expect them to sinter at a lower temperature than that of conventional powders. Thus, our main study was of relatively low temperature sintering (maximum temperature of 1500°C) of these powders by isostatic pressing and heating in a globar-heated alumina tube furnace. Experimental details have been reported elsewhere.[8, 9, 18] Most of the oxide powders and spinels densified close to theoretical density at or below 1500°C. Investigation of the densification characteristics of refractory nitrides and carbides is underway and preliminary results indicate that at least some of these powders will also sinter to reasonable densities.

RESULTS AND DISCUSSION

Results from the powder preparations and characterizations are given in Figure 1. Transmission electron micrographs of the powders showed their extremely fine particle size. Their high surface area and surface reactivity, produced a very high aggolomerative property which is clearly observed in the micrographs.

Chemical composition and crystallinity were determined by x-ray diffraction analysis and the results are shown in Table I. along with the temperature of formation of the respective materials. Powder density of materials was determined using methods described elsewhere[7, 8] and the results showed significant variation with change in the washing medium and reaction temperature.

Our sintering studies showed that powders with lower density have higher surface reactivity and hence a greater densifying property. The densities of the sintered materials (oxides) listed in Table I reveals that these ultrafine powders can be sintered close to theoretical density by isostatic pressing and heating at or below 1500°C. Microstructures of metallographically polished and thermally etched samples demonstrated that they are uniform in structure and have fine grain size with relatively few or no void spaces (Fig. 2). Sintering studies of other oxides, nitrides, and carbides are progressing and will be reported elsewhere.

Overall, our experiences indicate that extremely pure (99.99%) and ultrafine (<100Å) oxide and non-oxide ceramic powders can be synthesized and handled through good laboratory technique, and that their high surface reactivity can be maintained during sintering experiments. The mechanism of that sintering is, however, apparently different than that of conventional ceramic materials[9, 19] and needs further detailed study.

TABLE I

CHARACTERIZATION OF SOME ULTRAFINE POWDERS

Material	Medium	Calcination or Reaction Temp. (°C)	Powder Density (g/cc)	Sintered Density (% Theor.)
$Al(OH)_3$	1	125	0.22	
	2	125	0.19	
	3	125	0.09	
γ-Al_2O_3		900	0.33	
α-Al_2O_3	1	550	0.30	87.9
$Y(OH)_3$	4	125	0.59	
	1	125	0.19	
	3	125	0.14	
Y_2O_3	4	600	0.80	
	1	600	0.12	
	3	600	0.10	
Al_2O_3-ZrO_2	1	125	0.14	95.4
	1	300	0.10	85.0
	1	550	0.10	86.0
	4	125	0.90	72.0
Zyttrite ®	1	550	0.30	97.2
	4	550	1.00	89.5
$3Al_2O_3$-$2SiO_2$	1	125	0.14	94.2
	1	550	0.10	99.2
	4	550	0.39	70.3
$MgO \cdot Al_2O_3$	1	1500	0.40	98.0
$ZnO \cdot Al_2O_3$	1	1400	1.01	
$CoO \cdot Al_2O_3$	1	1500	1.50	
AlN		1200	0.40	
Si_3N_4 (amorphous)		1000	0.03 [from $Si(NH_2)_2$]	
Si_3N_4 (amorphous)		1000	0.84 [from amor. "SiO_2" + NH_3]	
αSi_3N_4		1350	0.16 [from $Si(NH_2)_2$]	
αSi_3N_4 (amorphous)		1350	0.26 [from amor. "SiO_2" + NH_3]	
β'-Sialon		1200	0.36	
		1350	0.76	
SiC (amorphous)		900	0.84 [from "SiO_2" + CH_4]	
SiC		1400	0.27 [from "SiO_2" + CH_4]	

1-isopropyl alcohol; 2-acetone; 3-benzene or toluene; 4-water

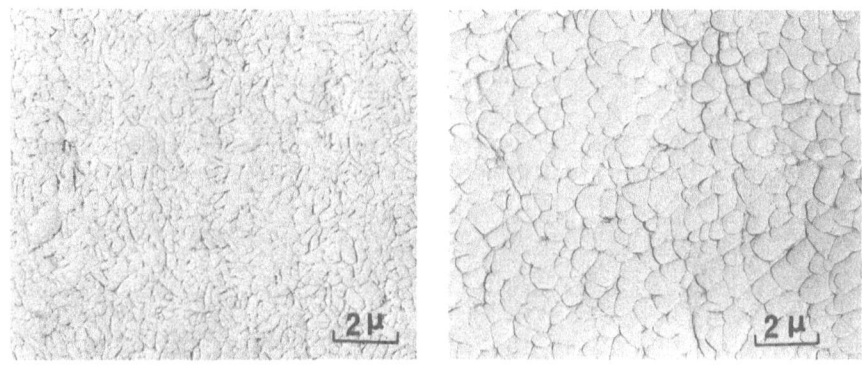

(a) mullite   (b) Al_2O_3-ZrO_2 eutectic

(c) ZYTTRITE ®

Figure 2. Electron micrograph of sintered metal oxides.

REFERENCES

1. K. S. Mazdiyasni, C. T. Lynch, and J. S. Smith, J. Am. Ceram.
    Soc., 50 [10] 532-37 (1967).
2. J. D. Crofts and W. W. Marshall, Trans. Brit. Ceram. Soc., 66
    [3] 121-26 (1967).
3. R. R. Neurgaonkar, T. P. O'Holleran, and R. Roy, Bull. Am. Ceram.
    Soc., 56 [3] 1-B-77, 289 (1977).
4. G. W. Brindley and M. Nakahira, J. Am. Ceram. Soc., 42, [7]
    311-24 (1959).
5. R. W. Rice, K. J. Wynne, W. B. Fox, and B. E. Walker, Bull. Am.
    Ceram. Soc., 56 [3] 4-B-77, 289 (1977).

6. K. R. Hancock, pp. 39-60 in <u>Ultrafine-Grain Ceramics</u>. Edited
    by J. J. Burke, N. L. Reed, and V. Weiss, Syracuse University
    Press, Syracuse, N. Y., 1970.

7. M. Hoch, Report No. AFML-TR-71-158 (1971).

8. M. Hoch and K. M. Nair, Ceramurgia Int., <u>2</u> [2], 88-97 (1976).

9. M. Hoch and K. M. Nair, Paper presented at Second International
    Powder and Bulk Solids Handling and Processing Conference/
    Exhibition, Rosemont, IL, May 1976.

10. M. Hoch and K. M. Nair, Universite Libre de Bruxelles,
    March 21-23, (1977).

11. M. Hoch and K. M. Nair, Bull. Am. Ceram. Soc., <u>56</u> [3] 120-B-77,
    300 (1977).

12. M. Hoch and K. M. Nair, ibid., <u>56</u> [3] 121-B-77, 200 (1977).

13. M. Hoch and K. M. Nair, ibid., <u>56</u> [3] 2-B-77, 289 (1959).

14. D. C. Bradley, Advances in Chem., Series <u>23</u>, 10-36 (1959).

15. D. C. Bradley, Progress Inorganic Chem., Vol. II, 303-361 (1960).

16. S. N. Ruddlesden and P. Popper, Acta. Cryst. <u>11</u>, 465-468 (1958).

17. K. S. Mazdiyasni and C. M. Cooke, J. Am. Ceram. Soc., <u>56</u> [12]
    628-33 (1973).

18. K. M. Nair and M. Hoch, Bull. Am. Ceram. Soc., <u>56</u> [8], 18-BN-77F,
    731 (1977).

19. W. S. Yound and I. B. Cutler, <u>53</u> []2] 659-663 (1970).

HANDLING AND GREEN FORMING OF FINE POWDERS

D. E. Niesz, L. G. McCoy and R. R. Wills

Battelle Columbus Laboratories

505 King Avenue, Columbus, Ohio   43201

INTRODUCTION

In its early stages of evolution, ceramic powder processing was entirely based on the use of natural minerals such as clay, flint and feldspar as starting materials. These materials were chosen not only because of their chemical compositions but also due to their physical characteristics. The introduction of synthetic starting powders such as alumina and the gradual evolution to materials made entirely of synthetic raw materials has required the development of a processing technology consistent with their physical characteristics. The technology for producing submicron, high purity oxide and nonoxide powders has been developing at a rapid pace in recent years and can be expected to continue. In order to take full advantage of these developments, processing science and technology for these advanced powders must also be developed. An important part of the required technology is the development of processing techniques for working with powders having bulk densities below 10 percent of theoretical without unacceptable chemical or tramp-particle contamination. Such techniques must be capable of producing materials with optimum microstructures that are free from singularities that limit mechanical and other properties.

PREPARATION OF FINE CERAMIC POWDERS

Fine ceramic powders are prepared by a variety of techniques. The most common technique is some form of precipitation followed by calcination. Gas-phase reaction has also become an important method of preparation especially for nonoxide powders. Other

41

techniques include size reduction, thermal decomposition, freeze drying and sol gel.

Grinding, attrition and impacting have the attraction that a wide variety of materials may be prepared in fine particulate form by essentially the same technique.[1,2]   Unfortunately no widely applicable, highly efficient grinding method is available to form high-volume quantities of submicron particles. The main types of mills used are ball mills, rod mills, fluid energy and shear mills and attrition mills.

A commonly used method for the production of metal oxide powders involves the decomposition of a metal salt by a calcination process. The metal salt may be a sulphate, nitrate, oxalate, or hydroxide. Organometallic compounds[3] can also be used.

The growth of ultrafine particles by precipitation reactions is a versatile technique because a wide range of particle sizes and shapes can be obtained.[2]   Precipitation can be carried out in water at room temperature or under high temperature pressure conditions (the hydrothermal process).[4] The hydrolytic decomposition of metal alkoxides in the nonaqueous media has also been used extensively to produce ultrafine oxide powders.[5]

In recent years increasing interest has been shown in the formation of ultrafine powders by gas-phase reactions in which the powder effectively precipitates out of the gas phase. Gas phase synthesis in a plasma has been applied to many nonoxide ceramics, and is particularly useful for endothermic reactions.[6]

Freeze drying is a salt decompositon process where droplets of metal salt solutions are rapidly frozen. The freezing produces spheres of frozen solution approximately 0.1-0.5 mm in diameter. To obtain the final oxide the frozen spheres are dried by a combination of vacuum drying and gentle heating so that all the ice is sublimed and the sphere is not melted. The resulting porous spheres are an anhydrous metal salt which are then calcined to form oxide powders.

In the sol gel[7]   process a colloidal suspension is first produced. Gel formation is controlled by either water removal (drying or solvent extraction) or by partial removal of the stabilizing anion. Droplets of the sol can be gelled to spheres with a controlled size within the overall range 5 to 1000 µ diameter which are then dried and calcined to obtain an oxide powder.

Characteristics of Fine Powders

Powder characteristics can be divided into chemical, physical, and structural characteristics.[8] Chemical characteristics are concerned with the concentrations of major and minor elements as well as their distribution in a powder. Structural characteristics deal with the phase composition of a powder and atomic-scale defects. Physical characteristics are those that describe the geometry of powder particles, powder agglomerates, and the bulk powder as well as the handling characteristics of the powder such as flowability.

Physical characteristics of powders play a major role in determining powder processing parameters, microstructure and properties of advanced ceramics fabricated by powder processing. The influence of physical powder characteristics on these factors is less recognized than the influence of chemical characteristics. In fine powders the physical characteristics are usually dominated by the agglomerate structure of the powder.[8] This factor is often not recognized in characterization of fine powders and not taken into account in attempts to relate powder characteristics to processing parameters, microstructure and properties of advanced ceramics.

Most fine ceramic powders are composed of fine ultimate particles which have sintered together during preparation or calcination to form agglomerates. These agglomerates can have a wide range of strengths depending on the material and preparation procedure. Often these powders have bulk densities of less than 10 percent of theoretical, making handling and processing by standard techniques difficult. The low bulk density is primarily a result of the presence of low-density agglomerates. Typically a fine powder is a random packing of low-density agglomerates rather than dense ultimate particles.

Direct use of such powders for mass producing advanced ceramics is often undesirable due to handling difficulties and microstructural considerations. In order to utilize these powders, they must be pretreated to develop the optimum physical characteristics. The pretreatment normally includes three steps. These are volume reduction, reduction of the powder to its ultimate particle size, and reagglomeration. Use of fine powders is further complicated by their reactivity with water which prevents the use of water as a medium for processing. Although not discussed here, processing of fine powders for advanced ceramics must be carried out in such a way as to prevent chemical and tramp-particle contamination.

POWDER PRETREATMENT

Many fine powders have bulk densities below 10 percent of
theoretical and often below 5 percent of theoretical. Such low
bulk densities lead to storage and handling problems and to compaction ratios of up to 10 to 1 if the powder is not pretreated
in some manner to improve its physical characteristics. Such high
compaction ratios are impractical for most commercial dry pressing
operations and lead to nonuniform green compacts and nonoptimum
microstructures.

A common technique for improving the compaction ratio of fine
ceramic powders is to prepress, granulate and repress. Although
this approach does improve the compaction behavior, it normally
leads to nonuniform porosity in a sintered compact. If the final
pressing pressure is an order of magnitude above the prepressing
pressure, the microstructural uniformity is acceptable for many
applications.

In working with powders that have a bulk density below about
15 percent of theoretical density it is often useful to isostatically prepress and granulate the powder to increase its bulk
density for handling purposes. This eliminates many of the
practical processing problems such as mill loading that result
from low-bulk-density powder.

Examples of the change in bulk density after prepressing at
various pressures and regranulating are given in Table 1.

For most advanced ceramics where microstructural uniformity
is critical, it is essential to reduce the starting powders to
their ultimate particle size.[8,9]   The most common technique for
doing this is ball milling, although several other techniques can
be utilized. Milling in a water medium has been standard practice
in alumina whiteware manufacture for many years. For nonoxide
ceramics such as Si_3N_4 where water reactivity is a problem and
for nuclear ceramics where nuclear considerations preclude the use
of a water medium, milling has been carried out in an organic
liquid such as hexane. An organic liquid is also required as the
milling medium for fine alumina powder if optimum microstructure
and sinterability is desired.[8]   In the last 10 years dry ball
milling has gained acceptance and is now routinely used by several
commercial alumina powder manufacturers. It is also being used in
the fabrication of several advanced ceramic materials and products.
Several other types of milling or grinding can also be utilized
for reduction of fine powders to their ultimate particle size.
Ultrasonic energy is also used for this purpose, especially in
the ferrite industry.

Table 1. Bulk Density of Selected Powders
After Prepressing and Granulating

		Density (% theoretical)		
	Bulk	34.5 MPa	69.0 MPa	138 MPa
Powder A	7	16	17	21
Powder B	4	8	11	11
Powder C	4	15	17	17

Ball milling is also used to achieve intimate mixing of
starting powders prior to calcining for solid-state reaction as
in ferrite manufacture or simply as a means of uniformly distributing additives prior to green forming.

For dry pressing, fine powders need to be agglomerated to
improve flowability, minimize lamination tendency, and reduce
dusting. Normally binders and lubricants are required to achieve
the desired agglomerate properties,[10] reduce die friction, and
improve green strength. Aside from the prepressing technique discussed earlier, spray drying and pelletization are the two
techniques employed. Spray drying is the most widely accepted
technique, but pelletizing technology is being developed and
should play a more important role in advanced ceramics fabrication in the future.

Figure 1 shows the compaction behavior[8,11]   of two Bayer-
derived reactive alumina (Powder A and B) which have been dry ball
milled by their suppliers. Also shown in Figure 1 is the compaction behavior of a typical 96 percent alumina body which has
been prepared by ball milling the starting powders followed by
spray drying (Powder C), a high surface area calcined alumina
(Powder D), and the compaction behavior of Powder D after dry ball
milling to its ultimate particle size (E).

Figure 2 shows the compaction behavior of a powder prepared
by gas-phase reaction (Powder F) and the same powder after prepressing, granulating and ball milling in a hexane medium (G).

A major factor to be noted in Figures 1 and 2 is the large
variation in the densities of the various powders at 0.0345 MPa.
The powders generally fall into two groups. All of the powders
in the as-calcined or as-prepared condition except Zittrite A[12]
have 0.0345 MPa densities between 8 and 12 percent of theoretical.
The milled powders and the spray dried powder have 0.0345 MPa
densities of 25 to 35 percent of theoretical. Zittrite A in the

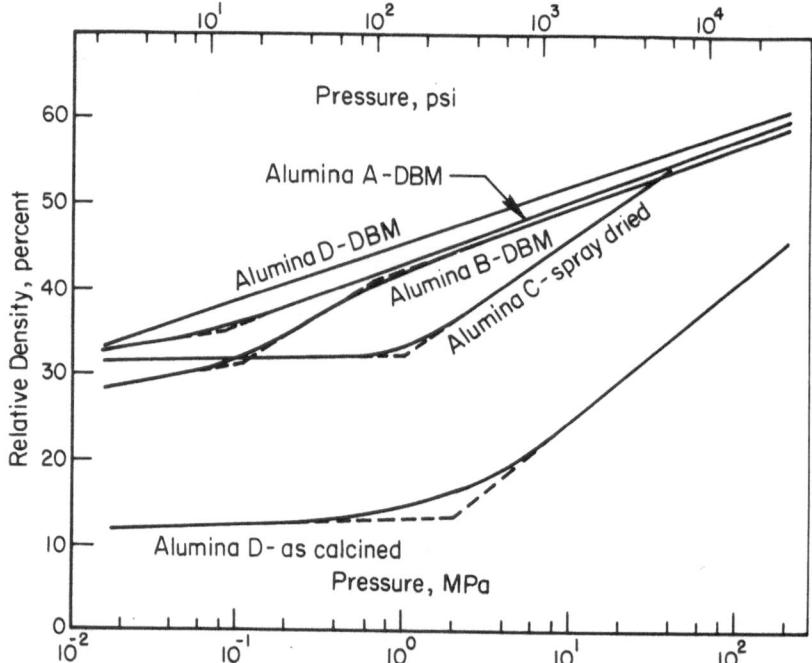

Figure 1. Compaction Behavior of Fine Alumina Powders

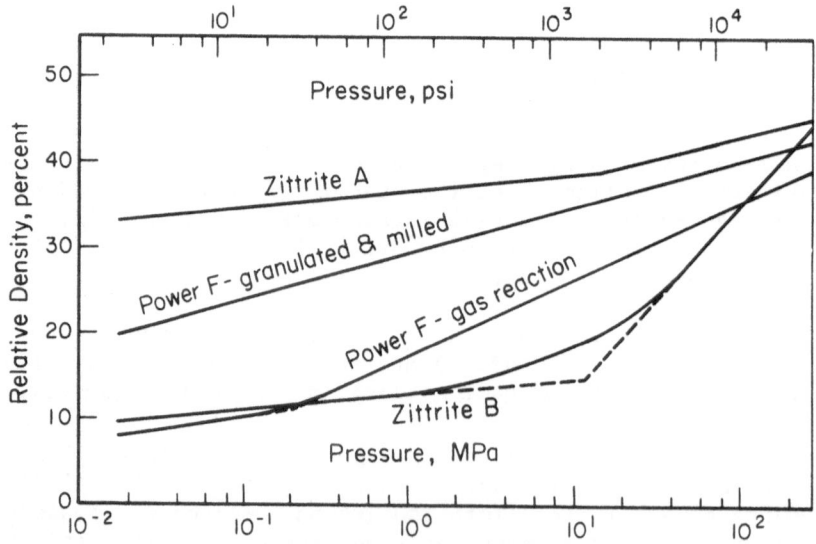

Figure 2. Compaction Behavior of Vapor Form and
Precipitated Fine Powders

as-prepared condition has an unusually high 0.0345 MPa density of 34 percent. This powder differs from Zittrite B only in that the alkoxide derived precursor was water washed prior to calcination rather than alcohol washed. This treatment apparently results in major differences in the agglomerate structure of the powders. The low densities (0.0345 MPa) are attributed to the presence of low density, porous agglomerates in the powders. Even weak agglomerates of this type are detrimental to the physical character of the powder unless they are broken up by ball milling or some other technique. The breakpoint in Curve C (spray dried alumina) is a measure of the compressive strength of the agglomerates in the powders, and the lack of such breakpoints in the milled powders indicates the absence of agglomerates that have strengths in the pressure range measured. Powders without breakpoints could contain agglomerates that are too strong to be broken down in the pressure range measured.

The breakpoint pressures for the various powders shown in Figures 1 and 2 are shown in Table 2. These pressures are useful for determining lot-to-lot variability of powders as well as to gain a better understanding of the characteristics of the agglomerates in the powder.

Table 2. Compaction Curve Breakpoint
Pressures for Powders Shown
in Figures 1 and 2

Powder	Breakpoint Pressure, MPa	(psi)
C (Spray Dried)	1.4	(200)
D (γ-phase in calcined alumina)	2.1	(300)
F (gas phase reaction)	0.21	(30)
H Zittrite A	15.2	(2200)
I Zittrite B	11.7	(1700)

References

1. Treatise on Materials Science and Technology, Vol. 9: "Ceramic Fabrication Processes", Edited by F. F. Y. Wong, Academic Press, 1976.

2. "Ultrafine Grain Ceramics", Editors: John H. Burke, Norman L. Reed, Volker Weiss, Syracuse University Press, 1970.

3. K. S. Mazdiyasni, C. T. Lynch and J. S. Smith, J. Am. Ceram. Soc., 48, 372–375, 1965.

4. R. Roy and O. F. Tuttle, pp. 138–80, Physics and Chemistry of the Earth, Vol. I, Edited by L. H. Ahrens, K. Pankama, F. Press, and S. K. Runoorn, Pergamon, New York, 1956.

5. K. S. Mazdiyasni, pp. 3–27 in "Fine Particles", Second International Conference, Electrochemical Society, 1974.

6. S. F. Exll, R. Roggen, J. Gillot, and B. Lux, pp. 165–186 in "Fine Particles:, Second International Conference, Electrochemical Society, 1974.

7. J. M. Fletchner and C. J. Masdy, Chemistry and Industry, 48–51, January , 1968.

8. Ceramic Processing Before Firing, Edited by G. Y. Onoda, Jr. and L. L. Hench, John Wiley & Sons, Inc., New York (1978), "Agglomerate Structure and Properties" (D. E. Niesz and R. B. Bennett).

9. H. Palmour III, T. M. Hare and M. L. Huckabee, "Optimal Densification of Ceramics by Rate Controlled Sintering", Final Technical Report, Contract Number N00019-73-C-0139, Engineering Research Services Division, North Carolina State University, Raleigh, North Carolina, March 1974.

10. S. J. Lukasiewicz and J. S. Reed, "Character and Compaction Behavior of Spray-Dried Powders, Paper No. 14-W-76, Presented at the Am. Ceram. Soc. Symposium, May 5, 1977.

11. D. E. Niesz, R. B. Bennett and M. J. Synder, "Strength Characterization of Powder Aggregates", Am. Ceram. Soc. Bull., 51 [9], pp. 677–680, (Sept. 1972).

12. M. Hoch, "Preparation and Characterization of High Purity Submicron Refractory Oxides and Mixed Oxides from Alkoxides", AFML-TR-71-158, Air Force Materials Lab., Wright-Patterson Air Force Base, Ohio, Aug. 1971.

THE POTENTIAL OF FINE PARTICLE TECHNOLOGY APPLIED TO CERAMIC RAW MATERIALS

N. O. Clark

Rutgers University, New Brunswick, N. J. and

Anglo-American Corp., Atlanta, Georgia

INTRODUCTION

Metalliferous ores and a number of industrial mineral sources consist of mixtures of crystals. Purification involves the preliminary separation by grinding in some form to the point at which the individuals are pure and separate grains. This is followed commonly by separation using methods which rely on different bulk or surface properties exhibited by the constituent minerals.

The comminution processes are not easy to control and adequate liberation usually results in a proportion of fines ("slimes"). Most mineral dressing plants are characteristically adjoined by slime ponds. These ponds characterize not only metalliferous ore dressing plants but also those preparing industrial minerals used in ceramics.[1] Slime ponds contain a wealth of mineral values in a form not easily accessible, and there for the time being the matter lies. On the question of economic value of a purified waste and the cost of a new extraction, it has been cheaper to mine more original ore than to push processing into the difficult region of fine particles.[2]

The application of mineral dressing techniques to ceramic raw materials, particularly clays and kaolins, has been virtually confined to particle size separation until comparatively recently. However, increasing commercial value of industrial minerals and consumption of reserves of better quality, particularly of kaolin for the paper industry, has provided a great stimulus to the application of mineral dressing technology to this mineral and hence to fine particle minerals generally, a trend of great interest to ceramists. Hand in hand with these developments has been the sophistication of solid-liquid separation applied to fine particles, since most mineral dressing of fine particles has to be carried out wet at appropriate dilution to allow overcoming of interparticle forces by appropriate physical chemistry.

Particle size separation is simple enough with the right physical chemistry and is effective if impurity minerals show that appropriate difference in size and density and, to a degree, shape to allow an elutriation or a settlement cut. The interesting new areas are, however, concerned more with other properties than size and density.

FLOTATION TECHNIQUES

Froth flotation has probably had the greatest impact in this area and is widely and increasingly used for purifying kaolins. There are two systems used: the "piggy back" system developed in the USA in which an ancillary mineral of normal flotation size is floated in anormal cell with the physical chemistry adjusted to attract the fine impurity mineral to the carrier, and the direct system developed in Great Britain which uses a special cell with a high shearing region to separate the finer minerals. Much of the development of fine particle flotation has been inhibited by the tradition of difficulty in normal size (i.e., 200 mesh) mineral flotation if the slimes (i.e. less than 10 microns) are not removed[3-6] see Table I).

Table I. Commercial Fine Particle Mineral Separation From Clays

Impurity	Separated Particle Size (microns)	Yield of Product	Color Change Reflectance At 4540Å. %	Analysis % Feed	Product
Blue Tourmaline	2.0	92%	78→81	1.42 B_2O_3	0.29 B_2O_3
Anatase/ Rutile	0.5	93%	87→91	1.80 TiO_2	0.60 TiO_2
Quartz	5.0	60%	75→77	5.0 SiO_2	1.0 SiO_2

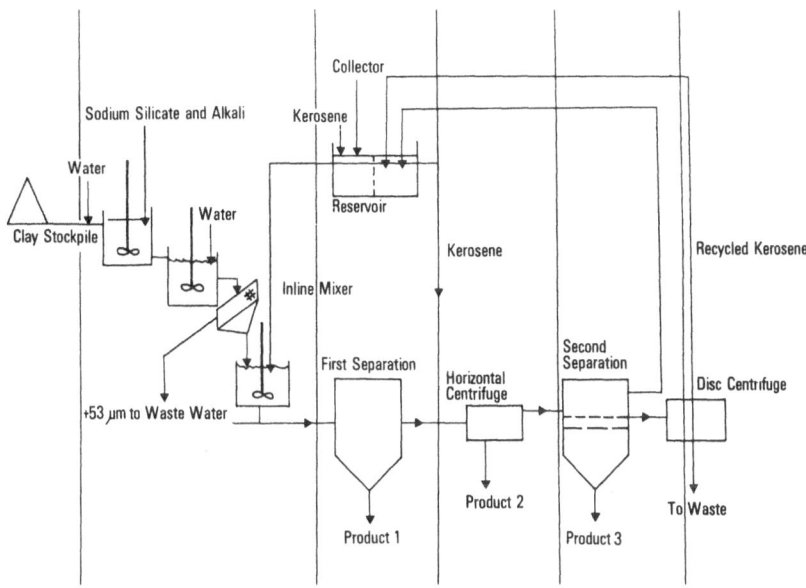

Figure 1. Flow diagram.

A closely allied process to froth flotation (which preceded it historically) is two-liquid phase separation in which the mineral is dispersed in a transient, unstable emulsion of oil and water and relies on retention of the floated solid at the interface. This process uses less energy than froth floatation and is very effective at the fine sizes met in clays and kaolins, but the problem area is reclamation of oil. However, this problem is disappearing with study and commercial processes are in operation on cassiterite slimes[7-9] (see Fig. 1 and Table II).

Table II. Two Liquid Process: Typical Results Obtained from Processing A Fine American Kaolin.

Sample	Recovery % wt	Tappi Brightness	Particle Size % Cum Finer 10 μm	2 μm	TiO_2 %
Feed	100	85.4	93	81	1.61
Product 1	73	91.2	97	83	0.45
Product 2	9	78.8	75	46	1.60
Product 3	6	89.3	100	99	1.10
Waste	12		88	75	10.8

SAND GRINDING

It has been tacitly assumed above that the mineral is naturally
of fine particle size as in clays; actually mineral separations are
normally carried out at the 200 mesh level, but if the mixture in
the rock formation is finer than the grinding has to be finer.
Very fine grinding of the order of 1-2 microns is expensive in
energy and capital cost and highly inefficient in equipment designed for normal ceramic raw material preparation.

In recent years there has been a substantial development of
sand grinders for superfine wet grinding applied to three areas:
delaminating kaolin for paper coating (using plastic "sand"),
straight grinding of minerals to these finer sizes for the properties
they develop and for liberation of minerals preparatory to
floatation. All should be of great interest to the ceramist but
have not yet been applied to any extent.[10]

MAGNETIC SEPARATION

Magnetic separation as applied to kaolin has developed substantially in the last ten years in the USA for removal of ferrous
titania (particle size 1 micron) and in Great Britain for removal
of mica (particle size 5 microns), largely to reduce iron content
from biotite but also to reduce alkali content from muscovite
(which contains 1-2% Fe_2O_3).

Magnetic systems employing the Jones reciprocating separator
were in use in Great Britain for ceramic clays in the early 1960's
(a paper was given to the American Ceramic Society on this process)
but the most widespread development came from application of the
Kohn magnet system for paper clays in the late 1960's.[11,12] The
Kohn type of magnetic filter succeeds by avoiding segregating a
large part of the field into the matrix in the separating chamber,
a fault of the Jones system.

The system should of course be applicable to ceramic bodies
but, unfortunately from an economic point of view, costs are high
if the magnet is small (by buying area and using volume), so that
it is to the raw material where treatment can be on a large scale
that the most attractive economics can be applied.

An interesting development of magnetic processing is the commercial design and application of superconducting systems which
present some considerable economic attractions, e.g., lower capital
and much lower operating costs and greatly improved accessibility.
Superconducting magnets of large size cannot be switched so that

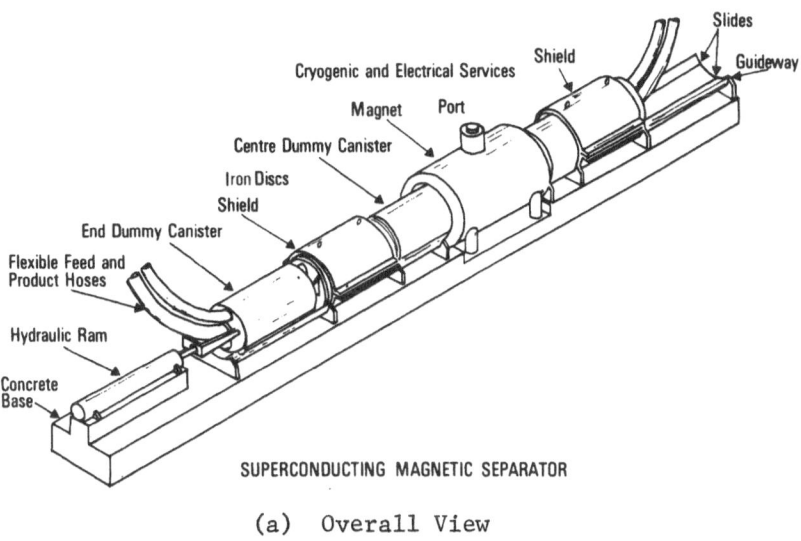

SUPERCONDUCTING MAGNETIC SEPARATOR

(a) Overall View

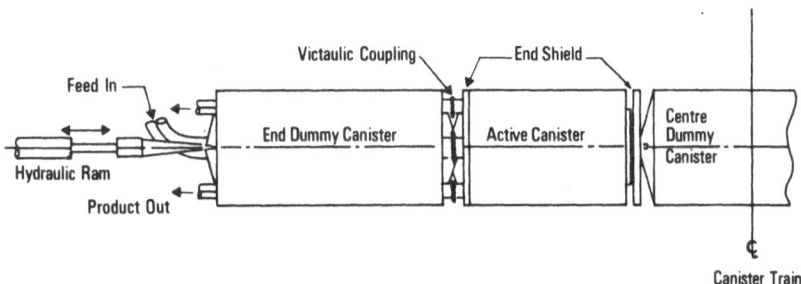

CANNISTER TRAIN ASSEMBLY SYMETRICAL ABOUT CENTRE
DUMMY CANISTER

(b) Detail of Canister Train Assembly

Figure 2. Superconducting Magnetic Separation.

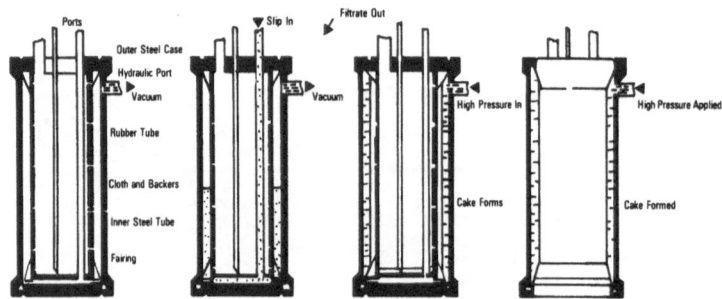

Figure 3. Tube-type solids separation (schematic).

the matrix must be reciprocated onto the field. Fields are of
course higher than in conventional solenoids by a factor of
1 1/2 - 2 to 1, and this makes for a small size. Moreover, no heavy
iron core circuit is required so that the matrix chambers are
virtually open for easy access [13] (See Figure 2).

All these systems for purification are run wet because the
forces between particles are lower or can be made lower and turbu-
lence is more easily avoided in a more viscous medium. As the
particles are fine in these minerals, the permeability is inherently
low and the volume occupied on settlement is high. In other words,
the viscosity of a suspension is high at a given concentration.

SOLIDS-LIQUID SEPARATION

Solids-liquid separation techniques are very relevant to any
purification system. This has led to a great interest in the
development of systems of thickening and filtration. There are two
solid forms of commercial interest for clays; a plastic solid for
ceramic working and a hard solid for transport in bulk. Two new
filtration systems directed towards the latter objective are
considered.

The problems of reducing water content to a low enough level
for direct transport implies pressures far above those in normal
use for clays, i.e., in the range 1000-2000 psi. There were severe
difficulties in the design of a plate filter for this range, but it
has been done in Great Britain. Not the least of the difficulties
was the design of a suitable pumping system for an abrasive slurry.
The problem has been solved by an ingenious system which isolates
the pump from the slurry by a free piston.

   An alternative is to use a tube as the containing vessel which
allows very high pressures and again isolates the slurry from the
pump.[14] The success of this machine comes from the simple design
of the seal at each end which uses the isolating diaphragm to form
the seals at each cycle - in fact a principle similar to that of the
cup washer. This machine can filter kaolin to 17% water content
which provides a hard solid for direct transport and avoids thermal
drying completely. A particular application has been in the
preparation of precipitated magnesite for refractories to give a
necessary high density to this material (see Fig. 3).

REFERENCES

1. C. L. Aplin and G. C. Argall, Tailings Disposal Today, Miller
       Freeman, publ. Pub. I. S. Book No. 0-87930-030-5.
2. N. O. Clark, Paper 14, 10th Int. Min. Proc. Congr., London, 1973.
3. E. W. Green, J. C. Duke and J. L. Hunter, U. S. Pat. 2990958
       (1961).
4. J. B. Duke, U. S. Pat. 3151062 (1964).
5. V. Mercade, U. S. Pat. 3331505 (1967).
6. E. K. Cundy, U. S. Pat. 3450257 (1969).
7. H. L. Shergold et al, Mtg. 5, Paper 3, 12 Int. Min. Proc. Congr.,
       Brazil, 1977.
8. Freeport Sulphur Co., U. S. Pat. 3432030 (1969).
9   (a) H. L. Shergold and O. Mellgren, Min. Metall., 78, 0121-32
       (1969).
    (b) Ibid., 79, C244-45 (1970).
10. English China Clays Lovering Pochin Co. Ltd., Brit. Pat.
       1469028 (1977).
11. H. H. Kolm, U. S. Pat. 3567026 (1971).
12. J. Iannicelli, Ger. Pat. 211986 (1971).
13. Watson, N. O. Clark and G. W. Windle, 11th Int. Min. Proc.
       Congr., Cagliari, 1975.
14. Filtration and Separation, March/April, 1971.

PARTICLE SIZE AND PERMEABILITY IN SLIP CASTING

G. W. Phelps

Rutgers University

New Brunswick, N. J.

INTRODUCTION

Ceramic slips are suspensions of polydisperse, lyophobic particles in a liquid, with particles ranging more or less continuously from an upper to some lower size limit. Colloidal lyophobic particles aggregate readily in suspensions, owing to Brownian motion and attracting van der Waals forces, unless these forces are offset by electric double-layers or liquid-solid affinity or both[1]. Casting slips are fluid high solids suspensions of one or more particulate materials deflocculated by one or more agents that cause development of diffuse electric double-layers. Certain other agents promote compression of double-layers and lead to coagulation. In practice, casting slip rheological and casting properties are controlled by a balance between coagulating and deflocculating agents[2].

Slip casting may be defined as consolidation of slip particles into a plastic mass through removal of part of the liquid by an absorbent mold, which brings the particles close enough for repelling forces to be offset by natural forces of attraction. The rate of coagulation rises sharply as the concentration of particles is increased[3]. Casting is a form of low pressure filtration[4] where the mold provides the driving force by suction and is a reservoir for separated liquid. Slip casting is an unsteady-state phenomenon with variable, increasing resistance to liquid flow as the cast develops. Resistance is related to a property called permeability.

Adcock and McDowall[4] developed an equation which permits determination of effective specific surface of a cast or filter-cake by a permeability measurement,

$$L^2/T = 2P_g E^3/5S_o^2 \eta (y-1) (1-E)^2 \tag{1}$$

where L is the cast wall thickness in time T, P the suction or pressure, E the voids fraction of moist cast or cake, η the liquid medium viscosity, S_o the effective specific surface, g the gravity constant, and y that volume of slip containing (1-E) volume of solids. The specific surface S_o is related to permeability K by the Kozeny equation

$$S_o^2 = E^3/5K (1-E)^2 \tag{2}$$

Commercial casting slips are usually moderately thixotropic, attaining a loosely gelled structure within a few seconds of cessation of shearing. The structure continues to strengthen only slowly from this point and stress provided by pouring is sufficient to break the gel. Ryan and Worrall[5] have demonstrated that the gelled structure plays a significant part in governing rate of cast.

EXPERIMENTAL

The experiments described here deal with the effects of degree of deflocculation, colloid content and modifiers, and size distribution on effective specific surface and permeability as determined by low pressure filtration.

(1) Porcelain Casting Slip

Fig. 1 shows particle size distribution curves for a blend of ball clays, a kaolin, and a mixture of ground quartzite and feldspar. A combination of these components constitutes the porcelain casting formula whose particle size curve also appears. Casting slips were prepared from this formula by dispersing 280 g of the ball clay in 370 ml of distilled water to which had been added 0.2 g of anhydrous sodium carbonate and either 0.2 g barium carbonate (for Series A slips) or 0.2 g precipitated calcium sulfate (for Series B slips). Fluidity was maintained during batching by drop-wise additions of an equal-volumes solution of distilled water and liquid sodium silicate ($Na_2O \cdot 3.22 \ SiO_2$, $40°$ Baume'). A 200 g portion of kaolin was next incorporated, followed by 520 g of non-clay powder. Finished slip was adjusted to a selected viscosity and thixotropy, as determined with a rotating disc viscometer, by further additions of sodium silicate solution. Thixotropy was

taken as the increase in viscosity over a 3 minute period. Three
progressively lower viscosity and thixotropy slips were prepared
for each slip series. These were marked, respectively, Slips AA,
AB and AC and Slips BA, BB and BC. The slip points were stored in
closed containers for 24 hours at room temperature, restirred and
again tested for viscosity and thixotropy. Each slip was cast for
25 minutes in a cup-shaped plaster test mold, drained for 5 minutes
and the casts removed for observation and measurement of wall
thickness with a vernier caliper. The several slip points were
each filterpressed in a laboratory press for 25 minutes at 35 psi
(2460 g cm^{-2}), drained of excess slip and the cakes re-pressed for
5 minutes at 35 psi. Moist cake thickness was measured with a
penetrometer and reported in cm. The cakes were weighed, dried 24
hours at $105°$ C. and reweighed. These data were used in equations
(1) and (2) to calculate S_0 and K values.

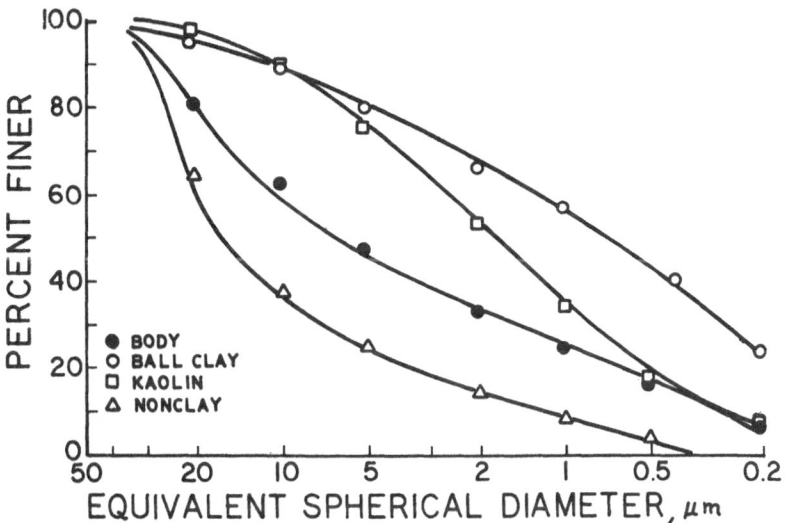

Fig. 1:- Particle size distributions of clay and nonclay constituents and of a typical porcelain casting formula.

(2) Cleaned Whiteware Slips

The experiment described here was performed by Casale[6] in a
demonstration of the effect of varying nonclay component particle
size on casting qualities of a sanitaryware-type slip. The constant clays component consisted of a combination of coarse-grained
and fine-grained Georgia kaolins, which had been subjected to
repeated washings to remove soluble salts, and by treatment with

an hydrogen-form cation exchange resin to assure a constant cation.
A series of quartz distributions served as the nonclay powder component of bodies whose particle size distributions appear in Fig.
2. The body materials were deflocculated in distilled water to
minimum viscosity (as determined with a rotating disc viscometer)
with sodium silicate ($Na_2O \cdot 3.22\ SiO_2$, $40°$ Baume'). The slips
were pressed at 35 psi for 16 minutes in a laboratory press,
excess slip drained and the cakes re-pressed at 35 psi for 4 minutes. Cake thickness was measured with a penetrometer and the
moist cakes weighed, dried at $105°$ C. for 24 hours, reweighed, and
the data used in calculating S_0 and K.

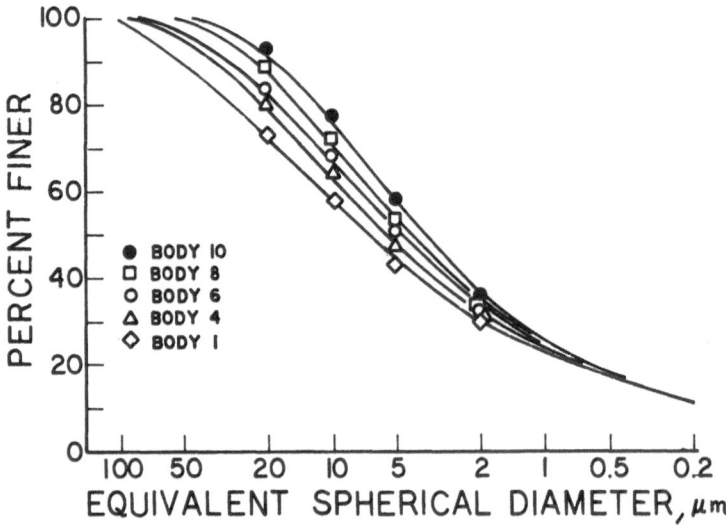

Fig. 2:- Particle size distributions of organic- and solubles free
sanitaryware casting bodies having a constant clays component and
variable nonclay component particle size distribution. Redrawn
from data of Casale[6].

(3) Narrow Distribution Clay Slips

The experiments outlined here were reported in detail by Vie
Wei Liu[7] in work designed to show the effect of removing the colloid fraction (minus 0.1 micron) and variable particle-size distribution on the rheological and filterpressing characteristics of
clay slips. Fractionated, cleaned, hydrogen-form kaolins were
blended to give the range of distributions lying between 44 and
0.1 micron shown by the curves of Fig. 3. The several distributions were deflocculated in distilled water with sodium hydroxide
at 67 percent solids and taken to minimum viscosity, as indicated
by a rotating disc viscometer. A 217 ml portion of each slip was

introduced into the barrel of a laboratory press and pressed at 40 psi until cessation of drip. Time was recorded and the cake was measured for thickness with a penetrometer, removed, weighed and dried at 105° C. The dry cakes were reweighed. These data were used to calculate S_o and K from equations (1) and (2).

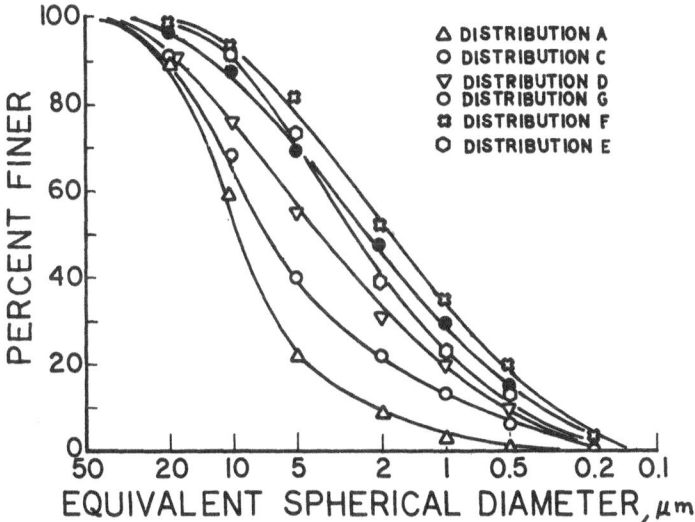

Fig. 3:- Particle size distributions of a number of cleaned, fractionated kaolins whose size limits arc 44-0.1 micron equivalent spherical diameter. Data taken from Vie Wei Liu[7].

RESULTS AND DISCUSSION

Case (1) experimental results appear in Fig. 4 and Table I. Increasing deflocculant was accompanied by progressive reduction of viscosity and thixotropy, decreasing rates of cast and filtration rates, decreasing voids fraction, decreasing permeability and increasing effective specific surface. The presence of higher sulfate and calcium ions of Series B slips coincided with higher rates of cast, improved mold release and improved feel. Equivalent viscosities and thixotropic buildups did not result in identical casting properties for Series A and Series B slips. The differences are probably due to (a) the extent of soluble humate formation and (b) the effect of exchangeable calcium in governing the formation of casts and filtercakes.

The sulfate ion reduces the production of hydroxyl ion from sodium carbonate and lessens neutralization of acidic groups of clay-associated organic matter. Alkaline earth humates are

insoluble at pH levels normally encountered in clay slip casting[8].
Hence, Series A slips with lower sulfate and calcium produced more
and finer humate particles[9], and responded more sharply to adjusting sodium silicate than did Series B slips. The less dispersed
humates and greater calcium ion content of Series B slips allowed
formation of a more open and stronger moist structure than found
for Series A casts and cakes, which accounts for the higher casting rates, greater permeabilities and lower effective specific
surfaces of the former.

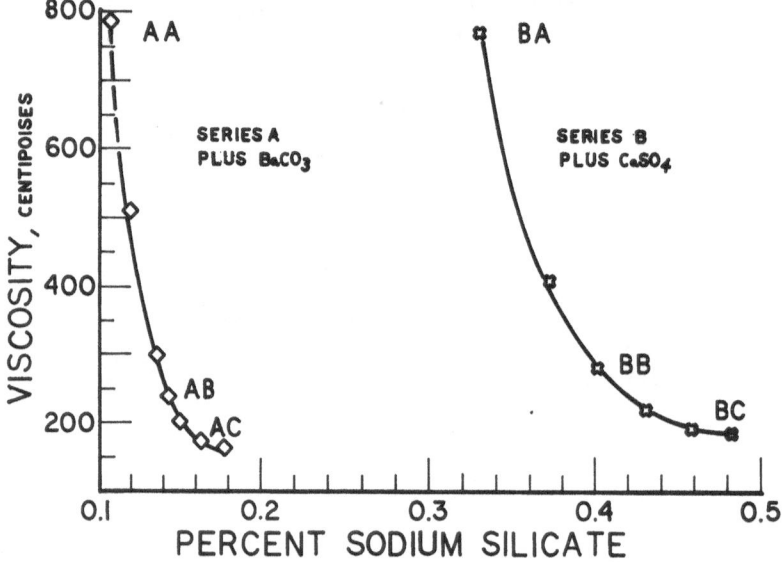

Fig. 4:- Deflocculation curves for a typical clay-based porcelain
casting slip at two levels of sulfate and calcium ions. Sulfate
of Series A slips was reduced by barium carbonate. Sulfate and
calcium levels of Series B slips were increased by adding calcium
sulfate.

Case (2) experimental data are listed in Table II. Contrary
to expectation, increasing intermediate-size particle concentration, shown by the curves of Fig. 2 for Body B1 to Body B10, did
not result in increasing deflocculant demand and increasing viscosity at maximum deflocculation. Instead, viscosity dropped
significantly from Slip B1 through Slip B8 then rose somewhat for
Slip B10. Deflocculant requirement dropped moderately from Slip
B1 through Slip B6, rose again for Slip B8 and dropped slightly
for Slip B10. Voids fractions for filter cakes from Slips B1-B8
did not vary significantly but rose slightly for Body B10 slip.
Permeability values rose steadily from Body B1 through Body B8,

accompanied by a corresponding drop in effective specific surface. However, Body B10 slip showed a drop in permeability and an increase in effective specific surface compared to Body B8 slip.

TABLE I

Porcelain Casting Slips

Slip, Cast, Cake Properties	Series A			Series B		
	AA	AB	AC	BA	BB	BC
Fresh , centipoise	810	240	164	807	260	186
Buildup, centipoise	160	90	4	160	90	24
Mold Cast:						
cm/25 min.	0.62	0.47	0.40	0.77	0.55	0.50
Release	Poor	Poor	Poor	Good	Good	Good
Feel	Soft	Soft	Hard	Good	Good	Good
Filterpressed:						
Voids Fraction	0.343	0.340	0.326	0.358	0.344	0.329
K, cc/sec^2 x 10^{10}	2.88	2.69	1.95	4.49	3.35	2.65
S_o, cm^2/cc x 10^{-5}	2.52	2.56	2.78	2.21	2.35	2.42

TABLE II

Organic- and Salts Free Slips

Slip, Cake Properties	Body Distributions				
	B1	B4	B6	B8	B10
% Sodium Silicate	0.152	0.146	0.142	0.147	0.143
Min. , centipoise	254	213	188	175	190
Voids Fraction	0.388	0.392	0.387	0.392	0.398
K, cc/sec^2 x 10^{10}	2.22	2.95	3.37	4.09	3.37
S_o, cm^2/cc x 10^{-5}	3.71	3.39	3.00	2.83	3.13

The apparently anomalous data of Case (2) are paralleled by
results obtained with calcined alumina slips having intermediate-
size deficits[10]. Casting was slower than expected and casts stuck
severely. Investigation revealed a large concentration of fine
particles at the mold face. It was postulated that the movement
of liquid into the mold carries fine particles through the coarser
material and concentrates them in a compact, low permeability
layer at the mold face. In the case under consideration, it is
possible to visualize an increasing concentration of intermediate-
size particles in the 5-50 micron region serving to trap mobile
colloidal and near-colloidal particles and so lessen buildup of a
low permeability zone at the mold face or filter paper. Once the
concentration of intermediate-size particles passes some critical
point the cake or cast loses permeability normally and effective
specific surface rises as expected.

TABLE III

Low Colloid, Clean Clays

Clay Distributions	Voids Fraction	Permeability cc/sec^2 x 10^{10}	Sp. Surface cm^2/cc x 10^{-5}
A	0.510	369.0	0.54
C	0.456	103.0	0.86
D	0.442	67.3	0.90
G	0.422	34.1	1.13
F	0.433	30.4	1.29
E	0.450	56.2	1.06

Case (3) data for the essentially colloid-free kaolin
distributions of Fig. 3 and in Table III are in accord with work
on permeability by Bo[11] et al, which showed that noncolloidal
glass bead distributions gained in permeability as deficits of
intermediate-size particles increased and voids fractions rose
correspondingly. However, the more extended the distribution
limits the greater the change in permeability and voids fraction.
An increase in intermediate-size particles beyond the optimum
distribution for minimum voids resulted in decreased permeability
and slightly increased voids fraction. Data for the sequence A,
C, D to G in Table III show steady decreases in voids fraction
and permeability and an accompanying increase in effective specific

surface. Distribution E has a superficial resemblance to Distribution G in Fig. 3 but its cakes had higher voids, were more permeable, had low effective specific surface. An inspection of Vie Wei Liu's data indicated that Distribution E constituted a more narrow-range system than the other distributions. The increase in voids volume and continuing reduction of permeability and effective surface area of Distribution F in the sequence A, C, D, G and F might be taken as indicative of an excess of intermediate-size particles.

In conclusion, it is evident that permeability and effective surface area of filtercakes and casts made from deflocculated ceramic suspensions depend upon (a) the size range and distribution of the suspended particles, (b) the amount and fineness of a colloid fraction, (c) the kinds and amounts of deflocculants, (d) the kinds and amount of coagulating substances, and (e) the liquid viscosity. Although the point has not been discussed here it has been shown that increasing pressure normally results in increased rate of cast or filtration[4]. However, under circumstances where there is a deficit of intermediate-size particles and the system is fully-deflocculated it might well happen that permeability would decrease owing to increased compaction at the mold-slip interface.

REFERENCES

1. D. J. Shaw, Introduction to Colloid and Surface Chemistry, pp. 164-71, 2nd Ed., Butterworth's, London, 1970.
2. G. W. Phelps, Am. Ceram. Soc. B., 38, 246-50 (1959).
3. Irving Reich and R. D. Vold, J. Phys. Coll. Chem., 63, 1497-1501 (1959).
4. D. S. Adcock and I. C. McDowall, J. Am. Ceram. Soc., 40, 355-62 (1957).
5. W. Ryan and W. E. Worrall, Tr. Brit. Ceram. Soc., 60, 540-55 (1961).
6. A. Casale, M. S. Thesis, Rutgers University, New Brunswick, N. J., 1977.
7. Vie Wei Liu, M. S. Thesis, Rutgers University, New Brunswick, N. J., 1976.
8. A. N. Puri, Soils: Their Physics and Chemistry, Ch. XXIV, Reinhold Publ. Corp., New York, 1949.
9. F. J. Stevenson and J. H. A. Butler, Ch. 22 in Organic Geo-Chemistry, Eds. G. Eglintonand M. T. J. Murphy. Springer-Verlag, Berlin-Heidelberg, 1969.
10. G. W.Phelps, A. Silwanowicz and W. Romig, Am. Ceram. Soc. B., 50, 720-22 (1971).
11. M. K. Bo, D. C. Freshwater and B. Scarlett, Trans. Instn. Chem. Engrs., 43, T228-32 (1965).

CHEMICAL PROCESSING FOR CERAMICS (AND POLYMERS)

P. E. D. Morgan

Department of Mettallurgy and Materials Science
University of Pittsburgh, Pa. 15261 USA

ABSTRACT

Using the specific chemistry of the elements involved, very
simple techniques for active powder synthesis can usually be devised. "Active" powders may need special precautions in the
handling and forming but, quite often, the required dense, fine
grained, ceramic, especially in complex systems, cannot readily
be achieved except by their use.

INTRODUCTION

The densification of "active" powders ($\ll 1\mu$ grain size) is
not readily describable by simple sintering models, nor can sinterability be quantitatively predicted through characterization (as
yet). Nevertheless, a set of ceramic principles is well established
for guidance in choosing the type of powder, sintering conditions
etc. for maximum densification with (usually) minimum grain growth.
Increasingly we run into the need to sinter more complex systems
and, as the number of elements involved increases, so also do
the potential for unwanted side effects. Particular problems are
discontinuous grain growth, which increases with the presence of
liquid phases or the possibility of evaporation condensation (EC),
and the problems of the latter removing the driving force for
sintering while not contributing to densification. In the case of
the more covalently bonded materials of current interest e.g., SiC
and Si_3N_4, the EC mechanism is a serious problem at the very high
temperatures needed for adequate diffusional mechanisms and the
introduction of liquid boundary layers will usually spoil the

very properties desired at high temperature. Systems containing
alkalies, (β-alumina, lithium ferrite), Pb(PLZT, etc.), Ti, Cr(La
CrO$_3$), chalcogens, and others all have problems related to EC.
Other problems arise in holding the required valence state especially at high temperatures of, for example, transition elements.
In many cases, however, (in fact in all that have been thoroughly
studied) the problems are most easily overcome by appropriate
chemical forming and handling of the starting powders. The best
starting powder often allows sintering (or hot pressing) at much
lower temperatures, where grain growth, EC, valence changes, are
minimal. "Active" powders often retain a "relic" structure depending on their origin and chemistry and so very different
behavior can be seen in powders that ought (on the basis of
simple characterization) to be similar. It is often easier to
change the starting powder than embark upon prolonged studies of
why a particular source material behaves as it does. This seemingly empirical approach has not been considered very de riguer
even though it is at the heart of ceramic technology.

SOME APPROACHES TO CHEMICAL SYNTHESIS OF POWDERS

A common ceramist's myth is that liquid chemical techniques
for powders are expensive and not easily adaptable to the large
scale. Two of the very largest fine powder techniques, i.e.,
rutile for paints and Bayer aluminum hydroxide for the Hall
process are, in fact, highly commercial processes.

The rather general methods of powder preparation for ceramics
or catalysts, e.g., coprecipitation, freeze drying-calcining,
spray drying-calcining, spray roasting, sol-gel processes, flame
spraying and use of citrates, lactates, polyelectrolytes, and
resins, have been well reviewed in recent years.[1,2] Many mixed
oxides, however, can be made by much simpler direct precipitation
routes; this is especially true of ferrites, titanates, zirconates,
and niobates, where the element does not form very stable hydroxides. Chemistry texts 130 years ago[3] mentioned the formation of
magnetite, Fe$_3$O$_4$, simply by pouring mixed ferric and ferrous
solutions into boiling alkali, e.g.,

$$FeCl_2 + 2FeCl_3 + 8NaOH \rightarrow Fe_3O_4 + 8NaCl + 4H_2O \qquad (1)$$

Ten years later air oxidation of suspended boiling ferrous
hydroxide was also observed to give magnetite:[4]

$$6Fe(OH)_2 + O_2 \rightarrow 2Fe_3O_4 + 6H_2O \qquad (2)$$

Many other processes were subsequently developed for preparing
this material.

Forestier and Longuet[5] prepared a second ferrite, $CuFe_2O_4$, by an aqueous method, although the method was subsequently disputed by Milligan and Holmes.[6] Schuele and Deetscreek[7] showed that nickel ferrite, $NiFe_2O_4$ (grain size $\sim$ 200 Å), could be similarly formed but required 50 to 100 h. in boiling solution. Sato et al.[8] similarly made $MnFe_2O_4$, $CoFe_2O_4$, $Ni_{0.5}Zn_{0.5}Fe_2O_4$, $ZnFe_2O_4$, $CuFe_2O_4$, and others and showed that the pH must be above 12 to obtain the compounds with a normal grain (or crystallite) size of 300 to 450 Å. It appears then that the pH was different in the Forestier and Milligan works.

Sato et al.[9] also produced a range of compositions $Mn_{1-x}Co_x$ Fe_2O_4 and were able to study the limiting particle size for superparamagnetic behavior. Takada and Kiyama[10] returned to the aqueous aerial oxidation technique to prepare a range of ferrites with a grain size of 500 Å to 1 μ m:

$$6(M,Fe)(OH)_2 + O_2 \rightarrow 2MFe_2O_4 + 6H_2O \qquad (3)$$

$M = (Mn^{2+}, Zn^{2+}, Co^{2+}, Mg^{2+}, Fe^{2+},$ etc. It was noted that barium hexaferrite could not be achieved in this simple way. Barium titanate has been made by direct aqueous precipitation[11,12] and by air oxidation using Ti^{3+} precursors.[13]

In attempting to study the hot pressing of lithium ferrite, $LiFe_5O_8$, it was natural to see if this compound could be made by the direct method; indeed it can.[14] A typical preparation is as follows: 420g (10 mol) $LiOH \cdot H_2O$ is dissolved in 3000 ml distilled water. A solution of 270g (1 mol) $FeCl_3 \cdot 6H_2O$ in 200 ml distilled water is added to the cold lithium hydroxide solution in a blender. This produces a uniform dispersion of gelatinous, foxy red $Fe(OH)_3$. The mixture is slowly heated with stirring, out of contact with air/CO_2. At $\sim$80C, chocolate brown magnetic lithium ferrite starts to form. The mixture is gently boiled overnight, then cooled. The container is placed between the poles of a very strong magnet. The $Li_{0.5}Fe_{2.5}O_4$ precipitate is pulled to the sides and the supernatant liquid poured off. The solid is quickly washed 6 times with distilled water, each time using magnetic separation. A final wash and rinse with acetone is followed by air drying. The yield is quantitative. It was noted that, in an aqueous suspension below $\sim$ pH 11, fine lithium ferrite reverts to ferric hydroxide. Thus filtering and washing of the ferrite are difficult (if not impossible); the magnetic separation method is therefore far superior.

The product has a particle size of 100 to 300 Å and gives an X-ray diffraction pattern normal for lithium ferrite but with considerable line broadening. Firing to 350°C in air produces a sharp X-ray diffraction pattern with additional lines due to ordering in the structure.[15] The cubic cell parameter is 8.327 $\pm$ 0.003 Å.

The material may be hot-pressed at 4000 psi $\rightarrow$ 900°C, giving a 99% dense body with a grain size 0.1 to 0.5 μm.

Obviously many variations of the basic preparation are possible: various iron and lithium salts plus various alkalis may be used, and air oxidation will work. Other ions useful for control of magnetic properties can be easily incorporated.

In a quick test similar to the above, a solution of titanium tetraisopropylate in isopropanol is blended into strong cold lithium hydroxide and the mixture then boiled: Li_2TiO_3 similarly precipitated, as evidenced by X-ray diffraction. Chromium and aluminum analogs of lithium ferrite are not produced by this technique. Attempts to produce orthoferrites, i.e., $LaFeO_3$ or the analogous chromium or aluminum compounds, by similar techniques failed. Also tried was to obtain beryllium ferrite (which is unknown), analogous to chrysoberyl, $BeAl_2O_4$; this too failed. Some quick tests have revealed that other simple compounds are obtainable in boiling alkaline solution thus: concentrated LiOH solution with silica gel slowly gives Li_2SiO_3, with cobalt solutions $LiCoO_2$ appears with air oxidation, with manganese both $LiMnO_2$ and Li_2MnO_3 have been seen as evidenced by XRD.

Analogous to the method of Flaschen,[11] it is possible to produce $KTaO_3$, a perovskite, by pouring liquid tantalum pentaethoxide into hot, pH 14, KOH solution. The material shows good crystallinity by XRD, with no visible traces of other phases. If tantalum pentachloride is used the reactions are more complicated and $KTaO_3$ was not initially formed. However, when a very strong solution of KOH (50% by weight) was used an unknown phase separated which upon dilution led then to the formation of $KTaO_3$. Thus a prediction made earlier[14] has been confirmed. With Niobium salts, including the pentaethoxide, $KNbO_3$, was never directly obtained, rather, complex sodium niobate hydrates are produced.

Another unknown compound, $BeSiO_3$, was attempted by direct methods but not obtained. However, when the washed beryllium silicate gel was heated to $\sim$ 300°C in an air oven, BeO was always the first crystalline phase to appear; phenacite, Be_2SiO_4, is apparently a metastable phase at low temperatures. I do not know if this has been noted before. One might add the low temperature regions of many ceramic phase diagrams are assumed rather than measured and here low temperature synthetic approaches may be helpful.

It is apparent that the higher valent states of e.g.,Mn and Co are more stable at lower temperatures and direct synthesis in water/air can be achieved for active powders (for catalysis, etc.) that otherwise can only be obtained at higher temperatures with oxygen overpressure. It is especially in these higher valent transition element compounds (with Ca, Sr, Y, Sc, etc.) that new types may be readily discovered.

Where direct synthesis is not immediately achieved, the use of co-precipitation, even if not completely homogeneous on a molecular scale, is so vastly superior for uniform powder preparation to the use of ball milled oxides, that it should be the method of choice. Special "tricks" are often possible. The oxides of Ca, Sr, Li, lanthanides, etc. are sufficiently basic to allow double decomposition reactions to produce gels without the use of other bases. If appropriate quantities of these oxides are added to hot strong solutions of transition metal nitrates or acetates, thick reaction pastes can be achieved. On drying, little phase separation occurs. A very light ball milling and precalination will generate fairly pure perovkites or potassium nickel fluoride types $La_{2-x}Sn_xXO_4$, $La_{1-x}Sr_xXO_3$, at temperatures around 600°C. These powders are sufficiently active that good sintering of isostatically pressed boules can be achieved at ~ 1200°C in air. We do find a definite advantage to isostatic pressing, at 50,000 p.s.i., as opposed to uniaxial pressing ~ 10,000 p.s.i., in this type of work. Presumably the higher pressure, allowed by the iso-pressing method, crushes the aggregates together quite efficiently avoiding non-uniform shrinkage effects. Sometimes the difference between use of the two techniques can be as much as 98% final density versus ~ 80%. It is remarkable that it has not been especially noted in the literature that with very fine, active, powders such large differences may be observed. Undoubtedly some observers have missed potentially good sintering character-istics by using insufficient pressing pressures.

With these perovskite types some caution is required in that sometimes the temperature required for good sintering in air causes some degradation through loss of oxygen and conversion of higher valent states to lower ones. If this is carefully con-trolled, a lower temperature anneal in air at ~ 900°C will convert the material to the proper pure perovskite or potassium nickel fluoride structure without loss of density. Coincident with this the electrical resistance of the sample may radically drop thus for $La_{0.5}Sr_{0.5}FeO_3$ the resistance drops from 2Ω cm as sintered to 7×10^{-2} Ω cm after the air anneal.

Much of the data in the literature on both pressed powders and sintered samples is highly suspect because of the problems of holding the ideal valences. Thus Gobalakrishnan et al.[16] give

values of $1 - 6_\Omega$, cm for $La_{2-x}Sr_xNiO_4$, our sintered and annealed
sample of $LaSrNiO_4$ runs $\sim 5 \times 10^{-2}{}_\Omega$, cm at room temperature
to $4\times10^{-3}{}_\Omega$, cm at $900^\circ C$.

As mentioned earlier $Al(OH)_3$, from the Bayer process, is readily
available cheaply in large quantities, and can be highly pure. Even
so, one of the major impurities may be sodium; however, for forming
β-alumina, for the Na/S battery for example, the soda content is im-
material. Precursor powders which satisfactorily can be sintered to
dense β-alumina (actually Li containing β'') can readily be made by
pouring appropriate water solutions of sodium bicarbonate (or carbonate
etc.) and lithium hydroxide into a blending slurry of Bayer $Al(OH)_3$,
C331, Alcoa, in isopropanol.[32] The slurry thickens to a viscous paste,
is then dried and lightly milled, calcined at $600^\circ C$ and isopressed to
boules for sintering to dense β'' at $1590^\circ C$. At $600^\circ C$ the precursor
powder is "η" alumina but the precalcination may be at sufficient temperature, $\sim 1100^\circ C$, to preform the β type. Again we believe that isopressing at 50,000 p.s.i. is an important necessary part of handling
these fine powders.

Ideally for a good powder preparation the specific chemistry
of the elements involved should be used. A good example of this
is the case of chromium which, because it forms the rather stable,
water soluble, higher oxide, CrO_3, allows some especially simple
procedures. In the preparation of $La_{1-x}Sr_xCrO_3$ one can completely
avoid the use of anions such as sulfate, nitrate, acetate, etc.,
by the following procedure:[17] to a strong solution of chromic
acid in water is added, in order, appropriate quantities of La_2O_3
and $SrCO_3$. The mixture quickly heats up in a blender as reaction
occurs to a thick paste of hydrated lanthanum and strontium
chromates (unidentified). After drying at $120^\circ C$ overnight, the
material is very lightly milled, the crystals are quite soft.
Heating to $600^\circ C$ in air shows only the perovskite phase, for
$\chi = 0 \rightarrow 0.16$, as the color changes from yellow to green through
shades of brown to chocolate depending on χ. Because the powder
is preformed as perovskite, with no Cr_2O_3 present, the volatility
of chrome is lower than in unreacted chrome systems and this type
of material has been isopressed and sintered in air to 98% theoretical density at $1500^\circ C$.

Pearlman et al.[18] who showed beautifully the power of chemical
techniques in producing transparent mixed chalcogenides, also
avoided the use of unwanted anions in their preparation of thiochromites, e.g., $ZnCr_2S_4$ spinel, by starting with chromate solutions and reducing with hydrazine to produce floculent mixed
hydroxides of Zn and Cr which were subsequently heated with
sulphur containing reducing gases H_2S, CS_2 etc. to produce the
active thiochromites. Pearlman et al.[18] have reviewed the various

methods of obtaining active mixed chalcogenides including the
reaction of H_2S with coprecipitated hydroxides first examined as
long ago as 1880.[19]   In addition, recent approaches have included
the addition of halide fluxes to stimulate the direct synthesis
from the elements.[20] However, simple precipitation techniques
can also be applied to the production of mixed chalcogenides:[21]
initially, in our work, in an attempt to prepare $CuCr_2Se_4$, a
stream of H_2Se gas (prepared by reacting water and Al_2Se_3) was
bubbled into a dry solution of Cu and Cr stearates in xylene.
The black amorphous precipitate produced could be converted into
crystalline $CuCr_2Se_2$ by heating it to $300^{\circ}C$ in evacuated tubes.
This result suggested that, if precipitation could be conducted
at ~ $300^{\circ}C$, the crystalline spinel might be produced directly.
Furthermore, the handling of H_2Se gas was unpleasant and awkward.
Since Se selectively will dehydrogenate cyclic 6 membered carbon
ring systems such as sterols at ~ $300^{\circ}C$, giving aromatic hydro-
carbons and H_2Se (a method used for their characterization, a good
example of very specific chemistry), an internal generation was
considered.

    23.32g (0.05 mol. on the basis of analyzed Cu) of Cu stearate
and 119.02g (0.10 mol. on the basis of analyzed Cr) of Cr stearate
with 35g sitosterol* and 23.69g Se chips (0.2 mol. + 50% excess)
were stirred into 350 ml of nujol in a 3-1 reaction vessel and
slowly heated to 330°C. A clear solution was obtained, since
water was driven off at ~ $160^{\circ}C$. After 1 hour at $300^{\circ}C$ the solution
had precipitated magnetic material. After a total of 8 hours at
$330^{\circ}C$, the reaction mixture was allowed to cool to 70° to $80^{\circ}C$
and diluted with 1-1 of benzene. The black solid was magnetically
separated and washed well with benzene. Separation and washing
by magnetic means were simple and greatly preferred to filtration.
The yield was 21.8g (90% of theoretical). The exclusion of air
from the reaction seemed unnecessary (although an N_2 atmosphere
was used initially) and the product was stable to boiling water.

    XRD and magnetic analysis proved that the material was pure,
stoichiometric, ferromagnetic $CuCr_2Se_4$ spinel with the correct
cell parameter, 10.33 Å,[22] composed of particles 200 to 300 Å in
size.

    Although Te is a much less active dehydrogenating agent than
Se, when Te in an active form+ was used, analogously, $CuCr_2Te_4$ was
produced, although impure and in poor yield.

    Undoubtedly, other complex sulfides, selenides, and tellurides
could be synthesized by this or a similar method including, perhaps,
compounds that have been impossible to prepare by the standard
processes. It is possible that this type of technique is adaptable
to the production of hitherto mixed chalcogenides and to types
such as LaS with unusual formal valence states. Recently new

promising techniques have been evolved for the control of stoichiometry in powders of LaS[23] and B_7O[24] using respectively PbS and
ZnO as the controlled sources of the anions and reacting with
elemental La or B while readily removing Pb and Zn as vapors.

Due to the extreme difficulty encountered with sintering of
pure covalent compounds such as SiC and Si_3N_4 attention has
turned to chemical techniques here also to produce "active"
precursors. Thus powders of Si_3N_4 have been recently produced by
reacting silicon tetrachloride and ammonia (liquid)[25] and from
silane and ammonia[26] and encouraging sintering (or hot pressing)
results obtained where coarse powders, as made by, for example, the
nitridation of silicon, are unsatisfactory. Again it is interesting
that very early work on the chemistry of silicon nitride in 1879[27]
and 1903[28] involved just the same techniques now being reevaluated.
Other work is reviewed in (25).

Powder forming techniques, completely analogous to ceramic
methods, have been applied to the production of solid polymer
bodies.[29] Polyimidazopyrrolones , complex, ladder like, high
temperature stable polyheterocyclics, were molded (reactively
hot pressed) from ball milled mixtures of organic tetracids and
tetra-amines. While the idea of solid state synthesis of such
polymers was well known, no one had previously tried in one step
to simultaneously synthesize, sinter and mold such materials.
Exactly as for ceramic forming, advantage was gained by studying
the starting powders.

It was discovered that previously unknown 1 : 1 crystalline
salt adducts could be formed from pyromellitic acid (PMDA) and
diaminobenzidine (DAB) by rapidly mixing separate water/alcohol
solutions of the reagents. The formation of the well crystallized
orange adduct was greatly facilitated by "seeding", without
which amorphous precipitates were more often obtained. A new
deep purple salt of napthalene tetracarboxylic acid (NTCA) and DAB
was also attained. The separate acids and amines are often quite
expensive in highly pure form but the salt preparations worked
just as well with commercial materials, the salt precipitation
serving to purify the product and leaving the impurities in
solution. Elemental analysis confirmed this and showed that the
salts were indeed 1 : 1 adducts of the correct stoichiometry .

---

* Sitosterol, readily available as a waste product of sterol
processing, was obtained from the Vitamerican Corp., Little
Falls, N.J.

+ Prepared by dissolving bulk Te in concentrated H_2SO_4, pouring
the wine -red solution into excess water, and washing and drying
the precipitated, fine, black, Te.

It is assumed from the polymerization characteristics, that the salts contain "end to end" linear molecules in a disposition suitable for good linear polymer formation (with little cross linking).

Just as ceramics may be reactively hot pressed from precursors, i.e., dense magnesia from magnesium hydroxide,[30] so the organic salts could be reactively hot pressed (allowing water to escape) directly to finished "Pyrrone" polymer in a one step operation. Previously these polymers were only formed by multistep "wet" synthetic techniques and, moreover, were difficult to mold anyway. The polymer properties obtained were greatly superior to previous achievements. (For example, 20,000 p.s.i. tensile strength in a bulk molding at a density of 1.3 g/cc with an unrivalled hardness for these systems). Elemental analysis on the products confirmed that, while we were attempting to get better moldings, we actually obtained polymer much nearer to the theoretical C:H:N:O ratios than ever achieved for these polymers by the exacting "wet" chemical treatments. Only by control of solid starting powder and a solid state synthesis was this possible.

Afterthought: the question of availability in large quantities of the starting materials for chemical preps. always arises. Sometimes the sources are surprising; Mitchell[31] describes a good simple way of making $MgAl_2O_4$ spinel by using aluminum chlorhydroxide which is available on the large scale for underarm deodorants.

REFERENCES

1. R. Roy, Powd. Met. Int.6, 25 (1974) & Keram. Z. 26, 386 (1974)
2. A. L. Stuijts: pp. 335-62 in Science of Ceramics, Vol. 5. Edited by C. Brosset and E. Knapp. The Swedish Institute for Silicate Research. Gothenburg, 1970.
3. W. Gregory, p. 336 in Elements of Chemistry. Edited by E. Turner. Philadelphia, PA, 1846.
4. F. A. Abel and C. L. Bloxam, Handbook of Chemistry: p. 344, London, 1854.
5. H. Forestier and J. Longuet, C. R. Acad. Sci., 208, 1729-30 (1949).
7. W. J. Schuele and V. D. Deetscreek, J. Appl. Phys., 32 (3) 235, (1961).
8. T. Sato, M. Sugihara, and M. Saito, Kogyo Kagaku Zasshi, 65 (11) 1748-53 (1962).
9. T. Sato, C. Kuroda, and M. Saito: pp. 72-74 in Ferrites. Edited by Y. Hoshimo. University Park Press, University Park, PA, 1971.

10. T. Takada and M. Kiyama: pp. 69-71 in Ref. 9.
11. S. S. Flaschen, J. Amer. Chem. Soc., 77 (12) 6194 (1955).
12. K. Kiss and J. Magder, U.S. Pat. 3,292,994, Dec. 20, 1966.
13. T. Takada, M. Kiyama, Y. Bando, and T. Shinjo, Bull. Inst. Chem. Res., Kyoto Univ., 47 (4) 298-307 (1969).
14. P. E. D. Morgan, J. Amer. Ceram. Soc., 57, 499 (1974).
15. M. Schieber, J. Inorg. Nucl. Chem., 26 (8) 1363-67 (1964).
16. J. Gopalakrishnan, G. Colsmann, and B. Renter, J. Solid State Chem., 22, 145 (1977).
17. P. E. D. Morgan and R. Staut, U. S. Patent No. 3,974,108, August 10, 1976.
18. D. Pearlman, E. Carnall, Jr., and T. W. Martin, J. Solid State Chem., 7, 138 (1973). Also U. S. Patent #3,773,909 (1973).
19. M. Groger, Ber. Kais. Acad. Wiss. 81 531 (1880) and 83 749 (1881).
20. W. Kwestroo, pp. 563-74 in Preparative Methods in Solid State Chemistry. Edited by Paul Hagenmuller. Academic Press Inc., New York, 1972. See also: Harry Hahn and Bernhard Harder, Z. Anorg. Allg. Chem., 288, 257-59 (1956) and Abraham Clearfield, Acta Crystallogr., 16 (2) 135-42 (1963).
21. P. W. D. Mitchell and P. E. D. Morgan, J. Amer. Ceram. Soc., 57, 278 (1974).
22. M. Robbins, H. W. Lehmann, and J. G. White, Phys. Chem. Solids, 28 (6) 897-902 (1967).
23. F. Ehrburger and R. Roy, J. Solid State Chem. 18 201 (1976).
24. C. E. Holcombe, Jr. and O. J. Horne, Jr., J. Am. Ceram. Soc. 55, 106 (1972).
25. P. E. D. Morgan, Office of Naval Research, Arlington, Va., Dec. 1973, AD-778 373.
26. C. D. Greskovich and S. Prochazka, NATO Advanced Study Institute "Nitrogen Ceramics," Canterbury, England, Aug. 16-27 (1976) (in press). Also: C. D. Greskovich, S. S. Prochazka and J. H. Rosolowski, Tech. Report AFML-TR-76-179, Nov. (1976).
27. M. P. Schutzenberger, Sur l'azoture de silicum, Comptes Rendus, 89, 644-646, (1879).
28. M. Blix und W. Wirbelauer: Ber. Deut. Chem. Ges. 36, 4220-28 (1903).
29. P. E. D. Morgan and H. Scott, J. Appl. Polymer Science, 16, 2029 (1972). Also, P. E. D. Morgan and H. Scott, U. S. Patent #3,923,953. December 2 (1975).
30. P.E.D. Morgan and E. Scala, Sintering and Related Phenomena, Ed. G.C. Kuczynski, Gordon and Breach (1967).
31. P.W.D. Mitchell. J. Am. Ceram. Soc. 55, 484, (1972).
32. P. E. D. Morgan, Final Report COO-2942-1 USERDA, May (1977).

DISCUSSION

David W. Johnson, Jr. (Bell Laboratories): Would you comment on
chemical preparation of β-Al_2O_3 and have you been able to form
stabilized β''-Al_2O_3 at low temperatures?
Author: The lowest temperature at which we have seen β-alumina
(a syntactical mixture of $\beta+\beta''$) is at $\sim750^\circ C$ by processing in sodium chloride melts. For other work on β-alumina (and a good literature review) see: P. E. D. Morgan, "Studies of Synthesis, Crystallography and Sintering of β-Aluminas," Final Report COO-2942-1
U.S. ERDA, May (1977).

Dennis J. Viechnicki (AMMRC): Do you have problems with anion impurities and gas entrapment in the final product when you use
chemically processed powders as starting materials?
Author: Anions used should be carefully selected according to the
specific chemistry and sintering properties of the system. If I had
to choose a general anion which might be good in most systems to
be sintered in air, I would suggest acetates.

Frank G. Recny (General Electric Company): You mentioned probable
breakdown of agglomerates resulting from the high pressure isopressing as a contribution to the high fired density achieved.
There may be good reasons why this high pressure isostatic pressing cannot be employed. Have you measured the degree of agglomeration and considered or tried other means to deagglomerate before
pressing?
Author: Our conclusion was as a result of the empirical observation that a particular powder ($LaCrO_3$) could be densified to 97%
density after isopressing when the same material, same sintering
conditions, only gave $\sim80\%$ final density on uniaxial pressing at
lower pressures. And we have observed this effect with other
"active" powders. We have not followed this up with rigorous
work, for which there is a great need.
    Certainly, other techniques of deagglomeration are available
but at the research level the use of isopressing is the easiest
first way to check this out. I was pleased to hear the progress
of Reed, Niesz, and McCoy in the area of fine powder handling and
forming at this conference which supports my suspicions.

CERAMICS SINTERED DIRECTLY FROM SOL-GELS

P. F. Becher, J. H. Sommers,* B. A. Bender**

and B. A. MacFarlane

Naval Research Laboratory, Washington, D. C. 20375

INTRODUCTION

Sol-gel technology has been extensively utilized for processing nuclear fuel pellets and powders. Currently, the direct firing of gels is being explored to produce ceramics without the use of any intervening powder steps as an extension of Youldas' work on glasses and polycrystalline oxides.[1] This is motivated by (1) the high purity and homogeneity available in sols; (2) the potential ability in a viscous liquid to minimize the sources of defects introduced in the processing of powders; (3) the ability to visually examine many gel products for defects after drying; and (4) the shaping potential offered by a "plastic" gel. Furthermore, much lower temperatures can be used to fire gels to a fully dense ceramic (e.g., ThO_2, Ref. 2) than are required for conventional powder processed bodies.

Key problems in the direct firing of gels are the determination of (1) how to control the large shrinkages involved in the gellation of sols, (2) the size and shape limitations necessary to avoid cracking or distortion, and (3) the range of compositions and materials to which direct gel processing can be applied. The present work focuses on gels for alumina-based ceramics both because of their technical importance and the understanding gained in previous studies.[1,2] Some assessment of the versatility of the process has been conducted with regards to fabricating fibers, coatings, and bulk pieces and to other ceramic systems.

---

*   Currently at Columbia University, New York, New York.
**   Currently at Lehigh University, Bethlehem, Pennsylvania.

PROCESSING

The sols* primarily employed yielded either alumina or alumina-zirconia ceramics with the composition altered by either use of soluble additives (e.g., magnesium-acetate), by mixing sols or by addition of solid particulates. Uniform mixing is quite readily achieved in the sol state; in fact, combination of sols which will react leads to molecular mixing[1] not unlike that achieved by co-precipitation processes. At the same time, uniform mixing of particulates in a sol is possible by controlling the viscosity of the sol-gel during mixing.

The sols are then partially gelled (volume reduction of ~50%) by evaporation of water at ~80°C thus controlling the gel viscosity to facilitate the extrusion of fibers or the casting of bulk pieces. The large final drying shrinkages needed in bulk pieces are accommo-dated by casting onto mercury (or similar high surface tension-low friction "substrate") or into a porous membrane (e.g., dialysis membrane). Use of the "mercury" casting method requires control of the atmosphere to regulate shrinkage (i.e., humidity, flow). This control is incorporated in the membrane technique due to the nature of the "pore" size. In addition, the original sol composition can be used to minimize the tendency for cracking during the final stages of gel drying by altering the viscosity versus shrinkage behavior. In fact, only very small variations in electrolyte content (e.g., acetic acid) may be achieved by altering the surface charge. Thus repulsions between particles are potentially significant in obtaining dry boehmite gels which are free of cracks, Fig. 1. This has also been noted by Youldas.[1b] As a result, the effects of all additions

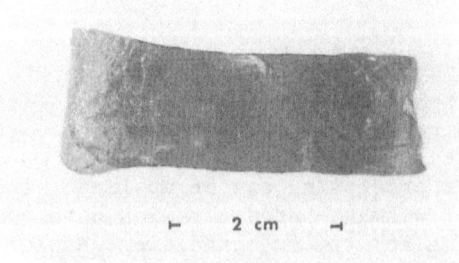

|← 2 cm →|

Fig. 1. Fully dried transparent boehmite gel "rod" formed by casting in dialysis membrane.

---

* Boehmite [AlO(OHO] with and without ZrO_2 colloidal suspensions with AlO(OHO derived by hydrolysis of aluminum sec − butoxide and ZrO_2 by thermal conversion of zirconium tetra-teriary butoxide (Ref. 1a).

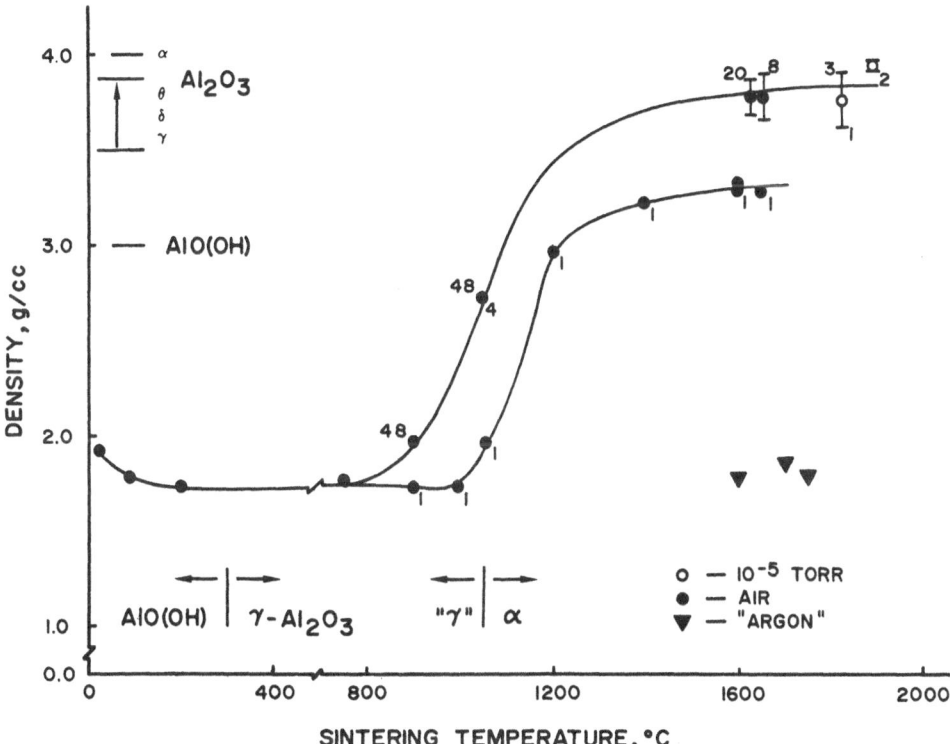

Fig. 2.   Sintering behavior of bulk ($\geq$ 0.3mm thick) boehmite
          (1/4 w/o magnesium acetate) gel.

to the sol must be considered; for example, magnesium acetate additions which are basic can promote premature setting and cracking of
the gels.

The sintering of the boehmite gels involves the desorption of
water, conversion to the γ-alumina phases at ~300°C and conversion
to α-alumina at ~1000°C. Little, if any, sintering occurs at temperatures below 900°C. In fact, there is a loss in density as water
is desorbed from room temperature up to about 200°C, as shown in
Fig. 2. For firing temperatures of <300°C, the alumina and alumina-
zirconia gels remain transparent. Sintering of these gels is
initiated at temperatures around 900°C, somewhat prior to the
γ- to α-alumina transformation, in agreement with studies of Badkar
and co-workers.[3] At this point, sintering is rapid compared to that
of conventional α-Al_2O_3 powders.

The microstructure of the initial boehmite gel is composed of
spherical particulates and is retained through the γ-Al_2O_3 conversion
and up to temperatures of around 1000°C, Fig. 3. However, the α

360°C;60m;AIR;39%   1025°C;180m;AIR;45%

1120°C;60m;AIR;61%   1500°C;15m;AIR;70%

Fig. 3.   Microstructure and Density (%) changes in gels due to
          temperature/time air-firing conditions. Sunbursts associated
          with α-Al$_2$O$_3$ nucleation at 1025°C.

conversion (~50% α-Al$_2$O$_3$ after ten hours at 1050°C, as observed by
x-ray techniques) is shown in the microstructure as a radial array
of particles nucleating from a center, i.e., sunburst-type patterns,
in the matrix (Fig. 3). Nucleation of α grains within the γ alumina
matrix is enhanced by increasing either temperature or time at
temperature. The α regions are subject to growth, yielding ~ 1/2
to 1 µm α particles at or above 1300°C; whereas, the γ matrix particulate size is $\leq$ 0.1 micron. The structure is now composed of two
interconnected phases; α alumina and continuous pores. Referring to
Fig. 2, one sees that at ~1300°C the sintering rates of the gels
decrease and, in fact, become comparable to those of cold pressed
powders. This is expected as the α-alumina grain size of the fired
gels is reaching the particle size of many α-alumina powders. This,
combined with the generation of porosity during the α transformation,
means that the alumina based gels will require temperatures to reach
final density that are not unlike those temperatures required for
cold pressed powder compacts, as pointed out by Badkar and
co-workers.[3a]

It is intriguing that significant sintering does occur at ~900°C in the extremely fine particulate size γ-alumina and some improvement in density (i.e., up to a 50% increase) can be achieved by sintering for long times (i.e., 1 vs 48 hrs) at temperatures below the α transformation. However, control of the grain growth and porosity generation during the α transformation is an obvious key to improving the sintering behavior of these gel materials. It should be noted that there is little difference in the sintering behavior for firing these gel materials in air or in vacuum. However, sintering under more extreme reducing conditions, e.g., an argon atmosphere combined with a graphite resistance heated furnace, results in the elimination of any sintering in these materials below about 1700°C, Fig. 2.

As in other alumina bodies, microstructural control via elimination of grain growth during the sintering of alumina based gels is required. This can be modified by addition of soluble additives such as magnesium acetate, Fig. 4A. In regard to microstructural control,

Fig. 4.   Ceramic microstructures by direct gel sintering. Fine,
          equiaxed grained aluminas with low porosity (A) or uniform,
          closed porosity (B) platelet grain-high porosity alumina
          (C) and dense Al_2O_3- 18 v/o ZrO_2 composite (D).

one can also take advantage of the unique porosity-grain structure developed in the alumina gels to obtain a microstructure with uniform distribution of fine closed porosity, Fig. 4B. This would be useful in applications where a lower thermal conductivity is desired. On the other hand, sintering in reducing conditions results in plate-type "grains" in an $\alpha-Al_2O_3$ body which is ~50% dense. Another approach has been to employ a gel containing both alumina and zirconia ($\leq$ 20 vol % ZrO_2) precursor particles. Because of the immiscibility of the zirconia, alumina results as a uniform dispersed second phase having a fine grain size in the sintered material. In fact, where uniform distribution of zirconia both along grain boundaries and triple points and in the alumina grains is observed, Fig. 4D, the alumina grain sizes are consistently uniform at about 5 microns or less in bulk materials that are sintered to temperatures of upwards of about 1700°C.

It is this extreme uniformity of porosity or second phase microstructural features, as in the case of alumina $\leq$ 20 vol % zirconia materials, that makes gel processing an attractive approach to composite materials. Comparable bodies made from either dry or wet mixing of conventional powders do not exhibit this high degree of uniformity of zirconia dispersion within the alumina matrix.

PROPERTIES AND APPLICATIONS OF GELS

The initial work has centered on the production of ceramic fibers using gels that can be extruded and subsequently fired to full density $\alpha-Al_2O_3$ quite readily. Fibers that have not been fully converted to α alumina with low densities (~45% of theoretical) had true tensile strengths ranging from ~69 to 104 N/m^2 (10,000 to 15,000 psi). Although not fully optimized, fibers that were fired to $\geq$ 95% density had the tensile strengths of ~416 to 455 N/m^2 (60,000 to 70,000 psi) when the grain size averaged 5 microns and fiber diameters were ~100 microns. As shown by Badkar and co-workers, fibers produced using gel precursors, can exhibit tensile strengths of over 690 N/m^2 (100,000 psi).[3b] Thus, there is potential for improvement in the present gel processing results mainly by reduction of both fiber and grain size. In addition, bend strengths of 416 to 455 N/m^2 have been obtained in samples cut from bulk pieces that were fabricated by sintering the alumina- 5-to-20 vol % zirconia base gels. These samples had grain sizes of 2-5 microns and exhibit a pore content of 2-5%. Here again, improved density and refined grain size would lead to increased mechanical properties for these bulk materials.

Other potential gel applications include fabrication of coatings. Our initial experience shows that dried gel coatings that are crack-free remain so after firing. For thin coatings, one does not see any polycrystalline microstructure in the gel coating after firing to temperatures up to ~1650°C in air. However, in the thicker region

of the coatings, a polycrystalline microst ructure is developed having
a grain size of about 2-5 microns after firing at 1650°C for an hour.
Sols can also be used as a composite matrix to add a particulate
second phase or to make ceramic compounds. In the composite case,
where a zirconia powder is mixed into the alumina sol, one also gets
a very uniform dispersion of the zirconia. This uniformity is much
more readily obtained in this manner than with wet or dry mixing
of the alumina and zirconia powders. In addition, preliminary
results show that a combination of alumina and silica-based sols
result in the formation of stoichiometric mullite compositions or
silica-mullite eutectic compositions that can be fired at 1600°C in
air for $\geq$ 8 hrs to densities of at least 90% of theoretical while
achieving bend strengths of ~20,000 psi.

SUMMARY

    As shown, the direct firing of gels can be applied to form
extruded fibers, and to fabricate coatings and bulk ceramic materials.
Size and shape limitations are related to shrinkage cracking during
the gellation process; however, this can be controlled to increase
the potential size produced by gel processing.

    The requirement that large volumes must be carefully dried to
overcome shrinkage cracking is somewhat compensated by the fact that
the gels generally remain transparent and therefore can be visually
inspected. Thus, when combined with the potential shaping capabilities, the ability to readily produce uniform composite microstructure
and ceramic compounds while avoiding the processing of powders make
the technique of direct conversion of sol-gels to ceramic pieces
an attractive fabrication method.

ACKNOWLEDGMENTS

    The authors wish to thank R. W. Rice, W. J. McDonough, and
B. Youldas who have contributed by their helpful discussions of this
work, Alberta Pellecchia and Roseann Gorman who have prepared the
manuscript, and L. Chuck and R. Ingel who aided during portions of
the experiments.

REFERENCES

1a.   B. E. Youldas, J. Mater. Sci., 12, 1203–1208 (1977).
  b.   J. Mater. Sci., 10, 1856–1860 (1975).
  c.   Bull. Am. Ceram. Soc. 54, 286–288 (1975).
2.   M. J. Bannister, J. Am. Ceram. Soc. 58 (1-2), 10-14 (1975).
3a.   P. A. Badkar, J. E. Bailey, J. Mater. Sci. 11, 1794–1806 (1976).
  b.   P. A. Badkar, J. E. Bailey, and H. A. Baker, Trans. Brit. Ceram.
        Soc. 71, 193–201 (1972).

CHARACTERIZATION OF CERAMIC MICROPROCESSING

P. F. Johnson and L. L. Hench

Department of Materials Science and Engineering

University of Florida, Gainesville, Florida 32611

I.   INTRODUCTION

Achieving improvements in the properties of many technical ceramics requires producing microstructures with uniform grains and an average grain size of <10 μm. Control of powders in the <10 μm size range is difficult and characterization of the powders throughout the processing cycle is essential. Control of such fine, active powders is an integral part of ceramic micro-processing.

Many instrumental methods are available for powder characterization[1,2], however, most methods yield only an average value of a single property of the powder. Identification of size or phase heterogeneities within a powder is often impossible to detect with most powder characterization techniques. Analysis of state of agglomeration, bonding mechanisms within agglomerates, and surface films or additives on particles within agglomerates are also formidable analytical problems. Likewise, many techniques cannot be used to follow changes in powder characteristics as a function of sintering after the particle necks become fused in a body. A previous publication of one of the authors (LLH) has shown some of these powder characterization requirements can be fulfilled by using the conventional transmission electron microscope (CTEM) containing a tilting stage[3]. The purpose of the present paper is to discuss possible application of the scanning transmission electron microscope (STEM) and the analytical scanning transmission electron microscope (ASTEM) in the characterization of several aspects of ceramic microprocessing.

II.   BACKGROUND

Use of the STEM as an analytical instrument was made
possible by developments in both electron sources and electro-
magnetic lenses. The key difference between the STEM and the
CTEM, are the addition of a strong, preobjective probe-forming
lense, beam scanning coils in the illumination system above the
sample, and a solid-state electron detector inserted below the
sample.

Figure 1 shows schematically some of the events which occur
during interaction between the electron beam and the sample.
Above the plane of the sample, secondary and backscattered electrons are available for detection and analysis. In addition,
characteristic x-rays can be analyzed using either solid-state
energy dispersive or crystal wave length dispersive detectors.

Below the sample, elastically and inelastically scattered
electron can be analyzed using energy loss, bright field and dark
field detectors, and photographic media.

All of these events occur through beam-sample interactions in
the CTEM as well as in the STEM or ASTEM. What then does the
ability to form a small (.5-2 nm) electron probe introduce into
the analytical process? Why were these various detectors not
commonly used in CTEM?

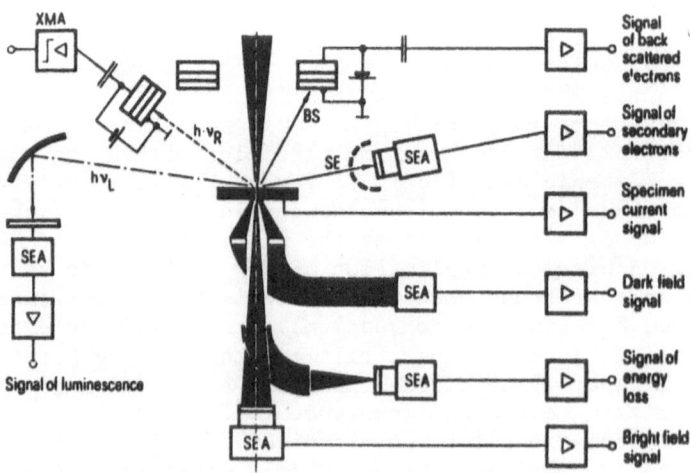

Figure 1. Schematic illustration of events which occur through
sample-beam interaction. The position of the detectors in the
column is indicated. Figure after Willasch[4].

First, in STEM, the sample can be analyzed in a point-by-point mode resulting in serial rather than parallel information processing. In CTEM, information from all of the illuminated sample is present all of the time. In STEM, only the sample area under the probe is illuminated. This, of course, implies that either more sensitive detectors or more intense electron sources must be used to achieve comparable detection limits. These are now available.

Secondly, the small probe size available in STEM means that very small sample areas can be analyzed. It is this feature of the STEM that is especially valuable in characterizing ceramic powders and the various steps.

The ASTEM in our lab is a Philips EM-301 equipped with scanning coils, a standard thermionic source, backscattered and transmitted electron detectors and an energy dispersive detector for compositional analysis. Modifications have been made which eliminate major sources of background x-radiation and sample contamination[5]. Compositional analyses of low atomic number elements is now possible with reasonable levels of sensitivity[6]. Thus, for the first time we have available a system which can provide a correlation of composition, morphology and ultrastructure from a single sample preparation.

Figure 2 illustrates the potential of the CTEM mode of operation for morphological analysis of powders. Shown is a very nice example of a clay particle. A number of beam-sample interactions give rise to the various areas of contrast seen in the particle. In general, contrast in mass, diffraction and phase interact to produce the complex ultrastructural morphology of the clay particle. The particle in Fig. 2 is about 2 μm across and appears to have many subgrain boundaries, surface steps or perhaps is even an agglomerate of many smaller particles.

In order to identify the ultrastructural features of the sample using CTEM, electron diffraction of the entire sample can be used. An image limiting aperture is inserted below the sample plane and the projection system adjusted to present the diffracted beams from the selected area at the viewing plane. Figure 3 shows that the clay particle has the diffraction characteristics of a single crystal. Controlled tilting of the sample yield a number of single crystal patterns such as that of Fig. 3. Analysis of these patterns, using standard techniques, provides a detailed description of the crystal lattice. The defect structure of the crystal can also be analyzed by performing microdiffraction of various parts of the crystal. Thus, the example illustrates use of CTEM for identification of single particles in a powder and distinguishing them from agglomerates.

Figure 2.   CTEM image of an isolated clay particle (20K).

Figure 3.   Selected area diffraction pattern from the clay
particle shown in Fig. 2.

III.   USE OF THE ASTEM

The clay particle shown was prepared for CTEM by simply spraying a dilute clay suspension upon a formvar coated copper EM grid. Often sample preparation is not so easy. Sample preparation for ASTEM analysis usually represents a major fraction of analytical time. Sample preparation is especially a problem when EM analysis of sintered compacts is attempted. To illustrate both the difficulties and opportunities ASTEM offers for studying powder changes in forming ceramic components, preliminary results from a study of Ta-Ta_2O_5 capacitor materials are discussed.

Ta-Ta_2O_5 capacitors are made by sintering Ta powder in a controlled low P_{O_2} atmosphere. Low density and uniform interconnected porosity are desired for the sintered compacts. Anodization in a 0.1% H_3PO_4 solution produces a film of Ta_2O_5 on the surface throughout the porous compact. The Ta_2O_5 film growth is 16 to 20 nm per volt of applied anodization voltage, yielding films typically 60-300 nm thick. Little structural characterization of electrochemically anodized thin films has been undertaken. For example, it is known that the P ion in the anodization solution results in a film with a higher dielectric constant than films anodized in H_2SO_4 or HNO_3. It is not known whether P becomes incorporated within the structure of the film, enhances cleaning the metal surface, controls film grain boundary structure or crystallinity, etc.

Figure 4 shows a standard scanning electron micrograph (SEM) of a lightly sintered compact of nominally "6 μm" Ta powder*. An apparent wide distribution of particle sizes are present. An ASTEM study was undertaken to understand the powder character resulting in the microstructure shown in Fig. 4.

The morphology of the unsintered Ta powder is shown in Fig. 5. By use of the tilting stage it was imaged at various angles, see reference 3 for the procedure. It was determined that the particles are initially aggregates of smaller crystallites. The aggregate morphology of the powder yields a much higher surface area and activity than if the powder were composed of solid particles of average 6 μm diameter.

The polycrystalline nature of the particles was confirmed using the selected area diffraction capability of the STEM. A microtome was used to obtain sections of the metal aggregates sufficiently thin for transmission of the electron beam. The

---

*Obtained from Norton Metals Division (now NRC Inc.) 45 Industrial Place Newton, Mass.   02164

Figure 4. Scanning electron micrograph of lightly sintered Ta compact (5K).

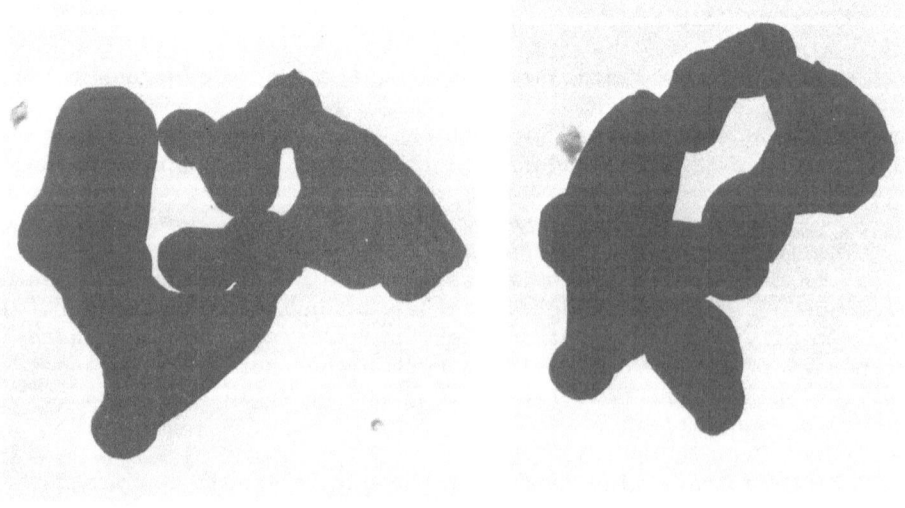

A B

Figure 5. CTEM image of unsintered Ta powder: A = 0° tilt; B = 45° tilt. The grains shown are the same, only the sample stage orientation relative to the beam was changed (6K).

lack of resolution of particle detail in Fig. 5 is because the
particles are too thick for the beam to go through them. However,
beam penetration of one of the microtomed aggregate fragments is
apparent in Fig. 6. The length of the fragment is ∿2 µm and the
width is approximately 0.7 µm. Selected area diffraction from the
entire fragment yields the polycrystalline pattern of Fig. 7.

Use of the C_2 limited diffraction conditions reduces the beam
diameter to 0.2 µm which covers only one tenth of the fragment.
The diffraction pattern obtained, Fig. 8, still contains multiple
spots, indicating that the size of the individual crystals in the
powder is less than 0.2 µm in size. Finally, STEM microdiffraction
conditions with a beam reduced to only ∿2 nm yields a single
crystal diffraction pattern characteristic of a single crystal,
Fig. 9. Thus, even the individual particles within the powder
aggregates, shown in Fig. 5, are polycrystalline. Control of
sintering and anodization behavior of this ultrafine, active
powder appears to require avoiding discontinuous grain growth during sintering which leads to the wide grain size distribution shown
in Fig. 4.

Figure 6. Fragment of Ta powder prepared by embedding in epoxy
and diamond knife microtoming (20K).

Figure 7. Selected area diffraction pattern of entire fragment shown in Fig. 6.

Figure 8. Diffraction pattern obtained from the fragment shown in Fig. 6 using the C_2 limiting aperture technique.

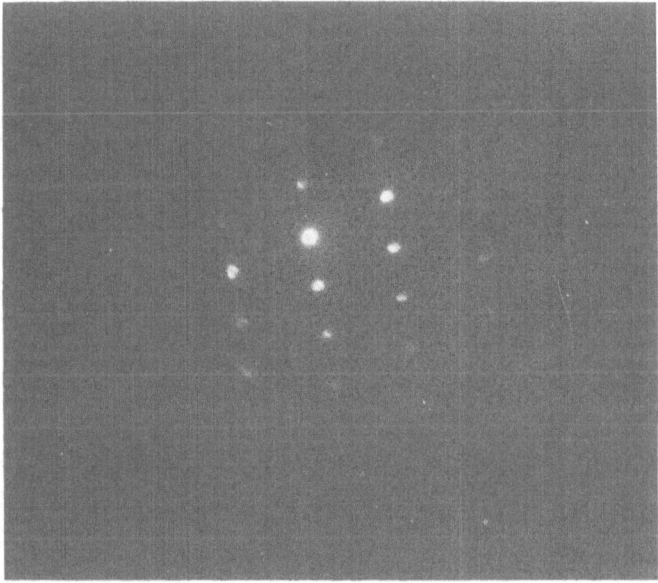

Figure 9.   STEM microdiffraction pattern for the fragment shown in
Fig. 6.

        The ASTEM was used to investigate anodization of this active
"6 μm" Ta powder. After sintering followed by anodization in
H_3PO_4 at 50 volts, the compact was sectioned by diamond knife
microtoming. Diffraction with a ∿200 nm beam on the anodized
layer showed only amorphous rings, Fig. 10. No microdiffraction
pattern could be recorded. Thus, the anodization process for this
powder results in a nearly amorphous Ta_2O_5 film. Additional
experiments must be conducted to determine whether the Ta_2O_5 layer
is amorphous throughout its thickness and whether anodization rate,
initial particle crystallinity, etc., affect its structural char-
actor and thereby the dielectric properties of the oxide film and
the capacitor made from the film.

        A preliminary effort to determine the role of phosphorus in
Ta_2O_5 film formation was also made. The energy dispersive spectrum
from the same region yielding the amorphous pattern of Fig. 10 is
presented in Fig. 11. The sample was supported on graphite EM
grids in a Be sample holder so nearly all of the systemic radiation
is eliminated from this spectrum[7]. A small amount of systemic Fe
is present from the Be sample holder.

Figure 10. Diffraction pattern obtained from the anodized Ta_2O_5 film using the C_2 limited aperture technique. No microdiffraction pattern could be recorded.

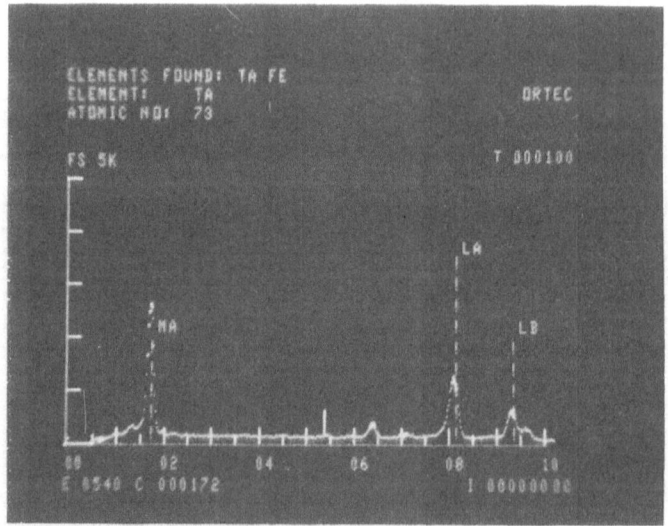

Figure 11. EDS spectrum from the same region as diffraction pattern shown in Fig. 10. The position of the Ta Mα, Lα and Lβ peaks are indicated. The small peak at 6.4 KeV is Fe in the Be sample holder.

Tantalum is the primary element detected in the analysis. A phosphorous peak is not observed. Thus, if P is built into the Ta_2O_5 film it is of sufficiently low concentration that it cannot be detected within the resolution of the ASTEM technique.

IV. CONCLUSIONS

The microfocus and capability for compositional analysis of structural features as small as 20 nm makes the ASTEM a powerful new tool for characterization of ceramic powders and microprocessing steps. Combined use of multiple modes of operation of the instrument is especially helpful when difficult characterization problems are encountered. At times, only by combined use of morphological examination, selected area diffraction, microdiffraction, and microcompositional analysis can one test various alternatives for changes in powder and agglomerate structures and surfaces.

V. ACKNOWLEDGMENTS

The authors acknowledge partial financial support of the University of Florida Engineering and Industrial Experiment Station, NINCDS Contract No. NIH 71-2286, and the assistance of J. J. Hren, E. J. Jenkins, and S. R. Bates.

VI. REFERENCES

1. L. L. Hench, pp. 457-597 in Characterization of Ceramics. Edited by L. L. Hench and R. W. Gould, Marcel Dekker, 1971.
2. C. Orr, pp. 39-59 in Ceramic Processing Before Firing. Edited by G. Y. Onoda, Jr. and L. L. Hench, J. Wiley & Sons, New York, 1978.
3. L. L. Hench and E. J. Jenkins, pp. 75-84 in Ceramic Processing Before Firing. Edited by G. Y. Onoda, Jr. and L. L. Hench, J. Wiley & Sons, New York, 1978.
4. D. Willasch, "Image Forming Systems", in Proc. Workshop in Analytical Electron Microscopy at Cornell University, Aug. 3-6, 1978, pp. 27-43.
5. J. J. Hren, S. R. Bates, P. F. Johnson, EMSA Proceedings, Claitors, Baton Rouge, 418-419 (1976).
6. J. J. Hren, P. F. Johnson, S. R. Bates, L. L. Hench, EMSA Proceedings, 29-291 (1976).
7. P. F. Johnson, J. J. Hren and S. R. Bates, "Spectral Contamination in the EM 301 ASTEM," submitted to 9th International Congress Electron Microscopy, 1978.

PACKING AND SINTERING RELATIONS FOR BINARY POWDERS

George Y. Onoda, Jr. and Gary L. Messing

University of Florida

Gainesville, Florida 32611

INTRODUCTION

Mixtures of coarse and fine powders are commonly used for processing refractory bodies. Compared with uniform fine powders, mixtures have higher green densities and less shrinkage when fired. The presence of coarse particles can provide a control for achieving fired bodies with a specific level of porosity. As a filler, the coarse particles may also reduce the cost of the raw material. A major disadvantage of mixtures is the limited strength of the fired body because of the large grains in the microstructure.

In this paper, theoretical aspects of the packing and sintering of binary mixtures will be reviewed. Compared with theories for the sintering of fine-grain powders, the theories for binary mixtures require far greater attention to the influences of the geometry of packing, to geometric constraints, to local stresses, and to the influences of compositional inhomogeneities within the body. The binary systems considered will be those for which the coarse particles are much larger than the fine particles (by more than an order of magnitude). Only solid state sintering processes will be considered.

We will be concerned with the volume changes due to sintering of two-component systems. Therefore, a useful mathematical relation is

$$v = \bar{v}_f w_f + \bar{v}_c w_c$$

where w_f and w_c are the weight fractions and $\bar{v}_f$ and $\bar{v}_c$ are the partial specific volumes of the fine and coarse components, respectively. The partial specific volumes are defined by $\bar{v}_f = (\partial V/\partial W_f)_{W_c}$ and $\bar{v}_c = (\partial V/\partial W_c)_{W_f}$, where V is the total volume and W_f and W_c are the weight of fine and coarse components. The specific volumes and partial specific volumes for fired and unfired bodies are designated by primes and no primes. Since $w_f + w_c = 1$, the relations for v and v' can be expressed as

$$v = (\bar{v}_f - \bar{v}_c)\, w_f + \bar{v}_c \tag{1}$$

$$v' = (\bar{v}_f' - \bar{v}_c')\, w_f + v_c' \tag{2}$$

and the volumetric shrinkage of the mixture in going from the unfired to fired state is

$$\left(\frac{\Delta V}{V}\right)_{mix} = \frac{v - v'}{v} = \frac{(\bar{v}_f - \bar{v}_f' - \bar{v}_c + \bar{v}_c')\, w_f + \bar{v}_c - \bar{v}_c'}{(\bar{v}_f - \bar{v}_c)\, w_f + \bar{v}_c} \tag{3}$$

The selection of the appropriate values for $\bar{v}_f$, $\bar{v}_c$, $\bar{v}_f'$, and $\bar{v}_c'$ is the essence of the theoretical considerations that follow.

HIGH CONCENTRATIONS OF FINE COMPONENT

When the weight fraction of coarse particles is small, the mixture can be visualized as a matrix of fine particles surrounding individually dispersed coarse particles. For this case, Furnas[1] assumed that $\bar{v}_f = v_f$ and $\bar{v}_c = v_o$, where v_f and v_o are the specific volumes for the 100% fine powder packing and the solid coarse particle, respectively. By analogy, $\bar{v}_f' = v_f'$ and $\bar{v}_c' = v_o$ for fired bodies. Equations 1-3 become[2]

$$v = (v_f - v_o)\, w_f + v_o \tag{4}$$

$$v = (v_f' - v_o)\, w_f + v_o \tag{5}$$

$$\left(\frac{\Delta V}{V}\right)_{mix} = \frac{(v_f - v_f')\, w_f}{(v_f - v_o)\, w_f + v_o}$$

$$= \left(\frac{\Delta V}{V}\right)_{fines} \left[\frac{v_f w_f}{(v_f - v_o)\, w_f + v_o} \right] \tag{6}$$

since $(\Delta V/V)_{fines} = (v_f - v_f')/v_f$. This shows that the shrinkage of the mixture is reduced from that of the 100% fine powder by the compositionally dependent factor in the brackets of Eq. 6.

The assumption that $\overline{v}_f' = v_f'$ may not always be valid. O'Hara and Cutler[3] suggested that a coarse particle may constrain the sintering shrinkage rate of surrounding fine particles. As the fine particles attempt to densify around the coarse particle, tensile stresses parallel to the surface of the coarse particle may develop which retard shrinkage. Their studies of sintering rates of binary alumina powders indicated that stress retardation occurs at lower temperatures and short times for sintering. At higher temperatures and longer times, the effect vanishes because stress relief through creep occurs. Unfortunately, no analytical solutions yet exist for the stresses and sintering retardation which could be incorporated into the present geometric model. The present model, therefore, is applicable only for those cases where sufficient stress relief occurs.

Compositional inhomogeneities (variations in composition at different positions in the body) can exist to a limited extent without invalidating Eqs. 4-6. The requirement is that all of the space not occupied by coarse particles is occupied completely by the matrix of fine particles. The equations would not be valid if the coarse particles clump together and the interstitial space between them is not completely filled with fine matrix. For this to happen, the local concentration of fines must drop below a certain critical value, to be defined later.

LOW CONCENTRATIONS OF THE FINE COMPONENT

A packing of coarse particles will have a specific volume of v_c. If fines are added into the interstitial spaces between the network of coarse particles without expanding the network, $\overline{v}_c = v_c$ and $\overline{v}_f = 0$ (Furnas[1]). During sintering, a reduction in volume of the mixture occurs only if the coarse particle network shrinks. Since the fines in the interstices sinter more rapidly, their volume remains smaller than the interstitial volume, so $\overline{v}_f' = 0$.

The value for $\overline{v}_c'$ depends on whether the fine particles influence the sintering rate of the coarse particles. If there is no influence, $\overline{v}_c' = v_c$, and Eqs. 1-3 become

$$v = v_c (1 - w_f) \tag{7}$$

$$v' = v_c' (1 - w_f) \tag{8}$$

$$\left(\frac{\Delta V}{V}\right)_{mix} = \frac{v_c - v_c'}{v_c} = \left(\frac{\Delta V}{V}\right)_c \tag{9}$$

where $(\Delta V/V)_c$ is the normal shrinkage for 100% coarse particles. Thus, the shrinkage rate of the mixture is independent of composition.

The maximum amount of fines, w_f*, that can fill the interstitial spaces without expanding the coarse network is equal to[1]

$$w_f* = \frac{v_c - v_o}{v_f + v_c - v_o} \tag{10}$$

This can only occur, however, if the fines are perfectly distributed through all of the interstices in the mixture.

If a mixture has a composition considerably less than w_f*, compositional inhomogeneities within limits can exist without affecting the relations expressed by Eqs. 7-9. The fines may be more concentrated in some interstitial spaces than others, as long as the coarse particle network is not expanded. If, however, the concentration of fines exceeds w_f* locally, the coarse particles are not in contact at that location, causing the overall value of $\bar{v}_c$ to increase above v_c.

The above discussion assumed that the fine particles do not affect the sintering rate of the coarse particles. Coble[4] has shown, however, that stress-enhanced sintering is possible for systems having variable particle sizes. The case of many fine particles in the interstices of coarse particles was not explictly considered. However, it was shown that a single fine particle that just fits into a square array of coarse particles would enhance the sintering rate of the coarse particles by around 25% because of compressive stresses developed at their points of contact. We can surmise that a similar enhancement could occur if the interstitial space is occupied by many fine particles, as long as the matrix of fines bridge at least two coarse particles. The more rapid shrinkage of the matrix would produce compressive stresses at the coarse necks. The greater the volume of the interstitial space occupied by fines, the greater should be the stress-enhanced sintering because more coarse particles are being bridged.

The stress-enhanced sintering may not occur if stresses are relieved. One possibility is that the small particles tear away from the coarse particles.[4] A second possibility is that stress relaxation occurs by creep or plastic deformation.

INTERMEDIATE COMPOSITIONS

The cases for low and high concentrations of fine component have been considered in the previous two sections. Different mechanisms apply for these two cases. A broad compositional region between these cases comprise the transition zone where a mixture of mechanisms is possible.

Homogeneous Mixtures

Case where coarse particle sintering is negligible. For this case, $(\Delta V/V)_{mix} = 0$ for the compositional range of $0 \leq w_f \leq w_f^*$, where w_f^* is given by Eq. 10. Above a certain composition w_f^m, enough fine matrix separates the coarse particles so that, even after complete densification of the fines, the coarse particles do not impinge. At w_f^m, given by

$$w_f^m = (v_c - v_o)/v_c \qquad (11)$$

the coarse particles pack together just at the moment that the fines completely densify. This value is the composition having the least shrinkage when fired to the dense state. Below w_f^m, the body cannot completely densify because of the constraint offered by the unsintering coarse particles.

At compositions between w_f^* and w_f^m, the coarse particles are either apart or in contact, depending on v_f'. For a given v_f', the composition dividing the two regions is[2]

$$w_{f_s}^m = \frac{v_c - v_o}{v_f' + v_c - v_o} \qquad (12)$$

The equations for each region are

$$\left(\frac{\Delta V}{V}\right)_{mix} = 0 \qquad 0 \leq w_f \leq w_f^*$$

$$\left(\frac{\Delta V}{V}\right)_{mix} = \frac{(v_f + v_c - v_o) \, w_f + v_o - v_c}{(v_f - v_o) \, w_f + v_o} \qquad w_f^* \leq w_f \leq w_{f_s}^m$$

$$\left(\frac{\Delta V}{V}\right)_{mix} = \left(\frac{\Delta V}{V}\right)_f \frac{v_f \, w_f}{(v_f - v_o) \, w_f + v_o} \qquad w_{f_s}^m \leq w_f \leq 1$$

These relations are expressed graphically in Fig. 1.

Case where coarse particles sinter. For compositions below w_f^*, Eq. 9 applies, and above $w_{f_s}^m$, Eq. 6 holds. The sintering behavior between w_f^* and $w_{f_s}^m$ cannot be simply described by the geometric factors. The reason for this is that $\bar{v}_c'$ is not simply

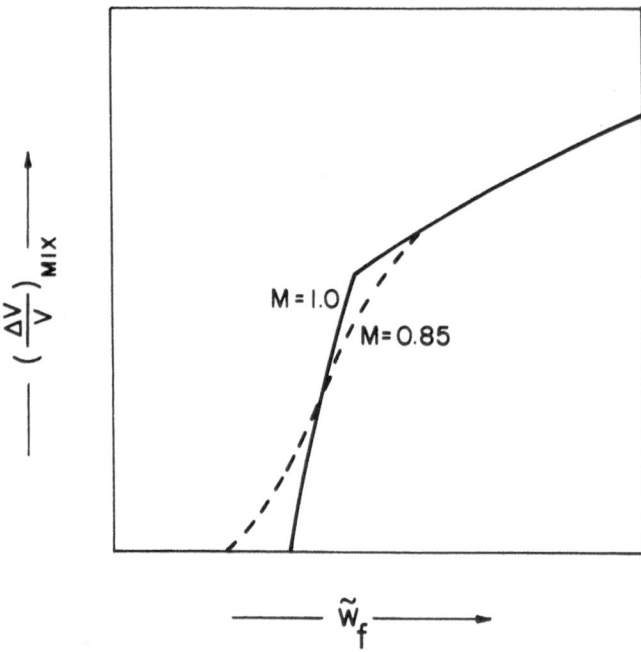

Figure 1. Shrinkage-composition relation when coarse particles do
not sinter. The curve for M=1 is for homogeneous mixtures; the
curve for M = 0.85 is for a mixture with compositional inhomogeneities.

related to v_c'. The value for v_c' as we have been defining it is
that it is the specific volume of the coarse that would exist
after the same sintering times as for v_f'. In the present case,
the coarse particles do not begin to sinter until after the fines
sinter sufficiently to bring the coarse particles together. Knowledge of the individual shrinkage rate kinetics for both the fines
and coarse particles is necessary to calculate the time required
for the fines to bring the coarse particles into contact, and to
use this time as the zero time for the sintering of coarse particles. In principle, this could be done and the results incorporated into the geometric theory. However, to date the analysis has
not been carried out.

    The variations in shrinkage with composition for the cases
where the coarse particles and fine particles sinter are shown in

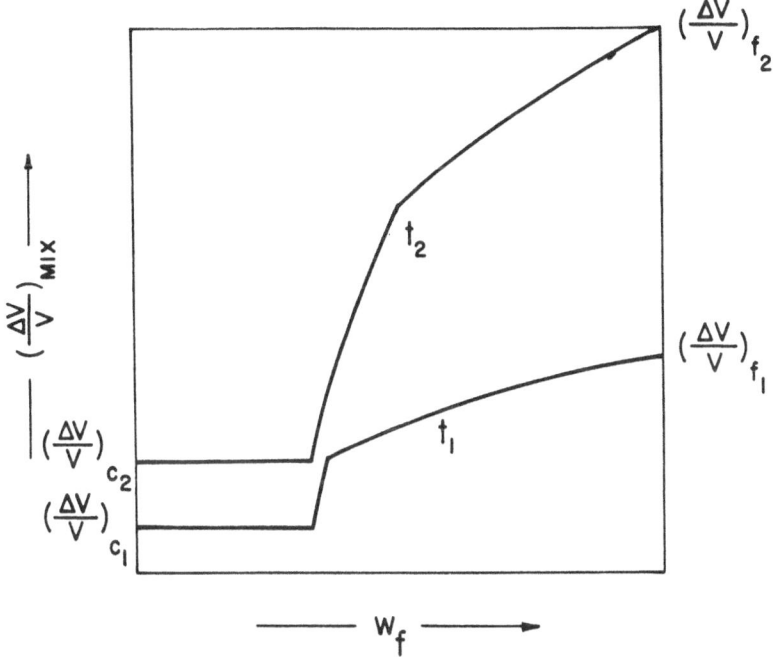

Figure 2. Shrinkage-composition relations when coarse particles
also sinter. The t_1 and t_2 represent sintering times, where $0 <$
$t_1 < t_2$.

Fig. 2. In lieu of an analytical solution for the region between
w_f^* and $w_{f_s}^m$, a straight-line approximation between v' and w_f is
assumed for the transition zone. The result of this approximation
is a slightly curved line for the transition zone in the shrinkage-
composition curve.

Inhomogeneous Mixtures

The specific volume and shrinkage of mixtures at intermediate
compositions depend very sensitively on the degree of compositional
inhomogeneities in the body. This problem can be treated analy-
tically if the compositional variations are known and if certain
simplifying assumptions are applied. One major assumption is that
the relations developed for homogeneous bodies are applicable for

small mass elements within the body, with each element having its own composition. Each element is treated as a minature body, and the overall specific volume and shrinkage for the entire body are obtained by combining the individual contributions. Using this approach, detailed analytical treatments for specific volumes and shrinkages of binary mixtures with compositional inhomogeneities have been presented in two recent papers.[5,6]

Unfired mixtures. Conceptually, the entire mixture is divided into a large number of small mass elements. The composition of each element is identified. The elements are divided into two population groups according to their composition. Population I contains those elements having compositions below w_f^*. Population II has elements with compositions above w_f^*. The total weight fractions of all elements in populations I and II are identified as q_I and q_{II}. The average compositions of populations I and II are $(\tilde{w}_f)_I$ and $(\tilde{w}_f)_{II}$. As was previously deduced, the overall specific volume (v_I) of population I can be calculated from Eq. 7, using $(\tilde{w}_f)_I$ as the composition. This is valid because compositional variations do not affect the calculation as long as they do not exceed the value of w_f^* in any mass element. Similarly, the overall specific volume (v_{II}) of population II can be calculated from Eq. 4, using $(\tilde{w}_f)_{II}$. It has been shown[5] that the specific volume $(\hat{v})$ for the original body is given by

$$\hat{v} = q_I v_I + q_{II} v_{II}$$

In practice, the compositional variations would be expressed by a continuous frequency function $f(w_f)$ or a cumulative weight less than w_f function $Q(w_f)$. The values for q_I, q_{II}, $(\tilde{w}_f)_I$ and $(\tilde{w}_f)_{II}$ needed to calculate the specific volume are obtained from the two functions by the following relations:[5]

$$q_I = \int_0^{w_f^*} f(w_f) \, dw_f = Q^*$$

$$q_{II} = \int_{w_f^*}^1 f(w_f) \, dw_f = 1 - Q^*$$

$$(\tilde{w}_f)_I = \frac{1}{q_I} \int_0^{w_f^*} w_f \, f(w_f) \, dw_f = \frac{1}{q_I} \int_0^{Q^*} w_f \, dQ$$

$$(\tilde{w}_f)_{II} = \frac{1}{q_{II}} \int_{w_f^*}^1 w_f \, f(w_f) \, dw_f = \frac{1}{q_{II}} \int_{Q^*}^1 w_f \, dQ$$

The parameter Q^* is the value of Q corresponding to w_f^*.

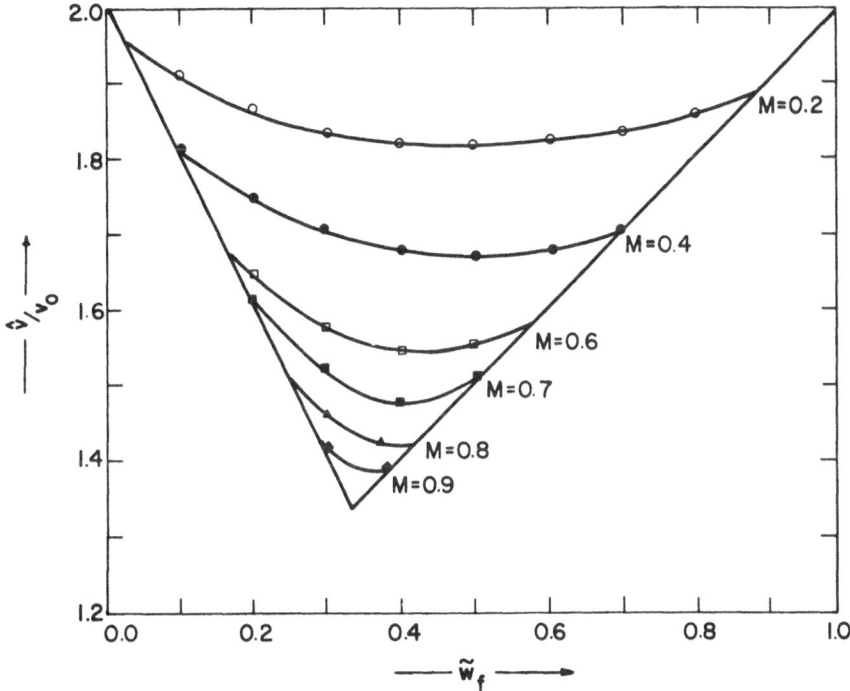

Figure 3.   The relative specific volume of binary mixtures of powders with different degrees of mixedness.

Theoretical specific volume-composition relations[5] for powders with different degrees of mixedness (M) are plotted in Fig. 3.   The parameter M is defined by M = $(1 - \sigma/\sigma_o)$, where σ is the standard deviation of the compositional distribution around the mean composition and $\sigma_o = [\tilde{w}_f (1 - \tilde{w}_f)]^{1/2}$ is the standard deviation of a completely unmixed powder of composition $\tilde{w}_f$.   Fig. 3 illustrates the changes in specific volume from a completely unmixed powder (M = 0) to a completely mixed powder (M = 1), assuming a model where the coarse and fine powders are placed side by side in a rotary mixer and interdiffusion occurs with time.   The deviations from the Furnas relation are similar to those found experimentally for real powders.

Studies[7] with glass beads have verified that these systems do obey the model which assumes the validity of the Furnas relations on a local scale.   Binary mixtures were prepared that had compositional inhomogeneities in the vertical direction in a column.

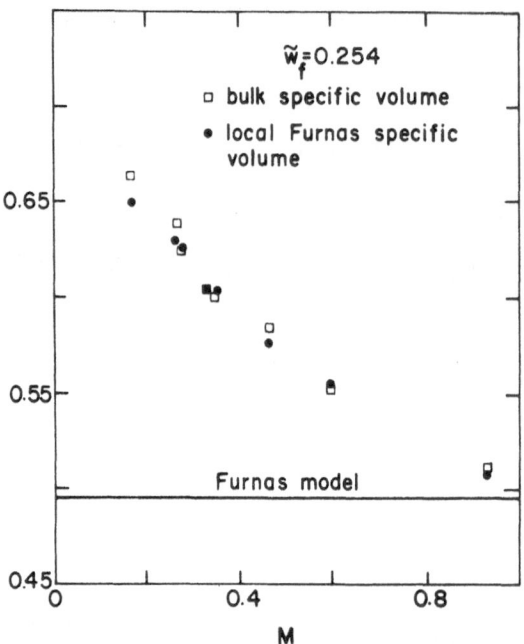

Figure 4. Comparison of calculated and measured specific volumes
of binary glass bead mixtures having different degrees of mixedness.[7]
Composition corresponds to w_f* for the system.

Compositional distributions were characterized and theoretical
specific volumes were calculated from the distribution data. The
agreement between theory and experiment is shown in Fig. 4.

Fired mixtures with no coarse particle sintering. As with
the unfired mixture, values for q_I', q_{II}', $(\tilde{w}_f)_I$ and $(\tilde{w}_f)_{II}$ are obtained by the same procedure used to obtain q_I, q_{II}, $(\tilde{w}_f)_I$ and
$(\tilde{w}_f)_{II}$, with one modification. Instead of w_f*, the composition
w_f^{ms} (Eq. 12) serves as the division between populations I and II.
The specific volume of the fired body is given by

$$\hat{v}' = q_I'v_I' + q_{II}'v_{II}'$$

and the overall shrinkage is described by

$$\left(\frac{\Delta V}{V}\right)_{mix} = \frac{\hat{v} - \hat{v}'}{\hat{v}} = \frac{q_I v_I + q_{II} v_{II} - q_I' v_I' - q_{II}' v_{II}'}{q_I v_I + q_{II} v_{II}}$$

For purposes of illustration, we will assume that a binary mixture has compositional inhomogeneities which can be characterized by a normal distribution around a mean, $\tilde{w}_f$. The frequency function is given by

$$f(w_f) = \frac{1}{\sigma\sqrt{2\pi}} \exp\left[-1/2\ (\frac{w_f - \tilde{w}_f}{\sigma})^2\right]$$

where σ is the standard deviation. Using numerical analysis techniques with the aid of a computer, analyses following the above procedures were carried out.[6] The calculated relation between shrinkage and composition is shown in Fig. 1 by the dashed lines for the case where M = 0.85. It can be seen that the effect of a small degree of compositional inhomogeneity is to broaden the transition zone at intermediate compositions.

Fired mixtures with coarse particles sintering.[6] Every small mass element is assumed to follow the relations developed for homogeneous bodies when the coarse particles sinter as well as the fine particles. Appropriate values of q_I', q_{II}', $(\tilde{w}_f)_I$, and $(\tilde{w}_f)_{II}$ are obtained from the functions $f(w_f)$ or $Q(w_f)$. Population I is defined by $0 \le w_f \le w_f^*$. Population II is for $w_{f_s}^m \le w_f \le 1$. A third population exists for the region between $w_f^* < w_f < w_{f_s}^m$, from which q_{III} and $\tilde{w}_{f_{III}}$ are obtained. The overall specific volume is given by $v' = q_I v_I' + q_{II} v_{II}' + q_{III} v_{III}'$, where v_{III} is obtained from $(\tilde{w}_f)_{III}'$ using the linear approximation discussed earlier. For a normal distribution of compositional variation, the theoretical relations are given in Fig. 5.

Data from O'Hara and Cutler for a binary mixture of Al_2O_3 powders is shown in Fig. 6. The experimental curves have an appearance more like Fig. 5 than Fig. 2. This suggests that their powders were not perfectly mixed but had an M of around 0.85. This might be expected since powders can never be perfectly mixed in practice. It is noteworthy that the shape of the experimental curves have the features expected theoretically. This supports the view that the Furnas relations are being obeyed locally.

Internal stresses. The above relations assume that every mass element can decrease in volume as predicted theoretically. However, different elements shrink different amounts because of compositional differences. The question must arise as to how the differential shrinkages can be accommodated. If the composition varies systematically in one direction in the body, the differential shrinkages could result in distortion or warpage of the body. With

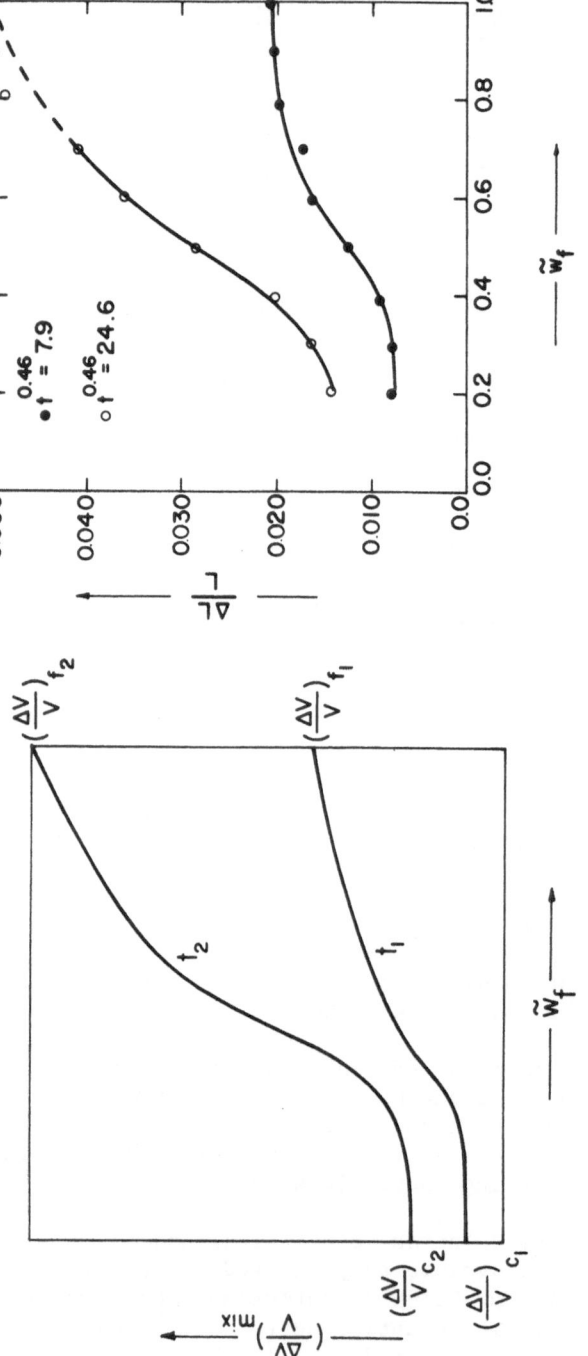

Figure 6. Replot of experimental data of O'Hara and Cutler,[3] expressing shrinkage versus composition for a binary alumina mixture.

Figure 5. Effect of compositional inhomogeneity on the shrinkage-composition relation when coarse and fine particles sinter. (Normal distribution, M = 0.85).

more random compositional variations, microscopic distortions
cannot relieve the stresses which arise from differential shrinkage.
To densify according to theory, permanent strains must occur (through
creep or plastic deformation) over large proportions of the body
and over long distances. Otherwise, the local stresses must cause
local cracking or either enhanced or retarded sintering. Cracks
and restricted shrinkage of porous areas are defects that can result. The magnitude of the problem depends on the degree of inhomogeneity and on the spatial distribution of the inhomogeneity.

IMPLICATIONS TO FUTURE RESEARCH

A geometric model for predicting the specific volumes and
shrinkages of binary powder mixtures, with or without compositional
inhomogeneities, has evolved in recent studies. The usefulness of
this model is that it predicts what should occur in the absence of
other complicating factors. In that regard, it can be considerd
as the "ideal case" against which real experimental systems can
be compared and "irregularities" recognized. Recognizing when
irregularities occur should be helpful in the future for studying
the effects of factors such as stress-enhanced or stress-retarded
sintering effects, local cracking, creep, plastic deformation, etc.

REFERENCES

1. C. C. Furnas, "The Relations Between Specific Volume, Voids and
    Size Composition in Systems of Broken Solids of Mixed Sizes,"
    Department of Commerce - Bureau of Mines, R. I. 2894,
    October 1928.
2. G. Y. Onoda, Jr., "Green Body Characteristics and Their Relation-
    ship to Finished Microstructure," in Ceramic Microstructures -
    '76, edited by R. M. Fulrath and J. Pask, Westview Press,
    Boulder, Colo., 163-81 (1976).
3. M. J. O'Hara and I. B. Cutler, "Sintering Kinetics of Binary
    Mixtures of Alumina Powders," Proceedings of the British
    Ceramic Society; Fabrication Science 2, No. 12, 145-54, (1969).
4. R. L. Coble, "Effects of Particle-Size Distribution in Initial-
    Stage Sintering,"J. Am. Ceram. Soc., 56 [9] 461-66 (1973).
5. G. L. Messing and G. Y. Onoda, Jr., "Inhomogeneity-Packing
    Density Relationships in Binary Powders," J. Am. Ceram. Soc.,
    61 [1] 1978.
6. G. L. Messing and G. Y. Onoda, Jr., "The Sintering of Inhomoge-
    nous Binary Powder Mixtures," submitted to the J. Am. Ceram.
    Soc. for publication, 1977.
7. G. L. Messing and G. Y. Onoda, Jr., "Inhomogeneity-Packing Density
    Relationships in Binary Powders - Experimental Studies sub-
    mitted to the J. Am. Ceram. Soc. for publication, 1977.

DISCUSSION

D. Lynn Johnson (Northwestern University): Have you examined actual binary powders in the SEM? It seems that the fine particles
should, to some extent at least, adhere to the coarse particles.
This would give rise to large-small-large particle contacts,
which seems to be different from the kinds of contacts you showed.
Author: We have not yet carried out experimental work on the sintering of binary powders. The presence of large-small-large
particle contacts, in what was described as region I in our paper,
would cause an increase in $\bar{v}_c$ above v_c. If this deviation from
theory is observed experimentally, the model must be modified
accordingly.

STRESS AND DENSITY DISTRIBUTIONS IN THE COMPACTION OF POWDERS

S. Strijbos

Philips Research Laboratories
Eindhoven - The Netherlands

P.A. Vermeer

Geotechnical Laboratory
Delft University of Technology
Delft - The Netherlands

ABSTRACT

Stress-strain curves for a compacting powder were obtained from tests in a specially developed high-pressure triaxial apparatus. A powder-wall shear apparatus was used to determine the stresses at a wall as a function of the powder-wall displacement. Calculations of stress and density distributions in a "two-dimensional powder compact" were based on mathematical relationships for the stress-strain response and for the powder-wall interaction. The results of these preliminary calculations show promising agreement with known experimental data.

1. INTRODUCTION

The technology of powder pressing in a rigid die has been highly developed, but is poorly supported by scientific understanding of the different process steps: filling of the die cavity, compaction of the powder mass, and ejection of the final compact from the die. A phenomenon which often gives rise to problems is non-homogeneous densification of the powder mass during the compaction step. This may cause, for example, warping of compacts during sintering.

A number of authors have studied experimentally the stress and density distributions, mostly in compacts of cylindrical shape.[1-4] Other authors made calculations on regarding the compacting powder

as a rigid-plastic material.[5-7] The present paper also deals with
the calculation of stress and density distributions.

In chapter 2, the basis of a new calculation procedure is briefly
discussed. Chapter 3 describes how relevant mechanical powder and powder-wall properties can be determined at the high stress levels
used in die compaction. The principles of the high-pressure triaxial test and the powder-wall shear test are explained. Some
test results for a very fine ferric oxide powder are presented
in chapter 4. This powder was chosen because data were available
on the measured boundary stresses.[1] The first test results were
used in a numerical procedure for calculating the stresses in a
"two-dimensional powder compact" (see chapter 5). It is shown that
the results of these preliminary calculations are in good agreement
with the boundary stresses measured along the surface of the compact.

2. MECHANICAL BEHAVIOUR OF POWDERS

When powder assemblies are considered as a continuum, powder
compaction can be defined as the induction of large strains which
by means of high stresses result in densification. This implies the
need to establish stress-strain relationships for powders. Such
relationships are discussed in the next section. One also needs an
expression for the interaction between the die wall and the powder
to be able to calculate the density and stress distribution during
compaction.

2.1. Stress-Strain Relationships of Powders

Schwartz and Weinstein were the first to introduce the principles of soil mechanics for calculating the stress pattern within
a compacting powder.[5] Their model calculations, and those in the
later work of Roman and Perelman,[6,7] are based on a "limiting stress
state" procedure, where it is assumed that the powder behaves in
a rigid-plastic way during compaction.[1]

For that procedure it is assumed that the failure properties
of a powder can be defined by a powder yield locus. This is a relation between the shear and normal stresses on a plane where slip
or failure is just about to occur; these stresses are called the
critical shear and normal stresses. According to the Mohr-Coulomb
criterion, as applied by the above-mentioned authors, the powder
yield locus can be approximated by the relation:

$$\tau = c_i + \sigma \tan \varphi_i$$

where c_i is a stress which is a measure of the internal cohesion of the powder, and φ_i is the angle of internal friction (Fig.1).

Assuming that internally the powder mass is everywhere in the failure state during compaction, numerical calculations can be made. However, these numerical calculations show large differences with the measured stress distributions.[1]   The most likely reason is that during the compaction of a powder, a failure state is not present everywhere, but that only minor parts of the compact will be in the limiting stress state, namely, only parts around corners and parts adjacent to the walls of the die.

A more promising model is the recently developed "double hardening model".[8]   This model accounts for the experimentally established fact that powders during compaction deviate strongly from rigid-plastic behaviour. It is found that stress states represented by a Mohr stress circle below the yield line, already cause an appreciable plastic deformation. The stress-strain relationship used is schematically drawn in Fig.2.

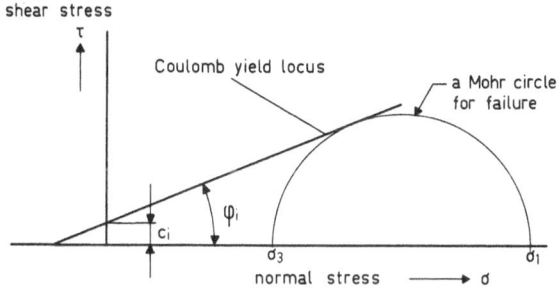

Fig.1. The Mohr diagram plot for the Coulomb yield criterion.

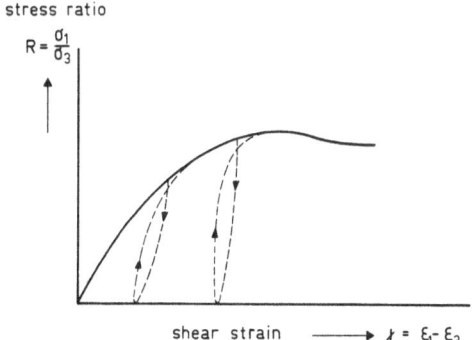

Fig.2. Relationship between the stress ratio $R = \sigma_1/\sigma_3$, $(\sigma_2 = \sigma_3)$, and the plastic shear strain $\gamma = \varepsilon_1 - \varepsilon_3$, $(\varepsilon_2 = \varepsilon_3)$, as used in the double hardening model.

2.2. Powder-Wall Interaction

By analogy with the introduction of φ_i as the angle of internal friction, we can introduce a friction angle φ_w to describe the interaction between a die wall and an assembly of particles[9]:

$$(\tau_w/\sigma_w) \leqslant \tan \varphi_w$$

where τ_w and σ_w are the shear and normal components, respectively, of the stress vector acting on the wall. The results of the powder-wall experiments, reported later in this paper, demonstrate that φ_w is a realistic quantity. The experiments lead to a relation of the type $\tau_w/\sigma_w = f(u)$, represented in a schematic form by the solid curve in Fig. 3. Here u is the displacement of the powder relative to the wall.

For the computer calculations reported in chapter 5, the boundary conditions from the experimental results are simplified as follows:

$$\text{if} \quad \delta u > 0 \quad \text{then} \quad (\tau_w/\sigma_w) = \tan \Phi_w$$

The symbol δ is used to indicate that a small increment of the displacement, u, is considered.

3. DETERMINATION OF POWDER PROPERTIES

A common test in civil engineering for determining the stress-strain response of soils is the triaxial test. A special triaxial test equipment was developed for testing small powder samples up to high stresses and strains, comparable to those existing during die compaction (section 3.1). Also a suitable powder-wall shear apparatus had to be developed, for studying powder-wall interaction at high stresses (section 3.2).

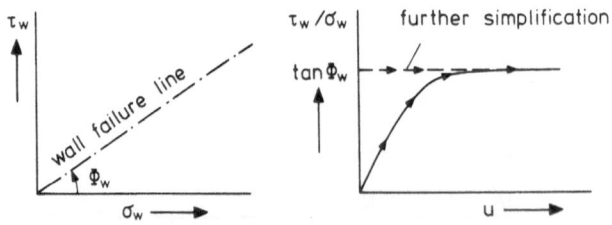

Fig. 3. Schematic representation of the results of powder-wall shear tests.

3.1. The High-Pressure Triaxial Test

A truly triaxial test would need independent control of the
three principal stresses, σ_1, σ_2 and σ_3. In the standard triaxial
test the two principal stresses σ_2 and σ_3 applied to the outer
surface of the sample have the same value. The cylindrical sample
is enveloped by a thin rubber membrane and placed inside a pressure
cell between two metal end platens. The outer surface of the sample
is initially loaded by a certain isostatic cell pressure ($\sigma_2 = \sigma_3$)
and then gradually compressed axially while the cell pressure is
kept constant.

Figure 4a shows a diagram of the high-pressure triaxial test
cell. It is a thick-walled cylindrical pressure vessel (2), incorporated in a frame. The dimensions are: 140 mm length, 20 mm inner
diameter, 60 mm outer diameter. The lower plunger (12) has a fixed
position, whereas the upper plunger (1) can be pressed downward by
the load pacer of a Tinius Olsen testing machine. Oil is fed into
the pressure chamber through a centre bore (13) in the lower plunger. Oil is supplied (14) from a hydraulic unit which allows us to
maintain the oil pressure at any desired level up to a maximum
pressure of 50 MN/m^2. The test specimen (4) is clamped between
the plungers, which are perfectly aligned (by the fits of the high-
pressure vessel) to prevent lateral forces.

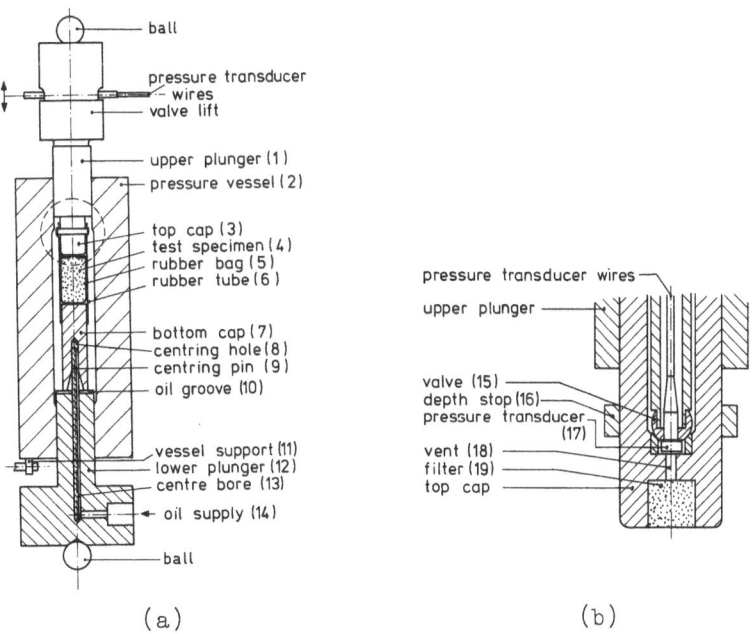

(a)   (b)

Fig.4. (a) Diagram of the high-pressure triaxial test cell, (b)
       detailed diagram of the top cap.

The axial load is measured externally. The measurement is affected by the friction between the upper plunger and the pressure vessel, but because the vessel is rotated and no seals are employed,the friction is negligibly small. A more uniform stress in the sample is promoted by reducing the friction between the ends of the sample and the rigid end platens. The length-to-diameter ratio of the samples is about 2. It is essential to make homogeneous powder packings.

Fig.4b shows some details of the top cap (3) which makes it possible to measure the volume change of the specimen during the compaction test. The measurement is done by determining the amount of air released from the sample during densification. The top cap is provided with a porous metal filter (19) and a vent (18). The vent is fitted with a valve (15) and a pressure transducer (17). When the valve is closed the pressure transducer permits measurement of the air pressure variations, from which the volume changes of the test sample can be determined.

A more detailed description of the preparation of the samples and the performance of the tests will be published elsewhere.[10]

3.2. The Powder/Wall Shear Test

The principle of the apparatus is shown in Fig.5. It consists of a carefully machined plane-parallel plate with known surface conditions, clamped by two cylindrical powder compacts of 15 mm diameter and about 2 mm height. The compacts are contained in holders (dies) from which they protrude about 0.3 mm. During a test the plate is pressed downward at a constant speed (range 0.01 – 1 mm/s), and the shearing force is recorded against the displacement.

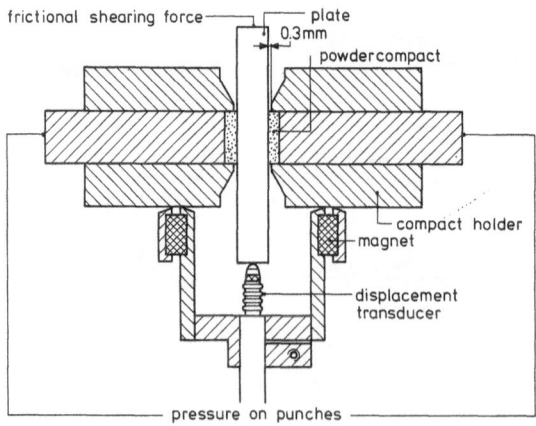

Fig.5. Schematic diagram of the powder/wall friction apparatus.

The plate is clamped between the two compacts by means of plungers exerting a constant load on the punches. Because the holders are strung along two guiding rods, with negligible friction, the load is fully transmitted to the powder-wall interface. The normal load can be varied between 1 and 40 MN/m^2. The construction ensures that pure powder wall displacement is recorded, not affected by the elastic compliances of the various parts of the equipment. Details of the equipment and the performance of the tests have been published elsewhere.[9]

4. EXPERIMENTAL RESULTS

The experiments were performed with a commercial ferric oxide powder, manufactured by BASF. The mean particle size was about 0.04 μm, the specific surface area ~ 40 m^2/g, the material density 5.27 g/cm^3, the bulk density 0.58 g/cm^3, and the tapped density 1.12 g/cm^3. The powder used for each experiment was dried for at least 48 hours at 120°C.

4.1. Standard Triaxial Tests

First an isostatic stress is applied to the outer surface of the sample, which results in a certain amount of volume strain $e_1=-\ln(V_o/V_i)$. V_i si the starting volume of the sample, and V_o the volume after isostatic compression. The tests continued by compressing the sample axially at a constant rate of 1 mm/min, while σ_2 and σ_3 are kept constant, and the average axial stress, σ_1, is measured. The results of all tests are summarized in Figs.6a and b, showing the stress ratio R (= σ_1/σ_3) versus the two strain components, namely the shear strain $\gamma = \varepsilon_1-\varepsilon_3$, and the volume strain

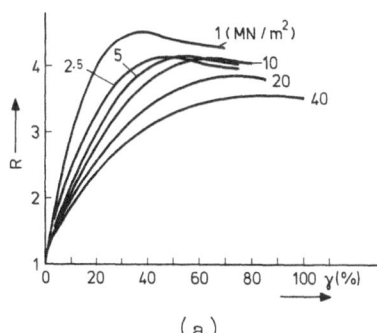 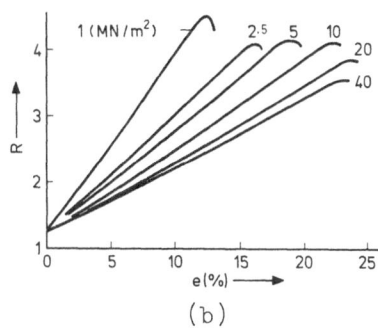

(a)   (b)

Fig.6. Triaxial test results. The principal stress ratio R = σ_1/σ_3, at various cell pressures σ_3, as a function of
a) the shear strain $\gamma = \varepsilon_1-\varepsilon_3$,
b) the volume strain $e = \varepsilon_1+2\varepsilon_2$.

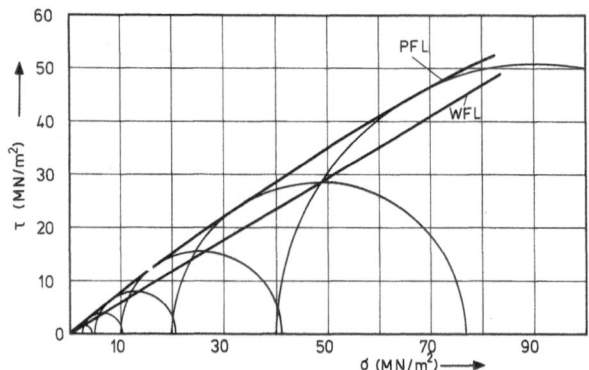

Fig.7. Powder failure line (PFL) of BASF ferric oxide derived from
        Fig. 6, and wall failure line (WFL) derived from Fig. 8.

e = ε_1 +2ε_3, respectively, at several cell pressures. The reproducibility of the results obtained with the test was good. It should
be noted that the volume strain e_1 caused by the isostatic compression is not included in the values presented in Fig.6b.
Figure 7 shows the powder failure line (PFL) which can be constructed
from the previous graphs. We chose the maxima of the graphs shown
in Fig. 6 as the stress ratio for failure.

4.2   Powder-Wall Shear Tests

    Tool steel walls with a Vickers hardness number of about 600
were used. The walls were ground so as to produce grooves on the

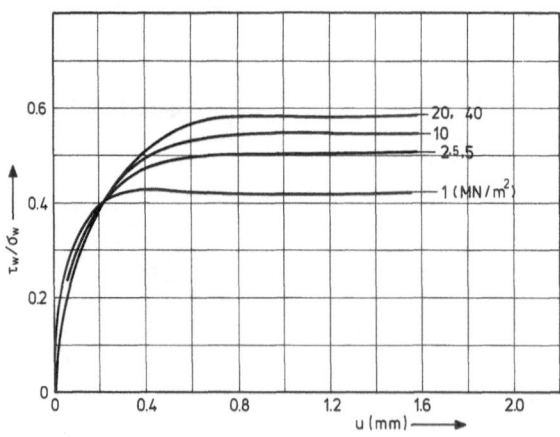

Fig. 8. Results of powder/wall tests at various loads, with BASF
         ferric oxide powder and a tool steel wall (V.H.N. 600)
         ground parallel to the sliding direction.

surface parallel or perpendicular to the direction of sliding. The
surface roughness, found by the Taylor–Hobson method, was 0.43 μm
measured perpendicular to the grinding direction, and 0.20 μm
measured to the grinding direction. Figure 8 summarizes the results
for the tests with the wall ground parallel to the sliding direction;
the results for the tests with the other wall show minor differences
and are not presented here. As can be seen, at low normal loads
the curve of τ_w/σ_w versus the axial displacement, u, is slightly
different from that at high normal loads. From the maxima of these
curves the powder–wall failure line has been constructed and is
shown in Fig. 7. One can observe that the powder and powder–wall
failure lines differ only slightly. An interpretation of this fact
is given elsewhere.[11]

5.   CALCULATIONS OF BOUNDARY STRESSES

     It has already been stated in Section 2.1 that numerical cal-
culations, based on the limiting stress state assumption, show poor
agreement between the calculated and measured wall stresses. As a
typical example, Fig. 9 shows the results of calculations made in
our laboratory,[13] using the rigid–plastic model and own experimental
data for the pressure distributions on the top and bottom plungers,
for a BASF ferric oxide powder at a compaction pressure of 31.8 MN/m^2.
As can be seen, the course of the wall stresses is quite different

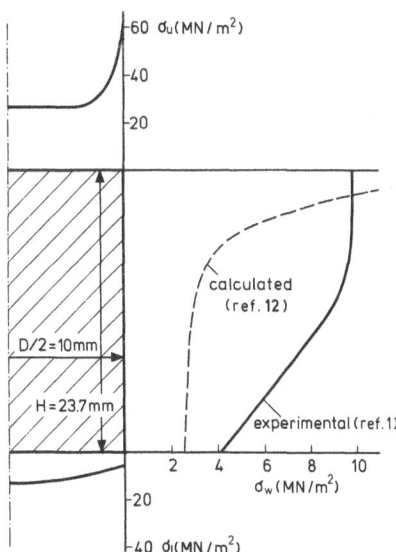

Fig. 9. Boundary stresses occurring during one–sided die compaction
        of 12 grams of BASF ferric oxide powder at a compaction
        pressure of 31.8 MN/m^2.
        σ_u is the normal stress on the upper plunger.
        σ_1 is the normal stress on the lower plunger.

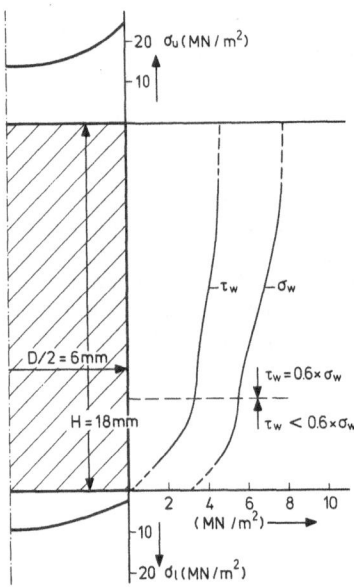

Fig. 10.   Calculated boundary stresses occurring during one-sided die
           compaction for a two-dimensional, theoretical case.

from that experimentally found.

The model recently developed by Vermeer[8] has been applied to the
problem of a "two-dimensional compact". A cylindrical compact could
not be treated since only a computer program for plane strain problems
was available. Another restriction was that the model was developed
for sand which is deformed at low stress levels. The parameters in
the model could not be chosen such that the stress-strain curves
shown in Fig. 6 were followed accurately in the calculations. Only
the dominant features of these curves were accounted for. The
computer program was based on the well-known finite element method.
The powder-wall interaction was taken into account by means of
special interface elements.

Figure 10 shows a typical result of the calculations at a compaction pressure of 17 MN/m^2. The curves of σ_w and of τ_w show good
agreement with those found experimentally. It can be seen that at
this stage of compaction the wall failure line is partly mobilized
along the wall, viz. where $\tau_w/\sigma_w = 0.6$. It is further worthwhile to
mention that our calculations of the internal stresses show that
indeed the powder failure state is reached hardly anywhere; this in
contrast to the assumption in the limiting stress state model.

The fact that these preliminary calculations show promising
agreement with the measurements of the stress distribution gives good
hope that it will prove possible to predict density distributions

in three-dimensional compacts of complex shapes. By means of such
calculations it will be possible to detect quantitatively the effect
of various powder and process parameters on the density and the density distributions.

ACKNOWLEDGEMENT

The authors are indebted to Professor A.L. Stuijts for his help in
preparing this paper and his stimulating interest in the present
study.

REFERENCES

1. S. Strijbos, P.J.Rankin, R.J.Klein Wassink, J.Bannink, and G.J.
   Oudemans, Powder Techn., 18, 187 (1977).
2. D.Train, J.Pharm.,London, 8, 745 (1956).
   D.Train, Trans.Inst.Chem.Engrs., 35, 258 (1957).
3. T.Takada, M.Kuramoto, T.Kunio and H.Kuno, Powder Techn., 14,
   51 (1976).
4. H.M.McLeod and K.Marshall, Powder Techn., 16, 107 (1977).
5. E.G.Schwartz and A.S.Weinstein, J.Am.Ceram.Soc., 48, 346 (1965).
6. V.E.Perelman and O.V.Roman, J.Powder Metall., 9, 692 (1971).
7. V.E.Perelman and O.V.Roman, paper presented at the ICC1, Brighton,
   England,Oct. 3-5, 1972.
8. P.A.Vermeer,Delft Progress Rep., 2, 303 (1977).
9. S.Strijbos, Sci.Ceram., Vol. 8, 415 (1976).
10. W.C.P.M.Meerman and A.C.Knaapen, A high-pressure triaxial tes-
    ting cell; to be published.
11. S.Strijbos, Boundary Phenomena During Die Compaction of a Fine
    Powder; submitted for publication in Ceramurgia International.
12. P.J.Rankin, private communication.

DISCUSSION

F. F. Lange (Rockwell Science Center): You have considered shear to
cause the "fracture" of interparticle bonds. Have other failure modes,
e.g., particle fracture by tensile stresses that can develop in these
tri-axial compressive states, been observed? For example, is the
particle size changed during compaction?
Authors: No, there has not yet been any electron microscopic examination of the sheared samples.

D. E. Niesz (Battelle Columbus): As a comment on the previous
question by Fred Lange, surface area data on calcined powders before
and after pressing indicate that little fracture of ultimate particles
occurs. Surface area shows little increase on pressing. However,
particle agglomerates are crushed.

PORE MORPHOGRAPHY IN CERAMIC PROCESSING

O. J. Whittemore, Jr.

University of Washington
Ceramic Engineering Division
Seattle, Washington 98195

INTRODUCTION

Most ceramics while being processed begin as an aggregation of particles and of pores. During processing the size and distribution of pores change. Some factors affecting pore size distribution (PSD) during processing are the particle size distribution, forming pressure, the volatilization of temporary binders and plasticizers, decomposition, and sintering. After processing many ceramics are porous and their PSDs can affect the properties in use. Mercury intrusion has been found to be the most useful method of characterizing PSD of porous ceramics. The method was first proposed by Washburn in 1921,[1] but not until 1945 did Ritter and Drake report applications.[2] Since then many investigators have used the method to characterize porous materials.[4,5] This paper will describe several applications of mercury porosimetry for the characterization of ceramics at various stages of processing.

THE MERCURY INTRUSION METHOD

Washburn gave the relation for intrusion of a pore of circular opening as: $pd = -4 \sigma \cos \theta$, where p is the pressure required to force liquid into a pore of entry diameter d, σ is the liquid surface tension, and θ is the liquid-solid contact angle. A non-wetting liquid is required (θ must be larger than 90°) and mercury is most convenient. The surface tension can be considered constant, and wetting angles are found to range from 130° to 140° for most materials. Therefore, only the pressure of intrusion needs to be measured to determine the pore diameter. When the volumes of

intruded mercury are determined at increasing pressures, a plot of
PSD can be constructed. Pores of diameters ranging from 200 μm
down to 3 nm have been measured, the latter requiring 414 MPa
(60000 psi). Additional information can be obtained by construct-
ing pore volume frequency plots. Also, by considering porosity as
a series of cylinders, the surface area can be calculated. Joyner
et al[3] showed good agreement of surface areas thus determined with
those found by nitrogen adsorption.

   Factors contributing to errors are: compressibility of mercury,
the glass vessel and the sample, the pore shape, the contact angle
assumed, surface tension variations, thermal expansion and sample
collapse. These errors have been discussed in several reviews.[2,4,5]

APPLICATIONS IN CERAMIC PROCESSING

Characterization of Diatomite

   The pore size distributions (PSDs) of powders are often determined. However, the distinction between interparticle and intraparticle pores can be difficult. Diatoms have intricate systems
of intraparticle pores, but their volumes are much smaller than the
interparticle spaces in loose diatomite. When compacted, the interparticle space was reduced and a steep break in the curve was found

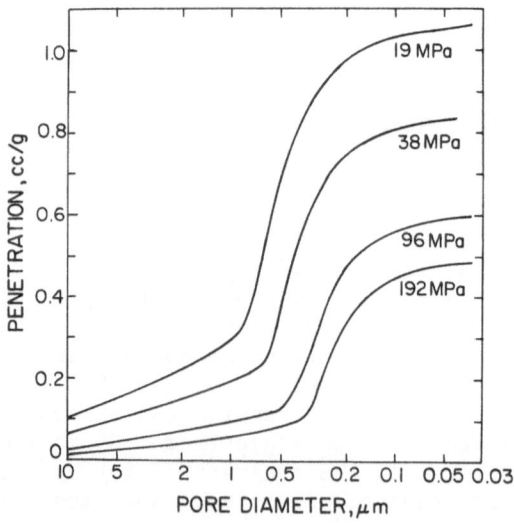

Fig. 1. Pore size distributions of Ceará Mirim, RN, Brazil
         diatomite after compacting at indicated pressures.

to agree with the pore diameters observed by scanning electron
microscopy.[6] Figure 1 gives PSD curves for one diatomite after
compacting at several pressures.

Compaction of Mono-Sized Glass Spheres

In a basic study of compaction, Calkins[7] studied carefully
sized 240 μm glass spheres. These were compacted by an Instron
machine in a steel mold which would fit into a mercury penetrometer
vessel. After compaction, the mold plungers were removed, screens
inserted for particle retention and the PSD determined. Figure 2
includes the distributions after compacting at various pressures
and the progressive decrease in pore size with pressure due to the
fracture of spheres and movement of the resultant fragments into
the pores. The pore frequency curve for the packed spheres before
compaction showed a peak at 94 μm, 0.4 of the initial sphere size.
This peak was progressively reduced and moved to a smaller size as
the pressure was increased and another peak at a smaller pore
diameter appeared. The curve of the 207 MPa compacted sample is
shown in Figure 3. After compacting to 345 MPa, only the small
size peak at 5 μm diameter was observed.

Calkins made careful calibration curves which, when subtracted
from the PSD curves, allowed him to calculate the surface areas.

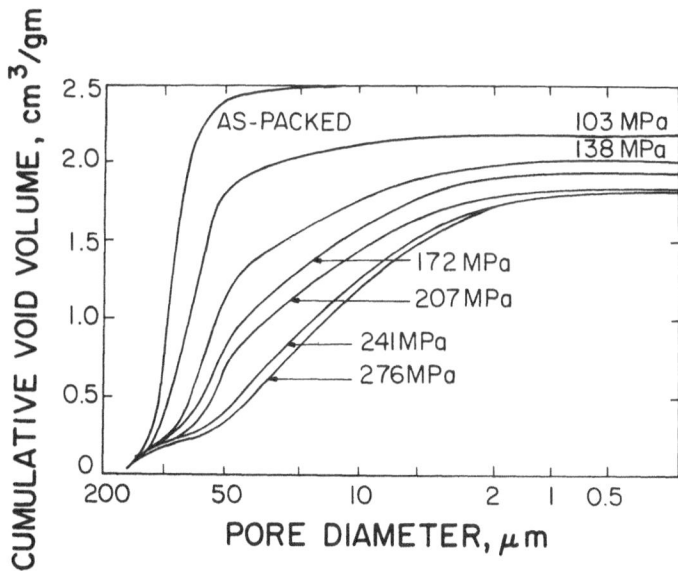

Fig. 2. Pore size distribution curves of glass spheres after
         compacting at indicated pressures.

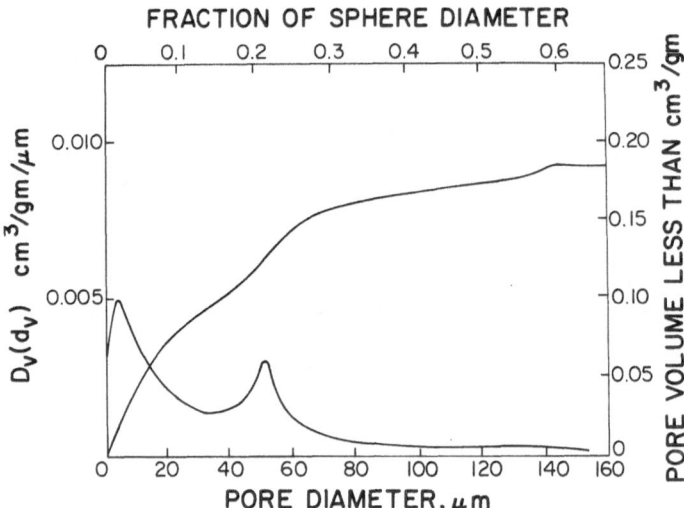

Fig. 3. Pore size distribution and frequency of glass spheres
         after compacting at 207 MPa.

He also calculated the work of compaction. A plot of surface areas
vs. work of compaction showed linearity, proof of Rittinger's "Law"
of comminution which states that the energy expended is proportional
to the new surface created.

Pore Sizes in Gypsum Plaster Molds

Gypsum plaster molds are commonly used for slip casting, and
they have been shown to act as filters.[8] A common water: plaster
consistency ratio for molds is 75:100 by weight although the strength
has been shown to be higher at lower ratios.[9] A series of samples
were prepared at consistencies from 40 to 85, dried and PSDs determined, from which surface areas, most frequent pore radii, and pore
volumes were derived. These are listed in Table I. In all cases,
the pore volume frequencies indicated only a narrow range of pores
in each sample. The relation between most frequent pore radii and
consistency was found to be nearly linear, with the relationship:

$$R = 0.46 \ W - 1.41$$

where R = most frequent pore radius and W = consistency or weights
of water per 100 weights plaster. One can thus predict the pore
size in a plaster mold from the consistency.

Table I.   Pore Properties of Dried Gypsum Plaster
Prepared at Various Water Consistencies

Consistency wt. water/100 wt. plaster	Pore Volume cc/g	Most Frequent Pore Radius μm	Surface Area m²/g
40	0.37	0.39	9.4
50	0.45	0.93	7.9
60	0.56	1.44	8.0
68	0.63	1.75	7.8
77	0.70	1.88	8.7
82	0.78	2.33	7.8
85	0.88	2.68	8.1

Pore Growth During the Firing of 60% MgO Basic Refractories

Many ceramics are prepared from a wide distribution of particle
sizes. During firing, bulk shrinkage is inhibited by the rigid
framework of large inert particles while the fine particles densify
in the interstices between the large particles, and pore sizes
increase. Watson et al[10] showed this effect in firing a wall tile
body. The effect has been compared with other causes of pore
growth.[11]

Refractories are commonly made of particle mixtures ranging
from as high as 5 mm down to 2 μm. The permeabilities of

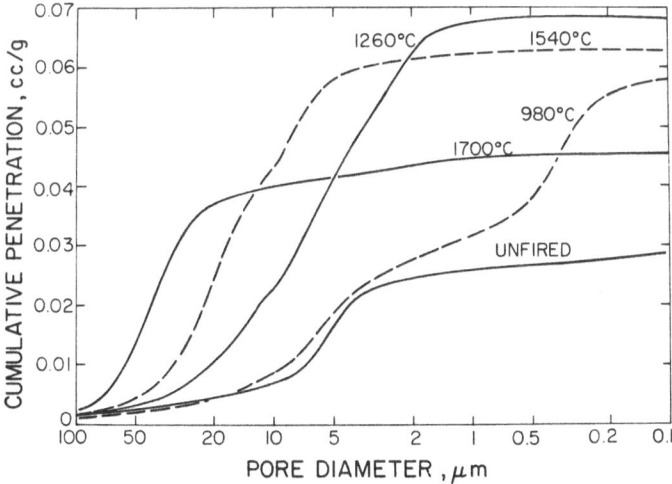

Fig. 4.   Pore Size Distributions of 60% MgO Brick Fired at Various
Temperatures.

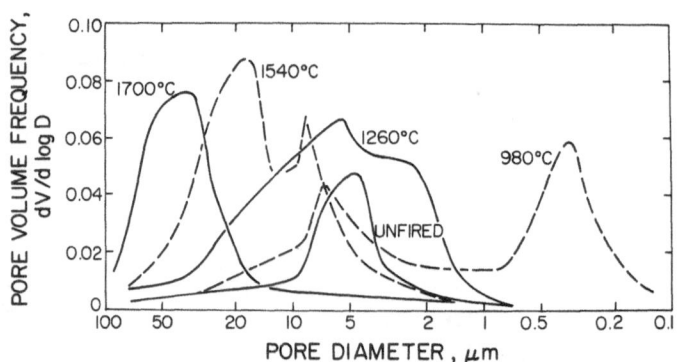

Fig. 5. Pore Volume Frequencies of 60% MgO Brick Fired at Various
         Temperatures.

magnesia-chrome brick containing 60% MgO were studied by Jeschke.[12]
A similar set of samples was prepared, fired at various temperatures,
and PSDs determined.[13] The data are plotted in Figure 4. Pore
volume frequencies derived from Figure 4 are plotted in Figure 5.

When the brick is first fired to 980°C, no change occurs in the
5 μm pores but the 0.3 μm pores appear, probably as a result of combustion of the temporary organic binder. Some pores are finer than

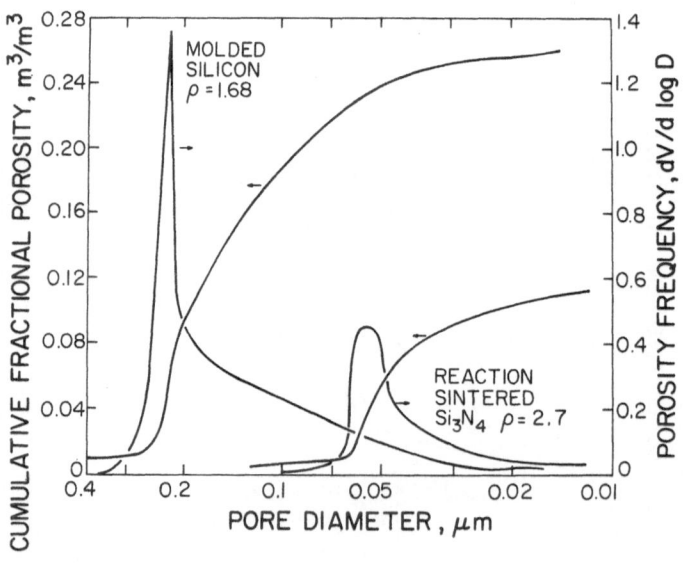

Fig. 6. Pore Size Distributions of Molded Silicon and of Silicon
         Nitride Resulting from the Nitriding of the Molded Silicon.

0.1 μm after 980°C, but after 1260°C all have grown larger than
1 μm. Pore growth continues through 1540°C to 1700°C. Porosity
decreases slightly after 1540°C and significantly after 1700°C.

     The pore volume frequency curves reveal additional information,
particularly the bimodal pore distribution in the samples fired at
1260 and 1540°C in addition to the obvious bimodal nature of the
980°C sample. The PSDs of several refractories were published by
Ulmer and Smothers.[14]

Reaction Sintering of Silicon Nitride

     An important method of manufacturing silicon nitride ceramics
is the reaction bonding process. Compacts of silicon particles are
heated slowly in nitrogen and are converted to dense strong articles
of silicon nitride with little exterior volume change. PSDs were
determined on a silicon compact of 1.68 bulk specific gravity and a
silicon nitride sample of 2.7 bulk specific gravity resulting from
the nitriding of a similar silicon sample. These data are plotted
in Figure 6 together with derived porosity frequencies. The process
can be seen not only to halve the porosity but also to reduce the
most frequent pore diameter from 0.2 to 0.05 μm.

The Initial Stage of Sintering

     During the initial and intermediate stages of sintering, pores
have been shown to grow in many materials.[11] Rarely have pores

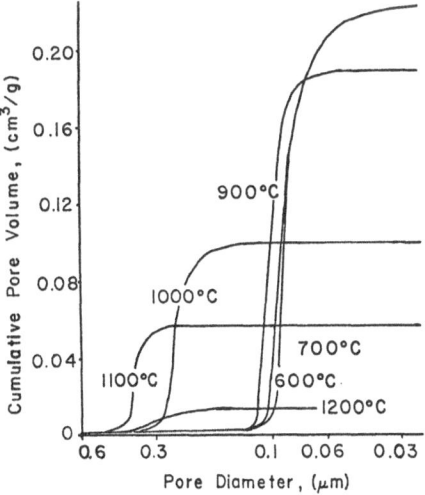

Fig. 7. Pore Distributions of TiO_2 Held at Temperature One Hour.

been noted to shrink until considerable reduction in porosity has occurred. The pore diameters of two types of Al_2O_3 showed little change until more than half of the porosity had been eliminated.

An intermediate group of materials, such as magnesia, iron oxide and titania, show pore growth occurring with shrinkage.[11,14] PSDs in titania samples sintered for one hour at various temperatures are displayed in Figure 7. Note that pore growth continued until only a quarter of the original porosity remained.

There also are a group of materials that show pore growth with little shrinkage. Sipe[15] found iron oxide behaved thus in the range 500 to 700°C. He proposed a "fast pairs" model where a small proportion of particle pairs coalesces within the mass without bulk shrinkage. His model, utilizing Coble's[16] relationship for grain boundary diffusion, fitted the porosimetry data. Sipe[17] also showed confirmation of fast-pairs by sintering monolayer particle films. Diffusion coefficients for particle pairs varied by a factor of 1000 at each temperature. He also found the resultant grains to be elliptical to spherical in shape. Joss found SnO_2 would show pores growing to 10 times their original sizes in the range of 680 to 1000°C.[14] His PSD curves for one hour at temperature are given in Figure 8. The porosimetry data also fit Sipes' model for fast-pairs coalescence by grain boundary diffusion.

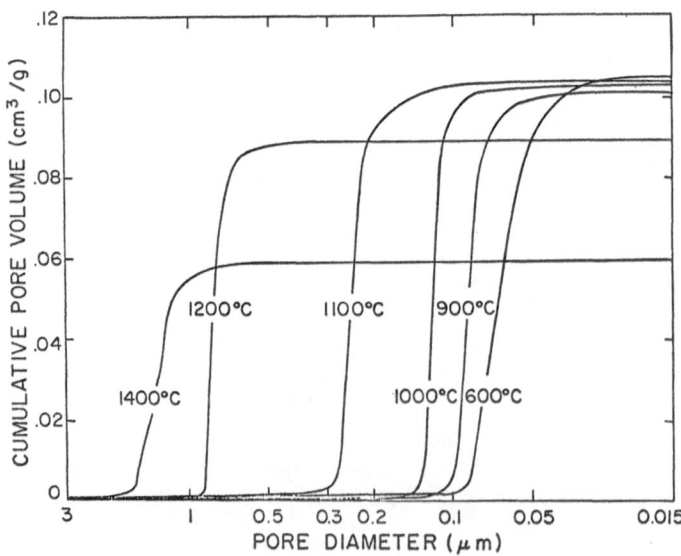

Fig. 8.   PSDs of SnO_2 Compacts Held One Hour at Temperature.

ACKNOWLEDGMENT

Much of the work described was supported by NASA Grant NGL-002-004.

REFERENCES

1. E. W. Washburn, Proc. Nat. Acad. Sci., 7, 115-16 (1921).
2. H. L. Ritter and L. C. Drake, Ind. Eng. Chem. Anal. Ed., 17, 12, 782-6 (1945).
3. L. G. Joyner, E. P. Barrett, and R. Skold, J. Am. Chem. Soc., 71, 3155-8 (1951).
4. H. M. Rootare, Aminco Lab. News, 24, 4A-4H (1968).
5. C. Orr, Jr., Powder Technol., 3, 117-23 (1969/70).
6. O. J. Whittemore, Jr., and J. H. C. Castro, Proceedings, Mat. & Equip. & Whitewares Div., Am. Ceram. Soc., 48-51 (1976).
7. D. B. Calkins, Ph.D. Thesis, Univ. of Washington, Seattle (1970).
8. D. S. Adcock and I. C. McDowall, J. Am. Ceram. Soc., 40, 10, 355-62 (1957).
9. C. M. Lambe, in Ceramic Fabrication Processes, ed. W. D. Kingery, 31-40 (1958).
10. A. Watson, J. O. May, and B. Butterworth, Trans. Brit. Ceram. Soc. 56, 37 (1957).
11. O. J. Whittemore, Jr., and J. J. Sipe, Powder Technol., 9, 155-64 (1974).
12. P. Jeschke, J. Am. Ceram. Soc., 49, 7, 360-3 (1966).
13. O. J. Whittemore, Jr., unpublished report to the Foundation in Refractories Education, 1974.
14. H. D. Joss, M.S. Thesis, Univ. of Washington, Seattle (1975).
15. J. J. Sipe, Ph.D. Thesis, Univ. of Washington, Seattle (1971).
16. R. L. Coble, J. Am. Ceram. Soc., 41, 2, 55-62 (1958).
17. J. J. Sipe and O. J. Whittemore, Jr., J. Am. Ceram. Soc., 53, 9, 525 (1970).

PART II

SOLID STATE SINTERING
AND GRAIN GROWTH

FUNDAMENTALS OF THE SINTERING OF CERAMICS

D. Lynn Johnson

Northwestern University
2145 Sheridan Road
Evanston, Illinois 60201

INTRODUCTION

Although sintering in some form is among the oldest of the technologies known to man, the level of understanding that he has of the problem is still in its infancy. Many interrelated and interacting phenomena occur as a compact of powder particles undergoes the morphological changes that are called sintering. The lack of understanding of sintering is manifested in the diversity of opinions as to the mechanisms of the process, and the inability to predict the sintering behavior of materials. The possible exception to this is in the case of simple glasses, for which the sintering mechanism is viscous flow.

The difficulty with predicting sintering behavior, and the key to the lack of understanding, is its complex nature. There is probably universal agreement that the driving force for sintering is the excess surface free energy present in a powder compact, but numerous interrelated processes act to bring about reduction in this excess energy. At elevated temperatures the surface tension of the particles gives rise to chemical potential gradients which, in turn, cause flow of matter to decrease these gradients and decrease the total free energy of the system. It is the purpose of this paper to describe some of the phenomena that occur in solid state sintering in response to these gradients, and to discuss experimental methods for quantitatively understanding their interrelationships.

DRIVING FORCE AND FLUXES

Surface tension in solids is not as simple a concept as it is in liquids, in which the surface tension is isotropic. In crystalline materials the surface tension often depends upon the orientation of the plane of the surface relative to the crystallographic axes. If the anisotropy is large, the surfaces found within a sintering compact will tend to be planar and faceted. If a surface tension is isotropic, and if the temperature is sufficiently high that relaxation can occur, (which is always the case if sintering is underway) the pressure beneath a curved surface is related to the curvature as follows:

$$\Delta p = \gamma K = \gamma \left(\frac{1}{r_1} + \frac{1}{r_2} \right) \tag{1}$$

where γ is the surface tension, K is the curvature defined in this equation in terms of the principal radii of curvature of the surface, r_1 and r_2. Compressive pressures will be positive and the radii of curvature will be positive if the center of curvature is on the solid side of the surface. Although the foregoing concept is not defined for particles with equilibrium flat surfaces and sharp edges and corners, nevertheless the excess surface energy is related inversely to the particle size in these cases as well. Edges and corners represent high energy sites, and the density of these increases as a particle size decreases. Thus there is a tendency for faceted, as well as smoothly curved, particles to sinter. An isotropic surface tension will be assumed for the present discussion.

The chemical potential of atoms under a hydrostatic pressure, p, is given by:

$$\mu_a = \mu_a^o + \Omega_a p + kT \cdot \ln(a_a) \tag{2}$$

where μ_a^o is the chemical potential of atoms at zero pressure and unit activity, Ω_a is the atomic volume, a_a is the activity of the atoms, and kT has its usual meaning. In the case of smoothly curved surfaces, Eq. (1) can be utilized in this equation to relate the chemical potential of atoms to the curvature. We thus see that atoms on concave surfaces have lower chemical potential than do those on convex surfaces, so that there will be a driving force to move atoms in such a way as to reduce the chemical potential gradients, and atoms will move from the convex to the concave portions of a powder compact.

The fluxes that arise as a result of the chemical potential
gradients can occur within the vapor phase, along the surface,
through the bulk of the particles, or along grain boundaries.[1-3]
In addition, if the local stresses are sufficiently high, plastic
flow is possible, although there is considerable controversy in
the literature as to how important this last mechanism is.[4-6]

Surface diffusion is perhaps the most universal mechanism
operative during sintering of crystalline materials. There is
apparently no difficulty in atoms being dislodged from convex
surfaces and diffusing to concave surfaces. Surface diffusion
causes a build-up of bridges or necks between particles, with a
concomitant loss in surface area, but cannot produce any densification of a three dimensional powder compact. It is observed usually that surface diffusion is important at low temperatures, in
that powder compacts can be strengthened at low temperatures without measurable densification. This is because surface diffusion
coefficients generally tend to have low activation energies and
higher values than other diffusivities at lower temperatures.
Surface diffusion can produce significant surface area loss and
contribute to particle coarsening during the earliest stages of
sintering of highly active submicron powders.[7,8] As might be sup-
posed, the mass flows driven by surface diffusion cause a more
rapid reduction in driving force than would be expected without
surface diffusion, and this will have a deleterious effect on densification rates and will influence the sintering kinetics, including the time dependence and other parameters used in identifying
sintering mechanisms.[9]

Vapor transport is another mechanism which is relatively
easily understood in a qualitative sense, at least, and which is
probably a fairly universal participant in sintering at the highest temperatures. The increased chemical potential under convex
surfaces gives rise to a higher vapor pressure, which in turn
causes a flow of material from convex surfaces to concave surfaces.
As with surface diffusion, vapor transport will cause surface area
loss without densification, and will alter kinetics and can inter-
fere with correct interpretation of sintering kinetics.

The role of plastic flow by dislocation motion has been perhaps the most controversial issue in sintering theory over the
past decades. There are those who have felt that the forces
produced by curvature gradients are sufficiently strong to cause
nucleation and glide of dislocations.[5,6] There are others who put
forth calculations which seem to indicate that the nucleation
stress is significantly greater than the stresses available, and
that, while the existing dislocations might move, there is insufficient force to cause additional dislocations to be formed.[4]

There is little unambiguous evidence that dislocation motion is an important factor in sintering. In submicron particles it likely can be neglected with impunity.

Volume diffusion can cause sintering both by atoms diffusing from the particle surfaces to the neck surface, and from the grain boundary that forms between particles to the neck surface. In the vast majority of cases of volume diffusion controlled sintering, the rate limiting diffusion will be by a vacancy mechanism, even though the material may have a Frenkel defect structure, since the slower diffusing species will be rate limiting.

The flow of atoms down a chemical potential gradient drives a flow of vacancies up a chemical potential gradient. This can be appreciated by recognizing that vacancies are created at the concave neck surface and are annihilated either at the convex particle surface or at the grain boundary. In order for a vacancy source to operate, the vacancy concentration beneath that source must be less than the equilibrium value, or, in other words, the vacancies must be undersaturated, that is their chemical potential is less than zero (it is convenient to take the chemical potential of vacancies as zero). In order for a vacancy sink to operate, the vacancies must be supersaturated; their chemical potential must be greater than zero. Thus the vacancies are being driven from a region of lower chemical potential to the region of higher chemical potential, or up a chemical potential gradient. This is despite the fact that they are moving down their concentration gradient. (Concentration gradients do not necessarily drive fluxes.) Herring[10] showed that the flux of atoms in a volume diffusion sintering situation is given by

$$j_a = -\frac{D_v}{\Omega_a kT} \nabla (\mu_a - \mu_v) \tag{3}$$

where D_v is the volume diffusion coefficient and μ_v is the chemical potential of vacancies. Thus a gradient in the chemical potential of vacancies appears as an impediment to the flow of atoms.

Despite Herring's comments and equation published in the first days of development of modern sintering theory, most people have chosen to discuss sintering in terms of the vacancy concentration gradient. Although the sintering equations derived on the basis of concentration gradients are the same as those derived from the chemical potential gradient viewpoint for pure metals and pure ionic compounds, one encounters serious difficulty in the case of extrinsic doped ionic compounds, where the dopant essentially eliminates the concentration gradients. The Schottky and Frenkel

constants are functions of the pressure, to be sure, but in a
doped case the vacancies or interstitials will be governed primarily by the concentration of dopant, and the vacancy concentration gradients, while not eliminated entirely, are greatly attenu-
ated, and one is led to conclude that doping should always retard
densification.[11] On the other hand, this difficulty does not
arise if one works from the chemical potential gradient point of
view, since the gradient in the chemical potential of atoms will
be independent of (uniform) doping and the dopant will influence
mass transport through its influence on the diffusion coefficient.[12]

In considering volume diffusion as a mechanism of sintering,
one must bear in mind that the diffusion path from the grain boundary to the neck surface is near and roughly parallel to the grain
boundary. If a space charge or extended disordered region exists,
the diffusion coefficient in this near-boundary region will not be
the bulk value; it is usually greater, but could be less. For submicron oxide particles this region could conceivably encompass a
significant volume fraction of the particle. Yan et al.[13] have
concluded that a simple space charge layer would likely not cause
significant enhancement of sintering of ionic materials, and the
wide regions of enhanced diffusion indicated for some oxide sintering and diffusion data must arise from other factors. A wide
region of enhanced diffusion at the grain boundary could cause
volume diffusion kinetics, but the diffusion coefficient would not
be that for the crystal interior. One should probably never fail
to consider the possibility of an enhanced boundary diffusivity in
seeking to understand sintering of compounds.

Grain boundary diffusion, in which the grain boundary between
particles is both the source of material for building the neck
between particles and the path of matter transport, has been proposed as the sintering mechanism of some materials, and is gaining
greater acceptance.[14] The scarcity of direct measurements of
grain boundary diffusivities in ceramic materials seriously ham-
pers the elucidation of the importance of this mechanism, but some
qualitative comments might be made.

Grain boundary diffusion is the simplest of the various diffusion mechanisms to model, and undoubtedly is a major contributor to
sintering of submicron ceramic particles. The ratio of grain
boundary and volume fluxes in the initial stage of sintering of
two identical spherical particles is given by[9]

$$\frac{J_v}{J_b} = \frac{aD_v}{2\pi bD_b} \cdot \frac{A}{X} \tag{4}$$

where a is the sphere radius, b is the thickness of the region of
enhanced diffusion at the grain boundary, D_b is the boundary diffusivity within this region, and A/X is the ratio of the neck surface
area to the neck radius, normalized to the sphere radius. The last
term has values up to unity for the initial stage of sintering.
Using this flux ratio, some interesting results can be obtained
from data in the literature. Various investigators have reported
that submicron alumina sinters by volume diffusion. If J_v/J_b is
to be greater than 10 for 1μm diameter particles, then D_v/bD_b must
be greater than $10^6 cm^{-1}$. If b is assumed to be in the range of
10^{-7}cm, then the ratio D_b/D_v must be less than 10. This is very
different from the case of metals, in which this ratio is on the
order of 10^5 to 10^8 at the melting temperature. Unfortunately,
there are hardly any data on grain boundary self diffusion in
ceramic materials. Cabané[15] reported $D_v/bD_b \cong 5000$ for iodine
diffusion in KI at 525°C, decreasing with temperature to 125 at
280°C. The autoradiographic technique he employed was not sensitive enough to measure the corresponding ratio for K diffusion; he
reported $D_v/bD_b > 2000$ at 500°C. While his data are inconclusive
for potassium, there is no reason to suppose that D_b/D_v should be
several orders of magnitude less for ceramics than for metals; the
opposite is more to be expected.

Volume diffusion compact shrinkage kinetics were reported by
Jorgensen[16] for calcia stabilized zirconia of 0.91μm radius and by
Hagel et al.[17] for Cr_2O_3 with a particle radius of 0.14μm. The
ratio of the boundary to volume diffusivities would have to be
less than 150 for the former and 20 for the latter in order to
achieve this. While it is true that ceramics are not like metals,
it is quite unlikely that the ratio of the diffusivities are as
small as these interpretations of sintering kinetics would imply.
One must then suppose that the sintering mechanism was misidenti-
fied. It is the opinion of this author that grain boundary diffusion is probably the major densification mechanism for submicron
ceramic powders, and that those cases which were identified as
volume diffusion were done so either because the region of enhanced
diffusion was larger than the neck width, or that the data were
simply misinterpreted. Bagley et al.[18] found that for larger particle size aluminas there was an apparent volume diffusion kinetics
portion of the sintering curves, followed by grain boundary diffusion kinetics, and that doping with Ti increased the extent of the
former. This was explained in terms of a wide region of enhanced
diffusion which increased in width as the doping was increased.
Finer particle size powders showed only volume diffusion kinetics,
which is contrary to expectation based on Eq. (4), but can be explained in terms of a wide region of enhanced diffusion which completely encompasses the neck between sufficiently small particles.

The diffusion sintering models are derived assuming rapid
annihilation and creation of vacancies at sinks and sources.

Impediments to vacancy annihilation at the grain boundary would
increase the magnitude of the gradient of the vacancy chemical
potential gradient and, as indicated in Eq. (3), result in a
lessening of the atomic flux. A coherent or epitaxial interface
between a powder particle and a second-phase particle incorporated
in the grain boundary between two sintering particles could be a
significantly less efficient vacancy sink than the grain boundary.
There will be a redistribution of the stress at the grain boundary,
with a concentration of stress over the regions of low vacancy
annihilation rate, and these regions will similarly have a high
vacancy chemical potential. Since vacancies must be annihilated
uniformly over the entire grain boundary in order for material
continuity to be maintained, the rate of densification will be
governed by the rate of vacancy annihilation at the second phase
particle surface under these circumstances.

Certain kinds of grain boundaries can also be poor sinks for
vacancies. It may well be that this is a contributing factor to
the failure of pure covalent solids to densify appreciably. Surface diffusion is not difficult in these solids, as demonstrated
by the coarsening of the microstructure and bonding between particles,[19,20] but if the grain boundaries will not absorb vacancies,
densification cannot occur. It may be that the action of the
dopants that are traditionally used is to disorder the grain boundary to the point that it can become a vacancy sink. Of course,
some additives cause a liquid phase with associated sintering
enhancement.

In a compound $A_\alpha B_\beta$ there will be a coupling of the fluxes,
since the driving force for sintering is not sufficiently great to
decompose the compound. Because of this coupling,[1,2] the fluxes of
A and B atoms or ions are given by the following:

$$j_A = - \frac{\alpha D_A D_B}{\Omega_s kT(\beta D_A + \alpha D_B)} \nabla (\alpha \mu_A + \beta \mu_B) \qquad (5)$$

$$j_B = - \frac{\beta D_A D_B}{\Omega_s kT(\beta D_A + \alpha D_B)} \nabla (\alpha \mu_A + \beta \mu_B) \qquad (6)$$

for volume diffusion acting alone, where Ω_s is the volume of a
formula unit, or "molecule". However, since both volume and grain
boundary diffusion are likely for each species, the appropriate
equation for the A species would be

$$J_A = - \frac{\alpha M_A M_B}{\Omega_s kT(\beta M_A + \alpha M_B)} \nabla (\alpha M_A + \beta M_B) \tag{7}$$

where $M = A_v D_v + A_b D_b$ and A_v and A_b are the areas across which volume and grain boundary diffusion fluxes flow at the neck surface. We see from this that the sintering rate is governed by whichever species has the lower net mobility considering both volume and grain boundary diffusion. It will move by a combination of volume and grain boundary diffusion and, if the sintering experiments are properly done, the volume and boundary diffusion coefficients obtained from the sintering experiment will be for the rate-limiting species.

Based on the above concepts, it is now possible to make some comments relative to the effect of atmospheres and dopants on sintering behavior. These can produce changes in the surface tension which will cause small changes in the sintering behavior. Large changes might be produced if the ratio of the surface to grain boundary tension changes, since this will govern the dihedral angle at the neck surface and can cause a drastic reduction in driving force. In fact, one of the early proposals on the influence of boron on sintering of SiC was that the dihedral angle was increased by boron, thus permitting sintering to occur.[21]   (There are other explanations that now better fit this particular situation.)   The atmospheres and impurities can also have an effect on the defect concentrations and will thus influence sintering through their control of grain boundary and volume diffusion. Nonstoichiometry, atmosphere control, and perhaps more particularly doping will govern the grain boundary diffusivities and the width of the region of enhanced diffusion and can cause large effects, although the current state of knowledge is not at a point to permit prediction of effects in most cases.

SINTERING OF REAL COMPACTS

The application of the above concepts to the sintering of an assemblage of powder particles in a real compact is the area in which the amount of knowledge is far overshadowed by the extent of ignorance. The various mass transport mechanisms can interact in ways that are more or less predictable, but as sintering proceeds, such things as particle coarsening and grain growth occur, which change the geometric picture and greatly influence the course of microstructural evolution, and the sintering kinetics. In general, one can say that the vacancy annihilation rate still must be uniform over the grain boundary surface that is developed in the compact, even in the intermediate and final stages of the process. Thus, as pores become isolated and subsequently disappear or are swept along by grain boundaries, the diffusion distances become greater and greater, and the resulting densification rate is reduced to lower and lower values. The well known observation of cessation of shrinkage of pores isolated from grain boundaries is proof of the importance of grain boundaries as a vacancy source, and is a strong case for the importance of grain boundary diffusion as a primary sintering mechanism. It is in the area of relative grain boundary migration rates and mass transport rates where much can be done to control the sintering behavior of ceramics. For instance, it may be that a particular dopant causes a lower grain boundary mobility because of impurity drag or particle drag. It may also cause a reduced rate of grain boundary and volume diffusion, but the final result in the last stages of sintering could be an enhanced densification rate because the distance between pores is kept smaller than would have been the case had greater grain growth occurred.

INTERPRETING SINTERING KINETICS

The complex nature of a real powder compact almost defies quantitative interpretation. In some very special cases, i.e., spherical particles with a very narrow size distribution, some success has been achieved[9,22] but in this author's opinion most or all conclusions about sintering mechanisms based on shrinkage· or surface area changes of compacts of real particles are subject to reappraisal. The models for initial stage sintering all depend upon having equal size spherical particles which do not undergo rearrangement or coarsening during sintering. Rearrangement and densification alone can cause new contacts between particles to be formed which will alter the over-all densification kinetics, even for compacts of spheres of narrow size distribution.

It is interesting to note that the isothermal shrinkage of powder compacts plotted on a log-log scale with time shows slopes that are in the range expected for various sintering mechanisms, and such agreement has frequently been used to identify sintering mechanisms. However, surface diffusion, particle rearrangement and particle coarsening all have an influence on this plot, the observed slopes will be a function of the sintering mechanisms and these complications, and therefore the log-log slope should never be used to identify sintering mechanisms or activation energies.

Constant rate of heating (CRH) methods have been employed recently.[23,24] Here again, the influence of surface diffusion can greatly change the slope of a log shrinkage or shrinkage rate versus reciprocal temperature plot. This is so because surface diffusion is more important at lower temperatures both because of its lower activation energy and also because it is more significant in the earliest stages of sintering. The use of cyclic heating and

cooling can shed light on the importance of surface diffusion for
compacts of spherical particles of narrow size distribution.[25]
This method has not been evaluated, yet, for typical ceramic powders. The so-called Dorn method[26] of determining the activation
energy for sintering also is subject to difficulties because of the
possibility of more than a single mechanism operating.

One of the biggest problems in interpreting sintering kinetics
is the lack of reliable grain boundary diffusivities in the literature. These are very difficult to determine with confidence because of the large influence of impurities, and it is not surprising
that few people attempt to measure grain boundary diffusivities.
However, it is an area of great need if the sintering of ceramic
materials is going to be quantitavely understood.

The intermediate stage model of sintering can be utilized,
provided the pores remain on the grain boundaries as assumed in
the model.[27] Initial and intermediate stage results give identical
answers in the case of the sintering of CoO,[28] but not in the case
of Al_2O_3.[29] In the latter case, grain growth and localized pore
shrinkage caused an increase in the distance between pores beyond
that expected on the basis of the grain size, with a subsequent
breakdown in the model assumptions. Nevertheless, meaningful
results were obtained by suitable modification of the model.

                            CONCLUSION

Sintering is such a complex process that only the origin of
the driving force, namely the excess surface energy of a powder
compact, and the list of possible sintering mechanisms, find universal acceptance. The final state of a sintered compact depends
in a complex manner upon not only diffusivities but also grain
growth and, finally, gaseous diffusion from closed pores. Under
some circumstances the influence of additives and atmosphere can
be predicted qualitatively. Considerable more research is needed
before sintering behavior can be predicted in detail.

The only unambiguous success in determining sintering mechanisms has involved measurements of both densification and suitable
microstructural parameters for systems which closely approximate
the assumptions on which the models are derived, namely, initial
sintering of a compact of spheres of very narrow size distribution
(standard deviation less than 5% of the mean) or that portion of
the intermediate stage of sintering in which the pores are uniformly distributed along three-grain edges. In particular, the log
shrinkage-log time plot is at best ambiguous and, more often than
not, misleading. The only non-isothermal technique useful for
determining sintering mechanisms involves cyclic heating and

and cooling; this has proved successful with spherical particles of a narrow size distribution, but has not been evaluated for typical ceramic powders.

Further work is obviously needed before the sintering of ceramics can be well understood.

REFERENCES

1. G. C. Kuczynski, Trans AIME, 185 169 (1949).
2. W. D. Kingery and M. J. Berg, J. Appl. Phys. 26 1205-1212 (1955).
3. R. L. Coble, J. Am. Ceram. Soc. 41 [2] 55-62 (1958).
4. K. E. Easterling, PHSNB6 4 [2] 75-86 (1972).
5. J. E. Sheehan, F. V. Lenel, and G. S. Ansell, Scripta Met. 7 [8] 809-14 (1973).
6. (a) C. S. Morgan, Mod. Develop. Powder Met. 4 231-40 (1971).
   (b) C. S. Morgan, Phys. of Sintering, 5 [1] 31-40 (1973).
7. (a) S. Prochazka and R. L. Coble, Phys. Sintering 2 [1] 1-18 (1970).
   (b) S. Prochazka and R. L. Coble, Phys. Sintering 2 [2] 1-14 (1970).
   (c) S. Prochazka and R. L. Coble, Phys. Sintering 2 [2] 15-33 (1970).
8. C. Greskovich and K. W. Lay, J. Am. Ceram. Soc. 55, 142-46 (1972).
9. D. Lynn Johnson, J. Appl. Phys. 40 192-200 (1969).
10. C. Herring, The Physics of Powder Metallurgy, ed. W. Kingston, McGraw-Hill, N. Y. (1951) p. 143.
11. D. Lynn Johnson, Sintering and Related Phenomena, eds. Kuczynski, Hooton and Gibbon, Gordon & Breach, N. Y., (1967) p. 393.
12. (a) Dennis W. Readey, J. Appl. Phys. 37 2309-12 (1966).
    (b) Dennis W. REadey, J. Am. Ceram. Soc. 49 366-369 (1966).
13. M. F. Yan, R. M. Cannon, H. K. Bowen, and R. L. Coble, J. Am. Ceram. Soc. 60 [3-4] 120-127 (1977).
14. R. L. Coble (This Conference).
15. J. Cabané, J. Chim. Phys. 59 1135-41 (1962).
16. P. J. Jorgensen, Sintering and Related Phenomena, eds., Kuczynski, Hooton, and Gibbon, Gordon & Breach, N. Y. (1967) p. 401.
17. W. C. Hagel, P. J. Jorgensen,a nd D. S. Tomalin, J. Am. Ceram. Soc. 49 23-26 (1966).
18. R. D. Bagley, I. B. Cutler,and D. Lynn Johnson, J. Am. Ceram. Soc. 53 136-141 (1970).
19. S. Prochazka, Bull. Am. Ceram. Soc. 55 397 (1976).
20. C. Greskovich and J. H. Rosolowski, J. Am. Ceram. Soc. 59 336 (1976).
21. (a) S. Prochazka, Spec. Ceram. 6 171-81 (1975).
    (b) S. Prochazka and R. M. Scanlan, J. Am. Ceram. Soc. 56 [1-2] 72 (1975).
22. D. Lynn Johnson and T. M. Clarke, Acta Met. 12 1173-1179 (1964).

23. Wayne S. Young and Ivan B. Cutler, J. Am. Ceram. Soc. <u>53</u>
     659-663 (1970).
24. J. L. W. Woolfrey and M. J. Bannister, J. Am. Ceram. Soc. <u>55</u>
     [8] 390-394 (1972).
25. (a) D. Lynn Johnson, 1970 Modern Developments in Powder Metal-
         lurgy, Vol. IV, ed. H. H. Hausner, Plenum Press, N.Y.
         (1979) pp. 189-198.
     (b) Stephen Brennom and D. Lynn Johnson, Sintering and Related
         Phenomena, Materials Science Research Vol. 6, ed. G. C.
         Kuczynski, Plenum Press (1973) p. 269.
26. J. C. Bacmann and G. Cizeron, C. R. Acad. Sci. <u>264C</u> 2077 (1967).
27. D. Lynn Johnson, J. Am. Ceram. Soc. <u>53</u> 574-577 (1970).
28. (a) P. Kumar and D. Lynn Johnson, J. Am. Ceram. Soc. <u>57</u> [2]
         62-64 (1974).
     (b) P. Kumar and D. Lynn JOhnson, J. Am. Ceram. Soc. <u>57</u> [2]
         65-68 (1974).
29. Y. P. Tsui, Ph.D. Thesis, Northwestern University, Evanston,
     Ill., June 1978.

ACKNOWLEDGEMENT

Supported by the National Science Foundation, Grant No. GH-41818.

DISCUSSION

L. De Jonghe (Cornell University): 1) To amplify on what you
said: it is important to realize that it is in principle *impossible* to conclude anything about mechanisms from kinetic measurements alone. 2) It is unfortunate that early stages of sintering
are examined to seek agreement with simple models, since in this
stage rearrangement is very important. 3) One should realize that
structurally grain boundaries are not wide; *chemically*, of course,
they may be. 4) Grain boundaries in Si may indeed be poor vacancy
sinks since current work (at Bell and at Cornell) indicates that
the covalent bonds tend to be satisfied at grain boundaries.
Author: 1) The only exception I would take to your first comment
is that sintering mechanisms can be determined if one measures
densification *and* appropriate microstructural parameters in com-

pacts in which the assumptions on which the models are based are satisfied, i.e., initial sintering compacts of spheres with very narrow size distributions (standard deviations less than 10% of the mean). 2) Apparently rearrangement was not a great problem in our studies of relatively large sphere size compacts of Ag, Fe, and LiF. 3) In general, I agree. 4) Thank you.

V. V. Boldyrev (Siberian Branch Acad. Sci. USSR): As I have understood, the model of sintering proposed by you is well accepted for systems with Schottky disorder (and simple substances, e.g., elements). Should we expect some different situation for the solids with Frenkel disorder, bearing in mind that the mobility of interstitial ions is usually higher than for vacancies and the algebraic sign of the derivative, dv/dc, for interstitials is opposite that for vacancies?

Author: Solid state sintering of compounds will be governed by transport of the slowest species. In the case of Frenkel disorder, with its highly mobile cations, the motion of the anions likely would be rate controlling. However, mass transport in sintering is not limited to vacancy mechanisms; atoms or ions may move by whatever mechanism is easiest.

CURRENT PARADIGMS IN POWDER PROCESSING

R. L. Coble and R. M. Cannon

Massachusetts Institute of Technology

The theoretical work and model experiments on sintering and hot pressing have provided a basis to qualitatively interpret much of the behavior observed in these processes as practiced in the technology. The data base and the modeling, however, are inadequate to quantitatively predict the kinetics of materials responses for these processes (including grain growth) for most systems of interest. In this paper the authors present their biases in a review of the paradigms (current "axioms") in the field, citing the literature sources which support our biases. Several paradigms have changed significantly in the past several years. As a consequence, new directions for the course of productive research are needed; we present our best guesses for them. We also give emphasis to some earlier theoretical and experimental work which we regard as important, the significance of which seems to us not to have been duly appreciated.

Complete characterization of powders, powder compacts, and their evolution during sintering has not been developed for any system and probably never will be on a routine basis. Hence, the framework for understanding is to connect the behavior and changes in behavior to controllable variables and operations, empirically, by measurements; and fundamentally, by theories and models and the materials properties data base. In order to illustrate the scope of the problem, each of these is expanded into a listing of items in Table 1. In what follows, we will emphasize the sub-topics independently in some cases, presuming that the relationships among the major topics are appreciated. For others, the major relationships will be emphasized; some topics will be ignored or given very limited attention.

Table 1: Sintering and Hot Pressing with/without liquid in metallic, ionic, or covalent systems

Behavior	Controllable Variables and Operations	Data Base Needed
General morphology	Powder preparation	γ_{sv}, $\gamma_{\ell v}$, $\gamma_{s\ell}$ and
Evolution of pores and grains	particle size,	$\gamma_b = f$'s(X_{dope})
Density f(T,t)	shape, distribut'n	D_ℓ^m D_ℓ^x
Grain size f(T,t)	Dopant distribution	D_b^m D_b^x
Dopant effects	2nd ϕ distribut'n	D_s^m D_s^x
	Fabrication	
Models	density distribut'n	D's in liquids
Neck growth	pore size "	$K_n^{m,x}$ $K_b^{m,x}$
Surface area change		P^m P^x
Shrinkage	**Firing**	
Densification in later	$\dot{T}$ and T_{max}	Phase equilibria
stages	$P_{hp}(t)$	Gas solubilities
defect reactions at	P_g, atmosphere	diffusivities
surfaces		Solute diffusivities
diffusive transport	**Characterization**	Creep behavior
evaporation conden-	**Measurements**	$T_y(T)$
sation		
plastic flow	Shrinkage, density	
Gas pressure effects	Neck growth	
Grain growth	Surface area change	
solute drag	Grain size, pore size,	
pore drag	and continuity	
breakaway	Permeability	
	Strength	
	Conductivity	
	Porosimetry	
	Dopant re-distribut'n	
	f(T,t)	

For the <u>behavior</u> to be understood, complete metrical and topological characterization is needed during sintering because the <u>behavior</u> simply connotes the evolution of the structure with (T,t). Complete characterization does not exist for any material undergoing densification - pore and grain size distributions and the relative locations of pores and shapes of pores during the course of densification is needed. Instead only <u>average</u> density and <u>average</u> grain and pore sizes are available from a few experimental papers - mean curvatures, etc. from a few others. Because of the difficulty in making a complete set of measurements we will probably continue with partial characterization to satisfy the minimal requirements to see what is taking place (density and grain sizes as f(t,T)). These <u>must</u> be supplemented by metallographic studies to look for the <u>usual</u> problem areas: low density areas due to poor agglomerate structures, pellet fabrication problems, or "accidental" large inclusions such as dirt, binders, dopant inhomogeneity, etc. In reviewing development programs on new materials, multiple phases have frequently been revealed by metallography, when only one solid phase was assumed (or hoped) to be present. Modern instrumentation has greatly alleviated this problem area because beam-probes can reveal gross compositional inhomogeneities in new materials for which chemical etchant techniques required to reveal multiple phases have not been developed.

The <u>operations and controlled variables</u> listed in Table 1 provide a set which are useful in mechanism studies, but are obviously even more important to control for reproducible production of high quality ceramics. A number of these are easily controlled to sufficient accuracy { $\dot{T}$, T_{max}, atmospheric species and pressure (P_g), and the applied pressure in hot pressing (P_{hp})} whereas powder preparation and fabrication present continuing problems for development programs and production processing.

<u>Powder preparation and fabrication</u> of powder compacts has long been recognized as important in controlling the densities achievable after heat treatment. The behavior of spray-dried powders, and of agglomerated powders after calcination of various precursor salts, has revealed that the agglomerates sinter to high density at relatively low temperatures, or at an early stage in the process, while large interagglomerate pores or large accidental pore inclusions remain as "the problems" to be removed at the final stages of the process. That the objective in fabrication would be to produce uniform density throughout the fabricated compacts with small pores of near-uniform size was illustrated by Rhodes[1] work on the fabrication of Zyttrite (Yttria stabilized zirconia prepared from an alkoxide pre-cursor developed by Mazdiyazni.) The agglomerates of $\sim$ 100 Å crystallites are difficult to fabricate into large samples but were found to be sinterable at $\sim$1400°C analogous to sol-gel ThO_2 and UO_2. Rhodes dispersed the agglomerates by grinding; he then removed coarse particles or agglomerates

by sedimentation. The suspended crystallites remaining were then centrifugally cast into a deposit of high green density (72%) with small and uniform pore sizes. These "samples" were sinterable to high density (99.5%) at $\sim$1100°C with a final grain size of 2000A. Twenty years ago, ZrO_2 was typically processed by grinding fused grain (to $\sim$1µm); samples at that time were sintered at $\sim$1800°C to achieve densities above 95% of theoretical. We conclude that greater effort is needed in learning to fabricate small crystallite sized materials into uniform, high density compacts of large sizes with the contained pores in narrow size distributions. Significantly lower firing temperatures are expected; greater possible benefits may be gained, however, by changing the ratio of densification rate/grain growth rate than to simply lower the sintering temperature. Perhaps discontinuous grain growth can be avoided without additive-engineering due to expected differences in the temperature dependencies for densification:grain growth.

Powders of small particle size have been recognized as being difficult to fabricate into large-sized specimens. The authors feel that new approaches are needed in fabrication. We have but one suggestion: that control of particle shape be pursued as well in developing techniques to produce controlled particle sizes and distributions in new materials and in the obdurate, old stand-bys. Because powders will not flow in die compaction to give uniform densities, modifications of slip casting or injection molding seem more promising to us for future developments - in general, inventors are needed! Elongated or platy particles present additional problems because preferred orientations may be developed locally and spatially by all of the fabrication procedures. Typically, platy particles do not sinter as readily as do equiaxed particles. Whether this is due to surface energy lowering (that is manifested in faceting), or to a shift in surface reaction controlled kinetics is an unresolved suspicion.

Characterization measurements typically used on powders and sintered or hot pressed structures are listed in Table 1 along with property measurements (such as strength) from which structural information might be inferred. Multiple measurements of the structure are needed at each stage as indicated above. In order to obtain information about all structural features of interest, specific groupings of measurements must be selected: (1) shrinkage or densification, (2) neck growth or surface area change or conductivity, (3) grain size, (4) pore size distribution, and (5) shape by metallography or porosimetry and permeability. The structure evolution as a function of temperature, time and dopants must be documented to improve the modeling in order to advance our understanding of the kinetics. For development programs, density changes and metallography to follow grain growth and to look for "the problems" are the minimum requirements.

The Data Base Needed, expanded in Table 1, includes all the parameters which appear in the theoretical models: the surface energies, diffusion coefficients, interfacial reaction coefficients, vapor pressures, phase equilibria, and solute diffusivities. The influences of dominant impurities or dopants on all of those properties are needed to make assessments of the relative contributions to changes in behavior. The creep behavior and yield stress are listed because of their obvious bearing on hot pressing, but also because of the analogue behavior with respect to the diffusion models for densification in sintering or hot pressing. Creep at low stresses in numerous dense polycrystalline metals and ceramics has been shown to take place by lattice or boundary diffusion.[2] The additional competing mechanisms (surface diffusion and evaporation condensation) are excluded in creep, and the geometric changes and net driving force are better described by the models. The fact that diffusional creep has been demonstrated in several oxides (MgO, Al_2O_3, UO_2, etc.)[2,3,4,5,6] showing:

1. dominantly viscous behavior at stresses up to 10^4 psi
2. conformance to the predicted grain size dependences
3. reasonable agreement between diffusivities calculated from creep data and the tracer values including effects of oxygen pressure with aliovalent dopants present,

lends confidence that the similar modeling (additive relations, geometric treatments and solutions for the diffusion fields for sintering) should also be applicable. Therefore, we conclude that most of the densification in sintering or hot pressing also takes place by diffusion in agreement with various conclusions from sintering studies.

It is partly the more complicated geometrical change occurring in sintering or hot pressing that has precluded precise quantitative understanding of the kinetics, particle size effects, temperature dependences, and diffusivity comparisons based on admittedly simplistic models. Simultaneous transport among competing paths, impurity effects, a failure to properly account for initial particle shapes and packing, space charge effects, grain growth occurring simultaneously with densification, and undocumented pore shape evolution are other recognized factors contributing to our inability to predict the kinetics from start to finish of the process.

In the modeling of diffusional creep for an elemental polycrystalline solid, Raj and Ashby[7] assumed that the lattice and grain boundary contributions are simply additive:

$$\dot{\varepsilon} = \frac{14\ \sigma\ \Omega}{d^2 KT}\ D_{eff} \quad \text{where } D_{eff} = (D_\ell + \pi\ \delta\ D_b/d) \tag{1}$$

Thus, motion along the fastest path controls the rate. For a binary compound, various combinations of ions and paths may control the process as Gordon[8] has discussed in detail, including the extrinsic

case. He extended Ruoff's[9] treatment of lattice diffusion controlled
creep in a binary compound displaying Schottky disorder and considered Frenkel disorder as well. He presumed that similar modeling
of the defect concentrations must also apply at the grain boundaries.
Gordon also followed Readey's[10] treatment to consider the extrinsic
case, assuming that the same principles apply at the grain bound-
aries. For the simplest case, that of a <u>pure</u> binary compound $M_\alpha X_\beta$,
D_{eff} for use with Equation 1 becomes

$$D_{eff} = \frac{(\alpha + \beta)(D_\ell^m + \pi \, \delta \, D_b^m/d)(D_\ell^x + \pi \, \delta \, D_b^x/d)}{\beta(D_\ell^m + \pi \, \delta \, D_b^m/d) + \alpha(D_\ell^x + \pi \, \delta \, D_b^x/d)} \qquad (2)$$

and Ω = molecular volume $\div (\alpha + \beta)$. Examination of Equation 2 shows
that whatever the relative magnitudes of the four diffusivities, it
is the <u>slowest ion</u> in <u>it's fastest path</u> that controls the rate. The
dominant terms change with grain size (d) as illustrated schematically
in Figure 1. The case shown is typical of MgO and Al_2O_3, for which
$D_\ell^m > D_\ell^o$ and $D_b^o > D_b^m$. The controlling species and path is indicated
by the dotted line; each species:path combination is controlling in
a specific grain size range. Changes in control from D_b^m at small
grain sizes to D_ℓ^m at intermediate sizes have been observed for the
behavior of both MgO and Al_2O_3.[3-5] The transition grain sizes will
change with temperature and with dopant effects from those illustrated
according to the relative changes in the diffusivities. Assuming
that an unknown impurity is present that could cause extrinsic behavior below an unknown temperature, there are at least eight possible activation energies which might be observed for creep at the
start of research on some new material.

The data on creep suggest an additional complication;[2,5] At low
stresses and small grain sizes non-viscous behavior is observed with
lower rates than those extrapolated from larger grain sizes. Defect
equilibration at the boundaries may then be rate controlling; this
would introduce two surface reaction coefficients: temperature and
impurity (or dopant) dependent (catalyze/poison?). These four
additional "steps" for possible control, added to the transport
processes give twelve possibilities for creep, the <u>simple</u> process
in comparison with sintering!

The grain shape changes in sintering or hot pressing can be
accomplished by any of the mechanisms of deformation normally considered for creep deformation. For the case of sintering, we presume that it is well accepted that the surface energy is the driving
force with the surface energy:curvature product giving an effective
pressure to evaluate the chemical potentials for diffusive transport.
They can also be used to evaluate the effective stresses to be compared with the yield stresses, to mark the dividing line between
plastic flow by dislocation motion and by diffusive transport. In
the initial stage of sintering the average boundary stress due to
the resolved surface tractions is $\sigma_n = 2 \pi X \gamma \sin\theta / \pi x^2$ but this

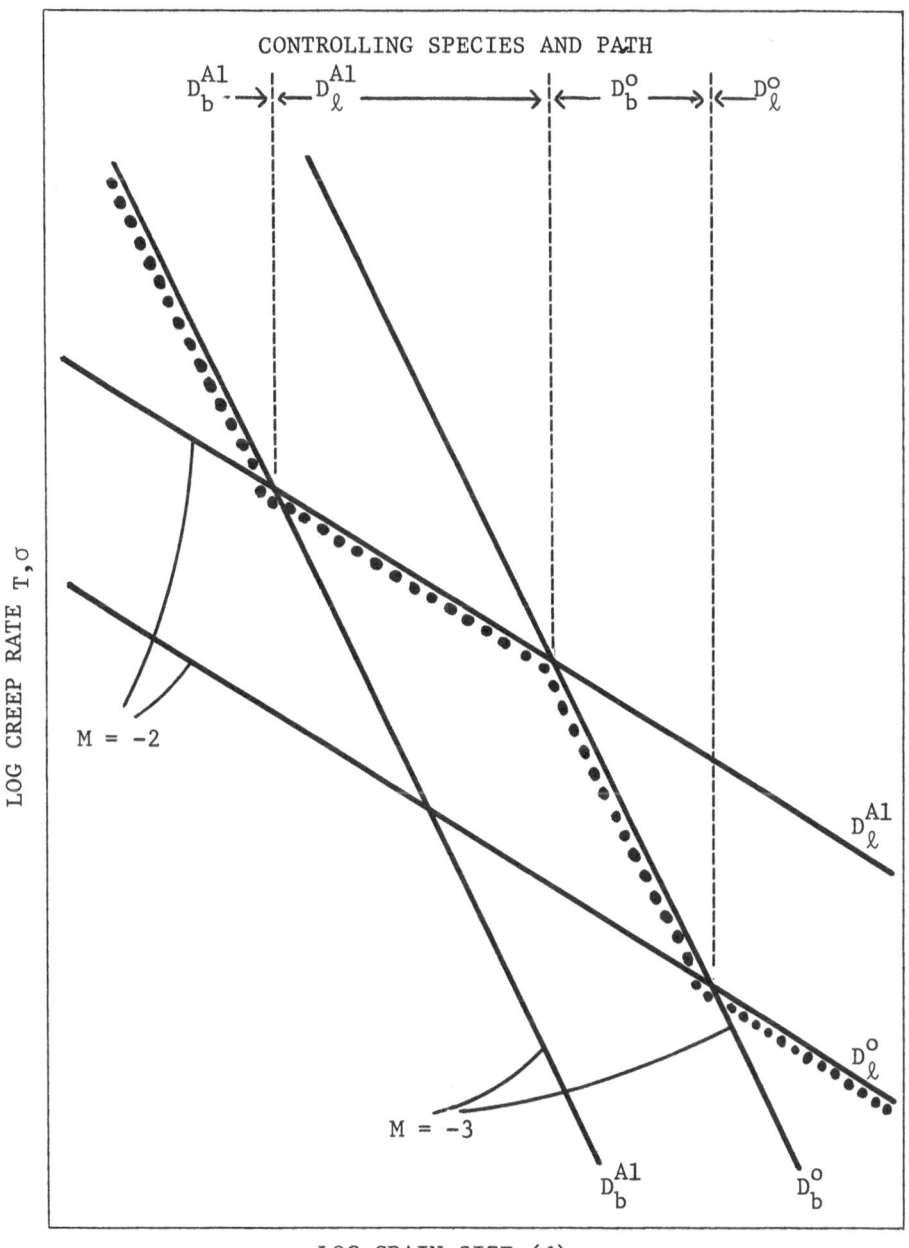

Fig 1: Diffusional creep in Al_2O_3 (schematical). Solid lines show the creep rate dependence on grain size for each assumed species and path controlling the rate. The dotted line traces the required controlling step: the slower ion in its fastest path.

term is smaller (and therefore ignored) than the local stress at the neck due to the curvature: $\gamma/\rho = 4R\gamma/X^2 = \gamma/RY$ where Y is the fractional shrinkage. In sapphire the yield stress for prismatic glide is > 40 KSI for T < 1400°C.[11] Thus for R = 10^{-4}cm, and $\gamma = 10^3$dynes/cm, $\sigma_n < \sigma_y$ for Y > 0.004; beyond the initial stage of sintering the stresses are below the yield stress and sintering is then presumed to occur only by diffusional processes. For hot pressing the range of densification in which plastic deformation is important is extended and can be evaluated based on the same principles[12,13].

Dopant effects on interfacial energies, etc. need to be documented independently of the diffusivities in order to separate the effects that are combined in the modeling. We assume that dopant effects on the interfacial energies and vapor pressures will be much smaller than the orders of magnitude changes in diffusivities or surface reaction rate coefficients. However, since the relative magnitudes of γ's (γ_b/γ_{sv}, e.g.) change the phase morphologies, small changes in the respective γ's can be very important.

The dopant effects observed on lattice diffusivities in oxides have revealed that the applicability of Schottky defect models combined with incorporation reactions for aliovalent dopants frequently do not predict the observed changes in diffusivities as well as in the alkali halides[14]. For example, both Ni and O show increases in diffusivity with increased oxygen pressure,[15] whereas the oxygen diffusivity would be expected to decrease. It is now conceded that impurity-defect complexes are more important than presumed earlier - more documentation is needed. There are also numerous examples in which the qualitative changes in diffusivities are predicted by the simple modeling, but in which the concentration dependences are not quantitatively explained {e.g.:$D_\ell^u(x)$ in UO_{2+x}}.[16] For lattice defects and diffusion we presume that further measurements of the diffusivities and characterization of the defects by optical, dielectric, ESR, EPR, and other measurements will provide the basis to establish the more complex impurity:defect and defect:defect equilibria and transport dependences on concentrations without requiring any major new concepts.

In contrast, knowledge about the structure and chemistry at grain boundaries is in a more primitive state particularly with regard to intrinsic or extrinsic diffusion. Whether the defects can be identified as vacancies or interstitials is not known, nor are their concentrations. For ionic materials, however, Yan et al.[17] concluded that the enhanced transport at grain boundaries cannot be attributed to space charge effects with/without dopants present. Thus, boundary transport and impurity effects on transport through the poorly understood boundary core region remain as major problems for focused research.

<u>Models</u>   Following Kuczynski's[18] pioneering work, there have been
numerous mechanistic models developed for the initial, intermediate
and final stages of sintering, and for hot pressing. It is the
authors' view that for the initial stage of the process, the modeling
is satisfactory for first order estimates of sintering if used in a
predictive mode. Neck growth can be approximated as an additive
function of all contributing mechanisms if the surface energies and
transport data are available for a particular material.

Taking the models from Coble,[12] Wilson and Shewmon[19] and
Kingery[20] for lattice and boundary, surface diffusion, and vapor
transport, respectively, the rate of neck growth between equal
sized spheres is:

$$\overset{\circ}{X} \simeq \frac{16D_\ell\gamma\Omega R}{X^3\ kT} + \frac{32\delta D_b\gamma\Omega R^2}{X^5\ kT} + \frac{4\delta D_s\gamma\Omega R^2}{X^5\ kT} + \frac{(m/kT)^{3/2}\gamma P^\circ R}{Xd^2\ (2/\pi)^{1/2}} \tag{4}$$

For the shrinkage rate:

$$\overset{\circ}{Y} \simeq \frac{8D_\ell\gamma\Omega}{X^2R^2kT} + \frac{16\delta D_b\gamma\Omega,}{X^4R^2kT} \tag{5}$$

which is useful if the neck sizes are measured. Equation 5 can be
reorganized in terms of $\overset{\circ}{Y}$ and Y alone if surface diffusion and evap-
oration:condensation are negligible. If all terms in Equation 4
are of the same magnitude, the neck growth from Equation 4 and the
shrinkage from Equation 5 can only be performed numerically.
Johnson[21] has done this for mixtures of surface and boundary diffusion. He showed that the course of densification was appreciably
altered for constant rate heating depending on the relative values
of D_s/D_b and their activation energies. The time dependence for
shrinkage was found to be variable and different from that found
for D_b dominant. Also the apparent activation energy for shrinkage
was different from that used for boundary diffusion in the analysis.
We regard Johnson's contribution as most significant because it shows
that when mixtures of competing mechanisms interact during sintering
no conclusion regarding the controlling mechanism can be drawn from
the observed time or temperature dependence for shrinkage. Even
if neck growth is measured along with shrinkage, there are many
combinations of ways in which the right hand sides of Equations
4 and 5 could interact to explain the observed behavior. Further,
consider that Equations 4 and 5 are analogous to Equation 1 and
for binary compounds require expansion as from Equation 1 to
Equation 2. Although Equation 5 appears to be a direct analogue
to Equation 1, we reiterate that it is only valid when the lattice
and/or boundary diffusivities are dominant.

In the extension to practical cases, i.e., when dealing with
impure and/or doped compounds, <u>all</u> transport coefficients must be
known for the compositions of interest as Readey's[10] work showed.

Also the particles employed are generally of distributed sizes and
not spherical. The modeling for angular particles has shown that
the time dependences are different than for spheres thus making
the kinetics even more difficult to understand. The size effects
for angular particles, however, are identical to those for spherical
particles when the same mechanisms of transport are considered,
which simply confirms (or further generalizes) Herring's[22] scaling
laws for additional geometries than those he considered. The
scaling laws are believed to be applicable, therefore, to equivalent
distributions of particles with different average sizes even if they
are not spherical, and could be used for limited ranges of sizes
for deductions of operative mechanisms if a sufficient number of
characterization measurements (shrinkage and neck growth etc.) are
conducted. But it should be noted, by reference to Figure 1, that
the mechanisms will change of necessity if there are very broad
differences in the average sizes of the different samples.

Increased densification rate with increased bulk density after
fabrication has been observed for many materials, but has not been
interpretable based on the first order models. The opening of
large pores observed in model structures (two dimensional particle
arrays) and the refined theories developed to explain that phenomenon are pertinent, but give no greater insight at the present time
than a restatement of the empirical facts: high green densities in
fabricated structures, without flaws, are the identified objectives
in fabrication.

Particle coarsening during the initial stage of sintering is
another phenomenon to consider as it may occur by coalescence of
small particles making only one contact with larger particles in
the compact. This is a different from "normal" grain growth which
is expected (and observed) at later stages in sintering, when all
grains have multiple contacts.

Another recent development with respect to modeling has been
the translation of neck growth models for various mechanisms to
the surface area changes that will be predicted in a powder compact,
by German.[23] Neck growth data on small particle size materials is
exceedingly difficult to collect whereas surface area measurements
can be easily conducted as a function of sintering treatments.
This provides a source of information regarding neck growth which,
coupled with shrinkage measurements enable improved assessments to
be made of the structure evolution. A precaution to be emphasized,
however, is that the models are dependent upon assumed monodisperse
particles uniformly packed. Most powder compacts do not meet these
boundary conditions. A second requirement, in order to predict
surface area change in a compact from predicted neck growth rates
among particles is that the coordination number of the particles
must be known; this, in general, is not known with precision.

The authors have derived a set of models for the initial stage of sintering based on the assumption that the interface reactions (by which defects are formed at the neck surface and annihilated at grain boundaries) are rate controlling instead of diffusive transport. They assumed that the defect reaction rates may be a power function of the driving force (ΔF): flux = $Ak(\Delta F)^{\alpha}$ where the coefficient (alpha) was assumed to take on values 1 or 2. The time dependences and size effects predicted for those models are shown in Table 2. In comparing these time dependences and size effects with earlier models for lattice and boundary diffusion controlled sintering it is seen that a greater spectrum of coefficients exists for time dependences and size effects which one might observe experimentally than has been presumed heretofore. The main conclusion which we would draw with respect to modeling is that the sintering data can be examined for conformance to one or another model or assumed mechanisms using the models in a predictive mode when enough other background data are available. But deductive conclusions regarding mechanisms from limited kinetic data are unwarranted.

For the intermediate stage of the process, reasonable agreement has been obtained between diffusion coefficients calculated from the data and models with independently measured values for copper and alumina.[24] The best data now available, collected by Wang,[25] in his thesis research with Dick Fulrath, and by Hare, Huckabee, and Palmour,[26] show that the intermediate stage models are too simple. Their results can be summarized as indicating that the sintering rates or densification rates when multiplied by the measured grain sizes cubed (or to the 4th power) do not yield constants at constant temperatures but variable densities as is predicted by the (simple) intermediate stage models. Continuous

Table 2: Time dependencies and size effects based on surface reaction controlled initial stage sintering compared with those from diffusion models.

$$(X/R)^N = Bt/R^M$$

INTERFACE CONTROL			DIFFUSION CONTROL	
$j = Ak_{n,b}(\gamma/\rho)^{\alpha}$				
α	N	M	N	M
AT BOUNDARY			BOUNDARY	
1	4	2	6	4
2	6	3		
AT NECK			LATTICE	
1	3	2	4	3
2	5	3		

tructure models of the kind Prochazcha[27] has described, or the statistical modeling that Kuczynski[28] has introduced may provide a suitable basis to improve the description of the evolution of the intermediate state of sintering. Not enough data have been generated on the structure evolution itself in order to properly guide thinking on model development; a more complete set of measurements and description of the total structure evolution is needed.

Modeling of the shrinkage of closed pores intersected by boundaries during the final stage of densification is regarded as adequate and the influences of trapped gases are understood. The major problems which appear at the final stage are, of course, pores trapped by discontinuous grain growth prior to complete densification, and other large holes (which originate from fabrication errors). Grain growth alone is not understood completely, although solute-drag effects on boundary mobilities are recognized to be important.[30] The transition in growth to the discontinuous mode will require more theoretical and experimental work to define the breakaway criterion. More experimental work is needed on pore mobilities and coarsening, and further testing of the models introduced by Brook,[31] extended by Carpay,[32] and further, by Cannon et al[33] to consider pore drag and the space charge effects on grain boundary drag. The current modeling is considered to be adequate to guide experiment design. The limited data base is a greater handicap than the modeling. More data are needed on the impurity effects or dopant effects on surface energy, the grain boundary energy, interphase energy, surface diffusion coefficients, boundary and lattice diffusion coefficients, all as functions of dopants, for a range of specific systems of interest.

Liquid Phase Sintering   In the authors' view, the most significant changes which have taken place in recent years regards the finding with respect to densification below the eutectics in numerous systems, notably tungsten-carbide:cobalt, tungsten:nickel,[34] and in a model system calcium fluoride:sodium fluoride which Skaar[35] investigated. The notable results are as follows: At high heating rates, typical of those used in laboratories in investigate isothermal sintering, the features of liquid phase sintering which have been classically described, are qualitatively observed. Hot-stage microscopy shows that liquid forms at the eutectic temperature, the particles rearrange rapidly in low viscosity liquids (seconds) and the structure undergoes additional rearrangement as shrinkage proceeds. The precise time dependences predicted by the modeling have not been observed.

Inhomogeneous liquid distribution after sintering is a feature observed that has not been modeled. Whether the liquid moves preferentially under the temperature gradients (during heating and cooling) is not clear. Skaar concluded that liquid would run to the more densely packed regions in the compacts - filling the small pores to lower the surface energy of the system. Skaar's results on the shrinkage of calcium fluoride doped with sodium fluoride as a function of heating rate are shown in Figure 2. It is noteworthy that at the heating rates which may be used commercially, ie ~5 degrees per minute, essentially all of the shrinkage has been completed below the eutectic temperature. In this system, heating rates $\geqslant 150°C/min$ are required in order to reach the eutectic temperature before shrinkage is complete.

There are numerous systems, of course, in which the sintering temperature is elevated above the eutectic temperature and the composition is in the two phase solid:liquid field (such as tungsten carbide:cobalt): therefore, grain growth and the final organization of the structure does take place with liquid present. We presume that the mechanism for densification relates to enhanced grain boundary diffusivity or enhanced transport in a carrier phase, an idea introduced by Brophy[34] et al. following studies of the enhanced sintering of tungsten with small additions of nickel. For calcium fluoride we expect that the lattice diffusivity of calcium would be decreased as sodium is added which further supports the conclusion that the process takes place by boundary carrier phase transport. A main point of interest is that a number of systems show that sintering rates are enhanced significantly by the addition of second components, indicating that in the extrinsic regime enhanced boundary transport can be very significant. Much more data is needed on dopant effects on boundary diffusivity (1) to provide a basis to analyze the structural and chemical models at boundaries, (2) to understand the sintering behavior, (3) to understand creep behavior for structural applications, and (4) to relate to electrical properties and potentially to conductivity degradation in electrolytes.

Another way to examine our overall understanding of the process and the relative importance of different materials properties, is to consider the behavior of the different materials classes. Most ionic and metallic materials readily undergo densification to ~95% of theoretical density if the particle sizes are sufficiently small. Control of discontinuous grain growth at the final stage of the process is necessary to achieve near-theoretical density. Although considerable grain growth or particle coarsening takes place early in the process, it does not present an over-riding restriction to densification. In contrast, covalent bonded materials were regarded as being non-sinterable (i.e. powder compacts did not show densification on heating) until recent efforts devoted to Si, SiC and Si_3N_4 were undertaken and sintering aids were found which would produce densification.[36,37,38]

Although there is a paucity of good data, several factors are suggested to contribute to the poor sinterability of covalent materials. The lattice diffusion coefficients are relatively low even at temperatures near the melting points. For instance for Si and

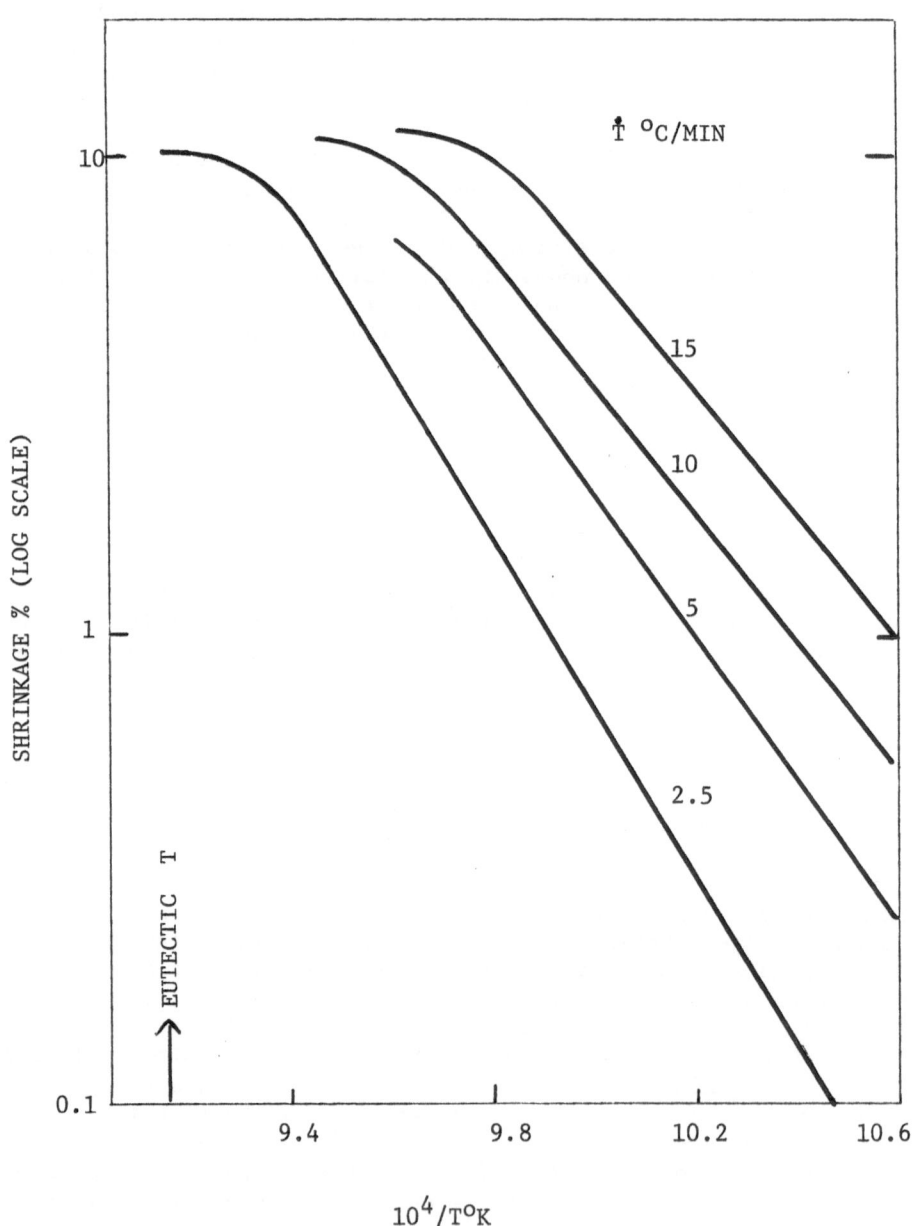

Fig 2: Sintering of 2μ CaF$_2$ with 3 mol % NaF at several constant
heating rates. Curves show log shrinkage versus reciprocal
temperature.

Ge at the melting point D 10^{-11} cm^2/sec, compared to 10^{-8} cm^2/sec for FCC metals and 10^{-9} for alkali halides.[39,40] The assumption of correspondingly low boundary diffusivities (poor data available) would account for low sintering rates. Further retardation or a complete lack of densification is apparently a result of the early grain coarsening which is observed; this eliminates the neck curvature and thus the driving force for densification. High vapor pressures (Si, SiC, Si$_3$N$_4$) and/or surface diffusivities (SiC) are probable causes. Low dihedral angles (i.e., high ratios of $\gamma_{gb}:\gamma_s$) that lower the driving force for densification[41] may be an additional problem for covalent materials. Although the dopants required to promote densification may enhance the lattice or boundary diffusivities, they are also apparently important in inhibiting coarsening, e.g., B in SiC.

Coarsening control may also require atmospheric control and a means to counteract the adverse effects due to the oxygen impurity contents in the original powders.

A counterpart among the oxides is Cr$_2$O$_3$, which was also "not sinterable" until Ownby[42] showed that Cr$_2$O$_3$ sintered readily if the oxygen pressure was controlled to $< 10^{-12}$ atm to suppress CrO$_3$(g) volatilization. For the silicon containing compounds, SiO(g) formed by reaction with oxygen also increases vapor transport from the relatively high values for silicon alone. Similarly (pressureless) sintering of Si$_3$N$_4$ can only be accomplished with high N$_2$ pressures[38,43] to inhibit vapor transport. In addition to reducing the driving force, vapor transport can also lead to excessive weight loss by decomposition.

The efficacy of many of the densification aids used for covalent compounds has been attributed to liquid phase sintering. For instance, from the observed hot pressing kinetics for Si$_3$N$_4$ + MgO, Brook[44] concluded that densification resulted from transport in a liquid film at the boundaries. By analyzing the creep, hot pressing, and grain growth (which is surprisingly slow) data for Si$_3$N$_4$;MgO, Cannon and Chowdhry[45] concluded that the mass transport rates are interface controlled and are much slower than would be anticipated for diffusion in the silicate liquids which are thought to be present. In general, these systems are poorly understood and enhanced (extrinsic) boundary diffusion may also be important.

For these materials we conclude that their poor sintering behaviors in comparison with metals and ionic materials are attributable to:
1. High vapor pressures relative to lattice diffusivities, and
2. Higher surface diffusivities relative to boundary diffusivities.

<u>Status and Needs</u> We have argued, perhaps too strongly, that sinter-
ing and hot pressing take place by diffusion, but that interface
control may be more important than has been realized under special
conditions in some materials. The theories and models now available
give reasonably good guidance for many of the operations and control
variables such as the size, temperature, and rate effects, based on
the application of the diffusion and vapor transport models to the
process. Various dopant and oxygen pressure effects give excellent
agreement with the expectations from independently measured diffusivities, etc. The models need refinements to better deal with
green density effects, particle shape, and grain/particle size
distribution effects.

The theories and models do not provide guidance for dopant
selection, nor will they for model refinements we envision may be
forthcoming. A rationale is needed for composition effects on a
range of properties that is much more precise than our current data
base provides. The area embraced by all materials of interest
would take decades to fill - perhaps systematic sets should be pursued on the covalent materials which seem technologically promising
and are least understood, or offer model systems for improving
understanding broadly.

REFERENCES

1. T. Vasilos, & W. H. Rhodes, <u>Ultrafine Grain Ceramics</u>,
    J. J. Burke, et al Eds., Syracuse Univ. Press, (1970)
2a. B. Burton, Diffusion & Defects Monograph, No. 5, (1977)
2b. B. Burton & G. L. Reynolds, Acta Metall <u>21</u>, 1073-78, 1641-1647,
    (1973)
3. R. T. Tremper, R. A. Gidding, J. D. Hodge & R. S. Gordon,
    J. Am. Ceram. Soc. <u>57</u> (10) 421-428, (1974)
4a. P. A. Lessing & R. Gordon, J. Mat'ls. Sci. <u>12</u>, 2291-2302 (1977)
4b. G. W. Hollenberg & R. S. Gordon, J. Am. Ceram. Soc. <u>56</u> (3) 140
    (1973)
5a. R. M. Cannon & R. L. Coble, <u>Deformation in Ceramic Materials</u>,
    R. C. Bradt & R. E. Tressler Eds., Plenum Press, N.Y. (1975)
    61-100
5b. R. M. Cannon, W. H. Rhodes, A. H. Heuer, J. Am. Ceram. Soc.
    (to be published)
6. M. Seltzer, J. S. Perrin, A. H. Clauer & B. A. Wilcox, Reactor
    Tech. <u>14</u>, 99, (1971)
7. R. Raj & M. F. Ashby, Met. Trans. <u>2</u> (4) 1113-27 (1971)
8. R. S. Gordon, J. Am. Ceram. Soc. <u>56</u> (3) 147-152 (1973)
9. A. Ruoff, J. Appl. Phys. <u>36</u>, 2903 (1965)
10. D. Readey, J. Am. Ceram. Soc. <u>49</u>, 366 (1966)
11. D. J. Gooch & G. W. Groves, J. Am. Ceram. Soc. <u>55</u>, 105 (1972)
12. R. L. Coble, J. App. Phys. <u>41</u> (12), 4798-4807 (Nov 1970)

13. D. S. Wilkinson & M. F. Ashby, 473-93 in Sintering and Catalysis, G. C. Kuczynski Ed., Plenum Press, N. Y. (1975)

14. L. C. Ianiello et al, Summary Report on Critical Needs & Opportunities in Fundamental Ceramics Research, ERDA 9, (April 1975)

15. M. O'Keefe & W. J. Moore, J. Phys. Chem. 65, 1438, (1961)

16a. H. Matzke, J. de Physique, 34, C9-317 (1973)

16b. F. A. Kroger, Zeit fur Phys. Chem Neue Fol 49, 178-97 (1966)

17. M. Yan, R. M. Cannon, H. K. Bowen, & R. L. Coble, J. Am. Ceram. Soc. 60 (3-4), 120-127, (1977)

18a. G. C. Kuczynski, Trans. A.I.M.E. 185, 169 (1949)

18b. G. C. Kuczynski, J. App. Phys. 21, 632 (1950)

19. T. L. Wilson & P. Shewmon,Trans. A.I.M.E. 236, 48 (1966)

20. W. D. Kingery & M. Berg, J. App. Phys. 26 (10), 1205-12 (1955)

21. D. L. Johnson in Ultrafine Grain Ceramics, J. J. Burke, et al Eds., Syracuse Univ. Press (1970)

22. C. Herring, J. App. Phys. 21, 301 (1950)

23. R. M. German, Sci. of Sintering 10 (1) 11-25 (1978)

24. R. L. Coble & T.Ҝ. Gupta, I & II, in Sintering and Related Phenomena, G. C. Kuczynski et al Eds., Gordon Breech (1965)

25. D. N-K. Wang. Ph.D. thesis, U. Cal., Berkeley (Dec 1976) (LBL report 5763)

26. (a) H. Palmour, T. M. Hare and M. L. Huckabee, Final Technical Report, Contract N00019-73-C-0139, Naval Air Systems Command, March 1974.
     (b) H. Palmour and T. M. Hare, Final Technical Report, Contract N00019-74-C-0265, Naval Air Systems Command, July, 1975.
     (c) M. L. Huckabee, T. M. Hare and H. Palmour III, this volume.

27. S. Prochazcha, Am. Ceram. Soc. Convention (1977).

28. G. C. Kuczynski, 325-339 in Sintering & Catalysis, G. C. Kuczynski, Ed., Plenum Press, N. Y. (1975).

29. J. E. Burke, J. Am. Ceram. Soc. 40 (3) 80-85 (1957)

30. J. W. Cahn, Acta. Metall. 10 (9) 789-98 (1962)

31. R. Brook, J. Am. Ceram. Soc. 52 (1) 56-57 (1969)

32. F. M. A. Carpay, J. Am. Ceram. Soc. 60 (1-2) 82 (1977)

33. M. F. Yan, R. M. Cannon & H. K. Bowen, in Ceramic Microstructures, '76, R. M. Fulrath & J. A. Pask, Eds., Westview Press, Boulder, Colo. 276 (1977)

34. J. Brophy, A. Prill, et al, Trans. A.I.M.E. 230, 769 (1964)

35. E. Skaar, Sc.D. Thesis, M.I.T. (1978)

36. S. Prochazka & R. M. Scanlan, J. Am. Ceram. Soc. 58, 72 (1975)

37. C. Greskovich & J. H. Rosolowski, J. Am. Ceram. Soc. 59, 336-343 (1976)

38. C. Greskovich, S. Prochazka & J. H. Rosolowski, Basic Research on Technology Development for Sintered Ceramics, Technical Report, AFML-TR-76-179, (Nov 1976)

39. A. Seegar & K. P. Chik, Phys. Stat. Sol. 29, 455-542 (1968)

40. M. Beniere, M. Chemla & F. Beniere, J. Phys. Chen. Sol. 37, 525-538 (1976)

41. W. D. Kingery & B. Francois, in Sintering and Related Phenomena, G. C. Kuczynski et al, Eds., Gordon & Breech, 471 (1967)
42. P. D. Ownby, in Materials Science Research, Vol. 6, G. C. Kuczynski Ed., Plenum Press, 431-37 (1973)
43a. G. R. Terwilliger & F. F. Lange, J. Mat. Sci. 10, 1169 (1975)
43b. M. Mitomo, J. Mat. Sci. 11, 1103-1107 (1976)
44. R. J. Brook, T. G. Carruthers, L. J. Bowen & R. J. Weston, presented at the NATO-ASI on Nitrogen Ceramics, Canterbury England (1976)
45. R. M. Cannon & U. Chowdhry, ibid.

The authors greatfully acknowledge financial support by DOE (ERDA) under Contract EY-76-S-02-2390 (RLC) and E(49-18)-2295 T. O. #8 (RMC).

DISCUSSION

C. S. Morgan (Oak Ridge National Laboratory): Another possible explanation of the sintering activation effect, such as the influence of Ni or Pd on the sintering of W, is dislocation slip. Very small quantities of surface additives often greatly increase the ease (or difficulty) of dislocation slip. At Oak Ridge, we have demonstrated that dislocations are generated during singering of tungsten.

Authors: For the initial stage of sintering, the calculated stresses based on the models are larger than the macroscopic yield stresses for small particle sizes and limited shrinkage, but decrease to less than the yield stresses within the initial stage. Although we question the appropriateness of using the macroscopic yield stress within the size of the regime in question, we concede the possibility, but note that the strong effect of particle size on the rate of sintering is predicted by the diffusion modeling and is counter to the expectations for control of the process by plastic deformation.

W. Rhodes, (GTE): It has been stated in describing the sintering kinetics of certain systems such as Al_2O_3 that D_b for the anion (or cation) is enhanced while the other ion is depressed. This is puzzling since it is known in several systems such as $SrTiO_3$ and Ca stabilized ZrO_2 that both anions and cations are enhanced at the grain boundary. It is really likely that these ionic systems would be so different?

Authors: In several instances direct diffusion experiments have failed to detect enhanced boundary diffusion for cations, e. g., Ni in MgO and Na and K in NaCl and KCl, in the absence of significant boundary impurity or second phase whereas the enhanced anion diffusion is easily observed. Creep measurements on fine grained materials can provide a more sensitive measurement of δD_b, and in the cases we have examined the values of δD_b for cations inferred from creep measurements are not in conflict with the detection limits for bicrystal or other self-diffusion measurements in which the results were negative. It is our judgement that the quantity $\delta D_b/\delta D_\ell$ is generally large for both ions in ceramics, although it is not necessarily the same for each, e.g., for undoped Al_2O_3 $\delta D_b/\delta D_\ell$ $\sim 10^9-10^{10}$ for O and $\sim 10^3-10^4$ for Al.

The space charge cloud near the boundary can, in principle, contribute to the total value of δD_b. Analysis[17] shows that the contribution from the charge cloud will be suppressed for one of the two species, which depends upon the dominant impurity. Calculations indicate, however, that δD_b is in general not significantly affected by these contributions from the charge cloud. We concluded that the enhanced D_v's must therefore arise from transport in the "core" of the boundary and expect that both species will be enhanced to different degrees in pure materials. No adequate theory, models, or experimental data exist to indicate the effects of solutes on δD_b for either species.

H. Palmour III, (North Carolina State University): Imposition of rate control on a near-optimal density-time profile[1] results in a "path of microstructural change"[2,3] which is likely to be rather different from that followed under conventional sintering conditions. Both the pore size and grain size distributions appear to be affected, with rate control tending to yield finer, much more uniform distributions. It is this greater degree of rate-related morphological uniformity (e.g., at $D \approx 0.95$) which appears to be in large part responsible for avoidance of breakaway grain growth during final stage densification under rate controlled conditions.[4] These findings are not inconsistent with expectations from the statistical distribution models for sintering being developed by Kuczynski.[5,6]

Authors: Certainly the microstructural evolution depends upon a competition between densifying processes (controlled by D_ℓ and D_b) and coarsening processes (controlled by wD_s and vapor transport) and grain growth (d). As the activation energies for these are different (e.g., "typically" $\Delta H(D_s) < \Delta H(D_b) \Delta\ H(D_\ell) < \Delta H(d,$ solute-drag) $< \Delta H(vap.)$), one can expect microstructural differences to result from different sintering time-temperature schedules. A recent analysis[7] shows that the critical porosity below which discontinuous grain growth can occur is very strongly dependent upon the grain size distribution, as has long been known, as well as on the average grain size and the solute content. Although we are intrigued by the results of rate-controlled sintering experiments which produce more desirable microstructures, we have been

unable to understand theoretically why the densification rate
per se should be the important variable to control in the densifi-
cation-coarsening competition.

References:
1. Hayne Palmour III and M. L. Huckabee, U. S. Pat. 3,900,542,
     Aug. 19, 1975.
2. R. T. DeHoff; pgs. 529-553 in Characterization of Ceramics.
     Edited by L. L. Hench and R. W. Gould. Marcel Dekker, Inc.,
     New York, N. Y., 1971.
3. R. T. DeHoff and E. D. Whitney, p. 81-100 in Ceramic Micro-
     structures - '76. R. M. Fulrath and J. A. Pask, Eds.,
     Westview Press, Boulder, Colorado, 1977.
4. (a) H. Palmour III, M. L. Huckabee and T. M. Hare, p. 308-319
     in Ceramic Microstructures - '76. R. M. Fulrath and J. A.
     Pask, Eds. Westview Press, Boulder, Colorado, 1977.
     (b) H. Palmour III, M. L. Huckabee and T. M. Hare, Presented
     IV International Round Table Meeting on Sintering,
     Dubrovnik, Yugoslavia, Sept. 5-10, 1977. To be published
     in the Conference Proceedings.
5. G. C. Kuczynski, p. 233-245 in Ceramic Microstructures - '76.
     R. M. Fulrath and J. A. Pask, Eds., Westview Press,
     Boulder, Colorado, 1977.
6. G. C. Kuczynski, Presented IV International Round Table
     Meeting on Sintering, Dubrovnik, Yugoslavia, Sept. 5-10,
     1977.   To be published in the Conference Proceedings.
7. M. F. Yan, R. M. Cannon, U. Chowdhry, H. K. Bowen, Bull. Am.
     Ceram. Soc., 56, 291 (1977).

Some Effects of Aggregates and Agglomerates in the Fabrication

of Fine Grained Ceramics

J. S. Reed, T. Carbone, Curtis Scott,
and S. Lukasiewicz

New York State College of Ceramics
Alfred University, Alfred, New York

INTRODUCTION

The processing engineer in production tends to equate processing with productivity. But viewed as a science, powder processing can be defined as the sequence of operations which systematically enhances the particulate character (presintering operations) and ultimately the composition and structure of the product material (sintering and surface finishing). Processing is a systems problem. Many functional operations and parameters are involved. But the continuous thread is the evolution of the chemical and physical character.

One very common cause of microstructural inhomogeneities (usually defects) is the inadequate control of aggregates[*] and agglomerates[*] in powders. Typical aggregates are crystallites held together by diffusion bonds. Agglomerates are more variable; the bonding forces may be surface charges, hydrogen bonds, van der Waals bonds associated with adsorbed organic molecules, and magnetic forces in ferrimagnetic powders. If moderate cementitious bonds exists, the distinction is less clear. Agglomerates are normally porous. Aggregates can be completely dense or porous.

_____

[*]Here a group of particles held together by strong and weak forces are termed aggregates and agglomerates, respectively.

Although the recognition of agglomerates and aggregates has increased in the past few years, research studies of their structure and their origin, persistence, and ultimate effects are rather sparse.[1-6] This paper discusses some effects of aggregates and agglomerates in the processing of technical ceramics.

Solid State Reactions

Calcined powders typically contain aggregates and a significant percent of these can escape/resist fracture on milling and/or high speed blunging. In studies of the mixing of suspensions of aggregated zinc oxide and unaggregated alumina, the size of the aggregates was <6 μm using a high speed impeller but sizes ranged up to 12 μm on using a low speed impeller. Ultimate crystallites were much finer but aggregates were ≥ size of the particles of alumina. On heating compacted mixtures at 1200°C, retarded reaction after 30 min. and incomplete reaction after 210 min. was traced to the presence of the coarse aggregates. This can be expected because the maximal diffusion path is a function of the size of the aggregates rather than the ultimate crystallites.

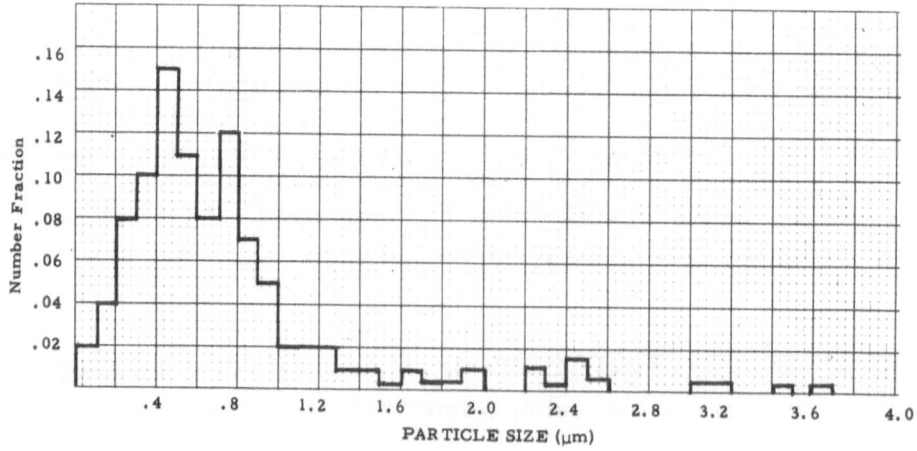

Fig. 1. Histogram for commercial, milled barium titanate powder

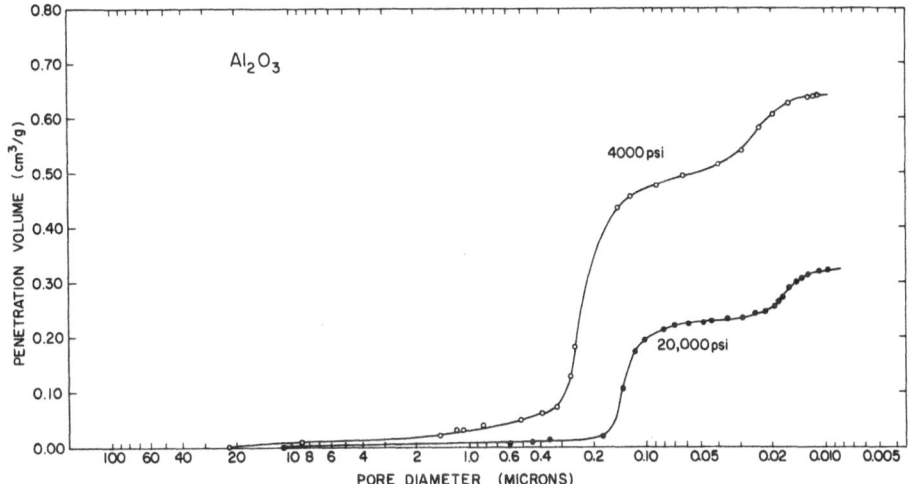

Fig. 2. Finer pores within aggregates persist in compacts
of Linde^R A alumina after isostatic pressing
at 140 Mpa (20 K psi).

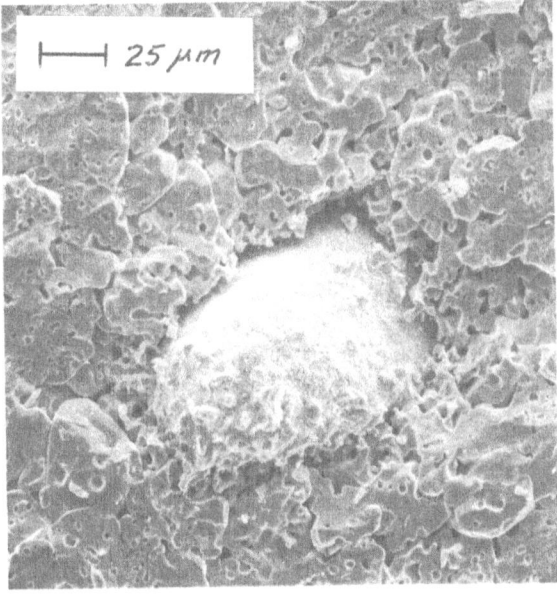

Fig. 3. Aggregate of differing BaO/TiO$_2$ in sintered, commercial CP barium titanate.

Microstructure Development in Barium Titanate

Many commercially produced milled barium titanate powders are log normal in size distribution to a good approximation if the coarser 2-4 percent is eliminated from the analysis (Fig. 1). The coarser fraction typically consists of aggregates $>8X$ the mean size. Aggregates have been observed to be dense in technical grade powders but porous in CP powder. Porous aggregates have also been observed in milled, chemically prepared and Bayer process aluminas.

These porous aggregates are observed to resist fracture into component crystallites even at high compaction pressures (Fig. 2); coarse interaggregate and finer intraaggregate pore inhomogeneities will be present in these compacts. In the CP barium titanate, aggregates appeared to be of higher BaO/TiO_2 ratio and sintered more rapidly than the bulk of the material (Fig. 3). When these aggregates were eliminated from the powder, the sintered microstructure was much more uniform and surface pores as large as 250 μm disappeared.

Ultrafine powders of $BaTiO_3$ with crystallites <25 nm are "active" in that initial stage sintering proceeds at an unusually low temperature (Fig. 4). However, aggregates as large as 500 nm greatly affect the evolution of microstructure and ultimate temperatures and grain sizes, required for densification, are rather typical.

Aggregates in Agglomerated Pressing Powders

Fine powders are purposefully agglomerated to achieve the flow response requisite for high speed pressing cycles. A survey of the characteristics of spray dried agglomerates of various compositions and their compaction response has been presented elsewhere.[7] Two characteristic (agglomerate) response parameters are the yield pressure (Py) and the joining pressure (Pj) when coarse pores $>$grain size within agglomerates and/or at interstices are eliminated (Fig. 5). Parameters affecting the response are under study. It is clear that "weak aggregates" in agglomerates are often a primary cause of an excessively high Pj (Fig. 6). Compaction results for bulky aggregated powders corroborate this interpretation.[5]

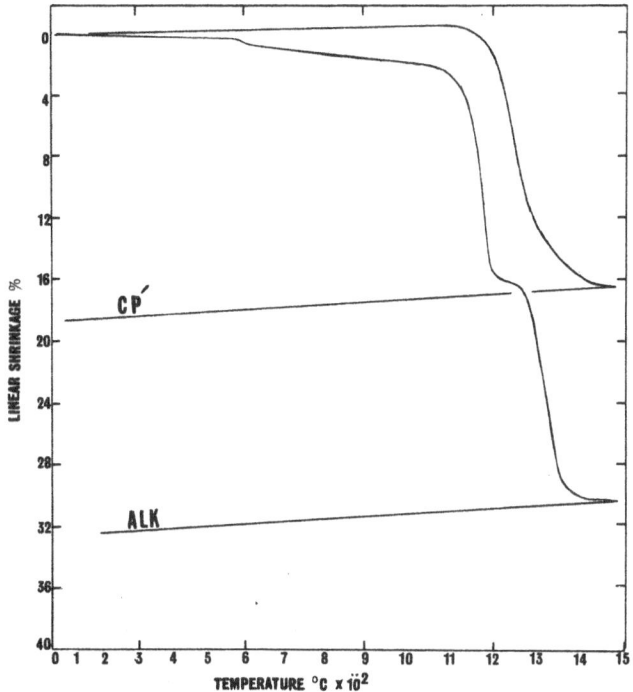

Fig. 4.   CRH shrinkage of deaggregated (CP') and alkoxide-
derived (active) $BaTiO_3$

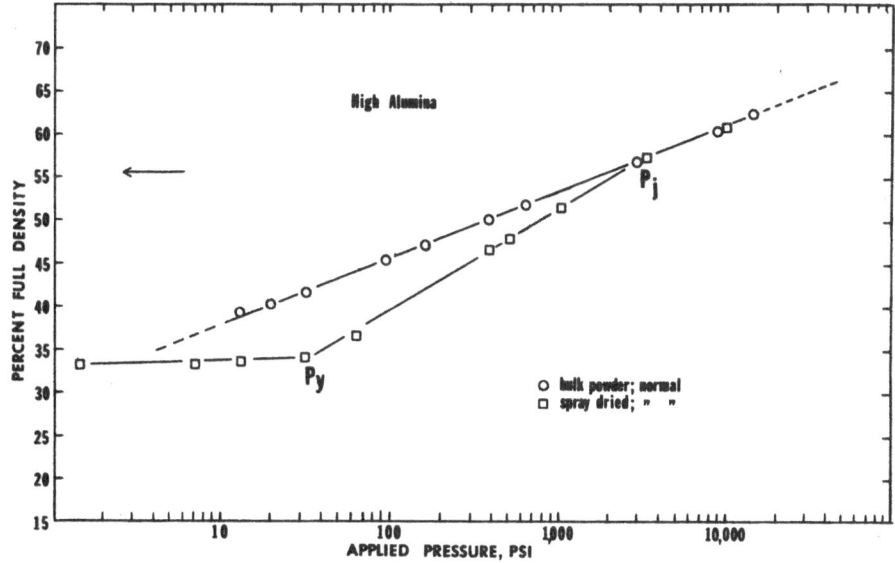

Fig. 5. Compaction diagram for spark plug body. (1 psi $\simeq$ 7000 Pa)

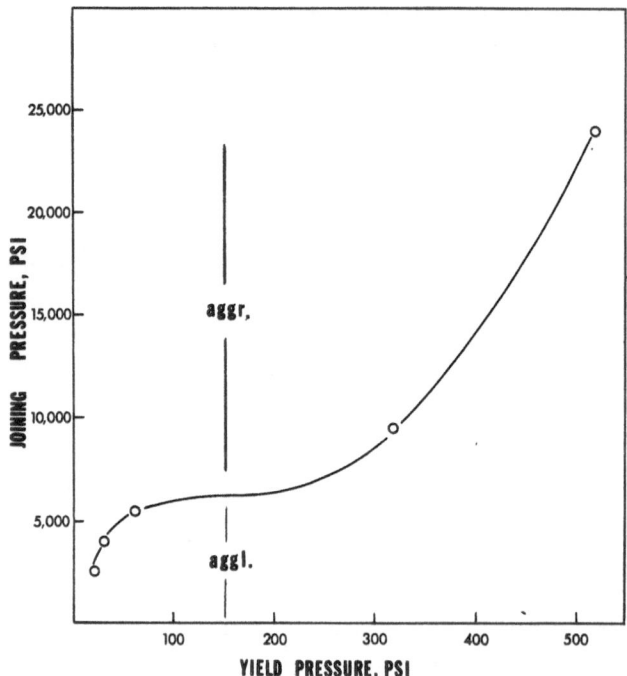

Fig. 6. Aggregates in agglomerate retard compaction.
(1 psi ≃ 7 K Pa)

Fabrication of Fine Grained Zirconia

Powders of 20-40 m^2/g and a mean crystallite size of ca.
50 nm have been used to produce cubic sintered material of >99%
density and <5 μm grain size.[8] Elimination of aggregates ≥ulti-
mate grain size and control of agglomerates was essential. As-
received milled powders contained aggregates generally <1 μm.
Wet milling using zirconia beads reduced both the aggregate and
agglomerate size but agglomerates reformed on drying. Agglomerates formed after milling in water and drying were noticeably stronger than those formed after milling in propanol and
drying. The former consistently survived handling in preparing
SEM specimens and incompletely collapsed on isostatic compaction at 20,000 psi (140 MPa). Similar results were obtained by
Neisz et al[5] for alumina. Compacts of powders milled in propanol and centrifuged to eliminate coarse aggregates, were consistently superior in densification response during CRH sintering.

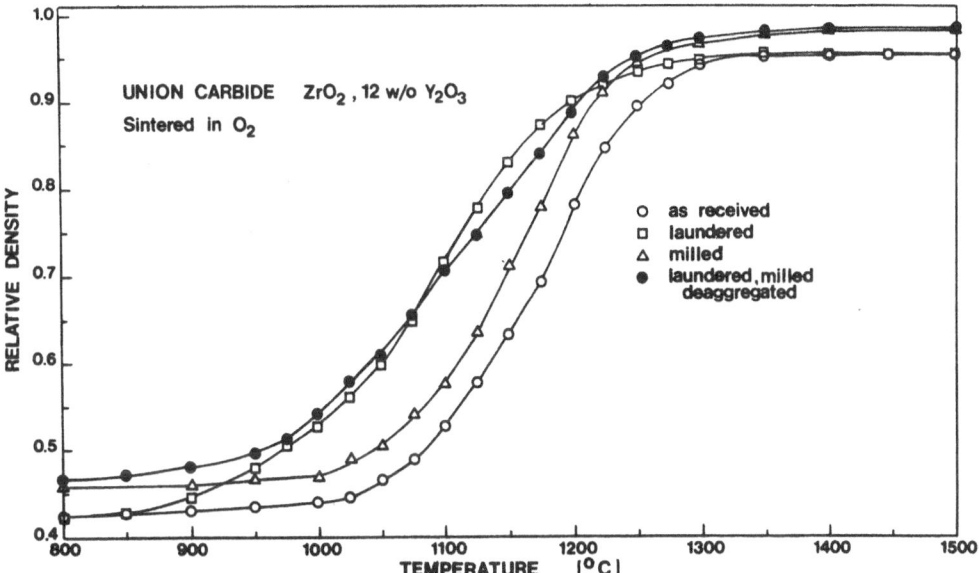

Fig. 7.   Densification response on CRH sintering of zirconia.

Maximal temperatures to attain 98% density were as much as
150°C lower and grain sizes were of smaller average size and
a more uniform size.   Microstructures in CRH sintered compacts of as-received powders contained inhomogeneous porosity
and exaggerated grains (Fig. 7).

Pressure Cast, Extruded Alumina

The pore structure in a slip cast (pressure cast) body is
more uniform if the casting slip is well deflocculated.   Partially
deflocculated (therefore partially flocculated) slips contain
agglomerates (flocs) and on deposition the pores in the cake
formed are less uniform in size and of greater average volume.
In fabricating whitewares by casting, slips are only partially deflocculated because the resultant cake is plastic rather than
brittle, and can survive stresses on handling and trimming.   The
range of pore size can be tolerated because sintering occurs in
the presence of a liquid phase.   However, for solid state sintering a cake with a uniform pore size is preferred (Fig. 8).

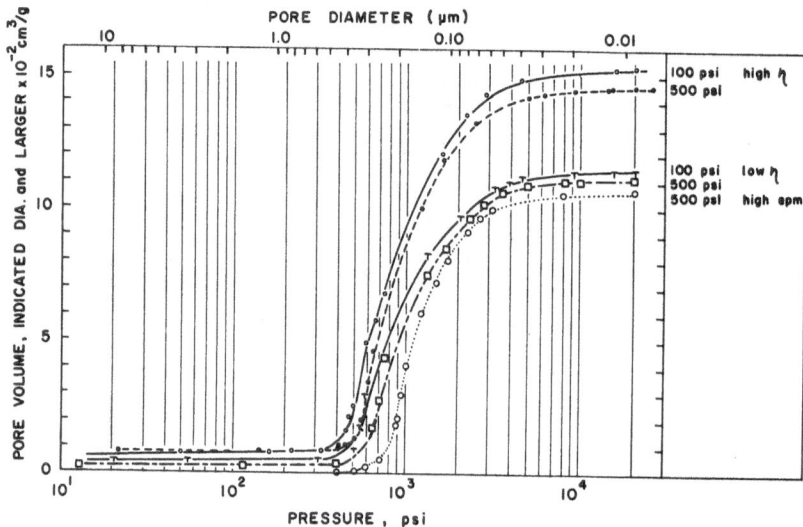

Fig. 8.   Dependence of pore size distribution in pressure cast
          body on deflocculation (indicated by viscosity (η) of slip).

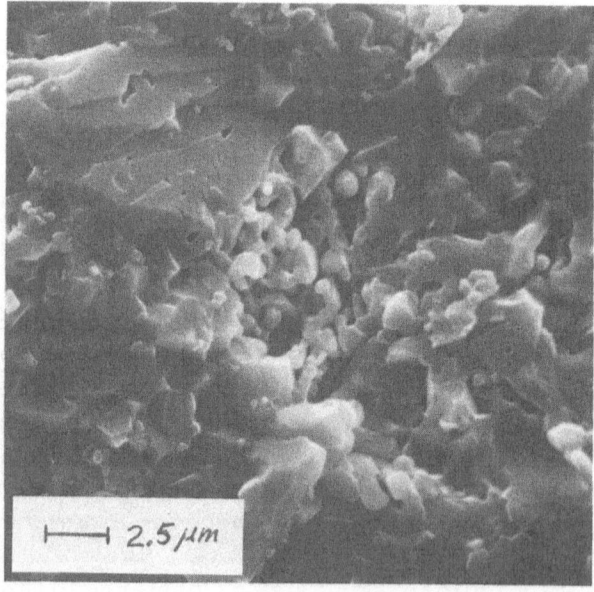

Fig. 9.   Exaggerated grains in extruded, sintered alumina.

Bodies for extrusion must be flocculated to exhibit psdudo-plastic flow behavior. If mixed in a deflocculated state and then flocculated uniformly prior to extrusion, the as-extruded micro-structure can be uniform. However, if mixed directly (concen-trated) in a flocculated state, it is very difficult to produce sin-tered material of uniform grain structure. Figure 9 illustrates exaggerated grains which originated from random agglomerates in directly mixed material.

Hard Ferrites

Sintered magnets of $SrO(5-6)Fe_2O_3$ and $BaO(5-6)Fe_2O_3$ are ideally of highest possible density and of an oriented $\approx$micron size grain structure. Initial powders must be of a submicron crystallite size. The use of ultrafine powders has not been par-ticularly beneficial because magnetic agglomerating forces pre-vent complete comminution of aggregates into individual crystal-lites. Aggregates are generally of a "random" grain structure and less porous than the bulk structure in compacts. On heating, densification and grain growth in aggregates can lead that in the bulk (inhomogeneous sintering). The mode of grain growth with-in aggregates is somewhat different than that between aggregates. Highly oriented grain structures cannot usually be achieved con-comitant with highest density and requisite grain size for maxi-mum intrinsic coercive force($_I$Hc). For sintered microstructures, the slope of the plot of residual magnetism (Br) against $_I$Hc is negative.

An alternative approach to minimize the effects of magnetic agglomerating forces is deformation processing at temperatures above the Curie temperature. Curves of Br vs. $_I$Hc for hot forged materials are of a positive slope and this method for cir-cumventing problems arising from agglomerating forces appears especially promising.

CONCLUSIONS

Material suppliers seldom report information about aggregates/agglomerates in powder specifications. But aggregates and agglo-merates are typically present in fine powders. Even if present in only a few percent, they may represent an extreme in the dis-tribution function of some powder characteristic. Better control of aggregates and agglomerates can often be the key to improving microstructures and properties of sintered materials. In fabricating dense ceramics with a micron grain size, the control of aggregates and agglomerates is especially important; "active" powders < 50 nm in size containing aggregates and uncontrolled agglomerates may be of no special advantage and even inferior to a coarser powder of uniform character.

Acknowledgments

Research discussed in this paper was supported by grants from the National Science Foundation (DMR 75-05445), The Edward Orton Jr. Ceramic Foundation, and Corporate Research, Union Carbide Corp.

References

1.   T. Vasilos and W. Rhodes, in Ultrafine Grain Ceramics, ed. by J. Burke, N. Reed, V. Weiss, pp. 137-172, Syracuse University Press, Syracuse, N.Y., 1970.

2.   C. Greskovich, in Treatise on Materials Science and Technology, V. 9, Ceramic Fabrication Processes, ed. by F. Wang, pp. 15-33, Academic Press, New York, N.Y., 1976.

3.   R. Katz, ibid pp. 35-49.

4.   Y. Kim, ibid pp. 51-70.

5.   D. Niesz, R. Bennet, and M. Snyder, Am. Ceram. Soc. Bull. 51 [9] 677-680 (1972).

6.   S. Lukasiewicz and J. Reed in Ceramic Microstructures '76, ed. by R. Fulrath and J. Pask, pp. 196-207, Westview Press, Boulder, Colorado, 1977.

7.   S. Lukasiewicz and J. Reed, (Character and Compaction Response of Spray Dried Agglomerates, submitted to Amer. Ceram. Soc., Oct. 1977).

8.   C. Scott and J. Reed (Sintering Response of Yttria Stabilized Zirconia: Agglomerate and Impurity Effects, submitted to Amer. Ceram. Soc. Oct. 1977).

THE SINTERING OF CONDUCTIVE RUTILE:

A MODEL SYSTEM FOR SINTERING ELECTRONIC CERAMICS[†]

G. R. Miller and O. W. Johnson

University of Utah

Salt Lake City, Utah   84112

ABSTRACT

Titanium dioxide in the rutile form, and doped with ions of normal and stable valence of +5, is a ceramic system which is useful in studying the effects of electron chemical potential on the solid state sintering kinetics of electronic oxides. Calculations of requisite defect concentrations as well as experimental sintering data from carefully characterized starting powders indicate a strong correlation between sintering kinetics and the value of the electron chemical potential in Ta-doped rutile. The correlation appears so strong as to suggest that monitoring the electron chemical potential would allow quantitative predictions as to sintering rates in bulk diffusion controlled densification.

INTRODUCTION

Sintering mechanisms have long been a topic of interest when discussing ceramic processing fundamentals. Over the years, many mechanims have been proposed and their characteristics documented both by calculation and experiment. In this paper we wish to investigate the sintering kinetics of at least a portion of the densification process where bulk diffusion of the cation and/or anion lattice plays a predominant role. Many will agree that volume diffusion controlled sintering often can be observed during the latter portions of the intermediate and final stages of the process. We wish to further limit our treatment to the case of ionic solids where the electrical neutrality condition is satisfied by the introduction of electronic defects such as

electrons and holes. One can easily surmise that the class of
solids covered by these limitations may still be very large.

We focus our attention on the fact that in volume controlled
diffusion process, one must consider three factors influencing
atomic migration, and hence, shrinkage. The first is the driving
force for diffusion, the second the defect mobility and lastly
the defect concentration. For this study we will assume that
powder compacts can be formed for study in which the driving
force is not significantly variable (from sample to sample) and
also that the defect mobility is concentration independent. Our
attention is thus directed to the effect of defect concentration
for each of the catonic and anionic sublattices. We must
recognize that local neutrality conditions will require movement
on all sublattices, but that not all are necessarily accomplished
at the same rate or by the same mechanism. For instance, one
might well observe ambipolar diffusion whereby one of the ionic
species moves long distances along grain boundaries while other
ions are transported by movement through the grains. We will
concentrate on the case in which volume diffusion of both ion
species is dominant, although many of the same conclusions
would follow for other cases.

It is well known that the change in the electronic chemical
potential of, say, electrons, (the electron Fermi level) can
drastically alter the solubility of lattice point defects, which,
when formed, contribute an electron or hole, or acceptor or donor
trap to the system.[1] Consider the simple case of adding a lattice
defect and q electrons to the conduction band of a solid whose
electron Fermi level is E_f, measured upward in energy from the
standard state of the pure, intrinsic crystal (at mid-gap if the
electron and hole effective masses are the same). Compared to
the intrinsic solid solubility, the equilibrium solubility of the
new defect and its electron is decreased by an amount ΔC, where

$$\Delta C = e^{-qE_f/kT} \tag{1}$$

With a solid of energy band gap of, say, 3eV and equal electron and
hole effective masses, ΔC will change by factors of exp − [57.4],
exp − [17.47] and exp − [10.2] at 300K, 1000K and 1700K
respectively for q equal to unity and the Fermi energy held very
near the conduction band edge. This simple model will only be
approximately valid if one assumes that no electron traps are
produced by the entry of the lattice defect. In this paper we will
investigate the usefulness of monitoring the electron Fermi level
in describing the densification process in a system made relatively
simple by chemical design. In other words, we wish to determine
the magnitude of the residual electron chemical potential effects
at high temperatures where sintering takes place and where nearly

intrinsic behavior is normally assumed.

THE MODEL SYSTEM

Titanium dioxide in the rutile phase is a system studied in enough detail[2] to be useful as a model system for these studies. Further, there is only one cation and one anion lattice that must move in the sintering process. (One could consider, as we will later, a more complex system with a double cation lattice, such as $SiTiO_3$. For the moment, the simpler case of rutile TiO_2 is most useful). In order to understand the effect of electron Fermi level, one must first consider the known defect structure of pure or slightly reduced TiO_2. For small changes in stoichiometry in the oxygen deficient direction it is rather clear that two lattice defects coexist--the titanium interstitial and the oxygen vacancy.[3]   Both of these defects are donors; each oxygen vacancy apparently produces one shallow electron trap. (We will assume that no trapping levels are associated with the Ti defect). One can easily surmise that the movement of these defects predominates in the solid state densification process in powder compacts in that both sublattices can move by virtue of the existence and mobility of their defects. It appears likely that these two defects are responsible for sintering and densification of relatively pure TiO_2 in air.

Rutile can be made conductive by the addition of certain dopants with a stable valence of +5 or greater in TiO_2, such as niobium, tantalum or uranium. Each of these elements will produce electron trap levels which are rather shallow, probably less than 0.030eV below the conduction band. Proper treatment of the chemical powders is a prerequisite for all studies reported here in that these dopants diffuse very slowly in rutile and we wish to study effects produced by their presence during almost all times for which sintering experiments are carried out. We cannot, for instance, allow homogenization of the dopant in the powders during major portions of the times over which sintering studies take place. Such powder preparation methods are available and have been reported elsewhere.[4]

One can, then, produce the rutile system with a reasonably stable and high electron Fermi level by the addition of one or more of the previously mentioned dopants. In this paper we will confine our attention to Ta-doped TiO_2 although preliminary experiments indicate that our arguments extend to TiO_2 with Nb and U dopants as well. However, one must consider the effects of such a high Fermi level in changing the nature of the pre-dominant defect structure. From the arguments leading to equation (1), the high Fermi level in the doped lattice should strongly suppress the solubility of both oxygen vacancies and titanium interstitials, the latter being suppressed more than the former due to the expected larger charge. With the large electron contribution to the free energy of the system, one must consider the possible existence of new point defects which will lower the free energy. For the case of rutile, both oxygen interstitials and titanium vacancies would lower the Fermi level, the titanium vacancy probably being most effective by virtue of its ability to reduce the electron density by as much as four electrons per vacancy. Thus one titanium vacancy can remove the electrons produced by four tantalum substitutional ions. The oxygen interstitial would be an unlikely defect from stereographic considerations. In another paper [5] we have investigated the plausibility of forming the titanium vacancy in TiO_2 heavily doped with Ta. From visual observations, weight change experiments, and high frequency complex impedance measurements, the titanium vacancy as been established along with a model for the formation of essentially undoped TiO_2 at grain boundaries via rapid diffusion of oxygen along grain boundaries. The result of a thermodynamic analysis for titanium vacancy formation gives the following relation for the concentration of titanium vacancies:

$$[V_{Ti}^{4-}] = F(T) \ p_{O_2} \ e^{+4E_f/RT} \tag{2}$$

where $F(T)$ is a temperature dependent quantity and contains the standard state for the electron chemical potential, and p_{O_2} is the oxygen partial pressure. At high temperatures, the tantalum ions are all ionized and thus for small tantalum concentrations, one can assume that the titanium vacancy increases with roughly the fourth power of the tantalum concentration. At high tantalum concentations, the picture will not be as simple, since the titanium vacancy concentration will be high enough to also effect the Fermi level. Further, at high vacancy concentrations strong electrostatic interactions between such highly charged vacancies will inhibit their formation although the remaining electrons will serve to screen these defects rather well.

Thus we have a picture in which rutile, sinterable by diffusion of titanium interstitials and oxygen vacancies in the pure state, becomes difficult to sinter as these defects are suppressed by the addition of tantalum and the coincident high Fermi level. However, as the tantalum concentration is increased, a new defect arises which allows movement of the titanium lattice via a vacancy mechanism. Further, the lowering of the Fermi level will allow the formation of some oxygen vacancies and the sintering kinetics would still be limited by anion lattice migration, unless rapid oxygen diffusion along grain boundaries is indeed allowed. In such a situation the oxygen lattice may be effectively transported by a grain boundary mechanism while the titanium ion lattice moves via a bulk mechanism (ambipolar diffusion). The existence of a rapid diffusion path for the oxygen sublattice will not, however, affect the arguments we have developed whereby the "sinterability" of rutile decreases and then increases as the tantalum concentration is raised. We now turn to experimental techniques to observe the predicted phenomena.

EXPERIMENTAL

In order that a measure of "sinterability" be made to observe these phenomena without (or with minimal) interference from other variables, the following approach was used: TiO_2 powders of pigment grade (containing negligible impurity contents of those elements known to "compensate" tantalum) were converted to the rutile form by heating to about $1100^{\circ}C$ for 3 hours. In this operation, the particle size was increased to about 0.4 to 0.5 microns. Scanning electron micrographs were taken of the powders after the converting and coarsening operations. Tantalum from an oxalic acid solution of tantalum oxalate was precipitated onto the powders by the addition of NH_4OH. Calcination of the loose powders was carried out for 2 hours at $1100^{\circ}C$. Added Ta concentrations varied from 0.1 cation pct to 3.0 cation pct. All samples were formed by uniaxially pressing disks measuring 3.18 cm (1.25 in) dia by about 0.32 cm (1/8 in) thick. Several pressing pressures were used for each dopant concentration, each sintering temperature and each sintering time. Pressing pressures were used which gave samples whose final sintered density was independent of pressing pressure, which occurred quite consistently for samples with a green density of 57 percent of theoretical or more. Samples were then sintered in air for times ranging from one to six hours and at temperatures ranging from 1200°C to 1400°C.

Mass densities were determined by hydrostatic weighing and were expressed in terms of the corresponding percentage of theoretical density. X-ray theoretical densities were corrected for the tantalum–titanium substitution. Scanning electron micrographs and optical micrographs were made to check for grain size changes and the position and distribution of pores.

EXPERIMENTAL RESULTS

Samples sintered at $1200^{\circ}C$ and $1300^{\circ}C$ were found in all cases to lack homogenization of the Ta dopant with the grains. Densities were always very low for Ta concentrations of 0.1 cation percent or greater. Only the "pure" TiO_2 sintered to densities as high as 85 pct of theoretical at $1200^{\circ}C$ or $1300^{\circ}C$. This is in line with observations of ourselves and others[6] that tantalum serves as a sintering and grain growth depressant in rutile and diffuses very slowly in the rutile lattice. However, sintering times of as little

as half an hour at 1400°C gives a characteristic blue-black
coloration in the compacts showing that homogenization is
taking place. As a result of these observations, all subsequent
sintering was carried out at 1400°C.

Measurements of sintered density as well as photomicrographic
examinations as a function of sintering time show that after
about one hour, the sintering of all samples is in the regime
whereby the porosity is isolated. Other than the samples of
"pure" TiO_2, the pores are primarily at the grain boundaries and
grain size is essentially constant at between 5 and 8 μm mean
diameter for all doped samples sintered at 1400°C in air, for times
of one to six hours (see Figure 1 for example). For the "pure"
samples, porosity is entrapped within the grains and extensive
nonuniform grain growth occurs, a characteristic not seen in the
doped samples.

For samples sintered for 3 hours at 1400°C, the results of
density measurements are shown in Figure 2 where percent theoretical density is plotted as a function of tantalum concentration
(closed circles). As a check that this effect is indeed due to
the presence of the tantalum, some powders containing about 0.3
cation percent aluminum were prepared in the same fashion as
earlier described. Since the aluminum should provide deep
electron traps for the electrons arising from the tantalum
addition, the electron Fermi level should be strongly depressed
until the tantalum concentration begins to exceed the aluminum
concentration. (At high temperatures, this "compensation" effect
should be one for one). Thus the curve obtained for otherwise pure
TiO_2 should be shifted to higher Ta concentrations when aluminum
has been added to the system as is observed in the second curve
(open circles).

DISCUSSION

Earlier a model was introduced which predicted a depression
in the sintering kinetics of donor doped rutile due to the
suppression of the concentration of the normal defects found in
TiO_2, namely the Ti interstitial and the oxygen vacancy. That
such a suppression is observed would not be so interesting at
lower temperatures since this prediction is well known when the
electron Fermi level is not seriously affected by intrinsic
processes (as at lower temperatures). Normally at higher temperatures, the generation of intrinsic electronic defects would
depress the Fermi level until the electron Fermi level reached mid-
gap and plays no part in the defect generation process. However
the exponential form is so strong that the Fermi level, when
maintained to even moderately high values up to the sintering
temperature, has a large effect. With density of states data for

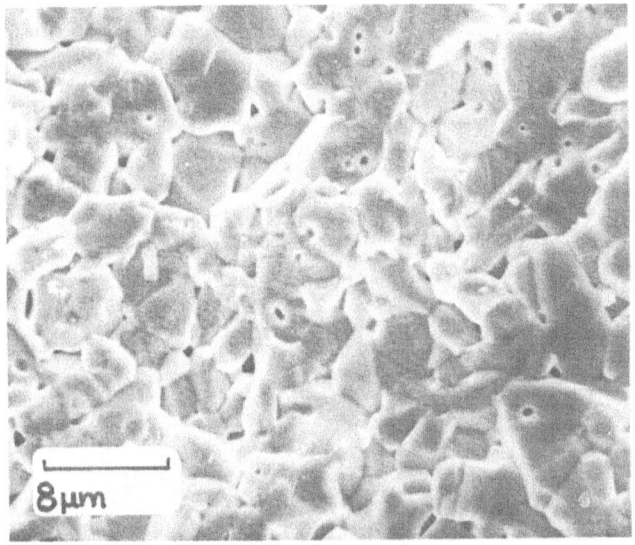

0.1% Tᴀ - 1400°C ᴀɪʀ sɪɴᴛᴇʀ - 3 ʜʀs.

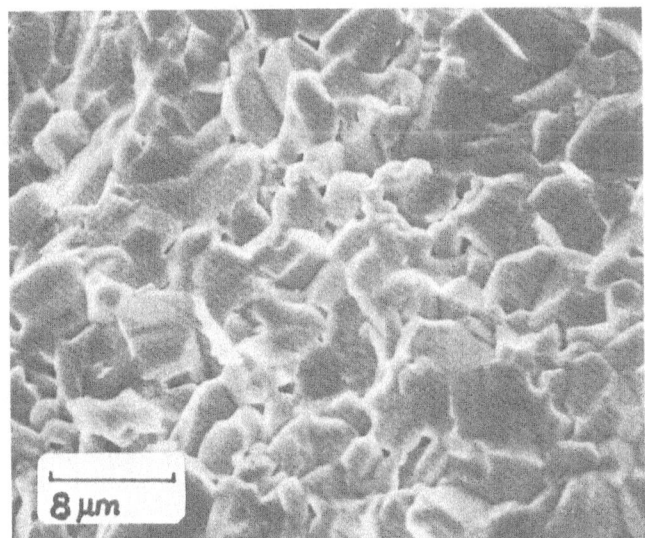

1.0% Tᴀ - 1400°C ᴀɪʀ sɪɴᴛᴇʀ-3 ʜʀs.

Figure 1:   SEM fracture surfaces of sintered, Ta doped rutile for
            Ta concentrations of 0.1 and 1.0 cation percent.

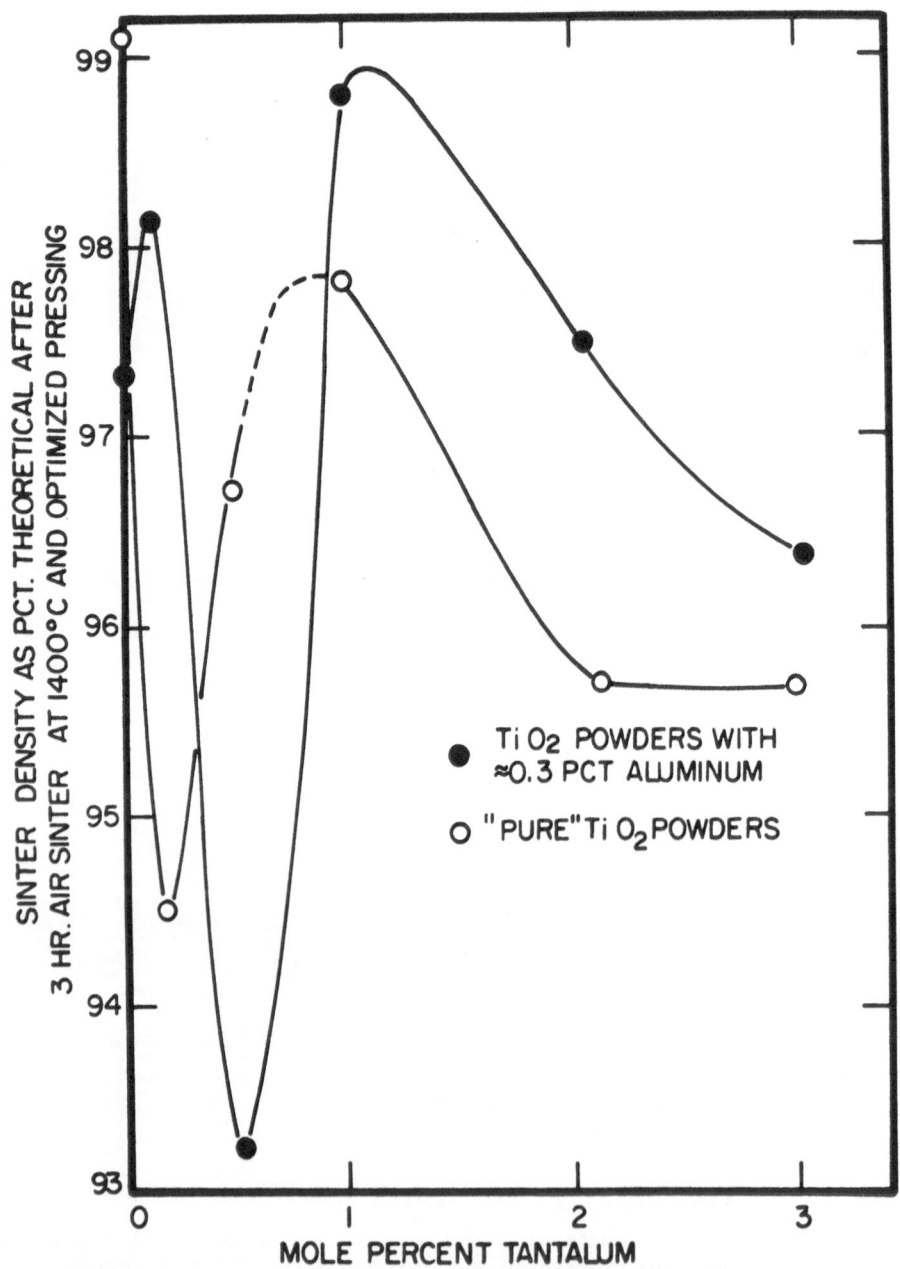

Figure 2:   Sintered density of Ta doped rutile and (Ta,Aℓ)
doped rutile samples as a function of dopant concentration.

rutile taken at this laboratory and assuming no other traps are present, the Fermi level will be approximately 0.6eV below the conduction band at 1400°C in air for material containing 1 pct Ta compared to about 1.5eV below the conductor band for undoped material. Thus, for example assuming that an oxygen vacancy traps one electron, the equilibrium concentration of oxygen vacancies in 1 pct Ta doped material is reduced from that expected in pure TiO_2 at 1400°C by a factor of about exp $[-(1.5-0.6)/KT \approx 2 \times 10^{-3}$. With no trapping of electrons by an oxygen vacancy the reduction in concentration is even larger. The reduction in concentration of Ti interstitials is even more dramatic--the corresponding factor is $\approx 10^{-11}$. On the other hand, introduction of 1 pct Ta tips the scales very strongly in favor of Ti vacancies, since their concentration is increased by a factor of 10^{11}. This analysis is obviously oversimplified and should not be expected to be quantitatively accurate; however, qualitatively these trends are correct. At dopant levels of 1 pct, then, we should expect the sintering kinetics to be limited by defect concentrations on the anion lattice. That the material obviously sinters astonishingly well at such concentrations may be due to rapid oxygen diffusion along grain boundaries in an ambipolar mode.

At very high defect concentrations, the sintering kinetics become quite complex. (There is no indication that the solubility limit has been exceeded). The electron Fermi level becomes very well pinned at a high value, even though according to equation (2) a large number of titanium vacancies should be produced. At such high concentrations, the large negative charge on a titanium vacancy, even though somewhat screened, will produce large electrostatic terms in the total internal energy as strong defect interactions come into play. Thus the concentration of titanium vacancies that can be formed is somewhat self limiting and the Fermi level is pinned at very high levels. The sintering kinetics again become severely limited by inability to form anionic defects.

It is interesting to speculate on the effect of Fermi level on grain growth and especially nonuniform and anomalous grain growth. Obviously for pure materials where volume diffusion is important in the grain growth mechanism, the tantalum dopant acts as an inhibitor. However, with inhomogeneously doped material, doped with say acceptor type impurities such as Al^{+3}, Fe^{+3}, Mg^{+2}, etc., as one might expect with contaminated material, the regions containing such foreign matter can be compensated by the tantalum additive, effectively pinning the Fermi level near mid-gap. As a result, the titanium interstitial and oxygen vacancy can form in large numbers since their numbers are now not even as self-limiting as in the case of pure material. Thus, grain growth can even be enhanced by the presence of tantalum when comparable numbers of impurities which produce deep traps are also present.

As a second, more complex example, consider a system such as $SrTiO_3$ and its double cation lattice. The oxygen vacancy dominates the defect structure under reduction.[7] Tantalum and niobium are soluble as substitutional donor defects on the titanium lattice as in TiO_2. Again the high Fermi level produced by the electrons from the dopant will suppress the oxygen vacancy and other donor defects such as the titanium (or strontium) interstitials. Again as in the case of TiO_2 where Ti vacancies exist at high stable Fermi levels, one expects that defects which lower a high Fermi level will become prevalent even though they might not be present without the donor dopant. In the case of $SrTiO_3$ lowering of the electron Fermi level can be accomplished by the formation of strontium vacancies or titanium vacancies but it would be unusual to observe both mechanisms at once. Hence, for instance if the strontium lattice becomes vacancy defective, the nondefective titanium lattice would remain relatively immobile. Thus, contrary to the results in TiO_2 where a minimum and subsequent maximum in sintered density is found as the donor dopant is increased in concentration in $SrTiO_3$ one might never observe the maximum in sinterability if only one of the cation sublattices becomes defective in order to reduce the electronic contributions to the energy.

SUMMARY

We have suggested a model for electronic ceramics in which one can qualitatively follow the late stages of sintering kinetics by rather simple arguments concerning movements of the electron Fermi level. Most of the requisite information for predicting relative sinterability of such solids comes from a knowledge of the defect structure or at least plausible defect structures. No information concerning the formation energy or mobilites of the defects is required as it appears that all of the data can be qualitatively accounted for by general consideration of the Fermi level alone. We acknowledge, of course that in complex systems, especially those containing trap-forming impurities, the situation regarding Fermi level position will be very complex. Nevertheless, from the fact that the effective formation energy of diffusing defects (lattice defect plus their electronic charges) is so strongly influenced by Fermi level movements, one must expect that the concepts shown here can be extended to many other ceramic systems. The essential point is that the effect of the Fermi level can be a major consideration even at rather high temperatures when the band gap is large enough.

[†]Work supported by the National Science Foundation under NSF-RANN C-805.

The authors wish to thank Mr. C.A. Venizelos and Mr. Paul Smith for their assistance in sample preparation and characterization.

REFERENCES

1. R.S. Brebrick, J. Phys. Chem. Solids 4, 190 (1958).

2. J.W. deFord and O.W. Johnson, J. Appl. Phys. 44 [7], 3001 (1973).

3. G. Levin and C. J. Rosa, Abstract 328, Meeting of The Electrochemical Society, Las Vegas, (1976).

4. O.W. Johnson, C.A. Venizelos, P. S. Beutler and G. R. Miller, submitted for review to J. Amer. Ceram. Soc. (1977).

5. O.W. Johnson and G. R. Miller, submitted for review to J. Amer. Ceram. Soc. (1977).

6. H. U. Anderson, private communication.

7. H. Yamada and G. R. Miller, J. Solid State Chem. 6, 169 (1973).

DISCUSSION

Man F. Yan (Bell Laboratories): How does the oxygen partial pressure in the TiO_2 sintering atmosphere affect the final densities? If so, can the effects of Ta and Al dopants be compensated by a change of the sintering atmosphere? Given the same sintered atmosphere or some dopant content, is there any difference in the microstructures (e.g., grain size)?
Author: Only very preliminary data on the effect of oxygen partial pressure on sintered densities and microstructures have been gathered and even these data only include oxygen partial pressures from 0.2 to about 10^{-3} atmospheres – very little effect has been observed. From our model on fully homogenized Ta-doped starting powders, one would expect a decrease in final sintered density as the oxygen pressure is decreased. This is followed by an increase in grain size and density as the pressure is substantially reduced to the region where reduced Ta-doped rutile is formed.

SINTERING OF MULLITE

M. D. Sacks and J. A. Pask

University of California

Berkeley, CA 94720

INTRODUCTION

Despite technological importance, limited information is available on the correlation of major processing variables with the sintering behavior and microstructure development of aluminum silicates containing mullite as the principal phase. Recent studies illustrate the excellent mechanical properties of mullite and the importance of controlled processing in achieving desired properties.[1,2]

The character of mullite bodies depends on the Al_2O_3/SiO_2 ratio, the degree of mixing achieved, the firing conditions (time, temperature, atmosphere), and the kind of impurities. Mullite is usually prepared by thermally decomposing aluminum silicates such as clay and sillimanite minerals. Alumina may be added to reduce the amounts of siliceous phase. Reviews on mullite formation and processing are given by Grofcsik,[3] Davis,[4] and Sacks.[2] Various kinds of Al_2O_3/SiO_2 mixtures are commonly sintered at high temperatures. Mixtures of α-quartz, silicic acid, β-cristobalite or fused silica with α-alumina, bauxite, diaspore ($HAlO_2$), gibbsite ($Al(OH)_3$) or AlF_3 have been reported.[2,3,4]

In these approaches, alumina and silica are not mixed on a molecular scale and, because of low diffusivities, normally result in imcomplete reactions despite high firing temperatures. Pressure sintering has been used to achieve higher densities and avoid the associated large grains.[5,6,7] In order to achieve better composition control, complete reaction, and a fine-grain powder, a number of chemical and gel preparation techniques have been developed to form mullite powder.[2,3,4]

193

In the present study, several processing schemes are utilized to
study the effects of overall Al_2O_3/SiO_2 composition, powder surface
area, and sintering atmosphere on the sintering kinetics and microstructure development of mullite.

EXPERIMENTAL

Processing

"Powder" Process. Starting materials were Alcoa Aluminum Co. A-
14 (α-Al_2O_3) and Illinois Minerals Co. silica flour (α-quartz) which
had purities of ~99.8% and ~99.5%, respectively. Compositions ranged
from 60 wt% Al_2O_3/40 wt% SiO_2 to 90 wt% Al_2O_3/10 wt% SiO_2.

Mixing and de-agglomeration were achieved by wet (isopropanol)
milling in (1) a porcelain ball mill (mixing condition "I") or (2) a
teflon-lined vibratory mill (mixing condition "P"). After stir-drying
and screening (-120 mesh), mixtures were calcined at 1700°C for 8 h
to form mullite, mullite and glass, or mullite and Al_2O_3 (depending on
the overall Al_2O_3/SiO_2 ratio). Experiments with the mixture 73 wt%
Al_2O_3 showed that calcination time and temperature did not affect subsequent sintering as long as the mullite reaction was complete.

After calcination, the mixture was subjected to coarse crushing
in a mechanically operated mortar and pestle and wet (isopropanol) vibratory milling (grinding condition "P"). Unless noted otherwise,
grinding time was 5 h. After stir-drying and screening (-120 mesh),
powders were calcined at 800°C for 1 h to burn out any organics. Powder mixtures were labeled according to composition and mixing/grinding
conditions, e.g., 73IP refers to a 73 wt% Al_2O_3 composition mixed by
condition "I" and ground by condition "P". X-ray diffraction and microscopy have shown the most significant contamination in processing to be
Al_2O_3 introduced by impact collisions during vibratory milling; thus,
PP compositions have the largest Al_2O_3 contamination.

Powder compacts were die formed under uniaxial pressure ($17MN/m^2$)
followed by isostatically pressing at ($170MN/m^2$).

"Gel" Process. The gel process utilized colloidal size reactants
to reduce the time and/or temperature required to complete mullite formation. Aqueous colloidal suspensions of alumina and silica were intimately mixed. To prevent segregation, mixtures were gelled by pH adjustment or by evaporation of water. Details of the process have been described by Ghate.[5] A 73 GP composition was used in kinetic studies.
Raw materials were Atomergic Chemicals Corp. GZ5, a γ-Al_2O_3 of ~99.95%
purity, and Ludox AS, a dispersion of silica particles of ~99.5% purity.
Complete reaction occurred on calcining the gel powders at 1450°C for
24 h.

Sintering Conditions

Compacts were fired under isothermal conditions at 1540°C to 1730°C for times of 0 to 100 h. The specimens were brought to temperature as quickly as possible (10-20 min); zero time indicates the point at which the sintering temperature is reached. Sintering atmospheres were air (quench-type furnace) and $\sim 10^{-6}$ torr vacuum (Brew furnace). The quench-type and Brew furnaces have been described in Ref.2.

Properties and Microstructure

Bulk density and open porosity were determined by the displacement method utilizing distilled water. Percent theoretical density (%ρth) was estimated by using the apparent true density as determined from green compacts. In green compacts, total porosity is $\approx$ open porosity; the latter quantity is measurable by the displacement method.

Standard ceramographic techniques were followed for preparation of specimens for viewing by optical microscopy, SEM, and TEM. Etching of 60 and 65 compositions was accomplished by brief (5-60 sec.) immersions in a weak (5-10%) hydrofluoric acid solution. Other compositions were etched thermally (20-45 min at $\sim$1550°C in air).

RESULTS

Effect of Al_2O_3 Content

The initial series was made to determine the sintering behavior at 1700°C of compositions that are in the mullite and liquid field (60 to 65 Al_2O_3), mullite solid solution range (71.8, 73 and 74), and the mullite and Al_2O_3 field (75, 80, 85 and 90) as indicated by the SiO_2-Al_2O_3 phase diagram.[8] The bulk density, % open pores, and %ρth for 1, 4 and 12 h firings in air are listed in Table 1; the estimated average %ρth of the unfired compacts was 57.5. Processing followed IP conditions. Photomicrographs in Fig. 1 show microstructures of representative compositions fired in air and vacuum for 12 h.

Each of the three groups showed characteristic behavior. The 60 and 65 compositions are at maximum density because of the presence of the liquid phase. Composition 60 in Fig. 1 shows that theoretical density has been realized; the %ρth values of $\sim$96 shown in the table are due to the presence of large pores which are in view at lower magnifications. The 71.8 to 74 compositions do not sinter to theoretical density and their microstructures at higher magnifications indicate the presence of glass; the densities are higher and less glass is visible on firing in vacuum, as seen in Fig. 1. The 75 to 90 compositions sinter less readily and are most sensitive to vacuum firing as easily seen on comparing the microstructures for the 75 and 80 specimens; α-Al_2O_3, which is characteristically present in these compositions, is most easily recognizable in the 80 vacuum-fired microstructure.

Table 1. Properties of Compositions Fired at 1700°C.

Composition, % Al₂O₃	Firing in Air, 1 hr.			Firing in Air, 4 hrs.			Firing in Air, 12 hrs.		
	Bulk Density, g/cm³	% Open Porosity	% Theor. Density	Bulk Density g/cm³	% Open Porosity	% Theor. Density	Bulk Density g/cm³	% Open Porosity	% Theor. Density
60 IP	2.91	0	97	2.90	0	96	2.89	0	96
65 IP	2.95	0	95	2.99	0	97	2.99	0	97
71.8 IP	2.51	20.6	79	2.78	8.7	87	2.97	0	93
73 IP	2.54	19.7	80	2.82	6.7	88	3.06	0	96
74 IP	2.56	16.6	80	2.72	11.1	85	3.00	0.5	94
75 IP	2.25	28.6	70	2.40	22.6	75	2.67	14.5	83
80 IP	2.43	26.7	72	2.64	17.6	78	2.83	11.9	84
85 IP	2.52	27.4	72	2.73	18.1	78	2.97	11.1	84
90 IP	2.64	26.3	72	2.82	21.4	77	3.34	2.2	91

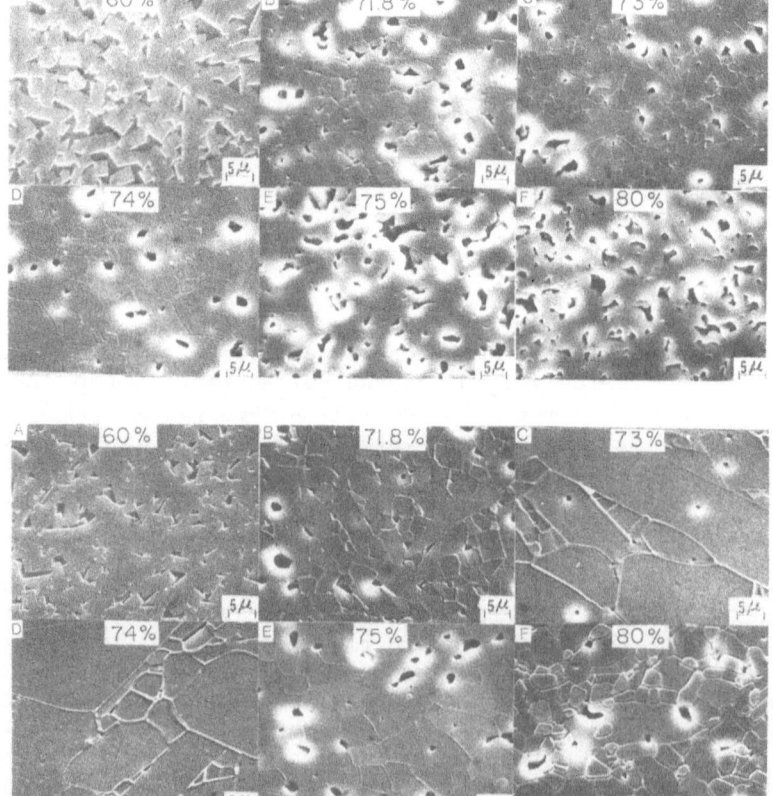

Fig. 1. Microstructures of representative compositions fired in air (top) and vacuum (bottom) at 1700°C for 12 h.

The contribution of etching to the characterization of these
materials is exemplified by the 71.8 composition. Figure 2 shows
photomicrographs of specimens heated at ~1710°C for 100 h in air after
etching for ~5 seconds with an ~5% HF solution (A left) and after thermally etching in air for ~20 min at ~1550°C (B left). It can be seen
that the glassy portions in (A) appear as pores in (B), and that the
thermal etching makes the grain boundaries visible. The effect of a
small amount of glassy phase on microstructure is shown in (A) and
(B) on the right of Fig. 2 for compositions fired at ~1710°C for 8 h
in air. Although thermal etching of air-fired specimens causes groov-
ing at grain boundaries, etching of vacuum-fired specimens causes
ridges as can be seen by comparing the photos for the 71.8, 73 and 74
specimens in Fig. 1. These ridges are removed on prolonged etching.

Kinetic Studies

The bulk densities, % open porosities, and %ρth for compositions
60 IP, 73 IP, 73 PP, 73 GP, and 75 IP sintered at a number of temperatures and times are listed in Table 2. The surface areas in m^2/g for
these compositions are 1.9, 1.8, 2.3, 6.8 and 1.7, respectively.
Representative plots of %ρth vs. ln t and rate vs. 1/T for composition

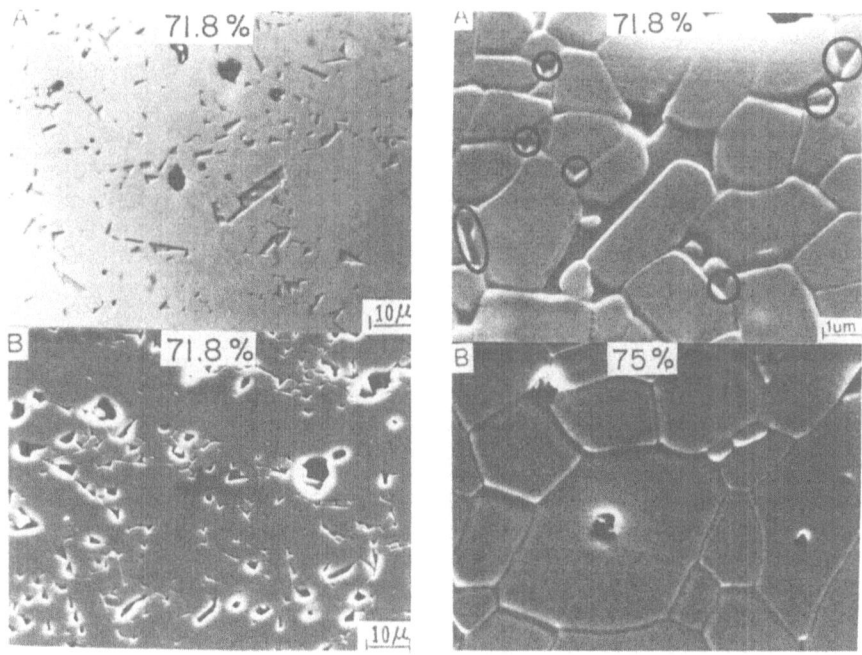

Fig. 2. Micrographs of chemical (A) and thermal (B) etching of 71.8
wt% Al$_2$O$_3$ fired in air (left). Effect of small amounts of glassy
phase (A) on microstructure (right). See text for details.

Table 2. Properties of Selected Compositions Fired in Air at Various Temperatures and Times.

Firing Temp. °C	Firing Time Hrs.	60/40 IP Bulk Density g/cm³	% Open Porosity	% Theor. Density	73/27 IP Bulk Density g/cm³	% Open Porosity	% Theor. Density	73 PP Bulk Density g/cm³	% Open Porosity	% Theor. Density	75/25 IP Bulk Density g/cm³	% Open Porosity	% Theor. Density	73/27 GP Bulk Density g/cm³	% Open Porosity	% Theor. Density
1540	0	1.93	35.3	64.1	1.91	39.6	60.1	1.94	38.9	60.6	1.90	40.2	59.0	1.74	44.7	53.4
	1	1.97	32.3	65.4	2.00	36.9	62.9	1.97	38.1	61.6	1.94	39.0	60.2	1.88	42.1	57.8
	4	2.04	31.7	67.8										1.97	38.1	60.5
	6	2.05	31.4	68.1	2.08	34.4	65.4	2.03	36.4	63.4	1.98	37.7	61.5	2.04	37.0	62.8
	9							2.05	35.8	64.1	2.00	32.7	62.0	2.12	34.9	65.2
	12	2.10	29.8	69.8	2.12	33.2	66.7	2.06	35.4	64.4	2.01	37.1	62.4			
	24				2.15	32.3	67.6	2.09	34.4	65.3	2.04	35.3	63.4	2.27	31.8	69.7
1580	0	1.99	32.7	66.1	1.97	38.3	61.9	1.95	39.2	60.9	1.90	40.3	59.0	1.82	43.0	56.0
	1	2.17	26.3	72.1	2.06	35.0	64.8	2.02	36.6	63.1	1.97	38.2	61.2	2.03	37.4	62.5
	2	2.30	22.5	76.4				2.05	35.4	64.1	1.99	37.2	61.8	2.09	33.6	64.3
	4	2.40	19.3	79.7										2.17	31.1	66.9
	6				2.18	31.2	68.6	2.09	34.1	65.3	2.04	39.9	63.4	2.28	29.2	70.2
	9	2.53	14.0	84.1				2.12	33.3	66.3	2.06	35.5	64.0	2.39	26.7	73.5
	12	2.57	11.0	85.4	2.24	29.2	70.4	2.14	32.7	66.9	2.09	34.3	64.9			
	24	2.75	6.6	91.4	2.29	27.8	72.0	2.19	30.9	68.4	2.12	33.5	65.8	2.52	19.0	77.5
1620	0	2.04	31.3	67.7	2.02	36.2	63.5	1.98	37.7	61.9	1.94	39.2	60.2	1.88	41.9	57.8
	1	2.42	17.1	80.4	2.16	31.7	67.9	2.09	34.3	65.3	2.03	36.3	63.0	2.21	3.03	68.1
	2	2.58	11.7	85.7				2.14	32.7	66.9	2.06	35.4	64.0	2.36	24.7	72.7
	4	2.69	5.6	89.4										2.48	22.0	76.4
	6				2.34	25.9	73.6	2.21	30.3	69.1	2.14	32.8	66.5	2.63	18.3	80.9
	6.5	2.78	.6	92.3										2.66	16.2	81.8
	8	2.84	0	94.3										2.75	13.3	84.8
	12	2.89	0	96.0	2.41	23.6	75.8	2.25	28.6	70.3	2.19	31.4	68.0	2.92	7.0	89.9
	24	2.93	0	97.3	2.50	20.4	78.6	2.33	25.7	72.8	2.26	28.9	70.2			
1660	0	2.32	20.4	77.1	2.10	34.0	66.0	1.98	37.6	61.9	1.98	37.8	61.5	2.18	31.9	67.0
	1	2.78	.3	92.3	2.34	26.1	73.6	2.22	30.1	69.4	2.12	33.3	65.8	2.56	18.3	78.8
	2	2.88	0	95.7				2.28	28.0	71.3	2.15	32.1	66.8	2.77	12.8	85.2
	4	2.91	0	96.6	2.54	19.1	79.9	2.34	25.6	73.1	2.25	29.8	69.9	2.90	6.5	89.2
	6													2.98	2.9	91.7
	8	2.92	0	97.0	2.67	13.6	84.0	2.44	23.0	76.3	2.33	27.1	72.4			
	9													3.06	0	94.2
	12							2.48	21.0	77.4	2.34	26.1	72.7	3.10	0	95.4
	16				2.81	9.2	88.4	2.54	19.6	79.4	2.41	24.4	74.8			
	18	2.93	0	97.3				2.56	18.5	80.0				3.10	0	95.4
	24				2.88	4.5	90.6	2.63	16.1	82.2	2.46	22.6	76.4			
1700	0				2.26	28.3	71.1	2.20	30.7	68.8	2.08	34.6	64.6	2.53	21.1	77.8
	1	2.91	0	96.6	2.56	18.4	80.5	2.38	24.4	74.4	2.25	29.4	69.9	2.82	5.3	86.6
	2	2.91	0	96.6				2.52	19.3	78.8	2.31	27.2	71.9	3.06	0	94.2
	4	2.90	0	96.3	2.82	6.7	88.7	2.60	19.0	81.3	2.41	25.2	74.8	3.08	0	94.8
	6							2.70	13.1	84.4	2.45	22.9	76.2			
	8	2.89	0	96.0				2.74	11.3	85.6				3.12	0	96.0
	12	2.89	0	96.0	3.03	.2	95.3	2.87	5.7	89.7	2.64	16.2	82.0	3.15	0	96.9
	24				3.06	0	96.7	2.90	5.0	90.6	2.71	13.3	84.2			
1730	0	2.90	0	96.3	2.46	21.6	77.4	2.34	26.0	73.1	2.19	31.0	68.0	2.84	9.6	87.2
	1	2.90	0	96.3	2.84	5.1	89.3	2.65	15.2	82.7	2.39	24.7	74.2	3.09	0	95.1
	2				3.03	.9	95.3	2.76	12.4	86.3	2.50	21.9	77.5			
	4	2.86	0	95.0							2.59	18.0	80.4	3.10	0	95.4
	6				3.04	0	95.6	2.89	5.2	90.3	2.64	15.7	82.0			
	9													3.14	0	96.6
	12				3.08	0	96.9	2.98	0	93.1	2.79	9.6	86.6			
	18	2.86	0	95.0	3.09	0	97.2							3.17	0	97.4
	24							3.06	0	95.6	2.90	2.7	90.1			

73 IP are shown in Fig. 3. For all cases at all temperatures, except at those temperatures at which maximum density was approached during the range of firing times, plots of %ρth vs. ln t approximate straight lines. The calculated apparent activation energies based on such data are approximately 70 kcal/mole for all mixtures except for 60 IP which was not determined. Relative rates of densification are illustrated in Fig. 4 by a comparison of 1660°C curves.

As expected, densification rates are fastest for 60 IP and slowest for 75 IP. Of interest is a comparison of the behavior of 73 compositions. It is seen that 73 IP densified faster than 73 PP which is attributed to its higher liquid content. It is believed that the higher

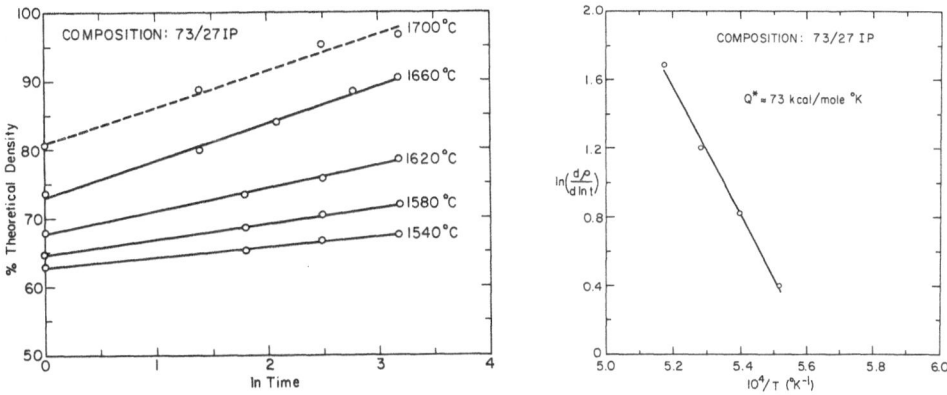

Fig. 3.   Plot of %ρth vs. ln t (left) and ln(dρ/dln t) vs. 1/T (right) for composition 73 IP.

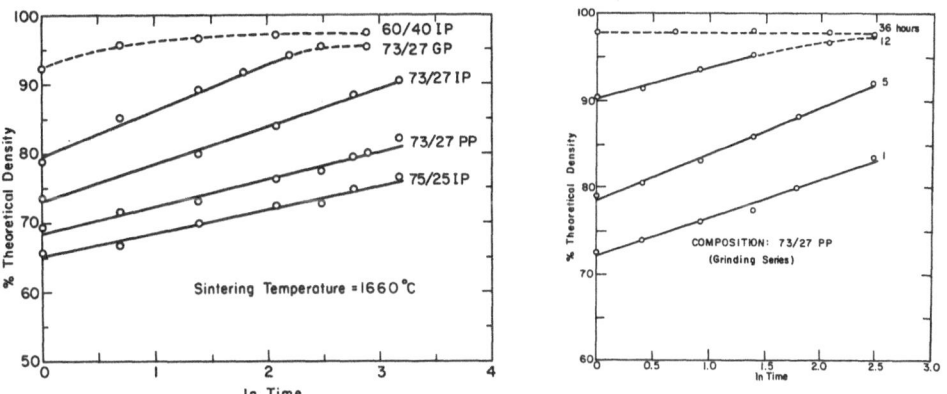

Fig. 4.   Plot of %ρth vs. ln t for a number of compositions at 1660°C (left).   Plot of %ρth vs. ln t for 73 PP ground for 1, 5, 12 and 36 h and fired at 1700°C in air (right).

Al_2O_3 contaminant in the latter reacted with some of the liquid.   A slightly higher total impurity content of the former specimen, 0.30 vs. 0.20%, and the higher surface area of the latter, 1.8 vs. 2.3 m^2/g, were not considered to be significant.   73 GP, however, densifies the fastest which is attributed primarily to the much higher surface area (~3.5 times greater) since its total impurity content was only 0.10%.

In accordance with sintering theory a proportionality between %ρth and ln t implies grain growth.[9]   Photomicrographs of 73 IP specimens fired at 1660°C for 1, 4, 12 and 24 h are shown in Fig. 5.   It can be

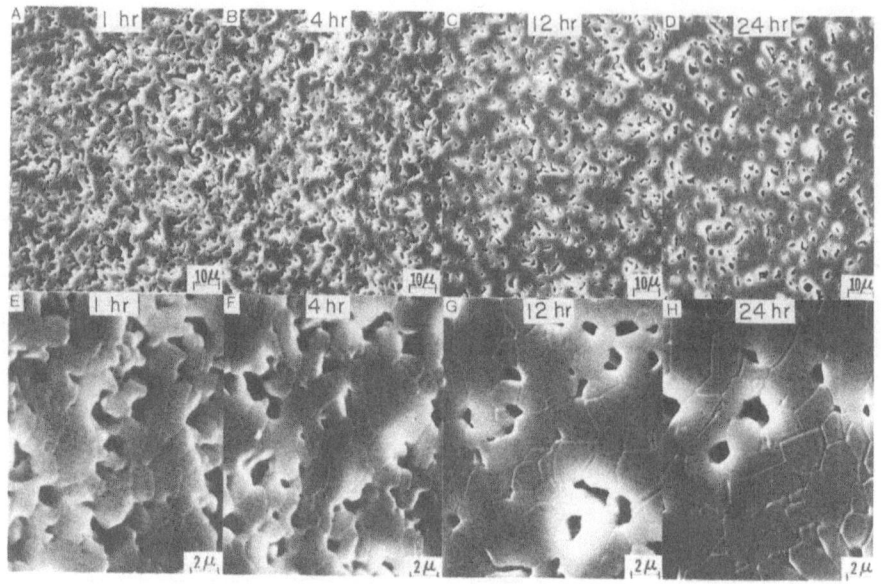

Fig. 5. Photomicrographs at two magnifications of 73 IP fired at
1660°C for 1, 4, 12 and 24 h in air.

seen that grain growth and a decrease of the number of pores have
occurred with time. This behavior was observed in all cases.

The powder equivalent to 73 PP, after calcining, was ground for
1, 5, 12 and 36 h. The surface area and green density are listed in
Table 3. The bulk densities, % open porosities, and %ρth after firing
at 1700°C for a number of times are also listed. Plots of %ρth vs.
ln t are shown in Fig. 4. Similar sintering behavior is observed as
before, but sinterability increased with increase in surface area.

DISCUSSION

In solid state sintering the surface areas of the particles
decrease (dA_{sv} is neg.) and the grain boundary areas forming at particle contacts increase (dA_{ss} is pos.). The free energy change for the
system at a given time is then equal to the instantaneous change in the
integrated interfacial energies and can be expressed by Eq.(1); as
long as δG is neg. sintering proceeds.[10,11]

$$\delta G = \delta \int \gamma_{ss} dA_{ss} + \delta \int \gamma_{sv} dA_{sv} \tag{1}$$

Therefore, a large γ_{sv} relative to γ_{ss}, or smaller ratio of γ_{ss}/γ_{sv},
is favorable. A minimum anisotropy of interfacial energies is also
favorable. Faster sintering in vacuum than in air could be attributed
to a favorable effect of this atmosphere on these factors. Lack of

Table 3. Effect of Grinding on Properties of 73/27 PP Composition
          Fired at 1700°C for a Number of Hours.

Firing, h	Grinding, 1 h			Grinding, 5 h			Grinding, 12 h			Grinding, 36 h		
	Bulk Density g/cm^3	% Open Porosity	% Theor. Density	Bulk Density g/cm^3	% Open Porosity	% Theor. Density	Bulk Density g/cm^3	% Open Porosity	% Theor. Density	Bulk Density g/cm^3	% Open Porosity	% Theor. Density
0	2.13	32.6	67	2.29	28.7	71	2.65	20.0	79	3.39	0	95
1	2.33	26.4	73	2.56	19.9	79	3.03	6.5	90	3.48	0	98
4	2.47	21.9	77	2.78	12.4	86	3.19	0	95	3.49	0	98
12	2.67	14.8	83	2.98	2.9	92	3.26	0.1	97	3.47	0	98
Unfired Compact, %pth	59.7			58.6			56.1			54.2		
Surface Area m^2/g	1.5			3.2			5.0			6.6		

thermodynamic equilibrium would lead to the presence of chemical potential gradients which are necessary for mass transport.

The structure of mullite leads to elongated prismatic crystals with anisotropy in surface energies since the surface structures are, and compositions can be, different for different crystallographic faces. This condition is emphasized in the presence of a liquid phase. Also, the presence of some liquid along grain boundaries in the 71.8 and 73 compositions permits the growth of larger crystals with straight edges in contact with several grains as seen in Fig. 1. In the absence of a liquid phase the anisotropy appears not to be as strong. In vacuum the apparent decrease of the γ_{ss}/γ_{sv} ratio can be due to a modification of the surface composition of the mullite crystals which would increase the surface energy. It has been observed that mullite surfaces decompose by losing SiO$_2$ according to Eq.(2) and leave a residue of Al$_2$O$_3$.[2,12,13]

$$SiO_2 = SiO + 1/2 \ O_2 \qquad (2)$$

The low partial pressure of O$_2$ in vacuum firing apparently also leads to an oxygen-deficient structure at the grain boundaries. This possibility is particularly true with the 71.8 to 74 compositions since thermal etching in air forms ridges at grain boundaries which can be explained on the basis of an increased volume on oxidation.

The large difference in sintering behavior in air between 74 and 75% Al$_2$O$_3$ compositions can be attributed to a liquid film at grain boundaries in the former specimen. The liquid film can provide a more favorable mass transport medium although the ln t relationship presumably indicates the same type of rate controlling step. Of interest is the fact that the film persists until the Al$_2$O$_3$ content of the composition is high enough to form free α-Al$_2$O$_3$, as in the 75 Al$_2$O$_3$ specimen.

A geometric analysis of sintering has indicated that the process can be divided into three stages: an initial stage with open porosity and no grain growth, an intermediate stage with open porosity and grain growth, and a final stage with closed pores.[10,11] When bulk diffusion is the rate controlling step during the intermediate stage, %pth is proportional to ln t as observed in this study. The extent of the initial stage is dependent on the range of particle sizes, agglomeration, and the degree of homogeneity in packing. In these cases the range and distribution of particle sizes is such that the initial stage of sintering appears to be over by the time the specimens reach or are at the test temperature for an hour.

It is known that finer grain sizes promote faster densification, but there is a question as to whether this behavior is due to the small size itself or some "reactivity" introduced into the particles by grinding in the form of defects. The grinding series behavior in this study can be attributed primarily to the size effect. In a model system composed of uniform size spheres, it can be visualized that the number of contacts in a unit volume will be determined by the size of the spheres. Small spheres thus will have many contacts from which material will be transported resulting in many more favorable short diffusion paths and faster sintering rates. The mass transport mechanisms, however, would still be the same. This argument can also be applied to a heterogeneous real system, such as this one, to account for the faster densification with increase in surface area by grinding.

CONCLUSIONS

Sintering of mullite bodies to near theoretical density can be realized by firing in air at temperatures above 1700°C for long periods of time. Enhanced densification can be achieved by utilizing completely reacted powders, increasing powder surface area, and sintering in vacuum. Mullite powders prepared by a gel process require lower temperatures for calcining and sintering. The presence of a liquid film along grain boundaries in the 71.8 to 74 wt% Al_2O_3 compositions affects the sintering and the microstructure.

This paper constitutes a report of preliminary studies that were made in an attempt to understand the factors and mechanisms that play a role in determining microstructure and kinetics of densification. Studies are continuing relative to "activity" of powders and their effect on sintering, phase equilibria under experimental conditions, and significance of thermodynamic considerations.

ACKNOWLEDGMENT

Grateful thanks are extended to the Division of Materials Research of the National Science Foundation for the grant in support of the study and to the Materials and Molecular Research Division of the Lawrence Berkeley Laboratory, supported by DOE, for use of laboratory facilities.

REFERENCES

1. P. C. Dokko, J. A. Pask, and K. S. Mazdiyasni, J. Am. Ceram. Soc. 60 [3-4] 150-55 (1977).
2. M. D. Sacks, (M. S. Thesis), LBL-6205, University of California Berkeley, 1977.
3. J. Grofcsik and F. Tamás, Publishing HOuse of the Hungarian Academy of Sciences, Budapest, 1961.
4. R. F. Davis and J. A. Pask, pp. 37-76 in High Temperature Oxides Part IV, edited by A. M. Alper, Academic Press, New York, 1971.
5. B. B. Ghate, (Ph.D. Thesis), Lehigh University, 1972.
6. R. A. Penty, (Ph.D. Thesis), Lehigh University, 1972.
7. K. S. Mazdiyasni and L. M. Brown, J. Am. Ceram. Soc., 55 [11] 548-52 (1972).
8. I. A. Aksay and J. A. Pask, ibid., 58 [11-1]] 507-12 (1975).
9. R. L. Coble, J. Appl. Physics, 32, 787 (1961).
10. J. A. Pask, C. E. Hoge, and B. Wong, pp. 246-254 in Ceramic Microstructures '76, edited by R. M. Fulrath and J. A. Pask, Westview Press, Boulder, Colorado, 1977.
11. C. E. Hoge and J. A. Pask, Ceramurgia International, 3 [3] 95-99 (1977).
12. R. F. Davis, I. A. Aksay and J. A. Pask, J. Am. Ceram. Soc. 55 [2] 98-101 (1972).
13. V. M. Grosheva, V. M. Pansevich, and V. Vu Boichun, Steklo i Keram. 25 [3] 35-36 (1968).

DISCUSSION

J. L. Pentecost (Georgia Tech): Did you have any problems with mullite formed from the gels? To give you an example of this problem: about 1970 Davison Division of W. R. Grace prepared about 10 kg of Al_2O_3-SiO_2 gel of mullite composition. This material was of high purity, less than 50 ppm Na_2O and other common impurities at a similar level. It was provided lightly calcined with about 200 M^2/gm surface area. Even after calcining at $1500^\circ C$ for 8 hr we were unable to achieve mullite grain growth in the calcine; the surface area remained above 40 M^2/gm, and the X-ray pattern remained diffuse with only broad brands for several major mullite peaks. Since we were unable to process the fluffy material ceramically, further work was abandoned.

Questions surrounding such metastability remain. It suggests that the presence of certain common impurities are essential for mullite crystallization or perhaps that some liquid phase is essential for mullite crystallization.

RATE CONTROLLED SINTERING AS A PROCESSING METHOD

M. L. Huckabee, T. M. Hare and H. Palmour III

North Carolina State University

Raleigh, North Carolina 27607

INTRODUCTION

Rate controlled sintering (RCS) has been under development at this University since 1965. It has emerged as an optimizable laboratory procedure which offers a rational alternative to the up-and-hold conventional temperature sintering (CTS) profiles commonly employed for the densification of kiln-fired crystalline ceramics.

Development of rate controlled sintering concepts, empirical optimization of a widely applicable, near-optimal three-stage density-time profile, related improvements in particulate processing prior to firing, and resultant microstructural refinements have been described elsewhere.[1-9] The roles of the different stages of the near-optimal three staged densification path in collectively influencing pore and grain size distributions, and the relative kinetics of densification and grain growth, have also been considered.[10,11] In this paper, we examine rate controlled sintering from the process viewpoint, both as it is presently practiced in the laboratory and as it scales up to conventional production kilns.

HOW RATE CONTROLLED SINTERING DIFFERS

Both the precedents of the past (implicit in the lore and in the literature) and the accessability and convenience of conventional thermocouple-based control instrumentation have conditioned most ceramists to think of optimizations of the firing process in terms of "time-at-temperature". In this "traditional" view of the CTS path, it is only during the soak time at the top temperature that the desirable characteristics of the ware being fired are developed.

205

The heating rates employed in reaching the soak temperature are generally considered to have only secondary importance, e.g., as being capable of influencing drying and binder burnout conditions, kiln furniture and refractory endurance, fuel consumption and overall kiln economics and the like. More often than not, the effects of the nonisothermal portion of the temperature-time path selected on the progress of the densification process itself have remained largely unknown, since in the usual CTS optimization of a firing process, only the final density and microstructure are considered.

Figure 1a represents such an up-and-hold curve initially developed in an industrial laboratory for chemically prepared 0.1% MgO doped alumina. It calls for a heating rate of ~8°C/min, with a 1 1/2 hr soak at 1550°C, yielding a fractional density ⩾0.99. Figure 1c illustrates the resultant (dependent) S-shaped density-time curve determined by dilatometry in our laboratory. Densification proceeds rapidly through the initial and intermediate stages but slows appreciably above D~0.90, even before the soak temperature is reached. The rate of densification during the soak period is still slower, and as a consequence, the volume change accomplished during the soak period is modest at best. This kind of firing profile, though relatively simple to achieve, has been shown to have undesirable overall consequences, since it facilitates (a) entrapment of occluded gases, (b) entrapment of pores within grains, and (c) excessive grain growth (see Figs. 3a, 3c).

By contrast, above the onset of shrinkage the independent variable for RCS firing is densification rate itself, i.e., the slope of the density-time profile (Fig. 1b). The near optimal density-time path developed[4]-- and patented[9]-- for high purity MgO-doped alumina comprises three different rate regimes: (1) a relatively fast initial linear region to $D = 0.75$ ($dD/dt = k_1$) (2) a significantly slower linear region $D = 0.85$ ($dD/dt = k_2 = \sim 1/3\ k_1$), and (3) a log decreasing region ($dD/dt \propto k_3/t$; $D \propto \ln t$) above $D = 0.85$. By means of feedback-controlled dilatometry, the sintering specimen is able to call for its own power needs in maintaining the preprogrammed density- (or shrinkage-) time profile. The resultant (dependent) temperature-time profile is commonly non-linear, and may well include negative slopes, as shown in Fig. 1d. Though typical of RCS firings for many materials, the reader is cautioned that any such actual curve is quite specific to a given material and process.

In RCS firing, densification proceeds at the lowest temperature which will maintain the desired density-time profile. With its controlled progression from fast to slow to still slower rates, the preplanned RCS density-time profile provides a much better opportunity than CTS for thorough outgassing during deliberately slow traverse of the permeable intermediate stages of densification.

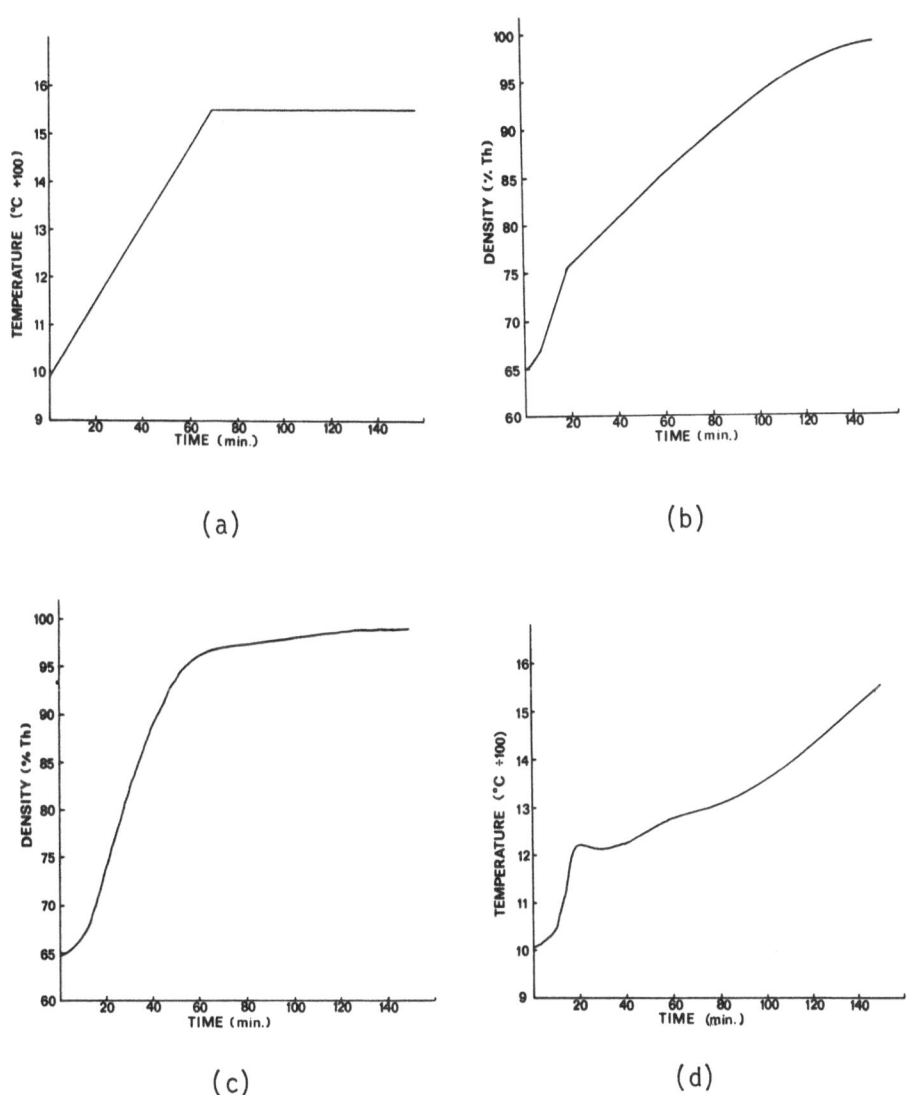

Figure 1.   Comparisons between CTS and RCS firings for chemically
            prepared 0.1% MgO doped alumina at 0.65 green density: (a)
            CTS temperature-time control profile, (b) RCS density-time
            control profile, (c) resultant CTS density-time profile and,
            (d) resultant RCS temperature-time profile.

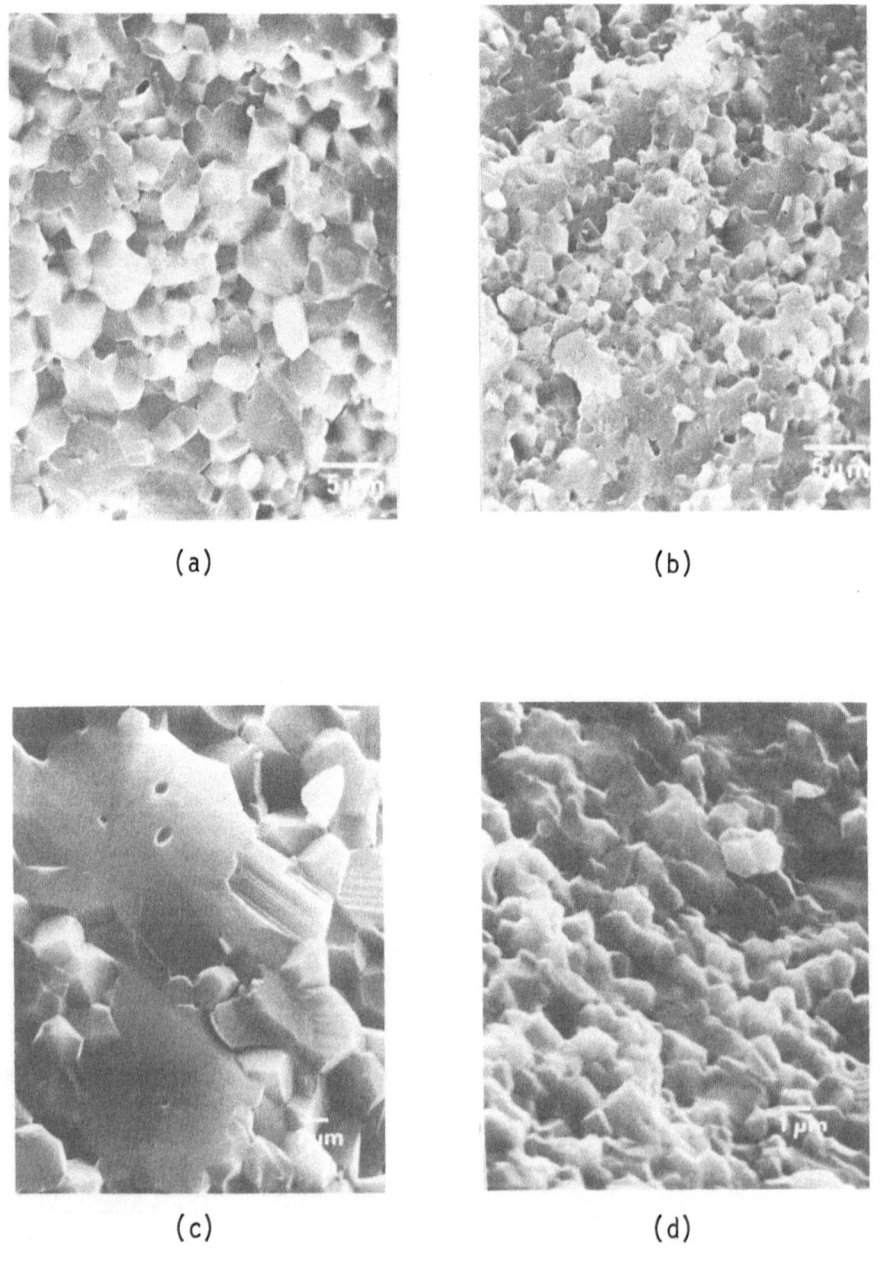

(a)   (b)

(c)   (d)

Figure 2.   Microstructural comparisons for 0.1% MgO doped alumina
            fired by different methods; (a) CTS, recent processing;
            (b) RCS, recent processing; (c) CTS, early processing; and
            (d) RCS, early processing.

Throughout the RCS firing, temperatures are kept as low as possible,
thus minimizing thermal excitation of undesirable transport processes which may accompany, but do not necessarily aid densification
(e.g., diminution of surface, pore coarsening and grain growth[10,11]).
The microstructures obtained by RCS firing are finer and much more
uniform than those characteristic of CTS firings (Fig. 2). In the
alumina examples illustrated, trapped pores within grains are
evident in the CTS specimens, but have been almost completely eliminated by RCS firing.

Figure 2 also provides two examples of the interactive nature
of particulate processing and subsequent thermal processing. The
alumina material was Kemalox KA-210*, with two very similar lots
being represented. The first lot, shown in the comparison 2c, 2d at
D ~0.994, was prepared early in our study by an existing industrial
laboratory procedure which involved wet milling in an organic
vehicle, with a Carbowax binder addition. The processed material
had a binder-free green density near 0.55. When fired in the CTS
mode, it displayed a marked tendency toward exaggerated grain growth
(Fig. 2c) which could be very effectively suppressed by firing in
the RCS mode (Fig. 2d). The second lot, shown in the comparison
2a, 2b at D = 0.99, was processed several years later by an improved
dry process[7,8] described in a following section; it had a binder-free
green density of ~0.65. In this instance, better particle compaction
helped reduce the tendency toward exaggerated grain growth in the
CTS mode (Fig. 2a). Even so, the RCS mode brought about significant
microstructural refinement (Fig. 2b).

HOW RATE CONTROLLED SINTERING IS ACCOMPLISHED IN THE LABORATORY

Specimen Selection and/or Preparation

For the dilatometric studies described here, relatively small,
easily characterized specimens are required; for a typical optimization study 25-30 such specimens having minimum variability may
be needed. They may be selected and cut from larger units, in
which case they must be treated as representative "volume elements"
of the larger whole. Since many forming methods impart a definite
orientation to the particles, causing anisotropy which in turn influences both linear shrinkage and physical properties, it is
important that the selection method translate that same sense of
orientation to the small dilatometer specimens. Commonly, auxiliary
experiments must be performed to establish these linear shrinkage
anisotropies and their respective relationships to volume shrinkage.
The feasibility of this approach has been successfully demonstrated
with "volume element" samples obtained directly from industrially
processed materials, shapes and sizes as diverse as building bricks,

* Product of W. R. Grace & Co., Washington Research Center, Columbia, Md

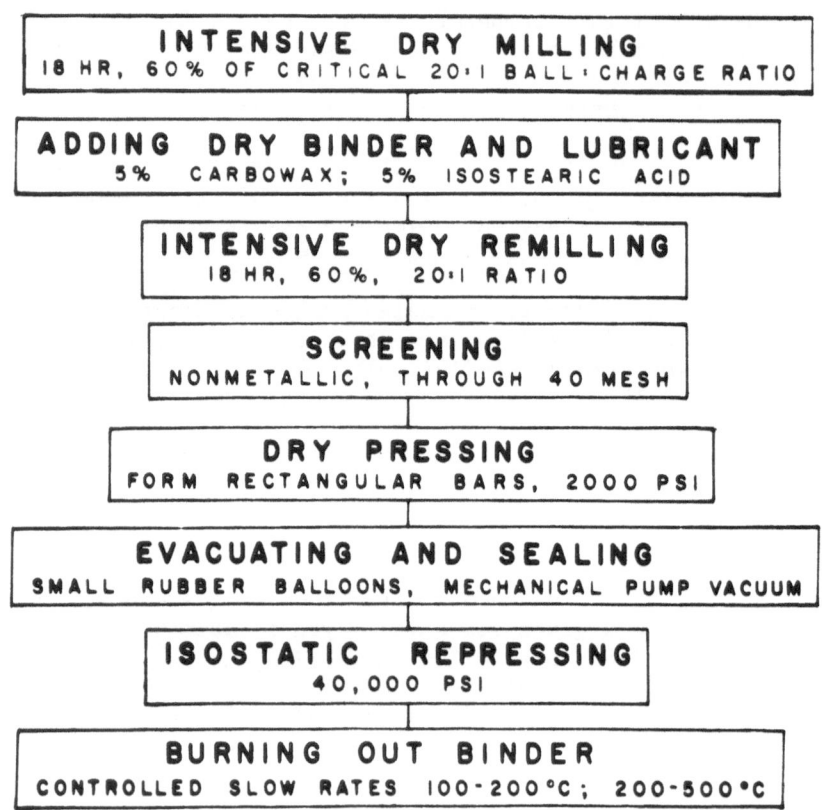

Figure 3. Flow diagram for optimized processing of alumina prior
to firing.

quarry tile, ferrite permanent magnets, and very thin tapecast high
dielectric capacitors (including multilayer types).

In other cases, it is possible to begin with the original particulate materials, allowing one to optimize both particle processing
and sintering steps as interactive components of an overall process.
This has been the case with much of our own systematic research with
alumina, where dry particle processing prior to firing has been
carefully optimized (Fig. 3) to achieve (a) uniform dispersion of
ultimate particles, (b) consistent mixing of binder and lubricant,
(c) good compaction, and (d) controlled burnout of organic additives.[7,8]
Conventionally processed, chemically prepared aluminas normally yield
fractional green densities near 0.50, but when processed in accordance
with the flow diagram of Fig. 3, the binder-free green density is
increased to 0.61-0.65, depending on particle characteristics of the
specific starting materials. At these higher green densities, the
coordination number is large, typically 7 contacts per particle for

D = 0.61, and 8 per particle for D = 0.65.[12] This improved uniformity
of compaction has a significant influence on sinterability, being
particularly evident in the RCS mode, where temperatures required to
attain given density levels with well processed material (D > 0.60)
are lowered by ~50°C in comparison to poorly processed material
(D~0.50).[7]

Dilatometry

Feedback-controlled dilatometry is the principal tool employed
in laboratory studies aimed at development of near-optimal RCS
profiles. Our dilatometers are of the vertical type and are
"homemade". It should be noted that at least two different
commercial dilatometer units, both horizontal, have been marketed as
being suitable for rate controlled sintering. The basic components
of our original one, still in use, were converted from another
application (TGA). Figure 4 schematically illustrates them: (a) a
small, watercooled, Pt-wound furnace, (b) c-axis-oriented sapphire
rods and sapphire stage which constitute the "hot end" of the
dilatometer, (c) the small specimen with its load-spreading alumina
shims, (d) a base plate and dove-tailed dilatometer support structure,
(e) displacement sensor (LVDT), and (f) a precision dial guage used
during set-up and calibration. In terms of components, the associated
instrumentation is quite conventional: the recorder-controller is of

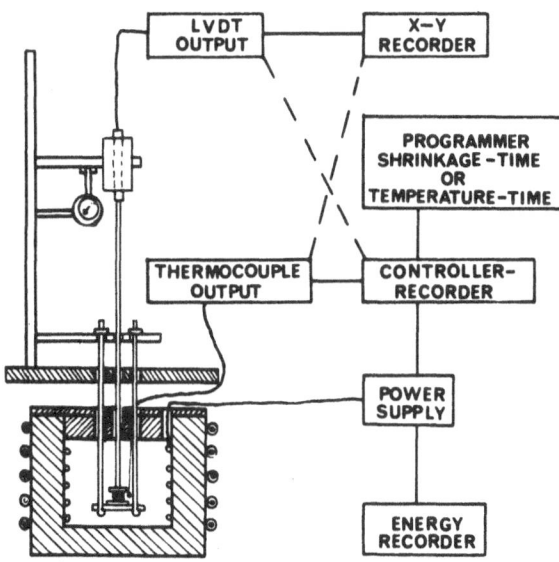

Figure 4.   Schematic illustration of basic dilatometer components
            for rate controlled sintering.

the exchangable range card, potentiometric type, with proportional
band, rate time and reset capabilities; the programmer is of the
curve-following, scribed conductive sheet type. Appropriate switches,
not illustrated, connect the recorder-controller to either the
thermocouple output (CTS mode, solid line input) or the LVDT output
(RCS mode, dashed line input) with due regard for proper polarity;
they simultaneously shift the displaced (now dependent) signal to
the x-y (or other suitable) recorder. For use in energy conservation
studies (to be reported elsewhere) the unit has also been equipped
with an optional power measuring/recording system. This system is
limited to specimens ~1.6 cm dia x ~1.6 cm high, and generally to
temperatures ⩽1550°C.

To minimize a variety of nuisance variables and/or experimental
limitations inherent in the older, retrofitted unit, a new dilatometer
has been designed and constructed in University shops. It features
(a) an air cooled, fast response furnace (1700°C max, using $MoSi_2$
elements and fibrous zirconia lining), (b) longer sapphire support
rods and larger sapphire stage to accommodate larger specimens, and
(c) a water cooled base plate with upper dilatometer components
constructed of Invar to minimize extraneous thermal drift. Its
instrumentation and power supply components are novel and largely
"homemade"; they provide for very rapid response, and also permit
direct programming and control of densification rate per se.

Data Analysis

Figure 5 schematically illustrates the input of data from dilatometry and auxiliary experiments to a computer-based data reduction
scheme developed and refined during almost a decade of rate controlled
sintering research. Temperature and shrinkage data are initially
extracted (typically at 5 min intervals) from recorder traces, and
are combined with dimensional and density measurements and data from
auxiliary experiments on preprinted "run sheets". The collected
data are transcribed to punched cards and fed along with a FORTRAN-
programmed control deck (which includes known instrument correction
factors in the form of frequently reviewed adjustable "constants")
to a local terminal for a large time-sharing computer (IBM 360,
Model 75). The output (not illustrated) is printed in standard
format containing all appropriate identification information and
the reduced data for each increment of time, including temperature,
specimen length, fractional shrinkage, fractional density, densification rate, heating rate, power level, and cumulative energy
required. From these printouts, the density-time and temperature-
time profiles routinely used to compare firings can be plotted, or
inputs can be extracted for other more complex computer analyses,
e.g., for densification kinetics (See Fig. 4, Ref. 10).

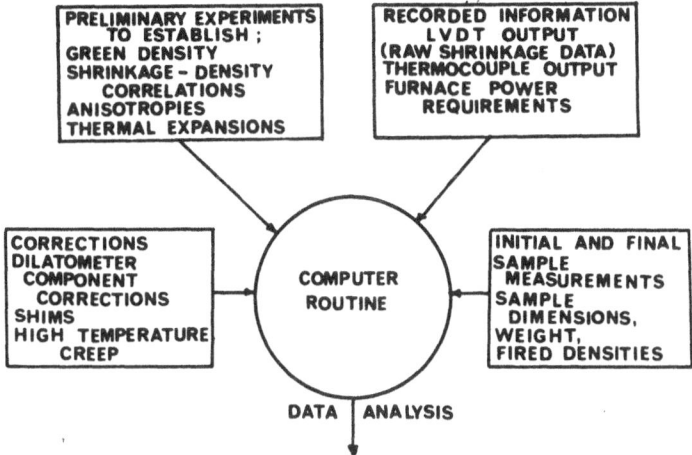

Figure 5. Data sources and analyses.

Optimization

Empirical optimization of RCS profiles has been discussed in part elsewhere;[1,5,8,10] it is an engineering "art form" based primarily upon integration of (a) systematically varied dilatometer studies in the RCS mode together with selected control firings in the CTS mode, (b) supporting microstructural analyses, typically by SEM fractography, and (c) related end-use property data (e.g., bend strength, dielectric strength, etc.). The near-optimal density-time profile originally developed for alumina[9] (Fig. 1b) appears to have certain "canonical" properties, because it also works well with only minor variations for a number of other rather dissimilar ceramic materials (e.g., $MgAl_2O_4$, $BaTiO_3$-$CaZrO_3$, structural clay products, etc.). Therefore, in a given situation one works to optimize both the specific shape of this generally applicable density-time curve and its time base (which combines with preheat and cooling portions to equal total firing time)[9] in achieving a finer, more uniform microstructure (or any other definable, rate-related technical objective) which in general, must also be capable of yielding "equal or better" end-use properties.

Process and Quality Control

RCS dilatometry provides a sensitive means for characterizing differences in sintering behavior attributable either to starting material (e.g., lot-to-lot variations) or to processing (e.g., changes in milling procedures, binders, lubricants, etc.).[5,7,8] For a given process, product and set of end-use properties, a properly optimized density-time path tends to remain almost invariant. Therefore, the corresponding RCS temperature-time profile provides a unique firing trace, or "fingerprint", over the entire shrinkage regime; such a "fingerprint" can be obtained in a single run for a given material processed in a given way. In general, the more active the material and the better the processing prior to firing, the lower the temperature required to reach any given density level along the fixed density-time curve. By establishing appropriate limits of variation at successive density levels, RCS dilatometry data can be utilized in an almost routine way to (a) maintain quality control over incoming materials and/or (b) establish and maintain process control over particulate process steps which precede firing. If, on the other hand, a given lot or process step falls outside the established norm, the RCS temperature-time trace can be used directly in bringing about appropriate changes in the firing profile which may be required to compensate for the abnormal variation.

Scaling Up

From Fig. 1, it is evident that there is an exact independent-dependent variable relationship between corresponding density-time and temperature-time paths for sintering of a given material; program according to one, obtain the other, and vice versa. This ability to exchange control variables was first demonstrated with laboratory dilatometers;[1,6,9] obviously, it also makes possible practical applications of rate controlled sintering in conventionally constructed, conventionally instrumented (but programmable) kilns and furnaces.[6,9] As an example, for alumina it has been shown that both the density and microstructure obtained with small "volume elements" in the RCS dilatometer can readily be reproduced in periodic gas kiln firings.[6]

In general, the larger the size of the ware (or assemblages of ware) being fired, the longer the total firing time must be to permit effective heat transfer, not only during sintering itself, but during preheat and cooling as well. For this reason, the RCS optimizations must be carried out over a suitably long time base, even though the small "volume element" specimen being used in the laboratory dilatometer could in fact be fired much faster. Thus, the dependent temperature-time profile must be generated with reference to an optimally shaped density-time profile having a time base appropriate to the scaled-up firing situation. By transcribing the resultant profile to the temperature-time programmer of a periodic kiln (or alternatively, to the temperature-length profile of a tunnel kiln), thereby transforming it to the independent control variable, one can recreate - and gain the benefits of - the initial laboratory rate controlled sintering conditions, but on a readily reproduced, much larger scale.

REFERENCES

1. H. Palmour III and D. R. Johnson, pp. 779-91 in Sintering and
    Related Phenomena. G. C. Kuczynski, N. A. Hooton, and C. F.
    Gibbon, Eds., Gordon and Breach, Science Publishers, Inc.,
    New York, 1967.

2. R. A. Lawhon, M. S. Thesis, N. C. State University, 1970.

3. D. R. Johnson, Ph.D. Dissertation, N. C. State University, 1970.

4. M. L. Huckabee and H. Palmour III, Amer. Ceram. Soc. Bull.
    51 (7) 574-576 (1972).

5. H. Palmour III and M. L. Huckabee, pp. 275-282 in Sintering and
    Related Phenomena. G. C. Kuczynski, Ed., Mat. Sci. Research,
    Vol. 6, Plenum Press, New York, 1973.

6. H. Palmour III, T. M. Hare and M. L. Huckabee, Final Technical
    Report, Contract N00019-73-C-0139, Naval Air Systems Command,
    March 1974.

7. T. M. Hare and H. Palmour III, pp. 307-320 in Ceramic Processing
    Before Firing. L. L. Hench and G. Onoda, Jr., Eds., John
    Wiley & Sons, New York, 1978.

8. H. Palmour III, T. M. Hare, Final Technical Report, Contract
    N00019-74-C-0265, Naval Air Systems Command, July 1975.

9. H. Palmour III and M. L. Huckabee, U. S. Pat. 3,900,542,
    Aug. 19, 1975.

10. H. Palmour III, M. L. Huckabee and T. M. Hare, pp. 308-19 in
    Ceramic Microstructures '76. R. M. Fulrath and J. A. Pask
    Eds., Westview Press, Boulder, Colorado, 1977.

11. H. Palmour III, M. L. Huckabee and T. M. Hare, "Rate Controlled
    Sintering: Principles and Practice", Presented at IV
    International Round Table Conference on Sintering,
    September 5-10, 1977, Dubrovnik, Yugoslavia. To be published
    in Conference Proceedings.

12. R. M. German and Z. A. Munir, pp. 249-267, in Sintering and
    Catalysis. G. C. Kuczynski, Ed., Mat. Sci. Research Vol. 10,
    Plenum Press, New York-London, 1975.

A MULTIPLE-LOGNORMAL MODEL OF NORMAL GRAIN GROWTH

F. M. A. Carpay and S. K. Kurtz*

Phillips Research Laboratories

Eindhoven, The Netherlands

INTRODUCTION

In the literature, grain growth models of two kinds are found: (a) "single grain" descriptions; grain growth is described for a single, freely growing grain, where the grain boundary movement resulting from the boundary curvature causes the change of the grain size. Simple arguments show that the diameter of a single free grain varies with the square root of the time. Sometimes with argument, but generally without, it is concluded from this single-grain behavior that the average grain size shows the same time dependence. Even modern textbooks[1] treat grain growth in this way, and (b) statistical models;[2-6] some authors have introduced the effect of the distribution of shapes and sizes. Smith[2] suggested a natural tendency of these distributions to be fixed. Assuming a lognormal distribution of shapes as well as of sizes, Feltham[3] was the first to present a quantitative statistical description of the grain growth process.

In this paper we will briefly outline a phenomenological, descriptive model of normal grain growth, based on an extension of the Feltham theory. A more extensive treatment of the model together with a comparison with the other existing models will be published elsewhere.[7] Here the emphasis will be placed on the experiments supporting the model proposed.

---

* Resident, permanent address: Philips Laboratories, Briarcliff Manor, New York, 10510.

THE MODEL

We will deal with three distinct types of parameters: dimension-less topological parameters (number of faces, number of edges), size parameters with the dimensions of length (diameters) and structural parameters which interrelate the first two types of parameters (grain boundary curvatures).

List of the most important symbols

D = three-dimensional equivalent spherical diameter.
d = two-dimensional equivalent planar diameter = s/π, where s is the perimeter of the grain.
F = number of faces of a grain (three-dimensional).
n = number of edges of a grain intersection (two-dimensional) counted experimentally as the number of neighbors; = number of the class to which the grains with n edges belong.
d_n = d in class n
σ_D, σ_d, σ_n, σ_{d_n} = standard deviation of respectively lnD, lnd, lnn and lnd_n.
N = total number of grains.
Subscript "med" stands for "median", defined in statistics.

Postulates

Starting from Feltham's model[3] we make a set of eight postulates which form the basis of the multiple-lognormal model:

Postulate 1. The three- and two-dimensional size parameters of the overall grain populations are lognormally distributed.

Postulate 2. The populations of the different three-and two-dimensional topological classes are lognormally distributed.*

Postulate 3. The grains within each distinct three- and two-dimensional topological class are distributed lognormally in their size parameter.

Postulate 4. In each of the above cases (1, 2, 3) the standard deviation of the logarithm of the variate is a constant independent of time.

Postulate 5. The two-dimensional topological class parameter n can be directly carried over to the three-dimensional case (parameter F) and used in lieu of F in developing the theory.

---

* Note added in proof: because of the discreteness of this distri-bution the mathematical similarity with Postulate 1 is not com-pletely correct; a better approximation would be the distribution of $(n + 1/2)$ instead of n; this will not be worked out here, but in another publication.[7]

Postulate 6. Grain growth occurs owing to the collapse of grains
in the lowest topological class accompanied by a
discontinuous transfer of grains (in time) from
the next higher to the next lower class.

Postulate 7. There is a power law relationship between the median
value of the size parameter of a specific topological class and the value of the topological parameter
for that class.

Postulate 8. The ratio of the equivalent spherical (or planar)
grain diameter to the mean radium of curvature,
averaged over the surface (or edge) of a grain is
a linear function of the logarithm of the topological class number F (or n) to which the grain
belongs.

Growth Equations

   With an iterative computer solution making use of an error
function (erf) subroutine the number of grains lost from the lowest
two-dimensional class ($n = 3$) in a fixed time interval Δt are calculated, while holding the grain size distribution for this class
constant. Then the new value of $N_3(t + \Delta t)$ is re-entered, to get
a new N_3 at which the iteration starts anew. Because the number
of grains lost from the lowest topological class is equal to the
decrease of the total number of grains in the same period of time
(Postulate 6) the variation of N and thus of d_{med} ($t = 0$) the values
of d^2_{med} computed are plotted in Figure 1. The curves in the d-plot
for the different "iteration" areas are slightly S-shaped as the

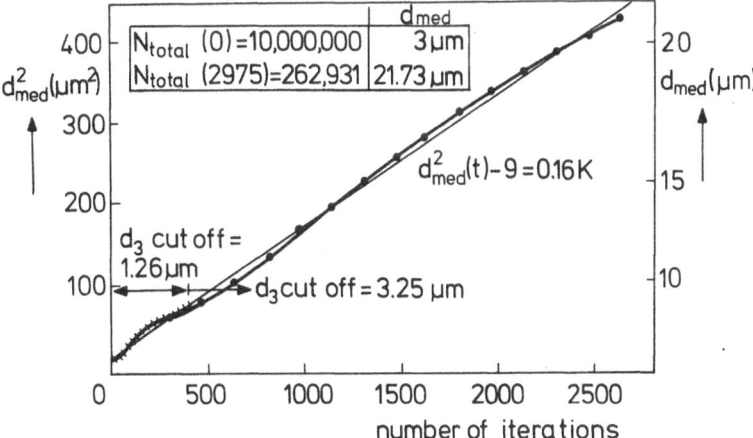

Figure 1.   Computer calculation of d^2_{med} as a function of the number
of iterations, which is equivalent to the time.   Two
iteration areas are used, resulting in a variation of 6.667
in the time scale.

value of d_{med} increases. Decreasing these areas (which costs much more computer time) straightens the curve and the fit approaches the equation.

$$d^2_{med} (t_2) - d^2_{med} (t_1) \text{ constant } (t_2 - t_1) \qquad (1)$$

With Postulate 7 this can also be written for the diameters in each topological class and using the two- to three-dimensional Schwarz-Saltikov transformation technique[8] this results into a general square law behavior for the three-dimensional diameter. Only the constants in the three distinct cases are different.

EXPERIMENTS

Procedure

The experiments were made on high-density nickel-zinc ferrites with composition $Ni_{0.36}Zn_{0.64}Fe_2O_4$. The samples were prepared by prefiring the pure starting materials $NiCO_3$, ZnO and Fe_2O_3 for one hour at 900°C in oxygen, followed by milling and deagglomerating, dry pressing and finally sintering at 1250°C (sample S8 at 1230°C). Table 1 gives the sample numbers and sintering times, and, for some samples, the end porosity.

Table 1.   Sintering Times

Sample numbers	S17	S18	S19	S20	S21	S22	S8
Sintering time (hrs)	1/4	1/2	1	2	4	8	48
End porosity (%)	--	1.0	--	0.9	0.4	--	0.1

All samples were cut, polished and thermally etched. Data on grain sizes were collected using a Kontron MOP-AM-01 semiautomatic image analyzer.

Results

Most of the experimental results are presented in this section in the form of tables or figures. Table 2 gives the results on the overall distribution of grain sizes for two- and three-dimensions and the two-dimensional topological parameters. The different standard variations are also included in the table.

Table 2. Overall Distribution

Sample number	Anneal- ing time (hrs)	Total number counted	Two dimensions		Three dimensions		Two dimensions*	
			d_{med} (µm)	σ_d	D_{med} (µm)	σ_D	n_{med}**	σ_n **
S17	$\frac{1}{4}$	2054	5.1	0.40	5.9	0.33	5.20	0.28
S18	$\frac{1}{2}$	2013	5.4	0.43	6.3	0.34	5.20	0.29
S19	1	2105	7.2	0.41	8.0	0.34	5.20	0.28
S20	2	2028/1284*	8.6	0.44	9.9	0.34	5.15	0.30
S21	4	2740/1440*	11.1	0.45	11.6	0.40	5.05	0.30
S22	8	2923/1329*	15.6	0.43	17.1	0.38	5.12	0.30
S8	48	1767/1135*	25.3	0.41	27.3	0.38	5.17	0.30

* The total number of grains counted for the n_{med} determinations is the second number given.

** Correction according to the postulate 2 footnote increases n_{med} and decreases σ_n both with about 10%.

Figure 2 shows for one sample (S22 as an example) the two-and three dimensional cumulative frequency of grain sizes on probability paper. Plots of this kind were used to calculate the median diameter and the standard deviation, while the final values were also obtained using a computer program. This program provided us with the Kolmogorov statistic as a check on the goodness of fit, which was generally above the "95% confidence" level, indicating that the lognormal presentation was permitted. Figure 3 gives the two- and three-dimensional mediam diameter squared as a function of time.

Table 3 summarizes the results on the two-dimensional diameter data of all samples (except S17) broken into classes, as well as their standard deviations.

Figure 4 shows for one sample (S19 as an example) the two-dimensional graphical plot of the cumulative frequency of grain sizes in the different classes on probability paper together with the cumulative frequency of the number of grains in the classes (topological parameter n). Figure 5 gives the two-dimensional median diameter (squared) for each class as a function of time.

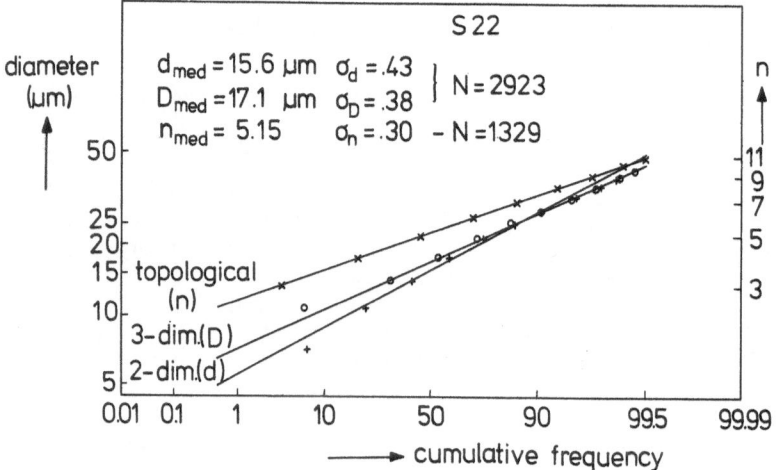

Figure 2. Cumulative frequency plot of the two- and three-dimensional
            diameters and of the topological parameter n for the
            sample S22.

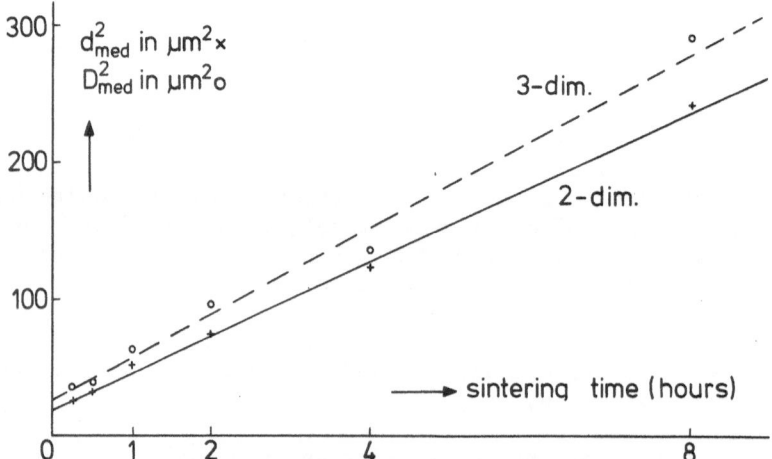

Figure 3. Experimentally determined two- and three-dimensional
            mediam diameter (squared) of the overall distributions
            as a function of sintering time.

Table 3. Two dimensional data per class

The values of d are mediam parameters given in µm and all
σ-values are standard deviations of the logarithm of these
d_n's (σ_{dn}).

Sample number	$n = 3$ d	σ	$n = 4$ d	σ	$n = 5$ d	σ	$n = 6$ d	σ	$n = 7$ d	σ
S18	2.5	0.35	3.1	0.30	4.4	0.27	5.7	0.23	7.0	0.21
S19	3.3	0.28	4.2	0.34	5.5	0.31	7.5	0.26	9.2	0.22
S20	3.7	0.32	5.2	0.30	7.3	0.27	9.3	0.24	11.5	0.22
S21	4.8	0.41	6.8	0.37	9.0	0.35	12.1	0.27	15.2	0.26
S22	6.7	0.38	9.5	0.40	13.0	0.34	16.9	0.30	21.3	0.27

Sample number	$n = 8$ d	σ	$n = 9$ d	σ	$n = 10$ d	σ	$n = 11$ d	σ
S18	8.8	0.20	10.0	0.19	11.7	0.19	13.4	0.18
S19	10.8	0.21	13.4	0.17	15.0	0.17	16.6	0.17
S20	13.2	0.20	16.7	0.20	18.4	0.16	20.0	0.14
S21	18.5	0.25	21.9	0.20	28.1	0.18	27.2	0.18
S22	25.2	0.25	30.2	0.22	32.9	0.22	39.1	0.18

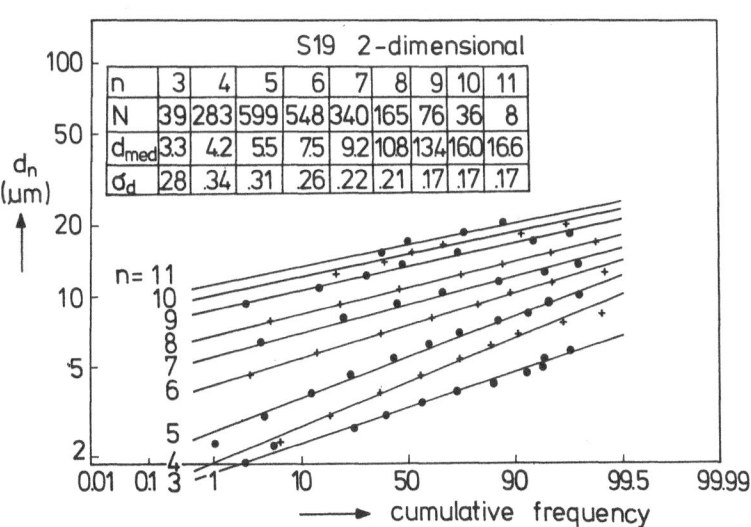

Figure 4. Cumulative frequency plot of the two-dimensional diameters
divided into classes for the sample S19.

DISCUSSION

Lognormality

The first three postulates concern the several lognormal distributions. As already mentioned in the introduction, lognormal distributions of grain sizes and numbers of sides have been reported before. We have also found, and have illustrated in Figures 2 and 4, that the lognormal distribution indeed fits very well the size and topological parameters of overall distributions as well as of distributions divided into classes.

The two- to three-dimensional transformation

Transformation of the two-dimensional overall size distribution to the three- dimensional one has been carried out using the Schwarz-Saltikov transformation.

As shown in Table 2 and in Figure 3 there is a slight change of the parameters: the three-dimensional diameter is somewhat larger,

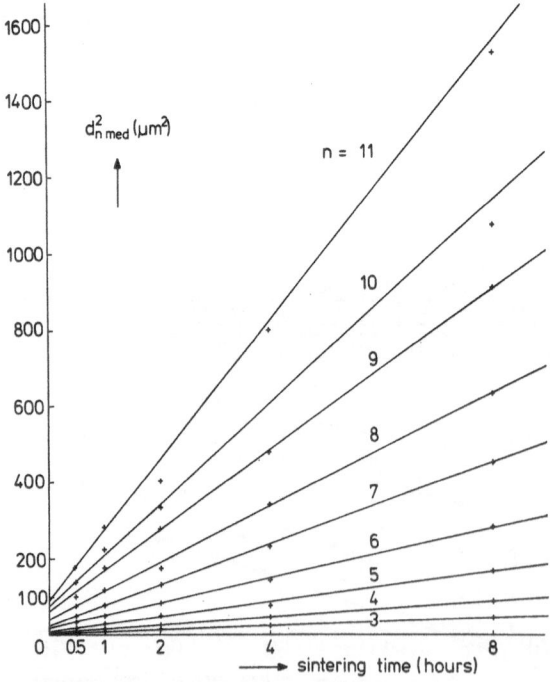

Figure 5.   Experimentally determined two-dimensional median diameter
            (squared) for each class as a function of sintering time.

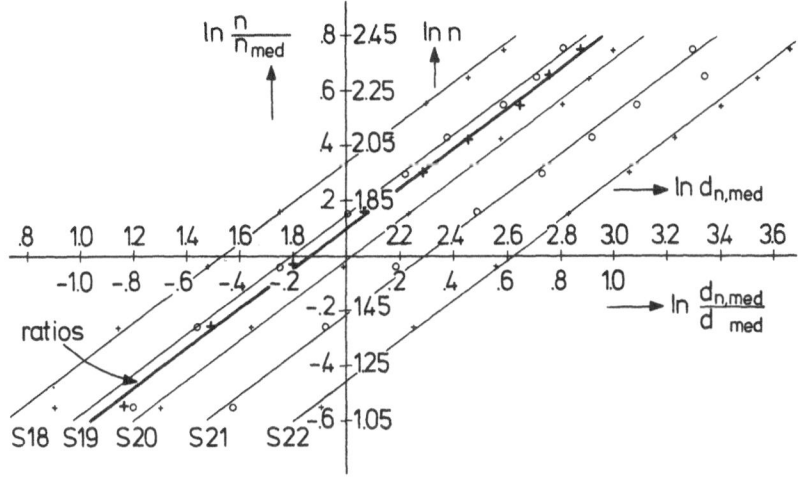

Figure 6.   A plot of $\ln d_{n,med}$ versus $\ln n$ of the different samples and a plot of the logarithm of the ratios: $\ln(d_{n,med}/d_{med})$ versus $\ln(n/n_{med})$ averaged over all samples. All curves show the same slope: $b = 1.4$.

while the value of σ decreases on transformation.   The small increase of the diameter can be explained using Spector's[11] analysis with the experimentally found σ.

Division into classes

According to postulate 7 the median diameter of a distinct class can be related to its topological parameter.   Figure 6 shows the results of Table 3 in the form $\ln d_{n,med}$ versus $\ln n$.   The result is $d_{n,med} = an^b$ with $b = 1.4$.   The constant a is a function of d_{med} and thus a function of time.   According to the same postulate $d_{med} = an_{med}^b$, which makes the following relationship hold:

$$\frac{d_{n,med}}{d_{med}} = \left(\frac{n}{n_{med}}\right)^b \tag{2}$$

A plot of these data, taken from Table 6, also gives $b = 1.4$ as shown in the same Figure 6.   (It should be noted that the curve does not go through the origin; it does when taking $n + 1/2$ instead of n, as reported above.)

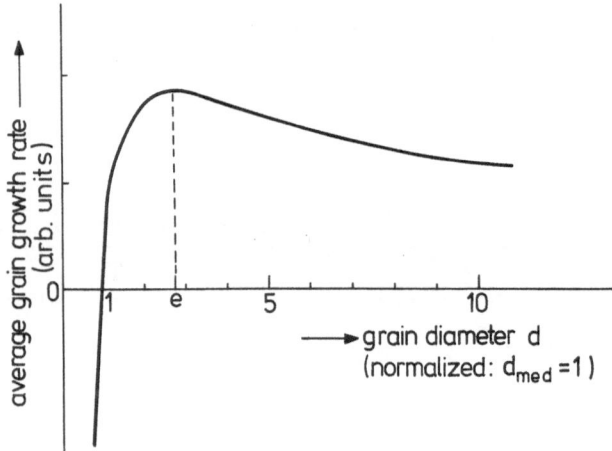

Figure 7.   Average grain growth velocity at a fixed instant of time
            (closed class: no grains enter or leave the class) as a
            function of grain diameter.

The maximum grain size

     With the postulates 7 and 8 the following average growth rate
of grains in a distinct class at a fixed instant of time (which means
that no grains enter or leave that class) can be derived

$$< \frac{d\, d_{n,i}}{dt} > = \text{constant } \frac{1}{d_{n,\text{med}}} \, ln \, \frac{d_{n,\text{med}}}{d_{\text{med}}} \tag{3}$$

     The relationship for fixed d_{med} is plotted in Figure 7.   The
curve shows a maximum at the corresponding value of $d_{n,\text{med}} = e d_{\text{med}}$.
which means that the average growth rate in class n with $d_{n,\text{med}}$
equal to $e d_{\text{med}}$ is the highest of the system.   As a result the values
on the curve to the right of the maximum are unrealistic in a steady
state system.   Therefore the ratio of the diameters of the maximum
size grains in a sample (with normal grain growth) to the mediam
grain size will never exceed e.

     For some samples we have measured the ratio taking for the
maximum grain size the median diameter of all grains with $n>11$.   The
values for S18, S21 and S8 were 2.46, 2.56 and 2.70 respectively.
The ratio tends to increase slightly when enough grains in the
ultimate class are counted.   The occurrence of the maximum in the
average grain growth curve for the largest grains implies a stability
criterion for normal grain growth.   The largest grains will always

grow the fastest. If for some reason a grain becomes larger than d_{max} its growth rate will become smaller than that of the grains with d_{max} and it will be overtaken.

The time-independent variables

According to postulate 4 the standard deviations of the several distributions should be independent of time. It can be seen from Table 2 that indeed σ_d, σD and σ_n are time-invariant. The same holds true for the σ_{dn} values as given in Table 3. Because the right hand term in Eq. (2) is time-invariant the left-hand term must be invariant too. Compilation of these data from Tables 2 and 3 shows that indeed the experimental $d_{n,med}/d_{med}$ value is independent of the sintering time.

The time-dependent variables

The most important time dependent variables are the median diameters. In Figure 3 the median diameters of the overall two- and three-dimensional distributions are shown, indicating a square law behaviour. (The value of the sample S8 has been omitted because of the lower sintering temperature. Correction of the parameter using an activation energy of 74 kcal/mole[12] raises the value almost exactly to the straight line. We feel, however, that this procedure is not very elegant and we will leave the sample out of consideration here. In fact a time interval up to 8 hours is already long enough.)

The growth laws from the plots are:

$$d^2_{med}(t) - d^2_{med}(t=0) = 27.5 \ t \ \mu m^2 \tag{3}$$

and

$$D^2_{med}(t) - D^2_{med}(t=0) = 32 \ t \ \mu m^2 \tag{4}$$

which is in complete agreement with the model. The intercepts indicate that at $t = 0$, $d_{med}(t = 0) = 4.4$ μm and D_{med} $(t = 0) = 4.6$ μm. It is clear that the value of the diameter at time zero cannot be omitted for the time intervals used. This starting grain size is greatly determined by the heating of the sample and the earlier sintering steps.

The data for the two-dimensional distribution divided into classes are plotted in Figure 5 in a similar way as for the overall distribution. Here too, the square law is obeyed:

$$d^2_{n,\text{med}}(t) - d^2_{n,\text{med}}(t=0) = C_{nt}$$ (5)

Combination of Eqs. (5) and (2) lets us deduce quite easily that

$$C_n = \text{constant} \left[\frac{n}{n_{\text{med}}}\right]^{2b}$$ (6)

where the constant is the one found experimentally to be 27.5 μm^2/hr (Eq. (3)). Comparison reveals that the slopes from Figure 5 show a nearly perfect fit to Eq. (6) when taking n_{med} = 5.18 (Table 2), b = 1.4 and using a minor correction factor as reported earlier in this paper.

CONCLUSIONS

The multiple-lognormal model separates the growth kinetics of and average grain at a fixed instant of time from the more complex behavior of the time rate of change of such parameters as the average or median grain size. It thus overcomes the problem of adequately averaging in a system of grains that are growing but necessarily decreasing in number. The model predicts a parabolic growth law of the form:

$$D^2_{\text{med}}(t_2) - D^2_{\text{med}}(t_1) = \text{constant} \ (t_2-t_1)$$

The experimental results completely confirm the model. However, more systems should be investigated. Literature data on some metal systems (e.g., Al, Sn[3,6]) are in agreement with the model as far as can be concluded from the scarce data. Preliminary experiments on MnZn ferrites carried out in our laboratory are also in good agreement. Though the measurements are very tedious, even with the semiautomated measuring techniques now available, a full analysis is needed to provide full confidence.

REFERENCES

1. W. D. Kingery, H. K. Bowen and D. R. Uhlmann, Introduction to
       Ceramics, 2nd ed., John Wiley and Sons, New York, 1976.
2. C. S. Smith, Metal Interfaces, p. 65, Amer. Soc. Metals, 1952.
3. P. Feltham, Acta Met. $\underline{5}$, 97, 1957.
4. M. Hillert, Acta Met. 13, 227, 1965.
5. F. N. Rhines and K. R. Craig, Met. Trans. $\underline{5}$, 413, 1974.
6. P. A. Comte and W. Form, Praktische Metallographie, $\underline{13}$, 9. 1976.
7. (a) S. K. Kurtz and F. M. A. Carpay, to be published.
     (b) S. K. Kurtz, D. Veeneman and F. M. A. Carpay, to be published.
8. E. E. Underwood, "Particle-size Distribution", Ch. 6 in Quanti-
       tative Microscopy, ed. R. T. de Hoff, McGraw Hill Co., 1968.
9. J. Aitchison and J. A. C. Brown, The Lognormal Distribution,
       Cambridge View Press, 1957.
10. D. McLean, Grain Boundaries in Metals, Ch. 3, Oxford Univ. Press,
       1957.
11. A. G. Spektor, Zevod. Lab. $\underline{21}$, 193, 1955.
12. Y. Suemune, Jap. J. Appl. Phys. $\underline{10}$, 454, 1971.

PART III

LIQUID PHASE SINTERING
AND POST-FIRING TECHNOLOGY

REARRANGEMENT DURING LIQUID PHASE SINTERING OF CERAMICS

W.J. Huppmann, S. Pejovnik*, S.M. Han**

Max-Planck-Institute für Metals Research
Institute for Materials Sciences
Powder Metallurgy Laboratory
Stuttgart, West Germany

ABSTRACT

Rearrangement in ceramic-glass systems was studied on models. When Al_2O_3-spheres of 50 µm diameter are sintered in the presence of an alkali-borate glass which does not dissolve any appreciable amount of Al_2O_3, large pores form during the early sintering stage and persist even after prolonged sintering at high temperatures. Contrary, sintering with anorthite glass, which in the molten state rapidly dissolves more than 5 wt.% Al_2O_3, leads to particle desintegration and increased densification. Similarly, using Al_2O_3-agglomerates instead of solid spheres leads to increased densification. It is shown that the solubility of Al_2O_3 in the glass-phase is more important for densification by rearrangement than large changes in the viscosity of the liquid phase.

INTRODUCTION

Many ceramic materials are produced by sintering in the presence of liquid phase, but still the understanding of the mechanisms governing densification and microstructural development is rather limited. This applies particularly to particle rearrangement, a process which according to Kingery [1] leads to rapid ini-

---

* Permanent address: University of Ljubljana, Faculty for Natural Sciences and Technology and Inst. "J.Stefan", Ljubljana, Yugoslavia
** Permanent address: Hanyang University, Research Institute of Industrial Science, Seoul, South Korea

tial densification by particles sliding against each other and
assuming a denser packing. Rearrangement will depend on many
parameters like the viscosity of the liquid or glassy phase and
its wetting characteristics, particle size and shape, packing
density, etc., i.e., it is likely to be a complex process which is
difficult to analyse. Therefore in the present work an experimental approach was chosen which had been successfully applied for
studying rearrangement in metallic systems [2,3,4], namely the measurement of microstructural changes of geometrically simple systems, e.g. spheres of solid particles uniformly embedded in the
glassy phase.

REARRANGEMENT IN THE SYSTEM Al_2O_3-ALKALI BORATE GLASS

A nearly ideal system for studying rearrangement of ceramics
is Al_2O_3-alkalie borate glass (Schott glass No. 8454; appr. composition in wt.%: 70 SiO_2, 8 B_2O_3, 8 Na_2O, 5 CaO, 5 Al_2O_3, 2 K_2O,
developed for glass-Al_2O_3 joining[5]) because this glass, although
it wets Al_2O_3 well[6], shows hardly any reaction with it. Al_2O_3
spheres of 50-56 µm diameter were prepared from fine (<1µm) Al_2O_3
powder by agglomeration[7] and sintering at 1800°C for 1 h in
nitrogen. To the almost dense (<1% porosity) spheres, 15 wt.% of
the finely ground glass was uniformly admixed by stirring and
drying alcoholic suspensions. The changes during liquid phase
sintering were measured on cylindrical pellets (diam. 10 mm,
height 5 mm) pressed with 200 MPa and presintered at 600°C in air.

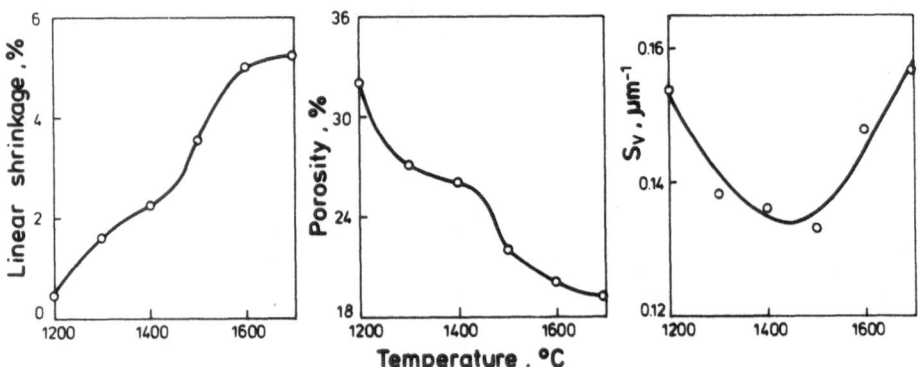

Fig. 1. Linear shrinkage, porosity and specific pore surface of
        Al_2O_3 spheres-15 wt.% alkali borate glass as a function
        of sintering temperature. Heating rate 20°C/min. Holding
        time 60 min.

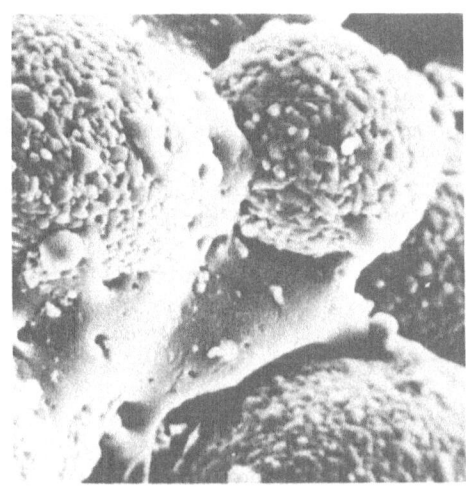

Fig. 2. 1 min. The glassy phase begins to flow into the interstices between the Al_2O_3 spheres.

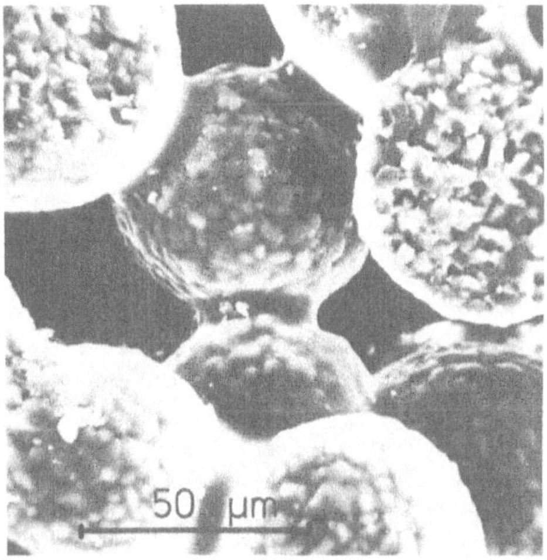

Fig. 3. 180 min. The glassy phase has reached the equilibrium shape, the Al_2O_3 spheres form stable configurations.

Fig. 2 and 3. Scanning electron micrographs of Al_2O_3 sintered in the presence of alkali borate glass at 1400°C.

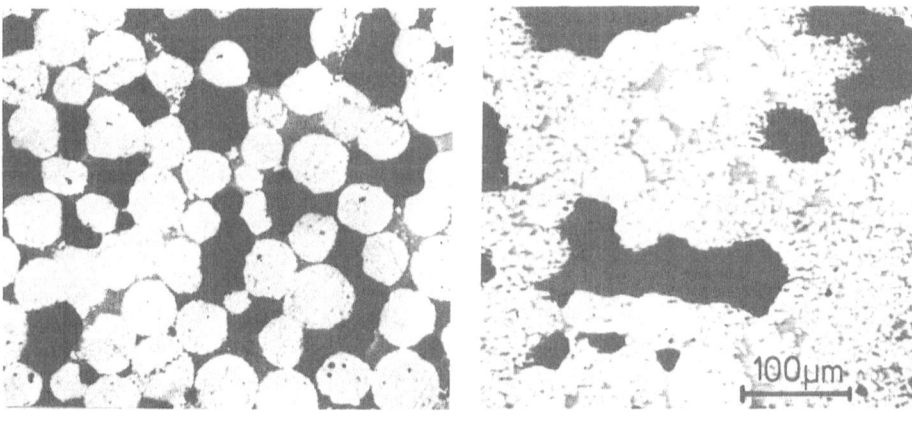

Fig. 4.   1400°C   Fig. 5.   1700°C

Fig. 4 and 5. Microstructure of Al_2O_3-alkali borate glass after sintering for 60 min.

Figure 1 shows the results of shrinkage measurements and quantitative microstructural analysis. As outlined earlier [6] these results indicate that rearrangement in this system takes place in two discernable stages, (I) below appr. 1500°C and (II) above this temperature. In stage I rearrangement is essentially rearrangement of the Al_2O_3 spheres. The glass phase first moves into the contact areas between the spheres (Fig. 2 shows an early stage of this movement) and then the glass bridges pull the spheres together until they form a rather stable array with large pores in between, Fig. 3 and 4. Rearrangement would thus terminate if stage II would not enable further shrinkage above appr. 1500°C. At this temperature effective penetration of the glass phase into the grain boundaries of the Al_2O_3 particles occurs [6]. Therefore rearrangement is now effectively the rearrangement of the small Al_2O_3 grains rather than the large spheres and shrinkage again increases, porosity decreases, and specific surface area of pores increases, Figure 1. Nevertheless densification is by no means complete even after sintering at 1700°C. The reason for this behavior is seen in Fig. 5. In a material sintered at 1700°C large pores as do normally form during stage I of rearrangement, are still to be seen. It is therefore concluded that during heating to this high temperature the large Al_2O_3 spheres rearrange and form a fairly stable microstructure with large pores which are stable even after complete penetration of glass into the grain boundaries of the spherical particles has occured. In summary it appears therefore that the initial packing of the Al_2O_3 spheres determines the final porosity in the system Al_2O_3-alkali borate glass because the glass penetration occurs only then when the spheres have already rearranged into a rather stable configuration.

THE INFLUENCE OF DISSOLUTION OF SOLID PARTICLES ON REARRANGEMENT

The results described in the previous chapter suggest that an effective possibility to speed up particle desintegration would be to enhance glass penetration into grain boundaries and thus to enhance rearrangement. On this basis larger shrinkage could be expected for a system in which the liquid phase dissolves substantial amounts of the solid particles because grain boundaries should then be attacked rapidly. From the Al_2O_3-anorthite phase diagram [8] it can be read that heating of an Al_2O_3-anorthite mixture to temperatures above 1550°C leads to dissolution of more than 5 wt.% Al_2O_3 in the liquid phase which exists above this temperature. In our experiments finely ground anorthite glass [9] was admixed to the Al_2O_3 spheres as described in the previous chapter. Compacted pellets were presintered at 900°C and then sintered at the temperatures shown in Fig. 6. Fig. 7 shows that glass penetration is practically complete at temperatures as low as 1550°C when shrinkage commences, Fig. 6. The stable large pores which form when the Al_2O_3 spheres move as a whole (as in the Al_2O_3 alkali borate glass system) do not occur here and porosity decreases to much lower levels when the sintering temperature is raised to 1700°C, Fig. 6 and 8.

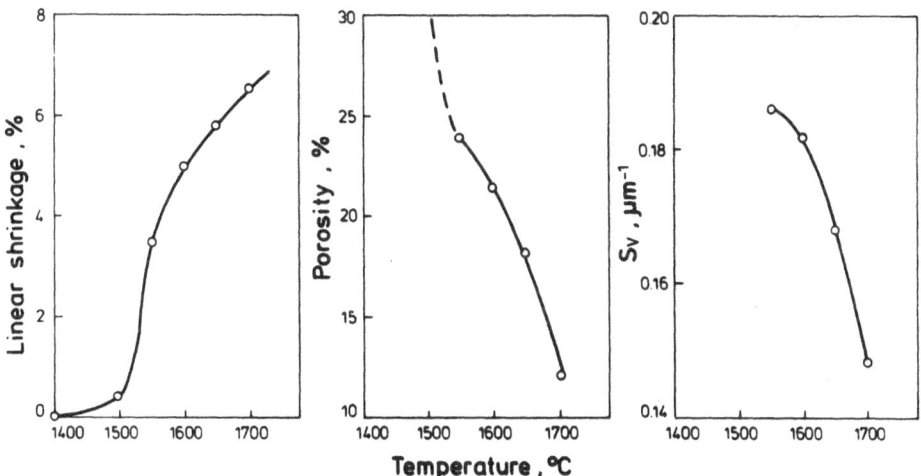

Fig. 6. Linear shrinkage, porosity and specific pore surface of Al_2O_3 spheres – 15 wt.% anorthite as a function of sintering temperature. Heating rate 20°C/min. Holding time 60 min.

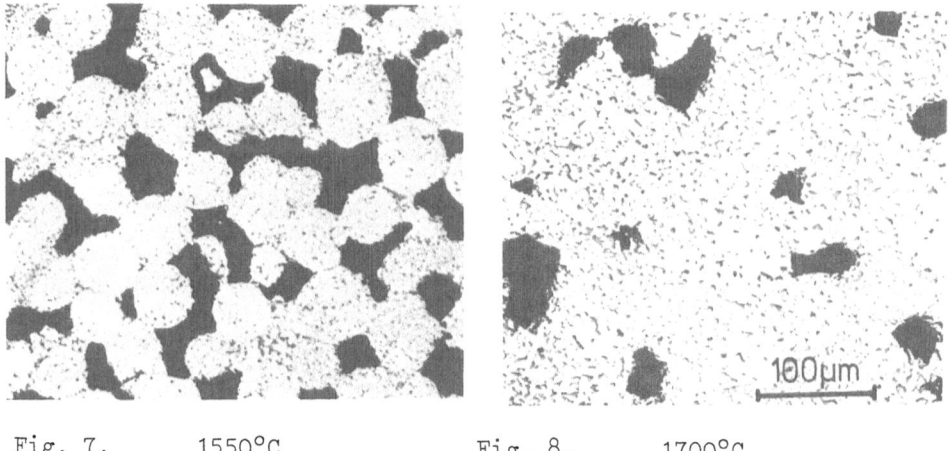

Fig. 7.   1550°C   Fig. 8.   1700°C

Fig. 7 and 8. Microstructure of Al_2O_3-anorthite after sintering for 60 min.

COMPARISON BETWEEN Al_2O_3-PARTICLES AND Al_2O_3-AGGLOMERATES

If rapid particle desintegration can enhance rearrangement then the use of only lightly sintered agglomerates instead of dense particles should also cause increased densification. This could be proven by sintering the Al_2O_3 shown in Fig. 9 and 10 [10] in the presence of the alkali borate glass described earlier. Figure 11 shows that with this material the two stages observed with Al_2O_3 spheres of comparable size, Fig. 1, are no longer present, which means that effective agglomerate desintegration occurs before the agglomerates are able to assume a stable configuration.

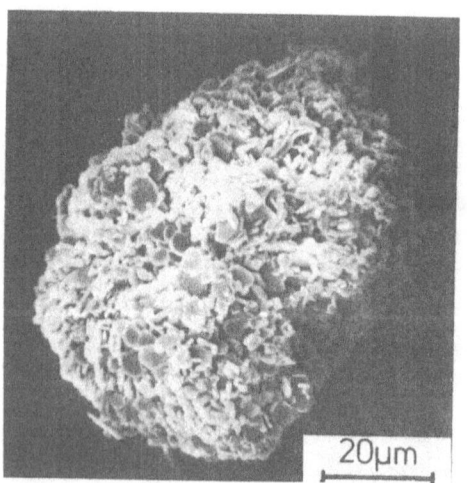 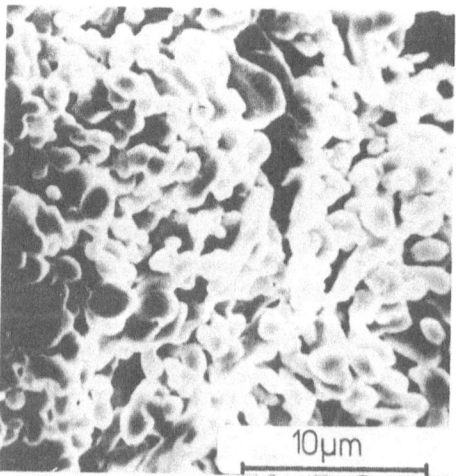

Fig. 9. Typical agglomerate   Fig. 10. Surface morphology

Fig. 9 and 10. Scanning electron micrographs of Al_2O_3 agglomerates.

THE INFLUENCE OF VISCOSITY OF THE LIQUID PHASE ON REARRANGEMENT

On the basis of the findings on Al_2O_3-alkali borate glass and Al_2O_3-anorthite the influence of the viscosity of the liquid phase on rearrangement must be considered as a minor one. At temperatures in excess of 1550°C the viscosity of the alkali borate glass is lower than the viscosity of the liquid phase in the Al_2O_3-anorthite system [6,11]. Therefore on the basis of viscosity alone shrinkage should be more pronounced in the Al_2O_3-alkali borate

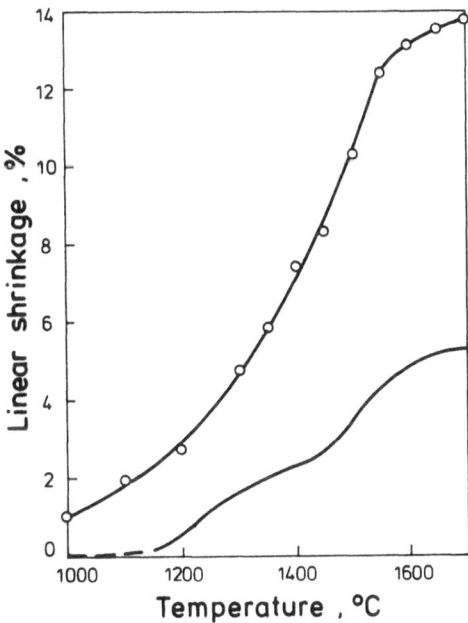

<u>Fig. 11.</u> Linear shrinkage of Al_2O_3 agglomerates - 15 wt.% alkali
borate glass as a function of sintering temperature.
Heating rate 20°C/min. Holding time 60 min. For comparison shrinkage using Al_2O_3 spheres (Fig.1) is also shown.

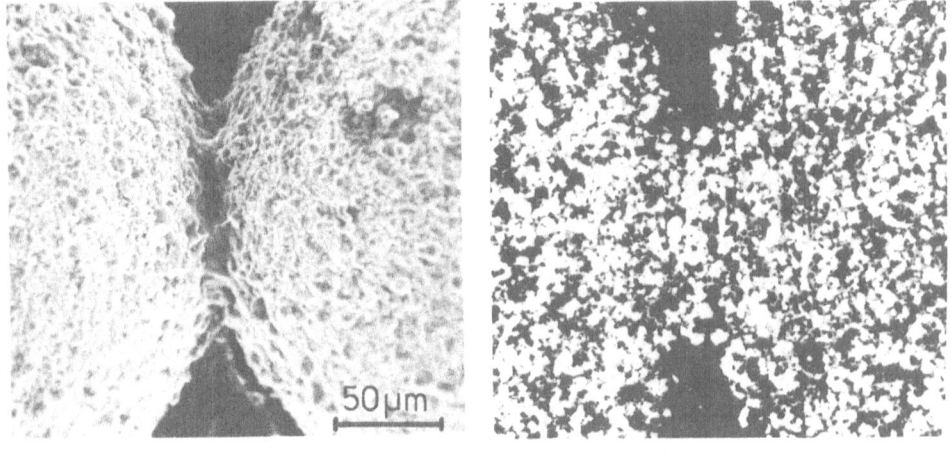

<u>Fig. 12.</u> Scanning electron mi-
crograph

<u>Fig. 13.</u> Section through neck
area   Ni   white
        ZrO_2 grey
        pores   black

<u>Fig. 12 and 13.</u> Sintering of 300 μm diameter spheres consisting of
finely interdispersed ZrO_2 and Ni. 1400°C, 60 min.

glass system contrary to our observations. As pointed out earlier [6] low viscosity may, however, accelerate particle desintegration.

The secondary role of viscosity on rearrangement is also evident from experiments with ZrO_2-Ni [12]. If agglomerated ZrO_2 spheres of 300 μm diameter are uniformly coated with Ni and then sintered above the melting point of Ni, i.e. in the presence of a liquid phase of very low viscosity, no densification and, more important, no rearrangement of the grains of the ZrO_2 spheres occurs. If ZrO_2-Ni spheres are prepared from uniformly mixed small ZrO_2 and Ni particles the structure of these spheres is comparable with the structure of a desintegrated particle (compare Fig. 7). When such spheres are sintered even below the melting temperature of Ni, that means at very high "viscosity" values of the second phase, densification occurs. Detailed analysis of pictures as shown in Fig. 12 and 13 strongly suggests that rearrangement of the small ZrO_2 grains causes this densification.

CONCLUSIONS

Desintegration of the solid particles into smaller grains must be considered a decisive factor for densification by rearrangement during liquid phase sintering of ceramics. The viscosity of the liquid phase seems of less importance. The parameters which influence particle desintegration primarily seem to be the ability of the liquid phase to take solid material into solution and the coherency of the solid particles.

ACKNOWLEDGEMENTS

The authors wish to thank Profs. G. Petzow and D. Kolar for their encouraging support of these investigations as well as Prof. L. Zagar and O. Lindig for repeated helpful discussions.

Part of this work was carried out under the German Yugoslavian Joint Agreement for Scientific Exchange and the administrative assistance of Dr. K. Scharmer is gratefully acknowledged.

REFERENCES

1. W. D. Kingery, J. Appl. Phys. 30 [3] 301-6 (1959).
2. W. J. Huppman and H. Riegger, Acta. Met. 23 [8] 965-71 (1975).
3. W. J. Huppmann and H. Riegger, The Int. J. Powder Met and Powder Tech. 13 [3] (1977) in press.
4. G. Petzow and W. J. Huppman, Z. Metallkde. 67 [9] 579-590 (1976).
5. O. Lindig (Jenaer Glaswerk Schott, Mainz) private communication.

6. S. Pejovnik, D. Kolar, W. J. Huppman, and G. Petzow, 4th Int.
        Round Table Conf. on Sintering, Dubrovnik, 5-10 Sept. (1977).
7. N. Claussen and G. Petzow, J. Mat. Technology 4 [3] 148-56 (1973).
8. E. M. Levin , C. R. Robins, and H. T. McMurdie, Phase Diagrams
        for Ceramists, p. 220, Am. Ceram. Soc., Columbus (1964).
9. L. Zagar (Technische Hochschule, Aachen) kindly provided the
        anorthite glass.
10. K. H. Friese (Robert Bosch Company, Stuttgart) kindly provided
        this Al_2O_3 powder.
11. G. Urbain, Rev. int. Htes Temp et Refract. 11 [2] 133-145 (1974).
12. W. J. Huppman and S. M. Han (Max-Planck-Institute for Metals
        Research, Stuttgart) unpublished work.

DISCUSSION

C. S. Morgan (Oak Ridge National Laboratory): What quantities of
glass are required to effect rearrangement and does this amount
decrease with increasing density of the Al_2O_3 particles? Is glass
permeation prevented in high density Al_2O_3 particles?
Author: Small quantities of liquid phase, say 1-5%, can affect rearrangement, and it is likely that with increasing density of the
Al_2O_3 particles less liquid is required to produce rearrangement
by particle disintegration. Glass penetration is not prevented in
high density Al_2O_3 particles. The spheres which we used in our
study were practically fully dense, and yet glass penetrated into
their grain boundaries. It appears, however, that in fully dense
particles grain boundary penetration is somewhat retarded in comparison with agglomerates of lower density (compare Fig. 11).

D. R. Clarke (Rockwell Science Center): 1) You are assuming glass
penetration occurs along a planar, flat channel and not a tube-
like channel which would be typical of penetration down a 3-grain
junction channel, are you not? Wouldn't this be more realistic
and affect the penetration times that you estimate? 2) The width
of the channel at $\sim$0.1 μm or so that you assume seems very large
for the width of a grain boundary before grain center separation
due to penetration has occurred. If r $\sim$ 3Å what would the rate of
penetration be?
Author: 1) The calculations which I presented here are described
in greater detail in the paper "Sintering of Al_2O_3 in Presence of
Liquid Phase" by S. Pejovnik, D. Kolar, W. J. Huppmann, G. Petzow,
4th Int. Round Table Conf. on Sintering, Dubrovnik, Yugoslavia,
September 5-10, 1977, and are for tubular channels. 2) The tube
diameter 0.1 μm represents the channel width when penetration is
completed. If, as you propose, a tube radius of 3Å is used, the
depth of penetration after 60 min would be in the range of 5 to

500 µm which would even better agree with our observations. We think, however, that more work is necessary in order to determine a meaningful tube diameter.

D. Lynn Johnson (Northwestern University): For calculating the penetration rate of the grain boundaries by glass, you used a planar capillary. It seems that a more realistic model would be a dihedral angle which goes to zero as the glass covers the surface of the polycrystalline alumina particles.
Author: Our calculations, described in more detail in ref. 6, are indeed a simplified description of glass penetration. However, the dihedral angle describes the equilibrium situation and cannot be used for deriving penetration kinetics.

Hayne Palmour III (North Carolina State University): One of our earliest experiments in rate controlled sintering was conducted with a classical polyphase body (BaSO$_4$, flint, clay) similar to Wedgewood's jasperware. To maintain a positive slope (constant densification rate) in the density-time plot at a fractional density near 0.70 as a very fluid liquid phase formed and spread, the feedback-controlled dilatometer generated a steep negative slope in the dependent temperature-time curve (see Fig. la,b). Both viscosity and solubility are known to be strongly temperature-dependent. This closely controlled densification rate experiment appears to confirm the significance you have attributed to viscosity and solubility in regulating boundary penetration and rearrangement during liquid phase sintering.

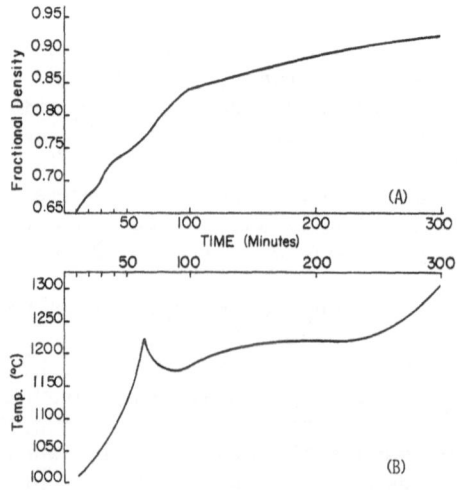

Fig. 1. Rate controlled sintering of Jasperware body. (After Lawhon, 1970).

REACTIVITY OF ALUMINA SUBSTRATES WITH HIGH LEAD GLASSES

William S. Machin   Robert W. Vest

Babcox-Wilcox   Purdue University

Lynchburg, Virginia 24505 W. Lafayette, Indiana 47907

INTRODUCTION

The substrate in a thick film resistor cannot be completely
inert because some degree of interaction with the glass in the
resistor is required for bonding, and this interaction is guaranteed
by the high temperatures used in processing. It has been determined[1]
that at least six properties of the glass contribute to microstructure development in thick film resistors, and that the presence
of 8 to 10 wt% substrate material dissolved in a resistor glass has
marked effects on microstructure development and electrical properties. Thus it is necessary to understand the kinetics of the glass-
substrate interactions before the properties of a thick film circuit
can be predicted for various processing conditions. This paper presents the findings of the initial studies of these interactions.

There are several possible rate limiting steps for the dissolution of an alumina substrate in a glass. At short times the dissolution rate may be controlled by the chemical reaction rate at the
interface. Such a process can be described[2] by:

$$Y = (KA_c/A_o)\, t \tag{1}$$

where Y is the linear recession of the surface, K is the phase
boundary reaction rate constant, A_c is the actual surface area of
the substrate and A_o is the geometric area of the substrate. In the
derivation of Eq. 1 it was assumed that both A_c and K are time
independent. As the solute concentration near the interface increases,
it becomes necessary to remove materials from the interface region
for the reaction to continue. In the absence of flow producing
hydrostatic instabilities, mass transport will be limited by

243

molecular diffusion in the glass. Cooper[3] has analyzed this case and
has shown that:

$$Y = 2\alpha \ (D*)^{\frac{1}{2}} \ t^{\frac{1}{2}} \qquad\qquad\qquad (2)$$

where α is a constant and D* is the effective binary diffusion coefficient. At longer times a boundary layer is built up at the
interface. Because of density, temperature or surface tension
gradients in the boundary layer, the region becomes unstable and
free convection occurs. The recession under these conditions is
given by:[4]

$$Y = (DC*/\delta*) \ t \qquad\qquad\qquad (3)$$

where D is the local binary diffusion coefficient, C* is a concentration parameter, and $\delta*$ is an effective boundary layer thickness.
A wide variety of work[2,4-10] has been reported on single and polycrystalline Al_2O_3 in many single and multiple compound melts, and
each of the three mechanisms has been found to control the kinetics
under certain conditions.

EXPERIMENTAL

Materials Preparation and Characterization

    The model lead borosilicate glass (70% PbO, 20% B_2O_3, 10% SiO_2)
used in these experiments was chosen because it is similar in composition to the glass used in earlier experiments at Purdue[1] and to
glasses used in the thick film industry. The glass was formed by
combining boric acid (H_3BO_3), red Lead (Pb_3O_4) and a lead silicate
glass frit (85% PbO·15% SiO_2) mixing on a roll mill for 1 hour, and
heating at 800-900°C until a clear, bubble-free, single phase liquid
was formed. The glass was then either used immediately or fritted
in deionized water and stored for later use (either remelting or
grinding and screen printing).

    Two types of polycrystalline substrates (AlSiMag 614-96%Al_2O_3
and AlSiMag 772-99.5%Al_2O_3 from 3M Company) were chosen because they
are both widely used in the thick film industry. Experiments were
also conducted with two orientations of single crystal sapphire. The
orientations selected were (0001), the basal plane, and (1123), a
plane tilted at 60° to the basal plane. The surface grain size of
the AlSiMag 614 substrates was 2 to 8μm, but the average grain size
in the bulk was considerably larger. After 3 minutes contact with
the glass at 800°C the etched surface grain size was observed to
range from 2 to 15μm. This grain size distribution did not change
with further etching up to 12 hours contact time with the glass,
indicating that it was characteristic of the bulk substrate.

Long Time Dissolution

Surface recession rates were determined by suspending AlSiMag 614 substrate chips in glass at times ranging from 30 minutes to 12 hours at temperatures ranging from 750° to 1000°C. Similar experiments were performed with AlSiMag 772 substrates at 800°C only, and with oriented single crystal sapphire chips at 800°C and 2 times only. The substrate chips were cleaned, dried and weighed to +0.1 mg. They were then secured in a platinum wire harness and immersed in the glass (already at temperature in a platinum crucible). At the end of the time period, the substrate chip was removed from the glass, air quenched, and transferred to concentrated HCl in order to decompose and remove any glass adhering to the substrate. After the glass was removed, the substrate was again cleaned, dried and weighed. The weight loss was then converted to surface recession. Blank experiments conducted on substrates having no glass contact and substrates immersed just long enough to be coated by the glass indicated that all glass was removed by the leaching process, and that the substrate was not attacked by the HCl to within measurement error. At least three experiments were performed at each time and temperature.

Short Time Dissolution

In order to determine the substrate recession at typical thick film processing times (3-15 minutes) a series of experiments were run at 800°C using an atomic absorption spectrometer to analyze for aluminum content in the bulk glass. Substrates of known geometry were suspended above the glass in order to reach thermal equilibrium before immersion. At the end of each immersion time the glass and substrate were poured into deionized water. The frit was treated with HCl to decompose the glass, and the filtrate plus washings was evaporated almost to dryness in the presence of methanol to remove the borates before the aluminum concentration was determined by atomic absorption. The Al concentrations were corrected for Al already present in the glass ($\sim$ 10 ppm), and the results transformed to substrate surface recession.

Concentration Profiles

In order to determine the distribution of substrate materials in the glass during typical thick film processing conditions experiments were performed by screen printing -325 mesh glass in an organic binder onto AlSiMag 614 substrates. These "resistors" were dried for 30 minutes at 120°C and fired for 4, 8 and 10 minutes at 800°C. The thickness of the glass after firing was about 11μm. After firing they were fractured to expose the glass-substrate interface, mounted on aluminum blocks and coated with gold to provide a conductive surface. Concentration profiles were determined using the energy

dispersive x-ray analysis (EDAX) capability of the scanning electron
microscope (SEM). The interface area was magnified 6500X and x-ray
spectra obtained at varying distances from both sides of the interface. All major elements of both substrate and glass with the
exception of boron and oxygen were used in the analysis. EDAX analysis of glass which had had no substrate contact showed a background
Al_2O_3 concentration of 2.5 wt% probably due to the aluminum mounting
block.

RESULTS AND DISCUSSION

Weight Change Measurements

Plots of surface recession vs time at a constant temperature
revealed a linear relationship with a change in slope at some time,
t_c. The least squares lines calculated from the results at 800°C are
shown in Fig. 1. The AlSiMag 614 substrates show a higher dissolution
rate than the AlSiMag 772 substrates for $t < t_c$, while the slopes for
$t > t_c$ are the same within one standard deviation. Single crystal
sapphire dissolution (also represented on Fig. 1) indicates that
attack on the basal (001) plane is less vigorous than that on the
$11\bar{2}3$ plane. This result is contrary to the results of Ryabov,
et al.[8], but agrees with recent studies by Becher and Murday.[11]

Experiments performed on AlSiMag 614 substrates at other temperatures also showed two linear regions. These results are presented as
recession rate as a function of 1/T in Fig. 2. Least squares fits to
these data gave:

$$Y = 8.6 \times 10^7 \, t \, \exp \left[\frac{-2.06 \times 10^4}{T} \right] \qquad\qquad (t < t_c) \quad (4)$$

$$Y = Y_c + 6.5 \times 10^6 \, (t - t_c) \, \exp \left[\frac{-1.85 \times 10^4}{T} \right] \quad (t > t_c) \quad (5)$$

These results are consistent with rate controlling mechanisms of
chemical reaction at the interface according to Eq. 1 for $t < t_c$
followed by free convection through a boundary layer according to
Eq. 3 for $t > t_c$. The difference between the recession rates for
AlSiMag 614 and AlSiMag 772 substrates for $t < t_c$ could be due to
preferred orientation. The AlSiMag 772 substrates have been shown to
favor grains with the basal plane (0001) parallel to the substrate
surface,[11] which would lead to a slower dissolution rate according
to the sapphire results. If the rate controlling mechanism for $t > t_c$
is boundary layer diffusion, preferred orientation of grains at the
substrate surface would have no effect, and the recession rates would
be the same as is observed.

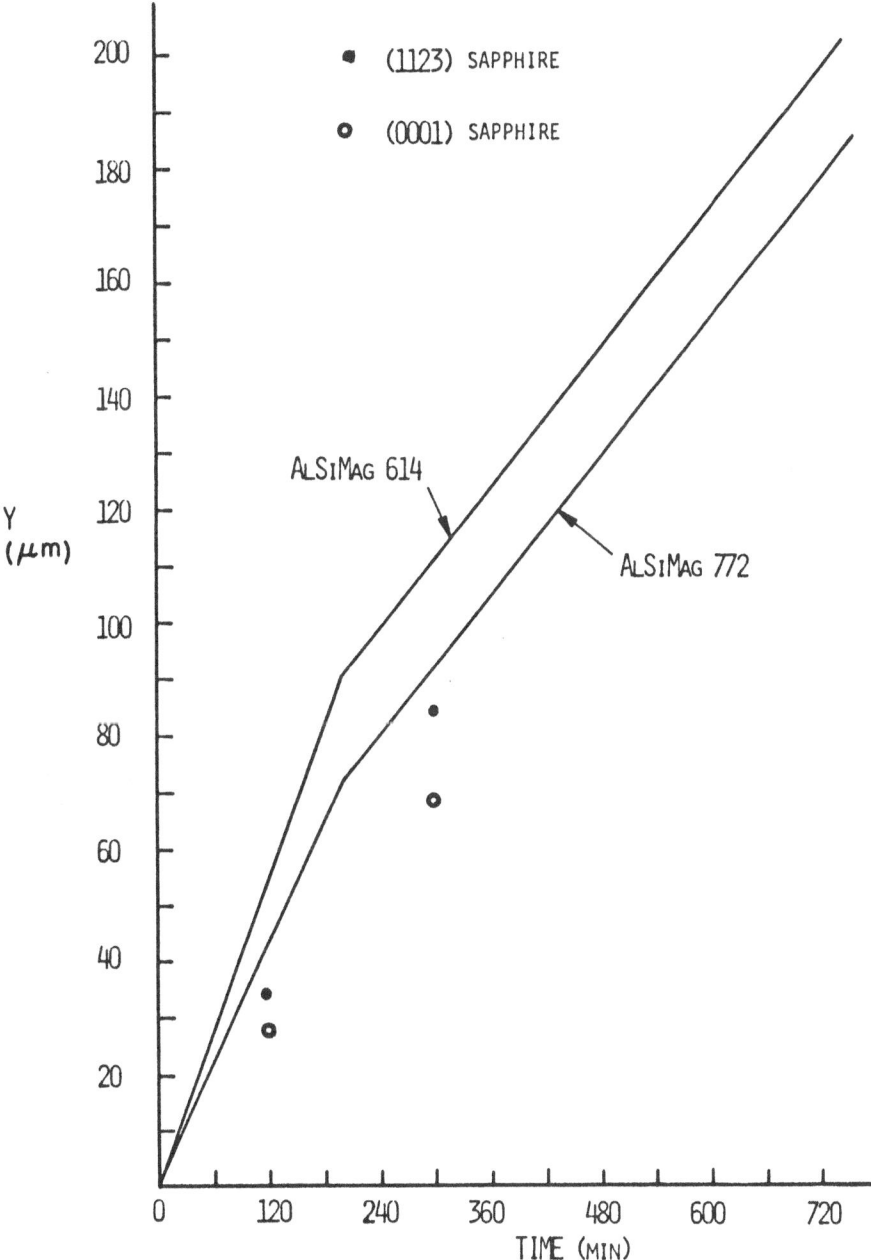

Figure 1.   Dissolution of Alumina Substrates at 800°C.

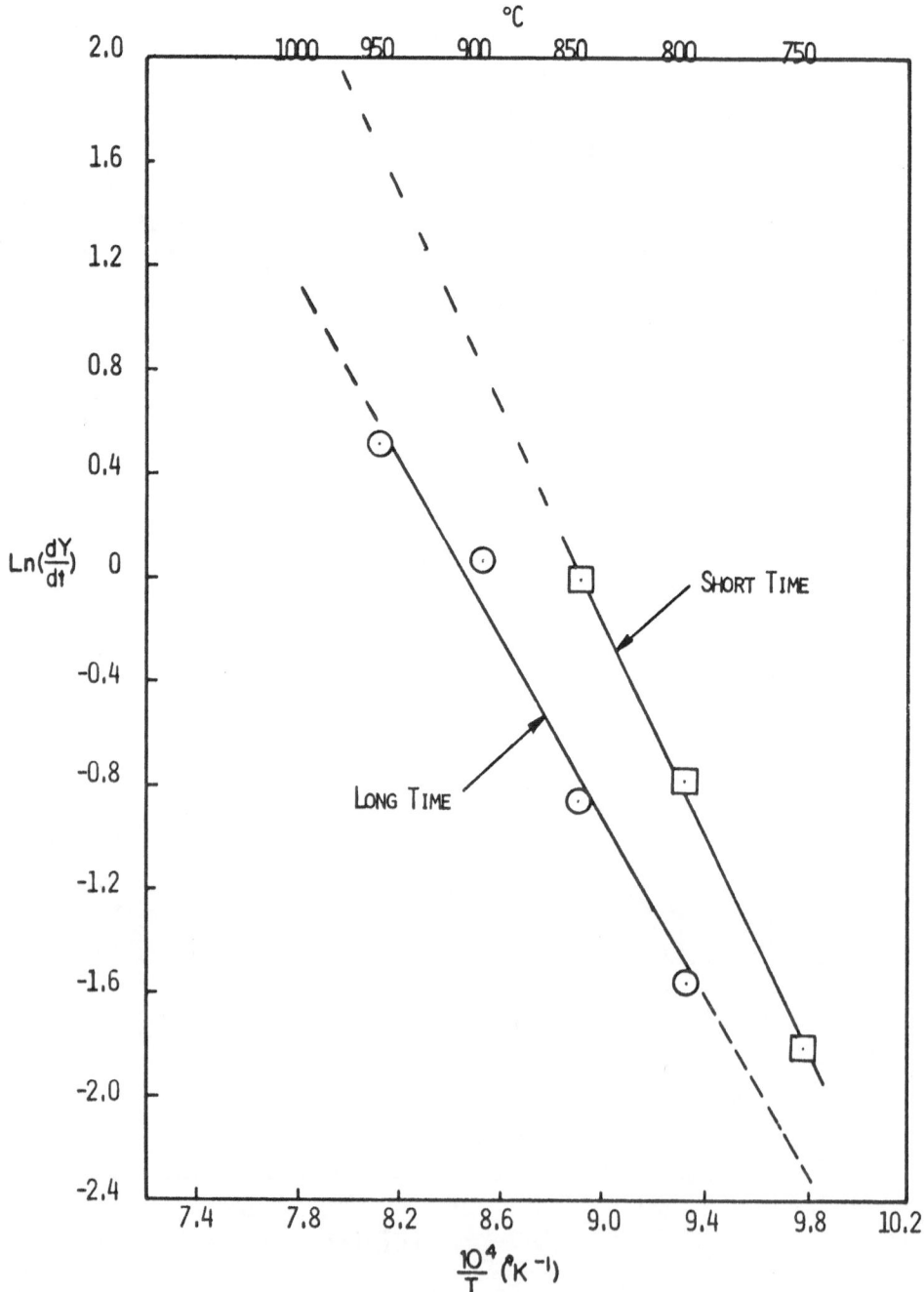

Figure 2.   Temperature Dependence of AlSiMag 614 Dissolution Rate

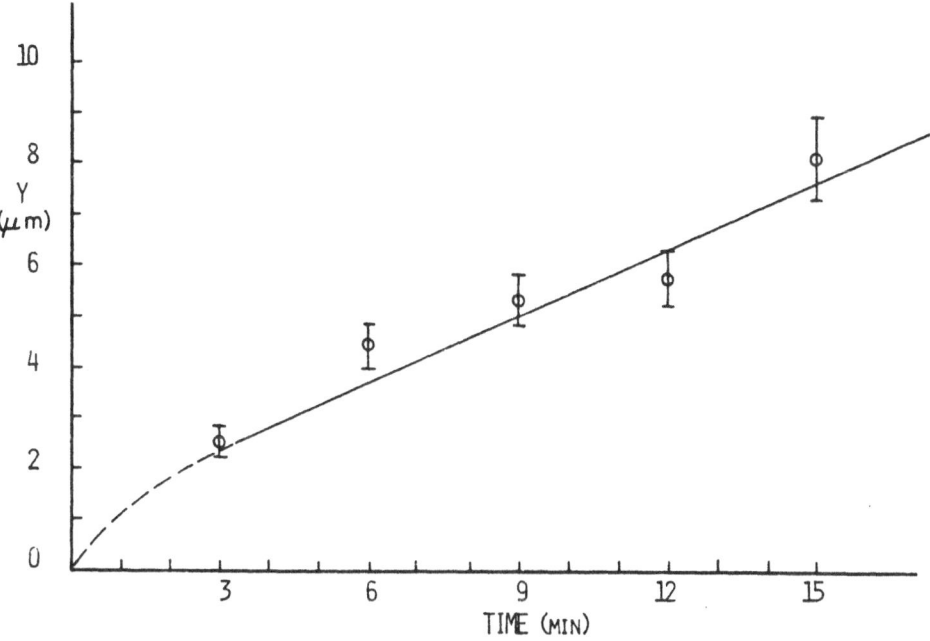

Figure 3.   Dissolution of AlSiMag 614 at 800°C.

Glass Analysis Measurements

        The AlSiMag 614 substrate recession data at 800°C calculated
from the atomic absorption analysis for aluminum in the glass are
shown in Fig. 3.   A least squares fit to the data yields Y = .44 t +
1.04, which is identical to the initial slope of the 800°C weight
change experiment.   A combination of the data from the two experimental techniques yields Y = .44 t + .83; the agreement between the
results obtained from the two techniques implies that the same
mechanisms is responsible for the entire range from t = 0 to t = t_c
for AlSiMag 614 at 800°C.

        Neither the line from the glass analysis data nor the line from
the combination of glass analysis and weight change data goes through
the origin as it should.   This may be due to some small systematic
error; however, it seems more likely that the effect is real and due
to a decrease in surface area (A_c in Eq. 1) during the first few
minutes of glass contact.   A decrease in A_c during the first three
minutes is expected based on the observed increase in surface grain
size.

Concentration Profiles

The concentrations of Al_2O_3 and PbO measured as a function of position relative to the substrate-glass interface after 10 minutes at 800°C are shown in Fig. 4. The beam size of the SEM is approximately 100 Å, but multiple scattering yields an x-ray source on the interface are influenced by the concentration of the element being

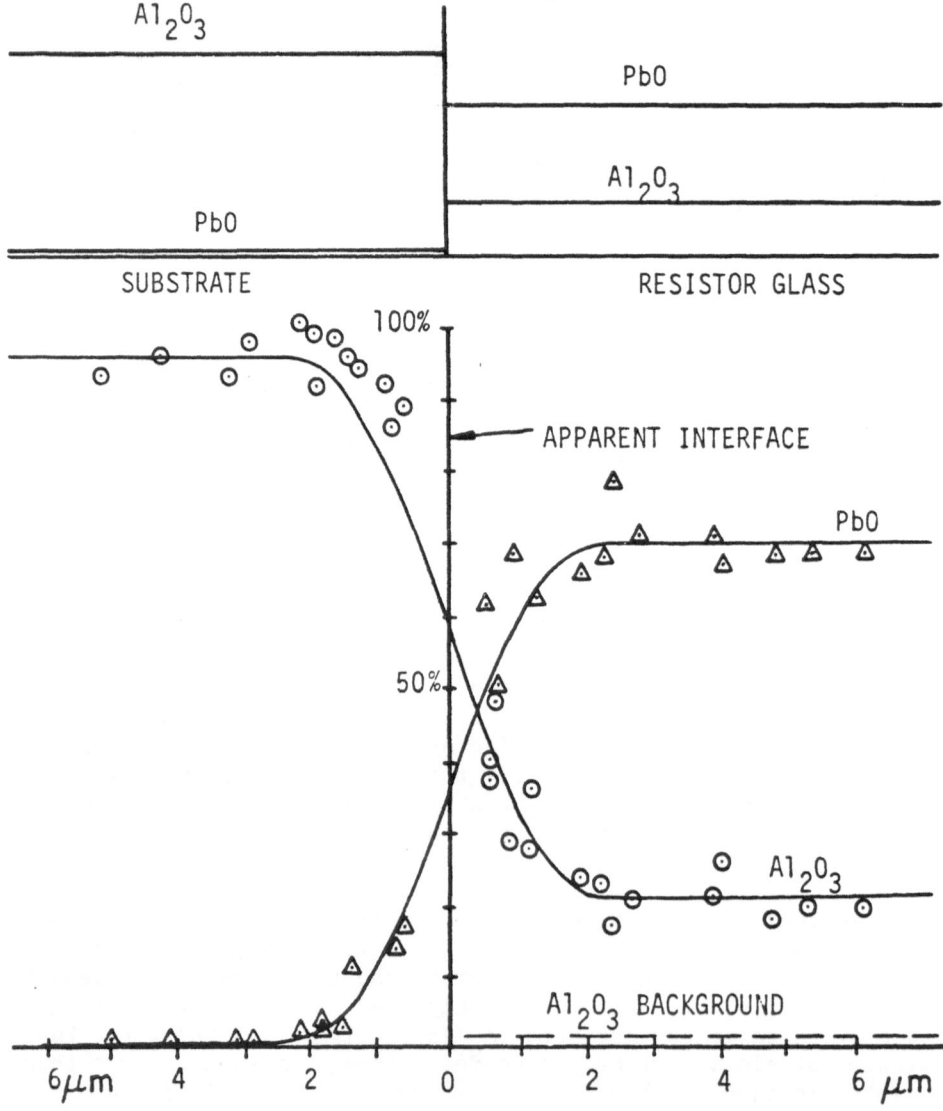

Figure 4.   Theoretical and Experimental Concentration Profiles After 10 Minutes at 800°C.

determined on the other side of the interface. The solid curves for
PbO and Al$_2$O$_3$ on Fig. 4 were calculated by assuming a concentration
step function at the interface as shown at the top of Fig. 4, and a
conical x-ray intensity distribution 2.4μm in radiu. The concentration step function would be expected if the kinetics are controlled
by the reaction rate at the interface. These assumptions lead to
close agreement between the apparent measured concentrations and the
calculated curves.

SUMMARY

     Of the three possible rate controlling mechanisms discussed, the
results are best interpreted on the basis of an initial dissolution
rate that is controlled by the chemical reaction rate at the interface followed by free convection through a boundary layer. For
typical thick film processing conditions *800°C, 10 minutes) the
experiments indicated that the glass in the fired film may be composed of as much as 20 wt% of substrate material.

ACKNOWLEDGEMENTS

     This work was supported by the Naval Air Systems Command under
Contract No. N00019-76-C-0354.

REFERENCES

1. R. W. Vest, "Conduction Mechanisms in Thick Film Microcircuits,"
    NTIS Report No. N76-13346, December 1975.
2. R. L. Reed and L. R. Barrett, Trans. Brit. Ceram. Soc., 63,
    509-534 (1964).
3. A. R. Cooper, Jr., Trans. Faraday Soc. 58, 2468-72 (1962).
4. A. R. Cooper, Jr., and W. D. Kingery, J. Amer. Ceram. Soc., 47,
    37-43 (1964).
5. R. L. Reed and L. R. Barrett, Trans. Brit. Ceram. Soc., 63,
    671-676 (1964).
6. N. McCallum and L. R. Barrett, Trans. Brit. Ceram. Soc., 51,
    523-548 (1952).
7. B. N. Samaddar, W. D. Kingery, and A. R. Cooper, Jr., J. Amer.
    Ceram. Soc., 47, 249-254 (1964).
8. A. N. Ryabov, T. I. Kiseleva, and L. B. Kulikova, ZH Prik Khim,
    48, 407-48 (1975).
9. M. Truhlarova, Silikaty, 18, 31-43 (1974).
10. M. Safdar, G. H. Frischat, and H. W. Hermicke, Ber. Dt. Keram.
    Ges., 51, 291-294 (1974).
11. P. F. Becher and J. S. Murday, J. Mat. Sci., 12, 1088-1094 (1977).

INFLUENCING OF THE PROCESSING PARAMETERS ON THE PROPERTIES OF RAPID FIRED PORCELAIN

H. Mörtel

University of Erlangen-Nürnberg

Erlangen, Germany

SUMMARY

The influence of the following parameters on the properties of rapid fired porcelain is discussed: green density of pressed and slip-cast bodies, grain size distribution of the raw materials, firing conditions (maximum temperature, firing time, rate of heating and cooling). The properties of the fired bodies which are considered are the microstructure, strength, linear thermal expansion coefficient, shrinkage, and the thermal shock resistance. It can be shown, that the processing parameters affect only the densification behavior of the body but scarcely the properties of the densified body, these are influenced mainly by the undissolved quartz. In this respect the firing schedule, especially the soak period, is of influence on the properties of the densified body in the sense that time is given for a complete dissolution of the quartz.

INTRODUCTION

Enormous increases in production costs are forcing the ceramic industry to intensify efforts to achieve cost savings. This is especially the case with energy related costs. Significant savings in the energy sector can be achieved with fast-firing programs. In our institute, a porcelain was produced, the properties of which were comparable or superior to those of normal hard porcelain, but whose firing required only two hours.[1] The present study now tries to shed some light onto the interdependence of such a porcelain and its manufacturing conditions.

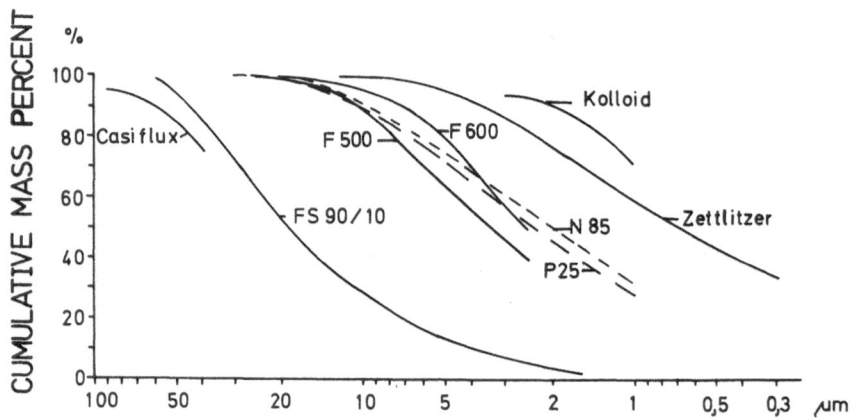

EQUIVALENT SPHERICAL DIAMETER MICRONS

Figure 1.   Grain size distribution of the raw materials. FS 90/10 =
            feldspar from Amberger Kaolin Werke, Hirschau, Germany,
            N 85 = Quartz flour from Hoffmann & Söhne, Neuburg/Donau,
            Germany, F 500 & F 600 = Quartz flour from Quarzwerke
            Frechen, Frechen, Germany, P 25 = Porcelain kaolin from
            Amberger Kaolinwerke, Hirschau, Germany, Kolloid and
            Zettlitzer = porcelain kaolin from Czechoslovakia.

RAW MATERIALS AND BATCH

        The following commercial raw materials were used: porcelain
kaolin of three different types, feldspar (not varied) and quartz
flour, also three different types. The chemical composition is
discussed in another paper.[2] The grain-size distributions are
shown in Figure 1.

        The batch components were milled for 20 hr in two different ball
mills (~10:1 and 2:1). The batch composition in each case was 50%
kaolin, 25% feldspar and 25% quartz. Figure 2 shows the resulting
grain-size distribution of two different batches with the same raw
materials but under different milling conditions. It is seen that
there is no serious influence of the milling conditions on the resulting grain-size distribution of the milled batch. All other
milled batches lie between these two distributions and are fully
reproducible. It is seen that, especially, the feldspar and the
larger fractions of the quartz flour were reduced by the milling
process.

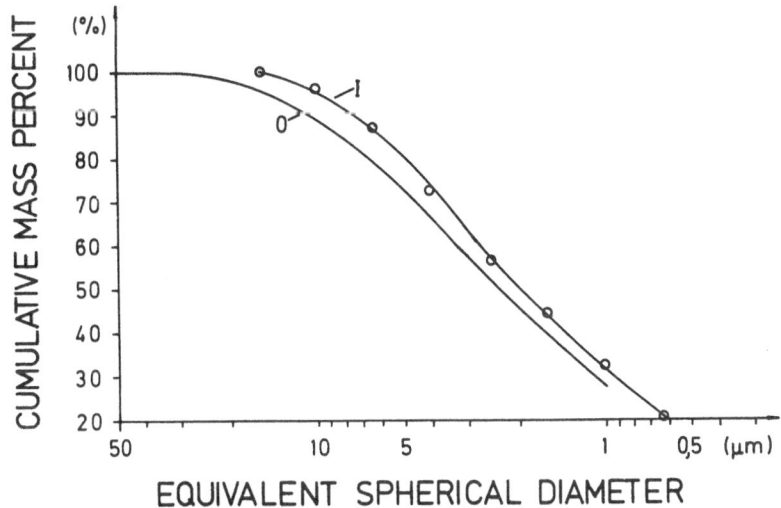

Figure 2.   Particle size distributions of two different batches.
             0 is milled 20 hrs in a 10:1 ball mill, I is milled 20 hrs
             in a 2:1 ball mill. The ratio of grinding balls to batch
             to water was 1:1:1.

The grain-size distributions for the raw materials were determined by using sedimentation methods (Andreasen) even in the case
of the batches (X-ray sedigraph).

The further characterization of the raw materials and the
batches is discussed in more detail in other papers.[1,2]

BODY PREPARATION

As discussed previously,[1] the preparation of a cast slip presented difficulties owing to the thixotropic nature of the largely
fine-grained body. Much lower water contents than normal are
sufficient to allow deflocculation to be achieved. Best results are
taken with DOLA-PIX PCN. Here more research is needed. The deflocculation behavior is mainly influenced by the kaolin.

Extrusion was performed using bodies prepared using 33-40% water.
A laboratory vacuum extrusion press was utilized to produce rods
having diameters of 11 mm.   Difficulties were present owing to the
thixotropic nature of the very fine grained bodies. But in all cases
a suitable extrusion rate was found. Texture was often seen in these

rods, which correspondingly affected the strength of the green and
the fired specimens.

Bodies suitable for dry pressing were prepared by first removing
the greater protion of the water by filter pressing. This was
followed by drying the filter cakes, pulverizing, and either granulating using a 2 wt% sulfite waste liquor addition or pulverizing
with no plasticizing agent added. Appropriate pieces (rods, tiles
50mm x 50 mm) were dry-pressed using a hydraulic press or a handpress.

CERAMIC PROPERTIES OF THE BODY

The plasticity of the bodies is mainly influenced by the raw
materials and shall not be discussed in this paper. The same holds
for the dry modulus of rupture. This modulus depends strongly on
the kaolin used.

The green density is influenced by the pressing method which is
connected with the water content. The influence of the batch composition, especially the type of kaolin used, is not so strong.
Table 1 shows the interdependences between the water content, the
pressing method and the resulting green density and dry shrinkage
in the diameter and the length.

Table 1. Some Ceramic Properties of the Bodies.

	dry pressed not plastified tiles	dry pressed plastified tiles	dry pressed plastified rods	wet pressed (extruded) rods
Water %	<1	2	2	35 ± 2
Green density c/cm^3	$1,12 \pm 0,04$	$1,62 \pm 0,2$	$1,66 \pm 0,02$	$1,42 \pm 0,07$
dry shrinkage ∅ %	0	<1	<1	3 ± 1
dry shrinkage l%	0	<1	<1	5 ± 1

From the data shown one expects that the higher the green density
the quicker the firing densification takes place. The high water
content of the wet pressed rods lowers the density of the dried
bodies.

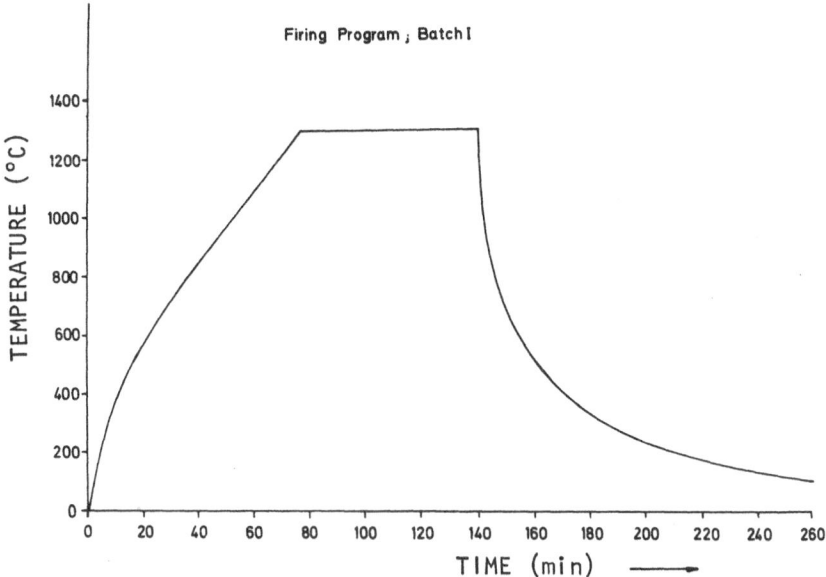

Figure 3.   Normal firing schedule. This schedule is modified by
            changing the soak period and the cooling rate up to a
            duration of 24 hours.

FIRING AND PROPERTIES OF FIRED SPECIMENS

Figure 3 shows the normal firing schedule. Quicker heating had
no influence on the properties of the densified bodies, even the
cooling rate had no significant influence on the properties of the
densified bodies. Even bodies without constitutional water (formulated
with meta-kaolin) had the same properties in the densified state.
For example, the modulus of rupture for a body fired according to the
schedule shown in Fig. 3 was 109 ± 8 MN/m^2 instead of 102 ± 6 MN/m^2
for a very slowly cooled body with the same batch composition.
Prefired specimens show a value of 103 MN/m^2 $\pm$ 3 MN/m^2, also no
influence.

That means the firing schedule has no influence on the properties
of the densified body. Only the soak period is of influence, but
this fact corresponds to the batch composition, in the sense that the
time required to dissolve the large quartz grains from a given grad
of quartz flour is longer. This is demonstrated for strength in
Fig. 4. In a similar manner the development of density and porosity
of the body is demonstrated for two different batches in Fig. 5.
Differences occur here between 1100°C and 1250°C according to the
reaction behavior of the kaolin used.

In these figures it is also shown that the sintering of the
porcelain can be subdivided into three steps. The first step is the

outgassing of the constitutional water followed by a solid state
reaction up to 1100°C.

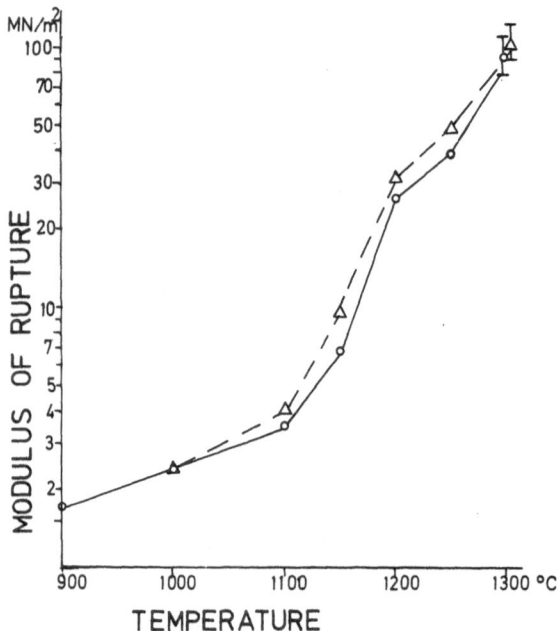

Figure 4. The development of strength during firing. The modulus
           of rupture for two bodies with different types of kaolin
           was determined on rods, which were cooled rapidly in air
           from the temperature shown in the plot.

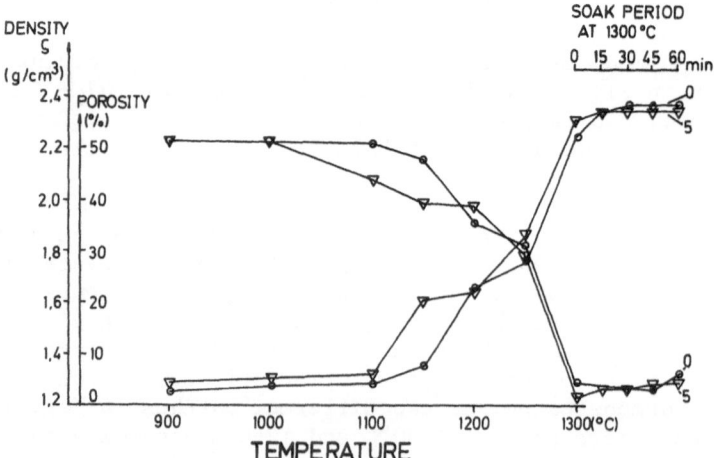

Figure 5. The development of density and porosity for two bodies with
           different types of kaolin. The parameters were measured
           on the same rods as in Figure 4.

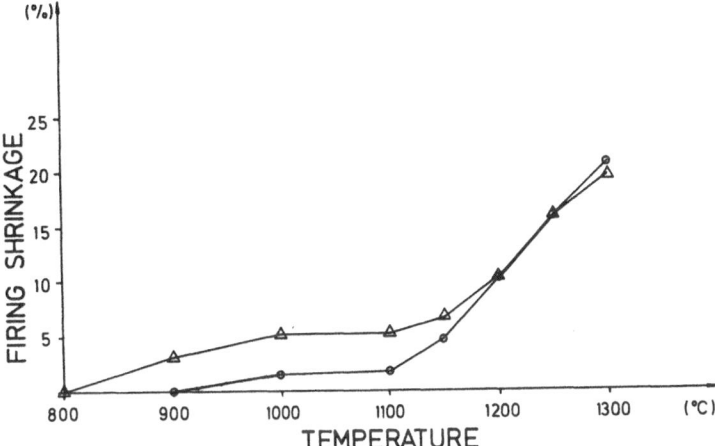

Figure 6.   The shrinkage of the two bodies during firing.   The values
were taken from quenched rods.

The second step consists of a quick densification with increasing
temperature.   The feldspar melts and quartz begins to dissolve.

The third step consists of small changes in the density and
related parameters, caused mainly by the development of porosity.
That means in the third step grain growth of the mullite needles from
the feldspar melt and pore size changes are the dominant processes.
In this step also the rest of the quartz dissolves.   These facts are
confirmed by analysing the microstructure with the SEM.

Firing shrinkage only occurs significantly in the first two steps.
It is shown in Fig. 6 again for two bodies.

Other properties are measured on tiles and rods with properties
of the green bodies as mentioned in Table 1 which were fired up to
1200°C and 1300°C respectively, according to the normal firing
schedule.   The results are presented in Table 2.   Comparing the results
given here with those in Figs. 4-6 some differences may be detected
but this is not very important, because the true temperature in each
case may have been different.   A modified furnace was used for
quenching and there the thermocouple could not be positioned exactly
between the samples.

The results must be discussed now in respect to the microstructure of the bodies.   These microstructures are shown for two bodies
at 1200°C and 1300°C in Figures 7-10.

Table 2. Properties of Fired Bodies.

Body	Density ρ [g/cm^3]	Modulus of Rupt. [MN/m^2]	Thermal Shock Resist [°C]	Shrinkage Dia. Total %	Shrinkage Length Total %	Therm. Expans. Coeff. α_{200} [10^{-6} K^{-1}]	Therm. Expans. Coeff. α_{600}
		Tiles not plastified, 1300°C					
0		83			16,8		
5		169			15,1		
		Tiles, plastified, 1200°C					
0	1,59± 0,04	32±4			7±0,4		
5	1,87± 0,04	74±6			9,8± 0,3		
		Tiles ,plastified,1300°C					
0	2,43± 0,02	119±13			14,5± 0,1		
5	2,444± 0,005	153±21			16,5± 0,3		
		Extruded Rods, 1200°C					
0	2,370± 0,006	85±4	240	20	18,1± 0,2	4,61	5,98
5	2,392± 0,002	73±10	200	15	16,1± 0,2	5,01	6,44
		Extruded Rods ,1300°C					
0	2,386± 0,01	85±4	210	23	18,9± 0,8	3,54	4,39
5	2,4 25± 0,01	104±8	260	18	17,5± 0,4	3,54	4,53

The micrographs are taken from extruded rods. The most important
parameter of a ceramic body is its strength. In the case of body 0*
there is no difference between the specimens fired at 1200°C and
specimens fired at 1300°C, respectively. From the micrographs it is
shown that the microstructure is very similar. In both cases small
grains of quartz remain undissolved, this affects mainly the strength.
In the specimens fired at 1200°C a higher porosity is present, this
affects the thermal shock resistance. The quartz present also affects
the thermal shock resistance but the amount in the compared specimens
is about the same. In the body 5 we detect important differences
between specimens fired at 1200°C and 1300°C respectively and the
micrographs explain that behavior. In the specimens at 1200°C a large
amount of quartz was not dissolved and the porosity is even higher.

---

* in the case of extruded rods

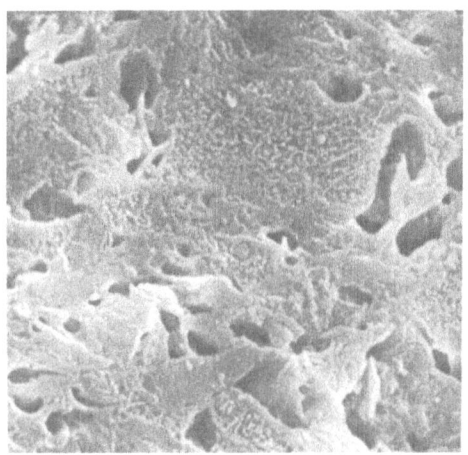

Figure 7. Body 0, 1200°C. Needle
          mullite in the feldspar
          melt, flaky mullite
          (dense gray areas) irreg-
          ular pores and sometimes
          quartz grains surrounded
          by SiO_2-melt.

Figure 8. Body 0, 1300°C. Needle
           mullite in the feldspar
           melt, flaky mullite
           (dense gray areas), fewer
           but bigger and more regu-
           lar pores and quartz
           grains surrounded by
           SiO_2-melt (dark gray).

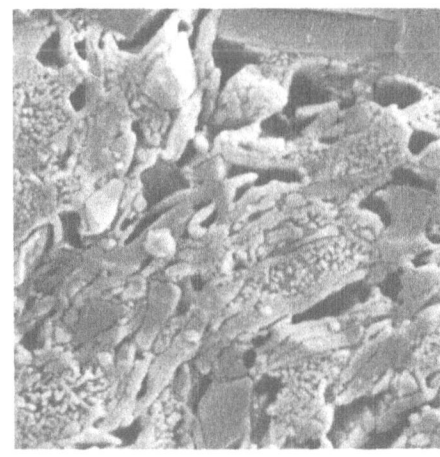

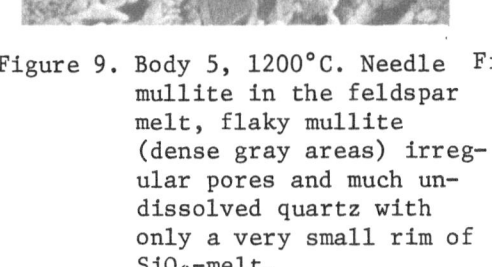

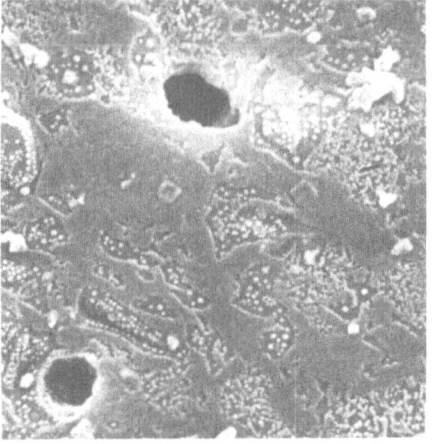

Figure 9. Body 5, 1200°C. Needle
          mullite in the feldspar
          melt, flaky mullite
          (dense gray areas) irreg-
          ular pores and much un-
          dissolved quartz with
          only a very small rim of
          SiO_2-melt.

Figure 10. Body 5, 1300°C. Needle
           mullite in the feldspar
           melt, flaky mullite
           (dense gray areas) few
           but big regular pores
           and nearly completely
           dissolved quarts i.e.,
           SiO_2-melt (dotted dark
           gray areas, cristobalite).

Therefore the strength is lower than in the compared specimens fired at 1300°C. For the thermal shock resistance the quartz has a more important influence than the porosity.

Discussing the rest of Table 2 it can be shown that the density of the fired specimens is influenced by the green density of the bodies. The same holds for the shrinkage. Comparing the results for the Bodies 0 and 5 it can be seen that the values tend in the same direction when comparing the data for specimens fired at 1200°C and 1300°C respectively.

CONCLUSIONS

It can be shown that the processing parameters of the production of fast firing porcelain have no important influence on the properties of the densified bodies. There is an influence on the firing behavior from the processing parameters during the densification process. The soak period was of special interest but in the cases studied this was connected with influences from the batch composition. In this respect the important factor is the dissolution of quartz during firing. The undissolved quartz determines mainly the strength, the thermal shock resistance and the linear expansion coefficient (to be discussed in another paper). In most cases (not all presented here) a firing up to 1300°C was needed. The soak period then was determined by the quartz used to give time for dissolution. Firing up to higher temperatures was not needed thus saving energy.

ACKNOWLEDGEMENTS

The author wants to thank his co-workers H. Lorenz, G. Braun and W. Donaubauer for sample preparation and measurements, and K. Piper for the time and nervous effort in the dark room.

LITERATURE

1. H. Mörtel, Ceramurgia International 3 65-69, (1977).
2. H. Mörtel, Science of Ceramics 9, 1977, in press.

STRENGTHENING OF LIME-STABILIZED ZIRCONIA BY POST SINTERING HEAT TREATMENTS

R.C. Garvie, R.R. Hughan and R.T. Pascoe[†]

CSIRO, Division of Tribophysics

Melbourne, Australia

INTRODUCTION

Two-phase ceramics formed from zirconia (ZrO_2) and moderate quantities (usually less than 15 mol%) of stabilizing oxides such as CaO, MgO or Y_2O_3 are known as partially-stabilized zirconia (PSZ). The matrix phase has a cubic, fluorite structure and is a solid solution of the stabilizer and zirconia. The dispersed phase has the structure of either the low or the intermediate temperature polymorph of zirconia, which have monoclinic and tetragonal symmetry, respectively. The stabilizer used in the present work was lime (CaO). The relevant part of the working phase diagram is given in Fig. 1.[1]

The transformation of the tetragonal phase to monoclinic at about 1000°C is fast and cannot be suppressed in bulk zirconia.[2] However, small domains with a tetragonal structure precipitated in PSZ can retain that symmetry on cooling to room temperature. Materials with such a precipitate are exceptionally strong and tough, compared to other oxide ceramics, when the precipitate domains approach a critical size; about 100nm when stabilized with lime.[3,4] These high strength materials are obtained by homogeneous precipitation from an originally supersaturated solid solution through post-sintering heat treatments. The effect of these treatments at different temperatures on materials in the composition range, 3.0-5.0 wt% CaO is described in this paper.

It will be seen that there are analogies with age hardening

[†] deceased

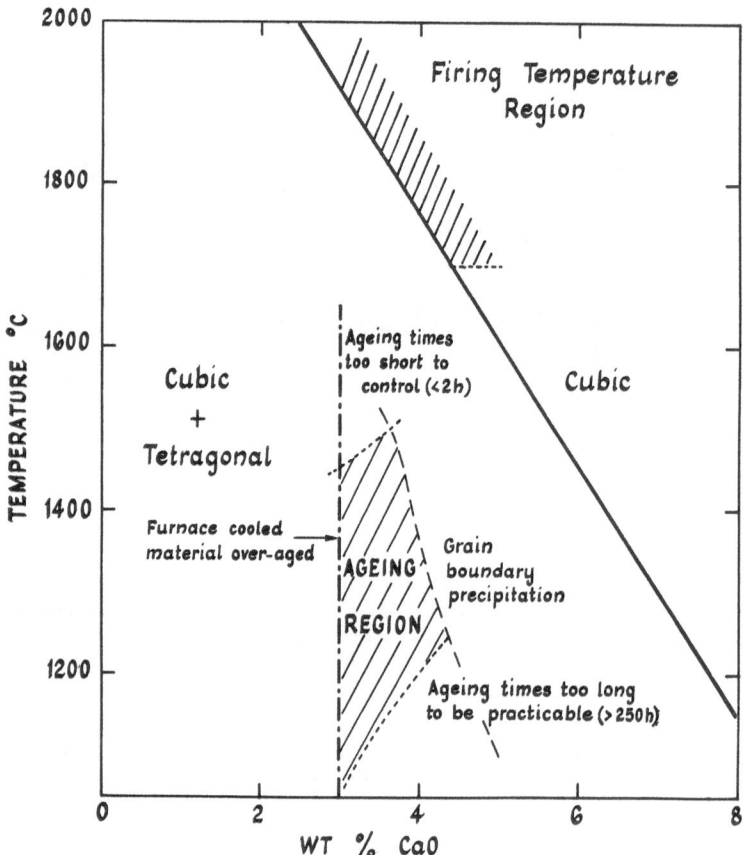

Fig. 1. Working phase diagram for the CaO-ZrO$_2$ system.

shown by some metallic alloys and the nomenclature of age hardening
will be adopted here. Precipitation and coarsening below the
solid solubility line (the solvus) will be referred to as ageing.
Materials with the highest strength will be called peak-aged and
those aged for longer and shorter periods as overaged and
underaged respectively.

FABRICATION AND CHARACTERIZATION

Powder Processing and Firing

The material was prepared by a conventional mixing and sintering process followed by a post-sintering heat treatment to optimize
the size and distribution of the precipitate phase.

The starting materials were commercial high-purity zirconia powder (99.7% excluding ~2% natural hafnia) and sufficient calcium acetate to give 3.0 to 5.0 wt% CaO in the final product. The stabilizer could be added as powdered calcium oxide but a more homogeneous solute distribution and a higher fired density was obtained when it was added as a compound that decomposed to the oxide on heating. Other compounds, such as the carbonate, hydroxide, nitrate or oxalate, could equally well have been used.

The zirconia powder and calcium acetate were blended under an inert fluid (1,1,1 - trichloroethane) and dried. The mixed powder was calcined in air at 1050°C for 24h to decompose the salt and then milled to break up coarse aggregates and mix in polyethylene glycol as a binder. The test bars used for the results reported here were die pressed and then isostatically pressed to 200 MPa pressure.

The sintering procedure was conventional except that the temperature was dictated by the need for a single phase micro-structure. The recommended firing temperature region, shown on Fig. 1, was maintained until solid solution was complete; this took 1 to 4 hours depending on temperature. The sintered body was then cooled comparatively rapidly. The rate must be sufficiently fast to achieve homogeneous rather than grain boundary precipitation during cooling, and yet be slow enough to avoid thermal shock cracking. Material described here as "furnace cooled" was cooled at the natural rate of a small furnace which took one hour to reach 1300°C from the firing temperature.

Ageing and Strength Changes

The sintered material was aged isothermally to give the required properties, usually peak strength. Changes in fracture strength during ageing were accompanied by structural changes that are to be described in detail elsewhere.[5] Nucleation and precipit-ation were very rapid, being largely completed during furnace cooling. Thereafter the precipitated particles increased in size and decreased in number, larger particles growing at the expense of smaller ones whilst maintaining the amounts of the two phases approximately constant.

Ageing can be combined with the sintering operation by arrest-ing cooling at a suitable temperature, or it can be a separate reheating process after first cooling to room temperature. Both methods have their merits but the second is preferred where the ageing characteristics of the material are not known.

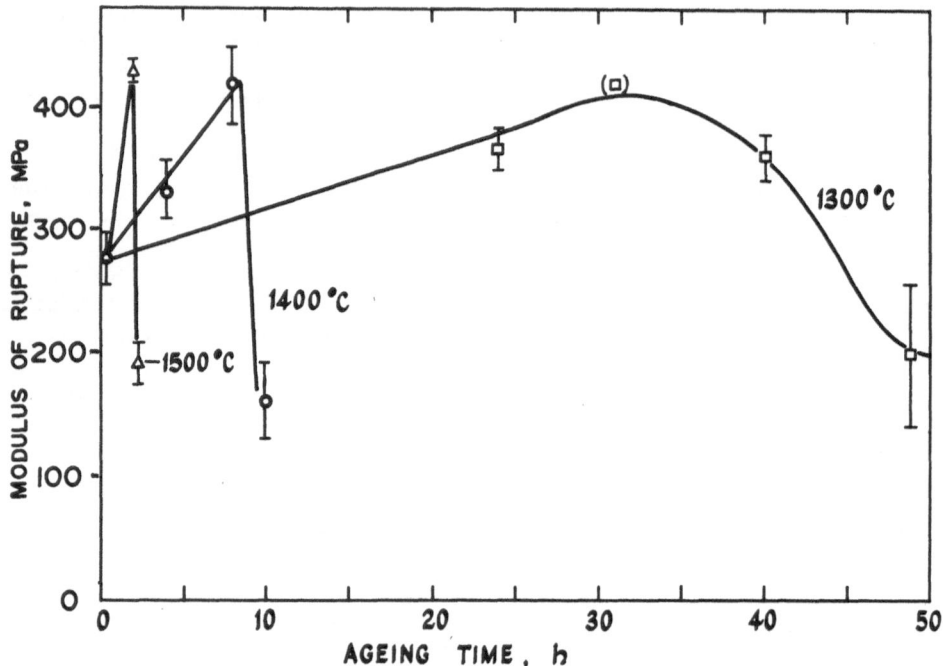

Fig. 2.   The effect of ageing temperature on the strength–ageing
          time curves for the 3.6 wt% alloy. Note: These specimens
          were annealed after surface grinding before the effect of
          grinding in increasing strength was discovered. Hence the
          discrepancy between values in Figs. 2 and 4. The point
          marked (□) came from a different batch of powder.

        The progress of ageing can be monitored by measuring the change
in fracture strength as a function of time at temperature.   The
flexural strengths presented here were determined in four point
bending on unnotched bars 3 x 3mm in section and 28.6mm between
outer knife edges. All bars were surface ground; this caused
metastable domains near the surface to transform to the stable
monoclinic structure to give a compressive surface layer and a
10–20% increase in fracture strength.[6]

        The effect of ageing temperature on the strength/time curves
for a fixed composition is shown in Fig. 2. For each temperature
the strength increases to a peak and then drops off rapidly. The
time to reach the peak decreases by a factor of about four for each
hundred degrees increase in temperature, corresponding to an
apparent activation energy of 300±50 kJ/mol. There is an upper
limit to the temperature at which useful ageing can be obtained.
For example, when the 3.6 wt% alloy used to determine the data

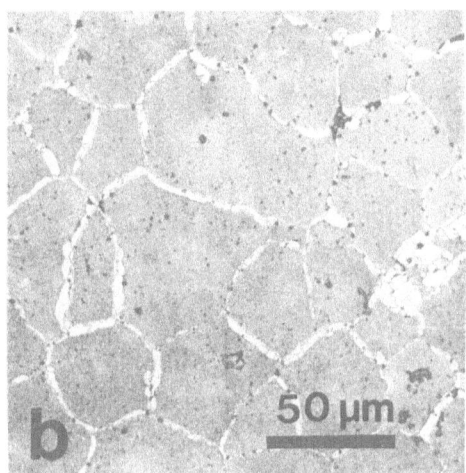

Fig. 3.   Optical microstructure of 3.6 wt% alloy.
(a) strong material; (b) weak material.

given in Fig. 2 was aged at 1600°C for a time calculated to yield
peak strength, it proved very weak, having a strength of only 160
MPa.   Ageing of the alloy at or beyond the upper limit results in
low under-cooling which favours grain boundary precipitation.   The
optical microstructures of strong and weak materials are compared
in Fig. 3.   Only small quantities of grain boundary precipitate
are visible in the strong material (Fig. 3a), electron microscopy
being necessary to reveal the 50% of tetragonal precipitate
present, while copious precipitation of monoclinic phase is seen
optically in Fig. 3b.

     The time to reach peak-strength after furnace cooling increased
as the CaO content increased, while the initial strength decreased,
as shown in Fig. 4.   All partially stabilized compositions were
stronger in the furnace cooled condition than fully-stabilized
zirconia, which has a strength of ∿170 MPa that is not strongly
dependent on composition.   It is apparent therefore that partial
ageing occurs during furnace cooling.

     The peak strength is roughly independent of composition up to
4 wt% CaO beyond which it decreases, probably as a result of
decreased supersaturation and therefore increased grain boundary
precipitation.   The 5.0 wt% alloy showed little strengthening at
1300°C so the data presented in Fig. 4 was obtained at 1200°C, and
the times corrected for temperature on the assumption that the
apparent activation energy is independent of composition.   The
lower temperature increased the level of under-cooling, favouring

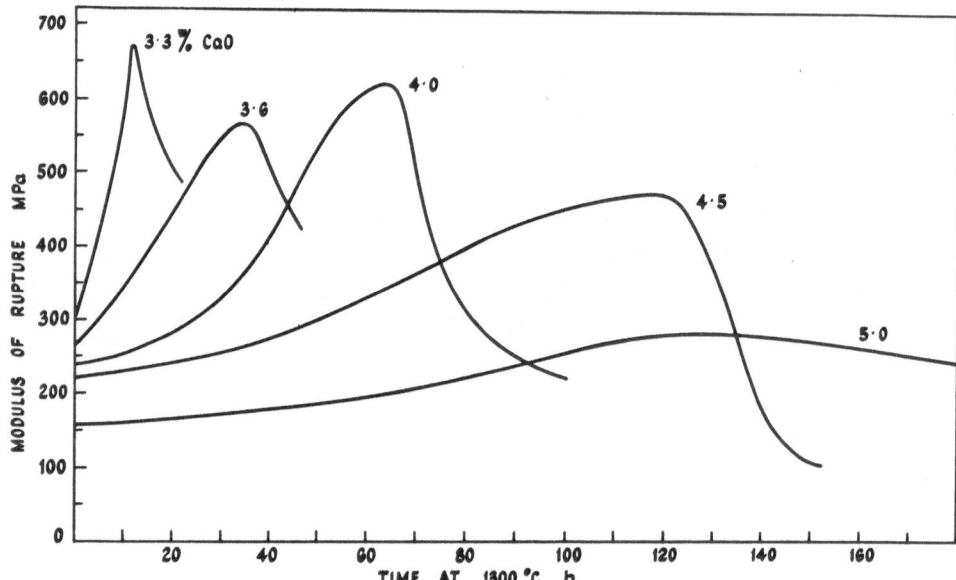

Fig. 4.   Strength–ageing time curves at 1300°C for various
          compositions. Note: surface ground specimens ⁻ see note
          on Fig. 2.

homogeneous precipitation.

     The temperature/composition region of optimum ageing, for the
cooling rate used in the present work, is shown on Fig. 1.   The
boundary dictated by the need to achieve homogeneous precipitation
is the curve roughly parallel to the solvus. Alloys containing 3.0
wt% and less are effectively overaged by furnace cooling so this
composition defines the lower composition limit to the field.   The
upper limit of composition is based on the arbitrary requirement
that ageing times should be less than 250 hours to be of any
practical use.

     Useful ageing can also be obtained by continuous cooling if
excessive time in the region of grain boundary precipitation is
avoided. Suitable cooling rates can be estimated from the time/
temperature relationships derived from Fig. 2 but control of the
process is more difficult.

                              Hardness

     There was little significant change in hardness during ageing
except that a drop occurred on overageing. The mean Vickers hardness

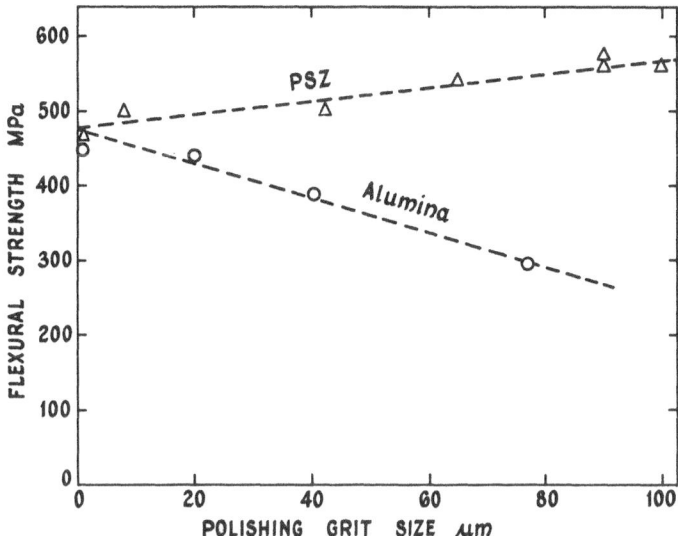

Fig. 5. Comparison of the effects of grinding with various grit
sizes on the strengths of Ca-PSZ and fine-grained aluminas.[10]

of a series of alloys aged up to the peak was 1320±80 kgf/mm^2 and
for a series of overaged alloys was 1180±50. A large part of the
scatter was due to the hardness impressions overlapping pores and
to hardness anisotropy in the various grains examined. Microhardness measurements were made in the hope of obtaining a more precise
monitoring technique but the scatter was similar and no significant
trends as a function of ageing time could be discerned.

The curves shown in Figs. 2 and 4 are similar to those obtained
for age hardening in metals. However, in the case of PSZ, it is the
resistance to fracture that increases to a peak on ageing and not
the resistance to plastic flow, as measured by the hardness. Therefore PSZ cannot be referred to as age-hardened or as precipitation
hardened. We suggest that a designation such as Transformation
Toughened Zirconia be used for materials of this type to distinguish
them from other forms of PSZ, such as those fired in the two-phase
region. We use the word 'toughened' because, as shown below, the
strengthening arises from an enhanced value of the work of fracture.

PROPERTIES

The equations of fracture mechanics show that a ceramic may be
strengthened by two routes: reducing the inherent defect size and
increasing the fracture energy.[7] Fracture energies of ceramics have

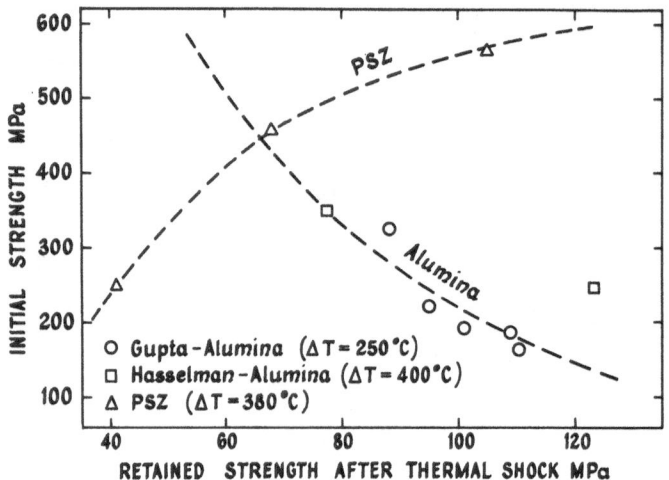

Fig. 6.   Comparison of the retained strength of Ca-PSZ after
          thermal shock with alumina.[11,12]

been measured on edge-notched bars using cut notches of various
depths.[8,9] We find that the strengthening of PSZ during ageing
results from an increase in fracture energy and that the inherent
defect size remains constant at ∿50μm. The fracture energy, G, at
peak strength is 500 J/m^2 corresponding to a fracture toughness,
K_{Ic}, of 10 MPa m$^{\frac{1}{2}}$ (Young's modulus 204 GPa). These fracture parameters are unusually high for an oxide ceramic material. For high
density alumina we measure G = 60 J/m^2 and K_{Ic} = 4 MPa m$^{\frac{1}{2}}$ in the
same test. Thus peak-aged PSZ is amongst the strongest and toughest
of all ceramics made by pressureless, solid-state sintering.

     As expected, most of the fracture-related properties of PSZ are
improved by the increase in fracture energy. The inherent flaw size
is large and similar to the grain size so abrasive processes that
introduce surface cracks smaller than 50μm should cause no degradation in strength. This resistance to abrasive degradation combined
with the surface transformation on grinding leads to a most unusual
effect for a brittle material.[6] As the grit size used in grinding
is increased, both the depth of damage and the amount of metastable
phase transformed to the monoclinic form increase. As a consequence,
the strength of peak-aged PSZ ceramics increases as the severity of
grinding increases, up to an abrasive particle size of 100μm (150
grit) at least. In Fig. 5 this behaviour is contrasted with that
of fine-grained alumina of a similar strength which is weakened by
abrasion.[10]

The resistance of PSZ to thermal shock is also improved by ageing. Peak-aged bars have to be water quenched from 300°C before a
catastrophic loss of strength occurs.[4] The corresponding critical
temperature for alumina is 220°C.[11] The strength retained by PSZ
after water quenching from higher than 300°C, increases with
increasing initial (unshocked) strength (Fig. 6). This behaviour
again contrasts with that of alumina where the retained strength
after shocking varies inversely with the initial strength. It can
be shown by an extension of Hasselman's[12] unified theory of thermal
shock that the behaviour of PSZ is typical of a material that is
strong because of a high fracture energy. The behaviour of alumina,
on the other hand, is typical of a material strengthened by reducing
inherent defect size.[11]

Variability of Properties of Commercial Materials

Partially-stabilized zirconia ceramics have been recognised for
some time as offering improved mechanical properties over fully-
stabilized zirconia and, in some respects such as toughness and
thermal shock resistance, over all other oxide ceramics. Commercial
production of PSZ has been undertaken but consistent performance has
rarely been achieved and the material has built up a reputation for
variability of properties. Thus, one set of extrusion dies for
copper alloys performed outstandingly well while some dies in a
subsequent batch failed before a single billet was extruded.[4]

The reasons for this variability are apparent from the ageing
behaviour described above. For any composition there is a narrow
"window" of time-temperature history that will give strong material.
When this is combined with the natural reluctance of ceramic
processors to cool at the necessary fast rates and risk high thermal
stresses, it is not surprising that material with the strength and
toughness reported here has not been made consistently.

A further problem arises when continuous cooling methods are
used to produce the necessary precipitation and strengthening. Once
conditions have been established that give peak strength any change
in cooling rate will lead to a reduction in properties. This is
particularly dangerous if a composition is established that gives
maximum strength in a small batch furnace. If a larger production
furnace is used for the same composition, or a larger batch is fired
in the same furnace, the cooling rate will almost certainly be
slower. As a result material that was peak-aged in a small furnace
will be overaged in a large one.

CONCLUSIONS

1. ZrO_2-CaO alloys containing 3.0 to 4.5 wt% CaO can be
    strengthened by isothermal ageing after rapid furnace cooling.

2. The rate of microstructural change that causes the strengthening
    increases with increasing temperature and decreases with
    increasing solute content.

3. The conditions under which useful strengthening occurs are
    limited.

4. The fracture energy is unusually high for a ceramic and results
    in improved fracture-related properties such as the resistance
    to thermal shock and surface degradation.

ACKNOWLEDGEMENTS

*We wish to thank Messrs. V. Gross and C. Urbani for sample
preparation, Dr. R.H.J. Hannink for microhardness measurements and
Dr. N.A. McKinnon for discussions and support.*

REFERENCES

1. R.K. Stringer et al, to be published.
2. E.C. Subbarao, H.S. Maiti and K.K. Srivastava, phys. stat.
    sol. A21 9 (1974).
3. R.C. Garvie, R.H.J. Hannink and R.T. Pascoe, Nature 258
    [5537] 703 (1975).
4. R.C. Garvie, R.H.J. Hannink, R.R. Hughan, N.A. McKinnon,
    R.T. Pascoe and R.K. Stringer, J. Aust. Ceram. Soc. 13
    8 (1977).
5. R.H.J. Hannink, K.A. Johnston and R.T. Pascoe, to be published.
6. R.T. Pascoe and R.C. Garvie, "Ceramic Microstructures '76"
    pp. 774-784, Eds. R. M. Fulrath and J. A. Pask, Westview
    Press, Boulder, Colorado, 1977.
7. F.F. Lange, "Fracture Mechanics of Ceramics". Vol. 1, 3,
    Plenum Press, New York, 1974.
8. R. W. Davidge and G. Tappin, J. Mater. Sci. 3 165 (1968).
9. C. Turner, Mater. Sci. Eng. 11, 275 (1973).
10. R.E. Tressler, R. A. Langensiepen and R. C. Bradt, J. Amer.
    Ceram. Soc. 57 226 (1974).
11. T.K. Gupta, J. Mater. Sci., 8 1283 (1973).
12. D.P.H. Hasselman, J. Amer. Ceram. Soc. 52 600 (1969).

DISCUSSION

R. N. Katz (AMMRC): The thermal shock data which you have shown is truly impressive. However, as the retained strength shown was due to the tetragonal → Monoclinic phase transformation is it possible that the first (or first several) thermal shocks will exhaust the sites for such transformation leading to an eventual reduction of strength in thermal fatigue environments?

Author: Limited repeated thermal shock tests (12 cycles, $400^\circ C$ to room temperature, water quench) showed an initial loss in strength to a level which then remained constant during the course of the experiment.

L. De Jonghe (Cornell University): Could you comment on why the modulus of rupture drops off so steeply upon high temperature versus $1300^\circ C$ aging?

Author: We think it is simply a question of kinetics. At high temperatures, coarsening proceeds at a faster rate. Consequently, at these temperatures the occurrence of the majority of the precipitate particles which exceed the critical size takes place in a shorter aging time interval resulting in a steep drop off in strength.

D. Viechnicki (AMMRC): What is the optimum or maximum use temperature of CaO-PSZ?

Author: $1000-1100^\circ C$. Steel has been extruded at $1000^\circ C$.

F. F. Lange (Rockwell International): With regard to the transformed layer on the surface due to surface grinding, we at Rockwell are using tetragonal ZrO_2 as a "tool" to measure damage depth due to particle impact. Damage depth (as measured by X-ray diffraction and calibration) correlate well with predictions obtained from single particle impact studies, viz. damage depth corresponds to the depth of lateral cracking during impact. Similar effects should be obtained during grinding.

Author: Your technique of using metastable tetragonal zirconia as an indicator of damage is interesting. We found the amount of transformed monoclinic material in a ground surface as a function of depth obeyed an exponential relation. See reference 6.

D. R. Clarke (Rockwell International Science Center): In your optimum aged material some discontinuous precipitation at the grain boundaries still occurs as seen in the optical microscope pictures. Does macroscopic fracture occur along these boundaries?

Author: The fracture mode in these materials is mostly transgranular. On this basis the small amount of grain boundary phase appears to have little adverse influence on the strength of peak-aged material.

Frans Carpay (Philips Research Laboratories): In your plot of the domain size versus the aging time you have found a combination of two curves, the one with a slope of 1/3, the other with a slope of 1. I would suggest not to plot the domain size (D) versus time (t), but D (time:t)-D (time:0) versus t. Kinetically this is more relevant, and I expect that it will change the slopes.

Author: No comment. These particular data will appear in another publication.

STRAIN AND SURFACE ENERGY EFFECTS IN CERAMIC PROCESSES

E. Dow Whitney

University of Florida

Gainesville, Florida 32611

ABSTRACT

Strain and surface energy effects often play important roles
in determining both the kinetic nature of phase transformations
and in the appearance of metastable high temperature phases.
Familiar examples are to be found in the effect of these parameters
on the kinetics of the monoclinic-tetragonal transformation in
zirconia, the occurrence of high temperature tetragonal ZrO_2 as a
metastable state at low temperatures, and in the appearance of
high cristobalite in samples of opal at room temperature. High
surface area powders possessing "excess" surface energy (relative
to a large single crystal) are called "active" and exhibit sometimes unusual properties.

INTRODUCTION

In the processing of crystalline ceramics the ceramist often
encounters first order phase transformations occurring within the
solid state where volume changes are not easily accomodated by flow
of the participating phase as in the case of transformations involving a fluid phase. Thus for phase transformations occurring
within the solid state the Gibbs free energy function $G = H - TS$
must include both surface and strain energy contributions, i.e.,
for the solid-state transformation $\alpha \rightleftarrows \beta$ the following relation is
established at the equilibrium temperature and pressure[1]

$$(G_\beta - G_\alpha) + (S_\beta \gamma_\beta - S_\alpha \gamma_\alpha) + (\epsilon_\beta - \epsilon_\alpha) = 0 \qquad (1)$$

where G, S, γ and ε denote the Gibbs free energy, molar surface area, surface energy and strain energy of the β or α phase, respectively. The first term on the left side of eq. (1) is the chemical free energy difference between the β and α phases in bulk, the second term the surface energy difference, and the third term the strain energy difference. Equation (1) shows that the transformation may have thermodynamic "permission" to proceed even though the chemical free energy term ($G_\beta - G_\alpha$) is positive provided that either one or a combination of the surface energy and strain energy differences is sufficiently negative.

These effects may, for example, play important roles in determining the kinetic nature of phase transformations and in the existence of metastable high temperature phases. Familiar examples may be found in the effect of these parameters on the kinetic nature of the monoclinic-tetragonal transformation in zirconia, in the existence of high temperature tetragonal ZrO_2 as a metastable state at low temperatures, and in the appearance of high cristobalite in samples of opal at room temperature.[2] High surface area powders possessing excess energy (relative to a large single crystal) are called "active" and exhibit sometimes unusual properties as in the variation of the heat of solution of MgO in aqueous hydrochloric acid by as much as 8.37 kJ mol^{-1} (2 kcal mol^{-1})[3] and the thermal decomposition of Al_2O_3 at relatively low temperatures.[4]

PHASE TRANSFORMATIONS IN ZIRCONIA

Metastable Tetragonal ZrO_2

Although many investigators have studied the mechanism of stabilization of tetragonal ZrO_2 at low temperatures and pressures the reason for the existence of this high temperature polymorph of ZrO_2 under ambient conditions is not fully understood. Mitsuhasi, et al.[1] have referenced much of the earlier studies on this subject. Subbarao, et al.[5] have published a comprehensive review of transformations in ZrO_2 including a discussion of the metastable high temperature phases. Universal agreement as to the mechanism of stabilization of metastable ZrO_2 does not exist, however. Stabilization of high temperature tetragonal ZrO_2 under ambient conditions has been attributed to (1) the presence of OH^- ions or other anion impurities,[6-9] (2) a size effect wherein fine particles of tetragonal ZrO_2 are considered stabilized as a result of the surface energy of the tetragonal phase being much lower than that of the monoclinic phase,[10-15] and (3) a martensitic-type transformation.[4,16-18] In addition to the size effect, Bailey, et al.[15] considered the influence of stored strain energy on the transformation but concluded it was too small to play a major role in determining the observed stability of the tetragonal phase at small particle sizes.

Pressure Induced Transformations

Both theoretical[19] and experimental[20] studies have shown that above 3.80 GPa (38.0 kbars) at 298 K, tetragonal ZrO_2 is thermodynamically stable. The transformation under these conditions is enantiotropic. Whereas tetragonal ZrO_2 formed at high pressures cannot be retained under ambient conditions, this modification can be obtained as a metastable phase at low temperatures and pressures from either the thermal decomposition of zirconyl salts[11] or via ball milling of monoclinic ZrO_2.[15]

The thermodynamic or mechanical work associated with the equilibrium high pressure monoclinic-tetragonal ZrO_2 transformation is $P\Delta V$. This quantity is negative since at 298 K the tetragonal phase, with a density of 6,100 kg m^{-3} (6.10 g cm^{-3}), is about 4.6% more dense than the monoclinic phase with a density of 5,830 kg m^{-3} (5.83 g cm^{-3}).[21] Corresponding molar volumes are 2.02×10^{-7} m^3 mol^{-1} (20.2 cm^3 mol^{-1}) for the tetragonal, and 2.11×10^{-7} m^3 mol^{-1} (21.1 cm^3 mol^{-1}) for the monoclinic phase. The mechanical work term for $P - 3.80$ GPa and $\Delta V = -0.94 \times 10^{-6}$ m^3 mol^{-1} is $P\Delta V = -3.57$ kJ mol^{-1} (-854 cal mol^{-1}) which represents the energy (mechanical work) loss associated with the formation of thermodynamically stable tetragonal ZrO_2 at 298 K under these conditions.*

DISCUSSION

From eq. (1) the condition for stabilization of metastable tetragonal ZrO_2 is given by the expression

$$G_t + S_t \gamma_t + \varepsilon_t < G_m + S_m \gamma_m + \varepsilon_m \tag{2}$$

where G, S, γ and ε denote the Gibbs free energy of zirconia in the form of large single crystals, molar surface area, surface energy and strain energy of the tetragonal and monoclinic phases,

---

*The enthalpy change ΔH associated with the monoclinic-tetragonal ZrO_2 transition at 1478 K is reported[22] to be 5.94 kJ mol^{-1} (1420 cal mol^{-1}) and thus the entropy change is

$$\Delta S = \frac{\Delta H}{T} = 4.02 \text{ J K}^{-1} \text{ mol}^{-1} \text{ (0.96 cal K}^{-1} \text{ mol}^{-1})$$

Assuming ΔH to be independent of temperature and pressure the change in internal energy ΔE associated with the monoclinic-tetragonal ZrO_2 transition at 298 K and 3.80 GPa is obtained from the expression $\Delta H = \Delta E + P\Delta V$, i.e.,

$$\Delta E = 9.51 \text{ kJ mol}^{-1} \text{ (2274 cal mol}^{-1})$$

respectively. Since the Gibbs (chemical) free energy difference
between tetragonal and monoclinic ZrO_2 in bulk is positive at
298 K $[G_t - G_m = 5.08$ kJ mol^{-1} (1213 cal mol^{-1})][19] stabilization
of the tetragonal phase must be the result of negative surface
energy and/or strain energy differences, i.e.,

$$S_t \gamma_t + \varepsilon_t \ll S_m \gamma_m + \varepsilon_m \qquad (3)$$

Garvie[11] has proposed that metastable tetragonal ZrO_2 occurs
as the result of a crystallite size effect and has correlated
this with those physical properties which are most important in
describing active powders namely, mean crystallite size, surface
area and excess energy. Starting with ZrO_2 of submicron crystallite size prepared by either precipitation or decomposition reactions, followed by heating at different temperatures, the crystallite size could be gradually increased with increasing temperature.
Garvie found the critical crystallite size for the tetragonal
monoclinic transformation to be about 300 m^{-10} (300 Å). Above
this critical size metastable tetragonal ZrO_2 could not exist at
ambient temperatures. Garvie postulated the two phases as being
in equilibrium at 300 K when the crystallite size is about 300 m^{-10},
and the sum of the free energies and surface energies are equal
as expressed by the equation

$$G_t + \gamma_t S_t = G_m + \gamma_m S_m \qquad (4)$$

This expression does not take into consideration the contribution
of strain energy to the overall free energy, however. Since $G_t >$
G_m the stability of the tetragonal phase according to this argument
is attributed exclusively to the smaller surface energy of tetragonal ZrO_2.

Bailey, et al.[15] found in their studies on phase transformations in zirconia that the tetragonal polymorph was formed from
monoclinic ZrO_2 at ambient temperature when the particle size was
made sufficiently small by vibration ball milling. Their x-ray
and electron microscopy studies indicated that the appearance of
the tetragonal phase was associated with a reduction in average
particle size and average crystallite size, and that milling introduced a degree of internal lattice strain. Utilizing the relationship that for cubic or spherical particles $A = 6/\rho D$, where ρ is
the density and D is either the cube edge or sphere diameter,
Bailey and his co-workers showed that eq. (2) can be written in
the form

$$G_t + \frac{6\gamma_t}{\rho_t D} + \varepsilon_t < G_m + \frac{6\gamma_m}{\rho_m D} + \varepsilon_m$$

$$\therefore \quad D < D_c = \frac{6}{[(G_t - G_m) + (\varepsilon_t - \varepsilon_m)]} \left(\frac{\gamma_m}{\rho_m} - \frac{\gamma_t}{\rho_t} \right) \quad (5)$$

Thus a critical particle diameter, D_c, can be calculated provided

$$\frac{\gamma_m}{\rho_m} > \frac{\gamma_t}{\rho_t}$$

Equation (5) was solved using values of $\gamma_m = 1.13$ J m^{-2} (1130 ergs cm^{-2})[11], $\gamma_t = 0.77$ J m^{-2} (770 ergs cm^{-2})[23] and $G_t - G_m = 5.08$ kJ mol^{-1}, giving a calculated critical particle diameter of $D_c \sim 100$ m^{-10} (100 Å). Although this was of the same order of magnitude as the crystallite and particle sizes measured by Bailey and his co-workers it is only 1/3 the critical tetragonal ZrO$_2$ particle diameter found by Garvie. It is important to point out that in calculating the above value for D_c the strain energy terms ε_t and ε_m in eq. (3) were ignored. The arguments given for this omission were (1) the strain energy terms were assumed to be very small, and (2) it is by no means clear what proportion of the stored strain energy is released on transformation.

The objective of this paper is to show that strain energy effects in this system cannot be neglected. Furthermore, the proposition is made that at the monoclinic-tetragonal ZrO$_2$ equilibrium pressure of 3.80 GPa at 298 K the release of PΔV mechanical work energy of -3.57 kJ mol^{-1} may be taken as a measure of the release of stored strain energy associated with the transformation. Substituting this value for the strain energy term $\varepsilon_t - \varepsilon_m$ in eq. (5) gives a calculated critical tetragonal ZrO$_2$ particle diameter of $D_c \simeq 330$ m^{-10} (330 Å), a value in much better agreement with the results of Garvie.

It would appear that when a crystallite size of $\sim$300 m^{-10} (300 Å) is exceeded, those factors counter-acting the increased strain in the system introduced by the nucleation and growth of monoclinic ZrO$_2$ in the tetragonal ZrO$_2$ host lattice become ineffective and transformation proceeds to completion.

Employing the Warren-Averbach method, Buljan, et al.[24] have measured the microstrain involved in the monoclinic-tetragonal ZrO$_2$ transformation. This study showed that initial room temperature microstrain in monoclinic ZrO$_2$ is almost completely relieved in the pre-transformation region ($\sim$1373 K), then increases to a maximum value when about 50% of the monoclinic phase has been transformed followed by a sharp decrease, the microstrain approaching very low values as the transformation progresses.

These studies are in agreement with observations by Hart and Chaklader[25] of superplasticity in pure ZrO_2 during the monoclinic $\rightleftarrows$ tetragonal transformation. Buljan, et al. also estimated the stored strain energy associated with the transformation at 1473 K from Faulkner's formula[26]

$$E_s = \frac{15E}{2\rho(3-4\nu+8\nu^2)}\left(\frac{\Delta d}{d}\right)^2 \tag{6}$$

where E is Young's Modulus, ν Poisson's ratio, ρ density and $\frac{\Delta d}{d}$ is the strain. Substituting the value of $\frac{\Delta V}{3V}$ for the theoretical microstrain, where ΔV is the difference between the volume of the monoclinic and tetragonal phases at 1473 K and V is the volume of the monoclinic phase at the same temperature, a value of 1.20 kJ mol^{-1} (286 cal mol^{-1}) for the stored strain energy was obtained. This is 1/3 the value obtained from $P\Delta V$ calculations. However, it should be realized that Faulkner's formula is valid only for isotropic materials and thus its application to the monoclinic $ZrO_2 \rightleftarrows$ tetragonal ZrO_2 transformation is only an approximation.

Strain energy as determined by $P\Delta V$ measurements is also in agreement with Garvie's estimation of the strain energy associated with the transformation from differential thermal analysis (d.t.a.) studies.[11] By comparing d.t.a. peaks of coarse-grained ($\sim$820 m^{-10}) and fine-grained ($\sim$232 m^{-10}) monoclinic ZrO_2 going through the transition Garvie estimated the strain energy associated with the transformation to be 3.35 kJ mol^{-1} (800 cal mol^{-1}).

One further comparison can be made. The above author also calculated the excess energy ΔE of active samples of precipitated zirconia from the derived expression

$$\Delta E = \frac{12N\ (n-1)^2\ d^2\ K\ \gamma}{n^3} \tag{7}$$

where N = Avogadro's number, d = the diameter of the oxygen ion, K = a factor to convert ergs to kilocalories and γ = the surface energy. It was assumed that ZrO_2 had the ideal fluorite lattice and that the powder was composed of cubes with n oxygen ions/edge. Equating surface energy and strain energy, i.e., expressing ΔE as 3.57 kJ mol^{-1} and solving for n gives n = 106 corresponding to a crystallite size of $\approx$290 m^{-10} (290 Å), a value again close to the observed critical tetragonal crystallite size of 300 m^{-10}. One may argue that strain energy is being consumed in the creation of additional ZrO_2 crystallite surface area analogous to the argument of relating surface energy and strain energy in the calculation of the theoretical strength of materials.

It would appear that at ordinary temperatures small crystallites of ZrO_2 prefer the tetragonal form whereas larger sizes prefer the monoclinic polymorph. This suggests that the portion of the crystal near the surface prefers the high temperature form. This is in general argument with the computations of Lennard-Jones and Dent[27] which showed that the surface of ionic crystals tended to have smaller lattice dimensions than the interior of the crystals. The contraction of the lattice at the (100) boundary of crystals of the NaCl type was found by Lennard-Jones and Dent to be of the order of 5 percent which is comparable with the $\frac{\Delta V}{V}$ contraction of 4.6 percent in transforming from monoclinic to tetragonal ZrO_2. It is to be expected, therefore, that the phenomena of surface contraction would favor the formation of tetragonal ZrO_2.

CONCLUDING REMARKS

The data of Garvie and Bailey and his co-workers on crystallite size effects in the stabilization of tetragonal ZrO_2 can be brought into much closer agreement by the introduction of previously neglected strain energy effects. A measure of this strain energy is taken to be the $P\Delta V$ work associated with the equilibrium monoclinic $ZrO_2 \rightleftarrows$ tetragonal ZrO_2 at 3.80 GPa and 298 K. Equating strain energy to Garvie's surface "excess" energy also leads to a calculated critical crystallite size for the existence of metastable tetragonal ZrO_2 in agreement with experimental observations.

ACKNOWLEDGEMENTS

The author thanks Robert T. DeHoff for his valuable discussions and critical review of the manuscript.

REFERENCES

1. T. Mitsuhasi, M. Ichihara and U. Tatsuke, "Characterization and Stabilization of Metastable Tetragonal ZrO_2", J. Amer. Ceram. Soc. 57 [2] 97-101 (1974).

2. M. J. Buerger, "Crystallographic Aspects of Phase Transformations", in Phase Transformations in Solids (R. Smoluchowski, J. E. Mayer and W. A. Weyl, Eds.), John Wiley and Sons, Inc., New York, N. Y. (1951) pp. 183-211.

3. D. K. Thomas and T. W. Baker, "An X-ray Study of the Factors Causing Variation in the Heats of Solution of Magnesium Oxide", Proc. Phys. Soc. (London) 92 [12] 673-79 (1959).

4. B. Arghiropoulos, J. Elston, P. Hilaire, F. Juillet and S. J. Teichner, "Reactivite des Alumines Noires Non Stoechiometriques", in Reactivity of Solids (S. H. DeBoer, et al., eds.), Elsevier Publishing Co., New York, N. Y. (1961) pp. 525-39.

5. E. C. Subbarao, H. S. Maiti and K. K. Srivastava, "Martensitic
    Transformation in Zirconia", Phys. Stat. Sol. (A) 21 9-40
    (1974).
6. G. L. Clark and D. H. Reynolds", Chemistry of Zirconium Dioxide
    X-ray Studies", Ind. Eng. Chem. 29 [6] 711-15 (1937).
7. R. Cypres, R. Wollast and J. Raucq, "Polymorphic Conversion of
    Pure Zirconia", Ber. Deut. Keram. Ges. 40 [9] 527-32 (1963).
8. A. Clearfield, "Crystalline Hydrous Zirconia", Inorg. Chem. 3
    [1] 146-48 (1964).
9. E. D. Whitney, "Kinetics and Mechanism of Transition of Meta-
    stable Tetragonal to Monoclinic Zirconia", Trans. Faraday
    Soc. 61 [9] 1991-2000 (1965).
10. A. Krauth and H. Meyer, "Modifications Produced by Quenching
    and Their Crystal Growth in the Systems Containing Zirconium
    Dioxide", Ber. Deut. Keram. Ges. 42 [3] 61-72 (1965).
11. R. C. Garvie, "Occurrence of Metastable Tetragonal Zirconia as
    a Crystallite Size Effect", J. Phys. Chem. 69 [4] 1238-43
    (1965).
12. K. S. Mazdiyasni, C. T. Lynch, and J. S. Smith, "Preparation
    of Ultra-High Purity Submicron Refractory Oxides", J. Amer.
    Ceram. Soc. 48 [7] 372-75 (1965).
13. M. Poleshaev, "Low Temperature Cubic and Tetragonal Forms of
    Zirconium Dioxide", Zh. Fiz. Khim. 41 [11] 2958-59 (1967).
14. I. A. El-Shanshoury, V. A. Rudenko and I. A. Ibrahim, "Poly-
    morphic Behavior of Thin Evaporated Films of Zirconium and
    Hafnium Oxides", J. Amer. Ceram. Soc. 53 [5] 264-68 (1970).
15. J. E. Bailey, D. Lewis, Z. M., Librant and L. J. Porter, "Phase
    Transformations in Milled Zirconia", Trans. J. Brit. Ceram.
    Soc. 71 [1] 25-30 (1972).
16. G. M. Wolten, "Direct High-Temperature Single-Crystal Observa-
    tion of Orientation Relationship in Zirconia Phase Transfor-
    mation", Acta Cryst. 17 763-65 (1964).
17. J. E. Bailey, "The Monoclinic-Tetragonal Transformation and
    Associated Twinning in Thin Films of Zirconia", Proc. Royal
    Los. (London) A279 [6] 395-412 (1964).
18. G. K. Bansal and A. H. Heuer, "On A Martensitic Phase Trans-
    formation in Zirconia (ZrO_2) -- I. Metallographic Evidence",
    Acta Met. 20 [11] 1281-89 (1972).
19. E. D. Whitney, "Effect of Pressure on Monoclinic-Tetragonal
    Transition of Zirconia; Thermodynamics". J. Am. Ceram. Soc.
    45 [12] 612-13 (1962).
20. G. L. Kulcinski, "High-Pressure Induced Phase Transition in
    ZrO_2", J. Amer. Ceram. Soc. 51 [10] 582-84 (1968).
21. J. D. McCullough and K. N. Trueblood, The Crystal Structure of
    Baddeleyite (Monoclinic ZrO_2)", Acta Cryst. 12 [7] 507-11
    (1959).
22. J. P. Coughlin and E. G. King, "High Temperature Heat Contents
    of Some Zirconium-Containing Substances", J. Amer. Chem.
    Soc. 72 [5] 226-65 (1950).

23. D. T. Livey and P. Murray, "Surface Energies of Solid Oxides and Carbides", J. Amer. Ceram. Soc. 39 [11] 363-72 (1956).
24. S. T. Buljan, H. A. McKinstry and V. S. Stubican, "Microstrain Measurements in Monoclinic Tetragonal Phase Transition of Zirconia", in PHase Transitions -- 1973 (H.K. Henisch, R. Roy and L. E. Cross, Eds.), Pergamon Press, New York, N. Y. (1973) pp. 307-15.
25. J. L. Hart and A. C. D. Chaklader, "Superplasticity in Pure ZrO_2", Mat. Res. Bull. 2, 521-26 (1967).
26. E. A. Faulkner, "Calculation of Stored Energy from Broadening of X-ray Diffraction Lines", Phil. Mag. 5, 519-21 (1960).
27. J. E. Lennard-Jones and B. M. Dent, "The Change in Lattice Spacing at a Crystal Boundary", Proc. Royal Soc. (London) A121, 247-59 (1928).

DISCUSSION

Dennis Viechnicki (AMMRC): In our work on growth of Al_2O_3/ZrO_2 eutectics we only saw monoclinic ZrO_2, never tetragonal ZrO_2, down to a rod size of 0.5 μm, which may support your strain energy hypothesis. To prove this one should try and splat cool this eutectic to get finer rod sizes to see the modification of ZrO_2. Also, what is the effect of impurities on stabilizing tetragonal ZrO_2, especially in light of the propensity of ZrO_2 to form cubic solid solutions with a large variety of dopants?

Author: Krauth and Meyer[1] melted 30-50 μm particles of pure ZrO_2 and of blends with various additives in a plasma jet. The molten material was severely chilled by spraying on a cold metal plate and then investigated by X-ray methods at room temperature. Whereas pure ZrO_2 gave only the monoclinic crystal form, additions of Al_2O_3 (as well as SiO_2 and ZrN) produced the tetragonal high temperature modification of ZrO_2. Interestingly enough, these investigators found in the ZrO_2/Al_2O_3 system a region in which the chilled specimens vitrified.

As discussed in this paper, there is no universal agreement as to the mechanism of stabilization of metastable tetragonal ZrO_2. The effects of impurities (primarily OH^- and other anions) in stabilizing high temperature tetragonal ZrO_2 under ambient conditions are discussed in references 6 through 9.

F. F. Lange (Rockwell International): Our views on this subject are that retention of tetragonal ZrO_2 (in a polycrystalline material) must be viewed in terms of the phase relations (monoclinic vs. tetragonal ZrO_2) in temperature, pressure, and compositional space. As shown by Whitney, pure tetragonal ZrO_2 can be retained at $25°C$ under ~37 kbar pressure. Constraint by one grain upon another resulting in $\sim$kbars is not likely, i.e., microcracks form and constraint is not maintained. On the other hand, investigators have shown that the T→M transformation temperature is lowered with additions of other oxides, e.g., the eutectoic is $\sim560°C$ for $ZrO_2 + Y_2O_3$. If it is assumed that $\partial T/\partial P$ is independent of composition, then the "constraint pressure" exerted by one grain upon another decreases with increasing Y_2O_3 (up to the eutectoid composition). Thus, for a given, small grain size, retention of tetragonal ZrO_2 can be achieved by constraint when a given amount of Y_2O_3 is added. Release of this constraint by the stress field of a crack "absorbs" work. Thus, such materials are tough.

---

[1] A. Krauth and H. Meyer, "Crystal Modifications Produced by Chilling (Quenching) and Their Growth Rate in Systems Containing Zirconium Dioxide," Ber. Deut. Keram. Ges. 42, 61-72 (1965).

DYNAMIC AND MATERIAL PARAMETERS IN BRITTLE FRACTURE IN CERAMICS

P.J. Gielisse, A. Choudry and T. Kim

University of Rhode Island

Kingston, Rhode Island 02881

I. INTRODUCTION

Some years ago when searching for an explanation of the behavior of materials, specifically brittle materials, under impact loading we considered the potential for designing a model which would incorporate the system's material properties as well as the impact dynamics. Our specific interests were directed towards the abrasive finishing or grinding process which is unique in that it requires consideration of the real time spatial stress distribution in a repetitive mode. Knowledge of this type is required in as much as it, directly or indirectly, relates to a wide variety of technologically inportant areas[1-4]; ballistic impact (armour), percussive wear, abrasive machining, two and three body polishing, grinding, erosion processes, general friction and wear, ultrasonic machining and rock excavation.

Modelling in these areas, but specifically in grinding of brittle materials, has been limited or has been directed at very special conditions only[5]. Most of the work has been done in the field of metal stock removal in which one heavily relies on concepts of plastic deformation. The models tend to be phenomenologically based, i.e., do not describe fracture phenomena and generally do not incorporate basic material characteristics. When material properties are introduced this is normally done with data that is statically derived at ambients not characteristic of the process. Most models are thus empirical, not based on first principles, and do not stress the dynamic (kinetic and vibrational aspects) but rather the static, quasistatic and local deformation concepts. If dynamic parameters are entered this is normally the result of volumetric or geometric considerations. Further developments in the field have been hampered not only by the relative complexity of the process and lack of reliable (materials) input data, but also, perhaps most significantly, by an apparent lack of need-to-know. A significant recent contribution has been the development of a geometrically based model with special application to very low speed single point deformation in ceramics, at the Philips Laboratories in the Netherlands[6].

We have presented at the last Alfred Conference the rudiments of a dynamic-elastic model of ceramic stock removal[7-8]. Since then we have found further support for the two basic points proposed at that time:

*In certain deformation processes, such as grinding, failure can be governed by transfer of energy via elastic waves even at subsonic impact speeds. This type of process must, therefore, be interpreted in terms of these dynamic aspects.

*In the stress-speed distribution there exist two discrete domains; one in which the stress field distribution is determined by kinetic considerations, the dynamic domain and another, the static domain, in which the determinant is load.

In the following sections we have refined and expanded our earlier concepts with particular emphasis on an expression for brittle fracture resistance. An evaluation of the model in terms of material property and system parameter variation, facilitated by computer analysis, has yielded insight into the relative importance of various parameters. Even in this early stage of development the model is capable of supplying data of importance in the screening of materials for ease of machining or resistance to brittle fracture, as the case may be. Such data should further afford the quantifying of grinding parameters, aid in designing brittle materials for ease of machining and supply input for adaptive or numerical control purposes.

II. THE ONE DIMENSIONAL MODEL

It shall be evident that the ceramic stock removal process in its most general form, is a complex process involving the simultaneous interaction of a large number of parameters. Simplification of the system to one in which a single point abrasive mounted at the periphery of a disc is impacted at high speed on a moving work piece, is often resorted to. In such a process one would like to know the energy expended in cutting the groove, the forces experienced by the abrasive (which lead to tool wear and eventual failure) and the relation between the observed depth and width of the cut to the (usually smaller) as-set experimental values. The problem plainly resolves to describing the well-known impact force versus speed curves which, after an initial precipitous drop in

force with increasing speed, go through a minimum and rise again.[7,9]
A theoretical model for even this simple process becomes quite complex and we shall therefore, as a first step, discuss a one dimensional
process as shown in Figure 1.

It is of importance to consider the various assumptions which
result from the simplification of the real system to the idealized
one. They should at once stress both the features and present limitations of the model.

A one dimensional constraint and centric impact is implied. A
very large driver impacts a small workpiece with a crack, inhomogeneity or in general a "trap" in which the transferred energy can
be dissipated. Fracture results when the energy couples with the inhomogeneity at or above the stress limit of the material.

Stress release is in zero time, the crack velocity is infinitely
fast or, in other words, crack propagation is instantaneous.

All energy is dissipated into fracture energy. The otherwise
very real thermal or chemical losses or dissipation of energy through
direct transfer to other points of the system, are not considered.

The stress wave propagation is assumed to be perfectly homogeneous without inelastic modifications.

The crack distribution is regarded as uniform in an otherwise
homogeneous material. A one or multiparameter crack distribution is
easily accomplished and thus forms no restriction.

The Griffith criteria apply. The material will fail, at the
inhomogeneity, at the maximum Griffith stress level.

Figure 1 shows an infinitely long rod of unit cross-section,
Young's modulus E and density ρ which is impacted at time $t = 0$ by
a driver D moving with a constant velocity V. A crack of length λ
is situated at a distance $L = ct$. A transducer is placed at the end
of the rod and will register the impact (paralleling an experimental
situation). It is desired to calculate the transducer response and
deduce from this the characteristics of the brittle fracture at L.
At impact a stress pulse $\sigma(x,t)$ travels down the rod for which the
space-time-stress distribution can, from theory of elasticity[10,11],
be expressed as

$$\sigma(x,t) = \frac{VE}{c} H(ct-x) \qquad (1)$$

in which the velocity of the stress wave, c, is expressed as

$$c = (E/\rho)^{1/2} \qquad (2)$$

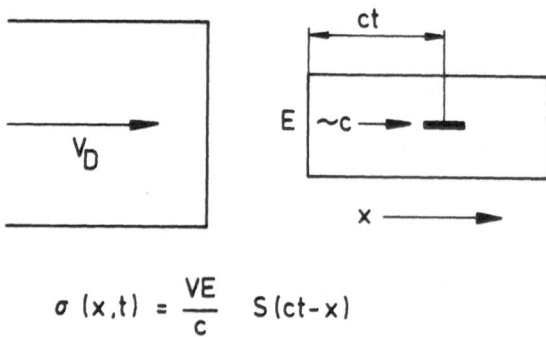

$$\sigma (x,t) = \frac{VE}{c} \ S(ct-x)$$

Figure 1.   Parameters of one dimensional model.

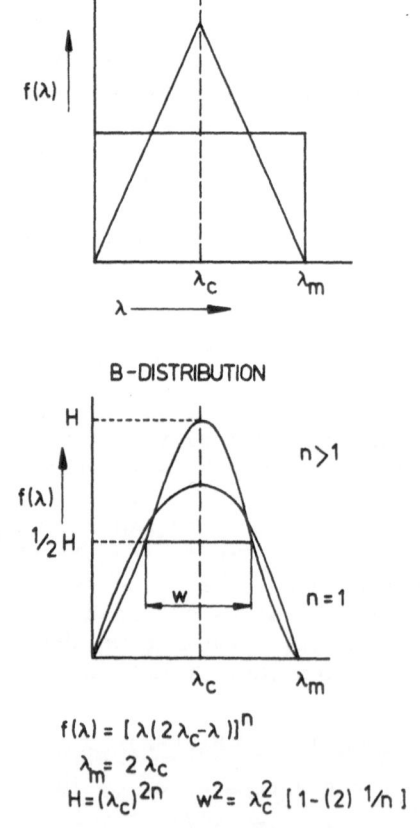

B-DISTRIBUTION

$$f(\lambda) = [\lambda(2\lambda_c - \lambda)]^n$$
$$\lambda_m = 2\lambda_c$$
$$H = (\lambda_c)^{2n} \quad w^2 = \lambda_c^2 \ [1-(2)^{1/n}]$$

Figure 2.   Various crack size distributions used in the modelling
process.

and

$$H(x) = \begin{bmatrix} 0 & x < 0 \\ 1 & x > 0 \end{bmatrix} \tag{3}$$

Assuming that Griffith's criterion is valid, a crack of length λ_o will propagate if,

$$(VE/c)^2 \geq \gamma E/\lambda_o \tag{4}$$

Rewriting the above where γ represents the specific surface energy we obtain,

$$\lambda_o = \gamma c^2 / EV^2 \tag{5}$$

This implies that at impact velocity V, all the cracks of length λ such that $\lambda_o < \lambda < \infty$ will undergo brittle fracture. If $n(\lambda)d\lambda$ is the number of cracks between λ and $(\lambda + d\lambda)$ length, then $N(\lambda_o)$ or the total number of cracks of length $\lambda > \lambda_o$ is given by

$$N(\lambda_o) = \int_{\lambda_o}^{\lambda_m} n(\lambda)d\lambda \tag{6}$$

in which λ_m is the maximum crack length.

The distance L which is now the mean distance between adjacent cracks, can be given as;

$$L = 1/N(\lambda_o) = [\int_{\lambda_o}^{\lambda_m} n(\lambda_o)d\lambda]^{-1} \tag{7}$$

The transducer response F is assumed to be proportional to the energy input from the driver which leads to fracture at L. If we further assume that the sample displays no internal friction then the entire energy input will manifest itself as elastic strain energy and kinetic energy must equal the kinetic energy under these circumstances. It can then be readily shown that the total energy input, W, from the driver just prior to fracture at L, will be given by,

$$W = \rho LV^2 \tag{8}$$

and thus

$$F = AW \tag{9}$$

where A is constant of proportionality. If the crack at L propagates instantaneously, i.e., with infinite velocity, then F should be zero since the stress pulse would not cross the crack prior to complete fracture and A should be zero. However it is well known that in brittle fracture, cracks propagate at a finite velocity and thus a fraction of W will reach the transducer (and hence $A \neq 0$). We assume that $A \sim V$ and to make it dimensionless we adopt that

$$A = V/c \tag{10}$$

It may be pointed out that this choice of A is not essential
to the present model and does not limit its generality. From eqs.
(8), (9) and (10) we get

$$F = \rho L V^3/c \qquad (11)$$

When this is combined with equation (7) we obtain as an expression
for the force

$$F = \rho V^3/[c \int_{\lambda_o}^{\lambda_m} n(\lambda)d\lambda] \qquad (12)$$

Writing equation (12) as

$$F = (V^3/c)\cdot\rho/[\int_{\lambda_o}^{\lambda_m} n(\lambda)d\lambda] \qquad (13)$$

separates the expression into terms which contain dynamic parameters
and material parameters only and facilitates interpretation of the
behavior of F in the speed regime. In the strictest sense and since
this is only a one-dimensional model, the impact dynamics are essentially contained in V.

To illustrate the implications of equation (13) we shall let
$n(\lambda)$ assume some analytical forms. The simplest one, the δ-function,
leads to,

$$n(\lambda) = n_o \delta(\lambda-\lambda_m) \qquad (14)$$

This corresponds to a sample with n_o cracks per unit length all of
the same length λ_m. Equation (13) will now yield

$$F = \begin{cases} \infty \quad, \quad 0<V<V_m \\ \rho V^3/cn_o, \quad V_m<V<\infty \end{cases} \qquad (15)$$

in which,

$$V_m = c(\gamma/\lambda_m E)^{1/2} \qquad (16)$$

As mentioned earlier F is essentially the energy input from the
driver between two successive fractures and thus the limit $F \to \infty$
for $V<V_m$ implies that the stress, VE/c, for such velocities is less
than the critical Griffith stress for the given crack size λ_m and
hence brittle fracture will not occur. In other words the system
just continues to get progressively strained without fracture and
the energy input from the driver becomes infinite, as expected,
since the specimen is of infinite length. However, if we have a
specimen of finite length L_m as discussed by Steverding[10], the
stress continues to increase with time contrary to the constant
stress given in equation (1) as,

$$\sigma(t) = EVt/L_m \qquad (17)$$

Thus even at a velocity $V<V_m$ failure will occur at a time τ after

the impact such that,

$$\sigma^2 = E^2 V^2 \tau^2 / L_m^2 = \gamma E / \lambda_m \qquad (18)$$

or

$$\tau = L_m / V [\gamma / \lambda_m E]^{1/2} \qquad (19)$$

In addition to the single parameter distribution we have considered and evaluated a two parameter distribution with a mean
(triangular) and several refined approaches of this type, see Figure
2. The so called β-distribution easily accommodates variation of
width at half height, generates discrete minimum and maximum crack
lengths and as a special case (n = 1) becomes a parabolic distribution which is very easily treated mathematically. Recent work indicates that grain size distribution in typical polycrystalline
bodies shows a log-normal behavior[12]. Little is known about crack
size distribution in typical ceramics[13]. The log-normal distribution
may be significant where fracture is predominantly intergranular.

III. APPLICATION OF THE MODEL

An initial evaluation and application of the model was accomplished with the aid of an extended Algol computer program. For
strict modelling purposes we have used the above mentioned parabolic,
beta, square and triangular crack size distributions. The actual
values of the materials variables included for the density (ρ) 2, 3
and 4 x 10^3 kg/m^3, for the surface energy (γ) 3 and 30 Nm/m^2 and for
the maximum or mean crack size 3 x 10^{-5}, 3 x 10^{-6} and 3 x 10^{-7}m.
The values of Young's Modulus (E) and Poisson's ratio were kept constant throughout the program at 10^{11} N/m^2 and 0.22 respectively.
The total number of cracks per unit length was more or less arbitrar-
ily set at 25.000, which comes down to one crack every 40 um, a not
unreasonable figure for a ceramic in which the grain boundaries could
be assumed to represent the cracks. The course of specific force
versus speed curve was, depending on the values of the particular
material parameters, evaluated to speeds well beyond the minimum in
the curve. Speeds of up to 70 m/sec proved insufficient to reach
this minimum, or to even supersede the Griffith strength for maximum
crack size, in those examples which combined the lowest crack size
(3 x 10^{-7}m) and highest surface energy (30 Nm/m^2) categories evaluated. In final analysis the speed range 0-50 m/sec proved sufficient
for property values of greatest interest. Evaluation intervals for
speed depended primarily on density of points required for smooth
curve tracing and ranged from as small as 0.01 m/sec at the onset of
significant data generation to steps of 5 m/sec at the highest speeds.
Other than a graphical presentation of specific force versus speed,
the program gave a running output of the values for critical crack
size, Griffith strength, the value of the integral (crack size distribution) and the specific force. In order to separately follow
the influence on the final result of the strict dynamic, i.e., speed,

aspects and the contributions resulting from the materials properties we have regarded equation (13) as to consist of a "macroproperties part" (MAP) and a "microproperties part" (MIP). The program
provided a running readout for both. The definitions are,

$$\text{Macroproperties part:} \quad V^3/c$$

$$\text{Microproperties part:} \quad \rho \left[\int_{\lambda_0}^{\lambda_m} n(\lambda)d\lambda \right]^{-1}$$

The actual results for a parabolic distribution are exempli-
fied by Figure 3. The input values were, $E = 10^{11} N/m^2$, $\lambda_m = 3 \times 10^{-5} m$,
$\gamma = 30 \ Nm/m^2$ and densities of 4, 3 and $2 \times 10^3 \ kg/m^3$, yielding separate curves running from left to right in that order. The res-
pective stress wave velocities were 5×10^3, 5.77×10^3 and
7.07×10^3 m/sec.

Figure 3 immediately calls attention to the effect of bulk
density on the process. A decrease in density shifts the force
curves towards higher speed values. The minimum speed for which
material can be removed in a brittle fashion from the higher density
material, is thus lower than that for a material of lesser density
all other factors being equal. A quantitatively similar observation
can be made for the force levels at any one speed up to a cross-
over region between 24 and 30 m/sec in Fig. 3. Beyond this region
the situation is reversed and the lower density material will generate the lower force levels for a given speed. The curves for the
lower density materials will also be broader around the minimum.
The higher the density the steeper the increase beyond the minimum.
This attitude can also be obtained by choosing materials with a
high surface energy, a small critical crack size or a combination
thereof. For a specific material the minimum in the curve will be
determined by the specific crack distribution. The speed at which
this minimum occurs has a physical significance in that it is the
point at which the energy for removal is the lowest which corresponds
to the minimum in the crack size distribution curve or the maximum
number of cracks of median size. Since the maximum in a parabolic
distribution will occur at $V_m/2$ we have for this case $V_{min} = 1.41 \ V_m$
(see equation 16).

It is instructive at this point to compare the behavior of the
MIP and MAP parts and their influence on the composite force curve,
shown in Figure 4. The attitude of the initial and sharply declin-
ing force curve up to the minimum is primarily determined by the
behavior of the MIP function, that is to say, the mechanical and
structural properties of the workpiece material such as crack size,
crack distribution and the number of cracks. Bulk density, which
is also part of the MIP function manifests itself, as mentioned
earlier, primarily in a shift along the speed axis. The behavior
of the curve on the high speed side of the minimum is virtually

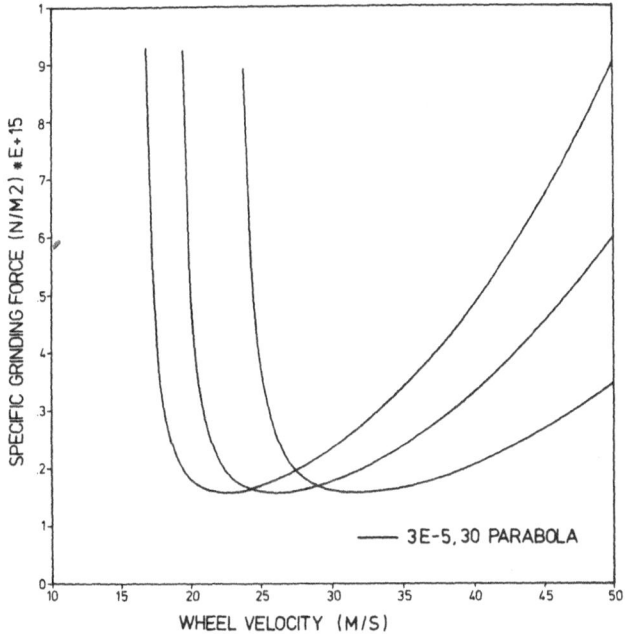

Figure 3. Results for a parabolic distribution with
$\lambda = 3 \times 10^{-5}$ μm and $\gamma = 30$ J/m^2.

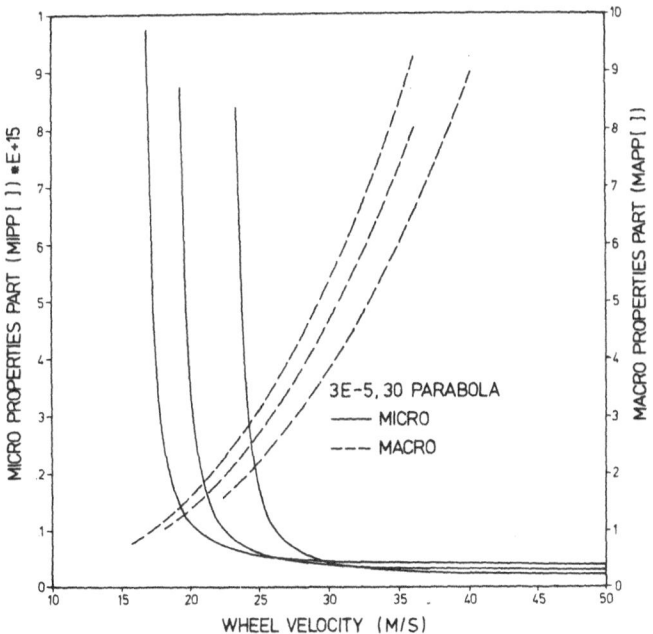

Figure 4.   MIP and MAP contributions to the force versus speed
curves.

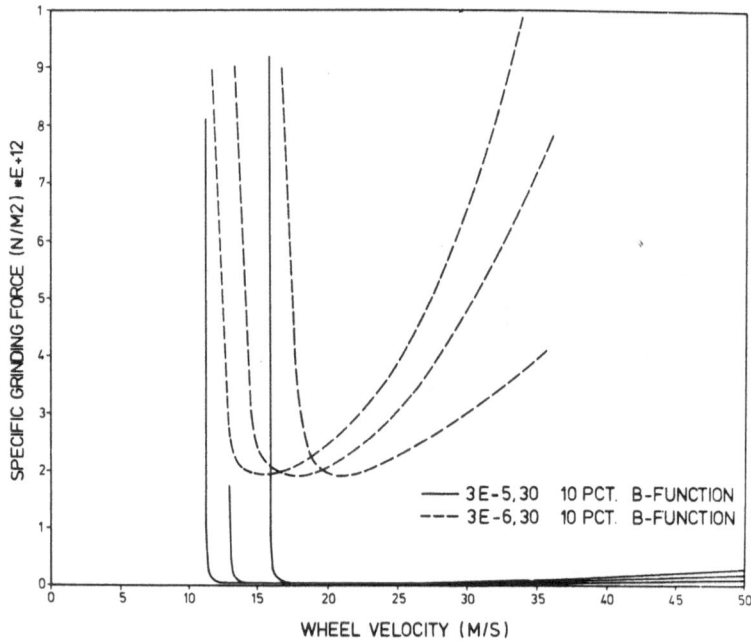

Figure 5.   Force distributions contrasting two grain sizes, 30 μm
             and 3 μm.

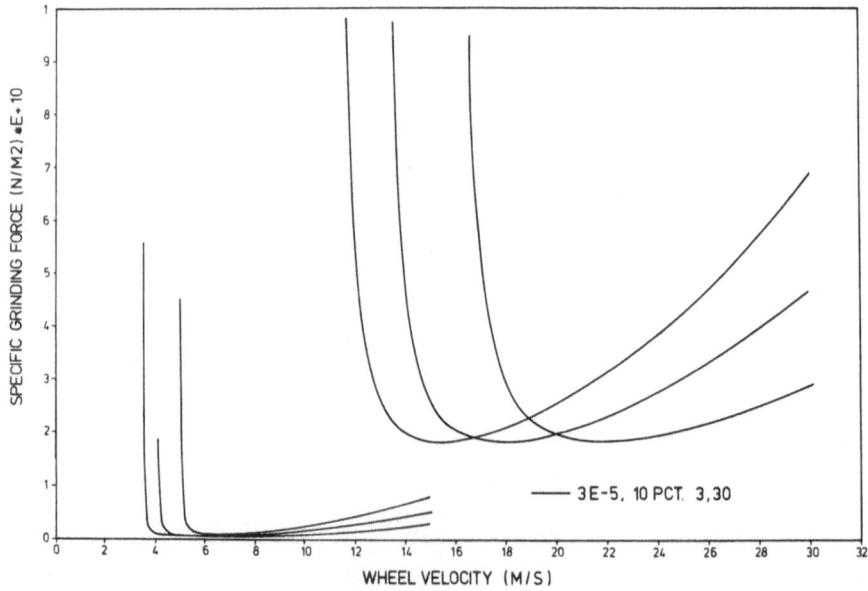

Figure 6.   Force distributions for two materials with 30 μm grain
             size γ = 3 and 30 J/m².

dominated by inputs of the MAP function or the dynamics of the
system. The rate of change for the region controlled by the material properties is much higher than that governed by the dynamic inputs. At $V > V_{min}$ this rate will increase with increase in density.
The lower density curves will tend to be flatter around the minimum
leading to smaller variations in force over broader velocity intervals. The adjustment and control of systems parameters for low energy grinding conditions should be less critical for materials of
low density. The overall important factor is thus the cause-duality
behind the attitude of the force curve. The material properties, at
least initially, make the greater impact. Removal processes at
$V > V_{min}$ would be influenced by fewer, more controllable, factors and
from a force or energy point of view be less sensitive to speed
changes particularly for the lower density materials. It will be
noted that even in the case of the examples of Figure 3, ΔV_{min} can
amount to as much as 50%.

Figure 5 shows the differences brought about by a change in
crack size from 30 μm to 3 μm in a sharp β-distribution. V_m will
shift by a factor $(30/3)^{1/2}$. Most important is the force level
shift by several orders of magnitude. This is not immediately apparent from the figure due to necessary scaling. The median crack
size is the most influential factor in setting the force levels.

Figure 6 illustrates the shifts in going from a surface energy
of 3 J/m^2 to 30 J/m^2, as in going from a typical glass to a ceramic.
Force levels increase by a factor of 10 to 30 depending on the exact location. The data for conditions used in Figure 6 but with
$W = 0.5\lambda_c$ (see Figure 2) rather than the 10% width, show a further
reduction in the force levels by somewhat less than one order of
magnitude. It reemphasizes the significance of crack size distribution. The median crack size does, however, remain the major determinant.

IV. ANALYSIS
   Brittleness reflects the way in which materials react to certain experimental conditions. It is a behavioral or state property
and not a material property. The ability of a material to resist
brittle stock removal, i.e., resist initiation, propagation and interaction of cracks is a measure of its work absorbing power. The
strain energy at failure for a brittle material, the well known
modulus of fracture or resilience Λ, can be expressed as,

$$\Lambda = \sigma_f^2 / 2E \qquad (20)$$

It plays a significant role in grinding brittle materials, see e.g.,
Figure 7. Rewriting the Griffith equation yields

$$\sigma_f^2 / E = \gamma_f / \lambda_m \qquad (21)$$

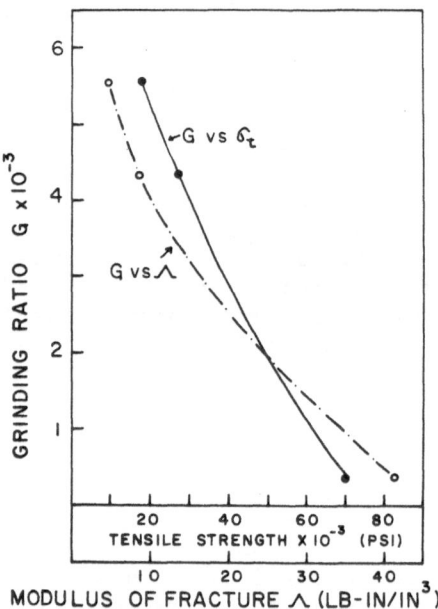

Figure 7.   Influence of modulus of fracture on grinding ratio.[14]

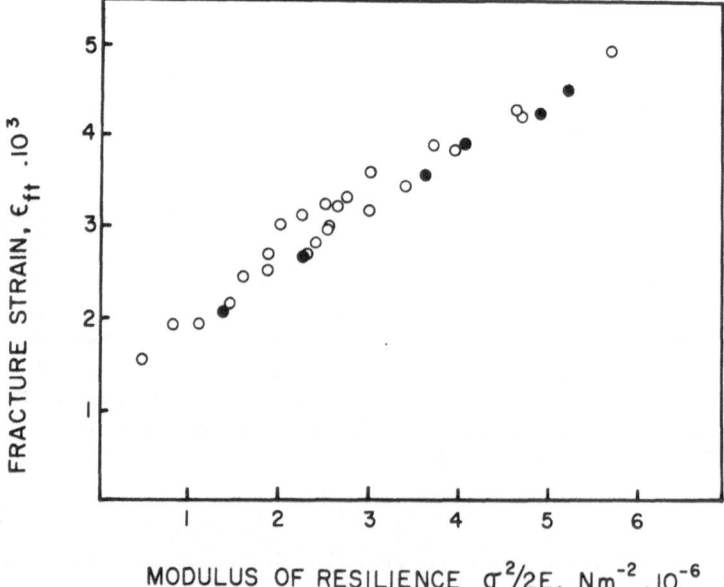

Figure 8.   Relationship between fracture strain and toughness for
some thirty carbide tool materials.[16]

in which γ_f = fracture initiation energy and thus σ_f/E, which represents the critical strain for brittle fracture, ε_f, will be

$$\varepsilon_f = (\gamma_f/E)^{1/2} \cdot \lambda_m^{-1/2} \qquad (22)$$

There should, therefore, be a direct relation between the strain energy Λ and the critical strain at fracture, ε_f. A number of experiences indicate that the ultimate strain at fracture, i.e., the maximum strain at the point of failure is an excellent criterion for brittle failure. Brandes[15] has shown, when investigating the strength of brittle materials under high hydrostatic pressure, that brittle failure, irrespective of the state of stress, occurs when the maximum elastic strain,

$$\varepsilon = 1/E \cdot [\sigma_1 - \nu(\sigma_2 + \sigma_3] \qquad (23)$$

where ν = Poisson's constant, reaches a critical value. The combined principal stresses rather than just the maximum uniaxial stress will determine where and when brittle fracture will occur. An important point is that although the specific rupture strength may depend on such parameters as loading rate, the ultimate uniaxial strain will remain the same. It can be shown that such quantities as impact strengths, grinding ratios (grindability) and grinding energies are a linear function of the ultimate strain at fracture. The higher the value of ε_f the more resistant the material will be to brittle failure. Figure 8 shows the relation between the fracture strain and the energy absorption characteristics of some 30 carbide tool materials[16]. Included are Co and (Mo-Ni) cemented WC and (TiC-TaC) as well as ceramic materials. The lowest point on the graph represents a 10% TiO_2-Al_2O_3 tool material. The solid circles represent data derived from measured plain strain fracture toughness (K_{Ic}) values[17].

When we equate the expression for stress at fracture from equation (1) with that from the Griffith criteria, equation (21), we obtain for the cut-off velocity

$$V_m = c/E \cdot (\gamma E/\lambda_m)^{1/2} \qquad (24)$$

or

$$V_m = (E/\rho)^{1/2} \cdot (\gamma/E)^{1/2} \cdot \lambda_m^{-1/2} \qquad (25)$$

This cut-off velocity, which can be regarded as a brittle fracture resistance parameter, is thus the simple product of two previously discussed parameters,

$$V_m = \varepsilon_f \cdot c_f \qquad (26)$$

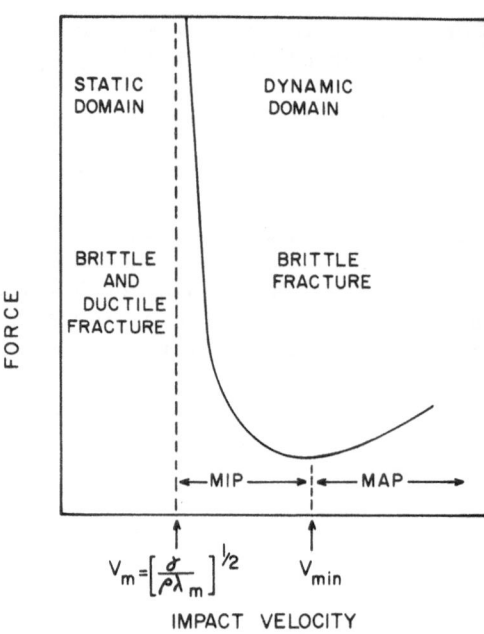

Figure 9.   Schematic representation of certain parameters in the
            stress-velocity distribution.

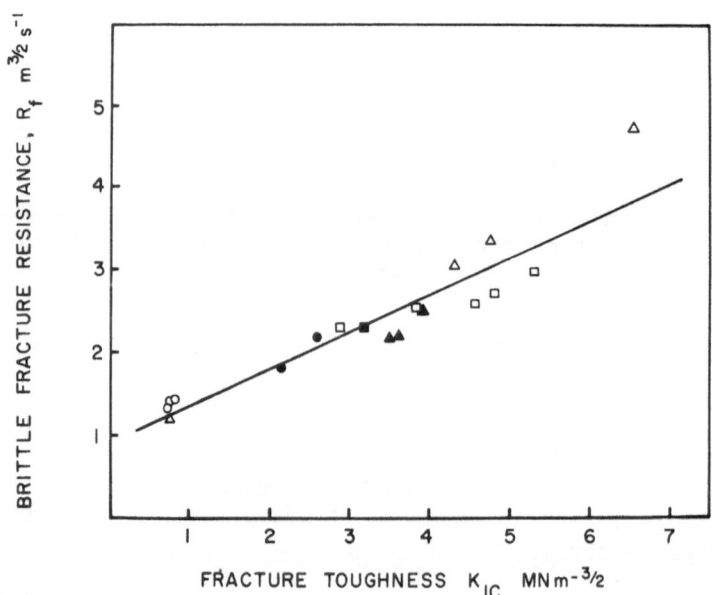

Figure 10.   Relation between brittle fracture resistance and
             fracture toughness.

where ε_f = ultimate strain at fracture
        c_f = strain wave velocity

The physical significance of V_m is further indicated in Figure 9 which, schematically, represents the overall results from equation (13) and the data from Figures 3 through 6. In the dynamic domain the stress field will be determined by kinetics and the deformation is velocity governed. In the static domain the nature of the stress field is primarily determined by local phenomenon and the deformation is load governed. A definition in this way should have significance in terms of the material involved, the system parameters as well as the specific operational ambients.

In Figure 10 we show the relation between a modified brittle fracture resistance parameter, R_f, and the plain strain fracture toughness, K_{Ic},

$$R_f = (\gamma/E)^{1/2} \cdot (E/\rho)^{1/2} \tag{27}$$

$$K_{Ic} = \sigma_f (\lambda_m)^{1/2} = (E\gamma_f)^{1/2} \tag{28}$$

for as many glasses and polycrystalline ceramics as appropriate data were available. The normalized brittle fracture resistance had to be resorted to in the absence of reliable experimental λ_m or V_m data.

Acknowledgement
      The authors should like to acknowledge the use of facilities and assistance of the staff of the Eindhoven University of Technology, the Netherlands, Division of Production Technology. We are particularly indebted to Dr. S.N. Touwen for his accurate and time consuming contributions to the computer work. Our work has much gained from many technical discussions with Prof. Dr. H.J.J. Kals, Twente University of Technology and the staff of the Ceramic Science Section, Philips Research Laboratories, Eindhoven, the Netherlands.

REFERENCES

1. M.L. Wilkins, C.F. Cline and C.A. Honodel, Light Armor, UCRL-71817.
2. A.G. Evans, M.E. Bulden, G.E. Eggum and M. Rosenblatt. Office of Naval Research N00014-75-C-0669 (1977).
3. D.R. Curran, L. Seaman and D.A. Shockey, Dynamic Failure of Solids. Stanford Research Institute Report #001-76 (1976).
4. D.A. Shockey, D.R. Curran, L. Seaman, J.T. Rosenberg and C.F. Petersen, Int. J. Rock. Mech. Sci. & Geomech. Abstr., 11 303 (1974).
5. M.C. Shaw, Mech. Chem. Eng. Trans. 8 (1), 73-78 (1972).

6. A. Broese van Groenou, N. Maan and J.D.B. Veldkanp, Philips Res. Repts., 30, 320–359 (1975).
   J.D.B. Veldkamp and R.J. Klein Wassink, Philips Res. Repts., 31, 156–170 (1976).
7. P.J. Gielisse, T.J. Kim and A. Choudry, An Experimental Investigation of the Dynamic and Thermal Characteristics of the Ceramic Stock Removal Process, in: Surfaces and Interfaces of Glass and Ceramics, V.D. Frechette et al. editors, 137–148 (1973).
8. A. Choudry and P.J. Gielisse, Dynamic Elastic Model of Ceramic Stock Removal, Ibid., 149–166 (1973).
9. P.J. Gielisse, T.J. Kim, A. Choudry, J.F. Short and E.J. Turker, Final Technical Report, N 00017-72-C-0202, Naval Air Systems Command, U.S. Navy, Washington, D.C. (1973).
10. B. Steverding, J. Am. Ceram. Soc., 52, 133 (1969).
11. Timoshenko and Goodier, Theory of Elasticity, McGraw Hill, N.Y. (1956).
12. F.M.A. Carpay and S. K. Kurtz, A Multiple-Lognormal Model of Normal Grain Growth, this volume.
13. H.A. Nied and K. Arin, Fracture Mechanics Model, International Symposium on Fracture Mechanics, Penn. State Univ. July 1977.
14. L. Goyette, Effects of Grain size on Grinding Forces in Ceramic Processing, M.S. Thesis, University of Rhode Island (1974).
15. M. Brandes, Int. J. Fract. Mech., 1, 56 (1965). See also, H.J.J. Kals and P.J. Gielisse, Annals C.I.R.P., 24 (1975).
16. P.J. Gielisse and H.J.J. Kals to be published, see also second part reference 15.
17. R.C. Leuth, Determination of Fracture Toughness Parameters for WC-Co Alloys, Thesis Michigan State University (1972).

DISCUSSION

V. Boldyrev (Siberian Branch, Acad. Sci., USSR): You are doing very interesting work and use good experimental methods for studying cracking of solids. What can you say about the chemical peculiarities of the processes which you have studied? How do you envision the chemical mechanisms of chemical reactions during the cracking of solids? In our laboratory we recently studied the chemical reactions on brittle solids (nitrate, nitrite, chlorate) during crack propagation in the origin near the tip of the crack. We observed mechanisms of decomposition reaction of alkaline nitrate, nitrite, and chlorate as distinct from thermal decomposition. For example, by cracking of sodium nitrate the main gaseous product is nitric oxide, but on thermal decomposition of this salt usually only oxygen is evolved. We think that in the crack region (or in the neighboring region of the crystal) we obtained the possibility to excite the high energy vibration levels of triangular

(three-fold coordinated) ions and as a result of this the change
in chemical mechanism. What do you think of the possibility of
similar effects on the system studied by you?
Authors: Your question points to a very fundamental aspect of the
fracture process, and it is not often that the results of such basic experiments are encountered. The question of chemical reactions during fracture and their relationship to thermal decomposition is directly related to the fundamental physics of the fracture process. We are generalizing our model to address such
questions. The underlying rationale is that in the thermal decomposition process in which the temperature obviously plays the dominant role, one can estimate the excitation density of the vibrational levels and then calculate the possible chemical reactions.
However, under fracture conditions it is the Hugoniot properties
and the chemical relaxation (or reaction) times that will dominate
the process results. Thus one must try to calculate the excitation of vibrational levels under the prevailing transient high
temperature as well as high pressure. This assumes that one can
first calculate the pressure and temperature distribution in the
impact zone. Chemical kinetics will indicate which reactions are
the dominant ones under such transient conditions. In most thermal processes there is essentially infinite reaction time available. Under fracture conditions very high pressure and
temperature levels are available for a very short time period only.
Such conditions might favor different chemical reactions.

Our recent spectrophotometric results[7,8] from microsecond discharges in individual grain impacts under typical grinding conditions in ceramics indicate true grinding temperatures between 1300
and $1600^{\circ}K$. We have calculated[8] pressure to reach levels of 10^{10}
Pa in about one microsecond unless failure occurs within a shorter
time interval. Discharges in different types of gaseous environments in grinding show dissimilar spectral distributions, and our
thermal emission spectra show structure indicating chemical reaction. Needless to say that elucidation of such experimental results is at the core of the fracture process.

D. R. Clarke (Rockwell Science Center): I do not see any momentum
terms of the impactor in your expressions. Are they hidden in
your "macroproperty" terms; and, if so, how does it affect the position of V_{min} in your Force vs. Input Velocity curves?
Authors: It is true that as such there is no momentum term in our
impact expression. The reason for this is that the momentum has
been treated as fixed by the boundary condition, provided by an
impacting driver which is moving with a constant velocity, even
after contact with the sample. This can further be interpreted as
an infinite mass of the driver. This is not as unrealistic as it
may sound since in practical cases of surface abrasion through
grinding the mass of the grinding wheel can be taken as infinite

as compared to the mass of the fractured part of the sample. The
important parameter is the energy, as transferred from the driver
to the sample. In ballistic impact this energy transfer occurs at
the expense of the momentum loss of the projectile and thus a momentum term is needed to calculate the energy transfer. In grinding or stock removal applications the grinding wheel velocity does
not or rather should not change during grinding. The drive motor
will supply extra energy for stock removal if needed. However,
our model is perfectly capable of being generalized to a finite
mass driver and then would include a momentum term. In this context the effect of a finite mass impact on V_{min} is rather interesting even though the final conclusion does not change qualitatively, i.e., below a certain velocity, V_{min}, an infinite one dimensional sample will not undergo brittle fracture irrespective of
the mass, and thus the momentum, of the projectile.

L. De Jonghe (Cornell University): Could you please clarify the
difference between micro and macro properties in the context of
this work?
Authors: The microproperties part (MIP) and macroproperties part
(MAP) terminology only has significance through their definition
in this paper. They do not refer to the micro- and macroproperties of a material as such. We intended to have these quantities further clarify the specific roles of the systems parameters in the force-velocity regime (see Figure 2) and to emphasize
the contributions of the "machine" conditions (MAP) on the one hand
and the material characteristics (MIP) on the other.

PROCESSING INDUCED SOURCES OF MECHANICAL FAILURE IN CERAMICS

R. W. Rice

Naval Research Laboratory

Washington, D. C. 20375

INTRODUCTION

Porosity, impurities, large grains, machining, and handling-induced flaws are recognized defects limiting ceramic strengths. This paper summarizes studies of the defect origins of room temperature fracture, and vividly illustrates that these origins are often (1) the actual sources of fracture and (2) the extremes of such defects that occur. Observations or plausible suggestions on the causes of such defects are given. Details of fractographic procedures are found elsewhere,[1] as are other examples of fracture origins from processing defects.[2-8] It should be noted that strengths in the absence of gross processing defects, i.e., strengths of the matrix of the laboratory test bars for the bodies considered in this paper containing only normal machining flaws are as follows: hot pressed Al_2O_3 520-820 MPa (75-125 ksi), commercial lead zirconate titanate (PST) 140 MPa (20 ksi), and hot pressed Si_3N_4 700-820 MPa (100-125 ksi). These are approximate guides for judging the impact of the grosser defects on strength.

FABRICATION INDUCED DEFECTS AS SOURCES OF MECHANICAL FAILURE

Pores as Sources of Failure

Two basic types of pores are collectively or individually common sources of failure. In addition they always lower Young's modulus and commonly lower fracture energy. First, collections of more, or somewhat larger, pores from the distribution of pores inherent in the densification of materials from powders, i.e., pores between grains,

303

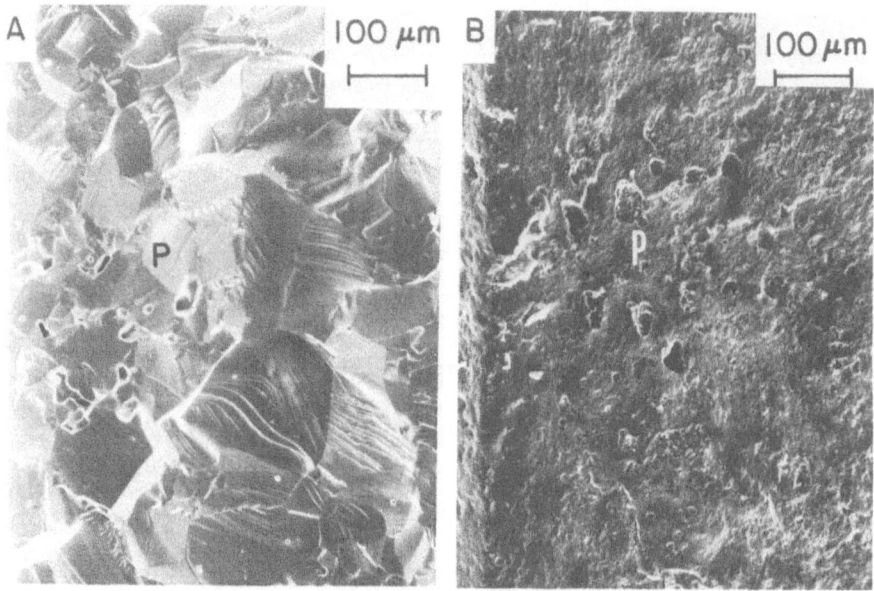

Figure 1.   Fracture surfaces of flexure bars of A) dense $MgAl_2O_4$
(σ_f ~210 MPa ~ 30 ksi) and B) reaction sintered Si_3N_4
(σ_f ~170 MPa ~ 25 ksi) from regions of higher than average
porosity (P). (Tensile surface at the left hand edges of
the photos). Whether the porous region in A) represents
an extreme of the normal distribution of porosity is
uncertain; however, the more porous region in the vicinity
of the fracture origin in B) is part of the normal distribution of porosity.

agglomerates or combinations of these cause failure, especially in
more porous bodies. On example is those bodies with >10% porosity,
where normal statistical variations of pores can result in a defect
more serious than other sources of failure, e.g., machining flaws.
Such origins, which may often cause only limited reductions in
strengths, are often not explicitly identifiable because they are
typically associated with lower strength failures having less definitive fracture markings and because there is typically no demarcation
separating the pore group causing failure and the surrounding "normal"
porosity. However, some fracture origins from regions of higher than
average porosity, e.g., Fig. 1, are observed.

The second, more commonly and clearly observed type of pore
fracture origin is associated with large pores or groups of pores
resulting from some irregularity in the processing operation. These
pores are usually distinctly different from those left between the
original particles or agglomerates. They can occur over any range
of porosity, but generally are more important in higher density,

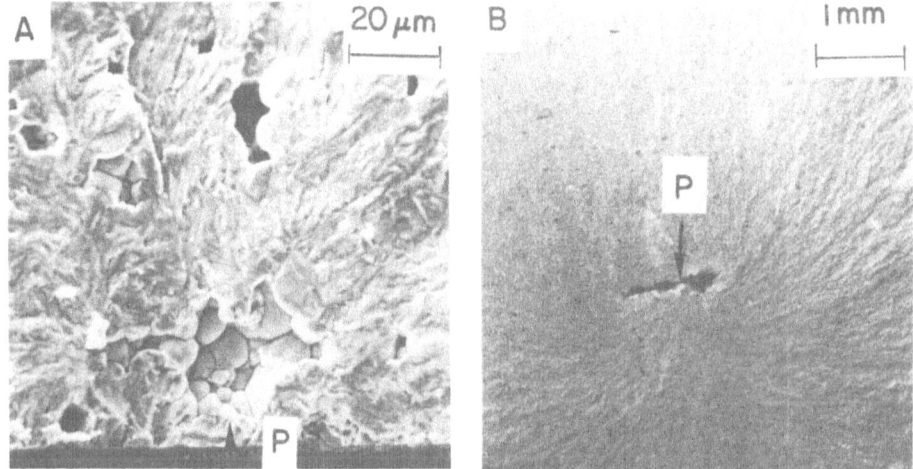

Figure 2.   Extremes of failure of commercial PZT sonar materials from
             isolated pores (P) larger than the grain size. A) Fracture
             surface of a flexure bar cut from one of the highest
             quality commercial materials available, (σ_f ~115 MPa
             ~17 ksi), tensile surface at photo bottom. B) Fracture
             surface of a poorer quality commercial sonar transducer
             ring failing under dynamic hoop tension loading at
             ~17 MPa (~2.5 ksi). Note the laminar and tear-like
             character to the pore (P) which is located approximatly
             in the center of the cross section.

higher quality bodies, because they tend to dominate less severe
flaws resulting from collections of normal porosity. Such origins
have been commonly observed in a variety of commercial PZT
bodies[6,9,10] (Figs. 2, 3); high strength partially stabilized
ZrO_2 bodies;[4] hot pressed (Fig. 4), commercially sintered,[2,3,6,13,15]
and a variety of experimental[2-6] α-Al_2O_3 bodies; hot pressed[11] or
sintered[12] β-Al_2O_3; reaction sintered Si_3N_4;[6-8] and occasionally in
commercially hot pressed MgF_2. Pores (bubbles) may also be the
sources of failures in commercially crystallized glasses[16] and glazed
electrical porcelain. Such pore origins often result in significant
(e.g., 25 to > 50%) reductions in strength.

     Such large pores can arise from a variety of sources, e.g.,
laminations from pressing or extrusion, or bubbles in slip or fusion
casting, or gas producing sources from binders, other purposely
introduced constituents, or foreign contaminants. For example,
fracture origins in the sonar rings shown in Figs. 2B and 3 are
believed to be caused by binder accumulations since; (1) their laminar
character is consistent with flow of a large binder particle during
cold pressing and (2) such large pores are typically near the center

of the body cross section where complete burnout would be most
difficult. Furthermore, the satellite pores are thought to result
from a binder agglomerate either having an irregular shape and
flowing from between other particles or agglomerates, or due to the
joining of the binder agglomerate with surrounding binder patches.
In Figures 3A, B it is hypothesized that burnout occurred at a low
enough temperature such that the gas pressure was kept below that
necessary to break the webs between the satellite pores. In
Figures 3C, D internal gas pressures apparently resulted in the
fracture of the webs between the satellite pores (with subsequent
rounding of the fractured webs leaving the stalagtite and the
stalagmite type appearance) and extended the pore, leading to its
crack-like character. Other important sources of material which
produce such relic pores include hair (human or animal, e.g.,
rodents), lint, dandruff, tobacco, cigarette ashes as well as bugs,
parts thereof, or bug excrement. Rhodes et al.[17] have shown that a
surprising amount of organic contamination can be screened from
even quite high purity Al_2O_3 powder.

Another example of pores resulting, at least in part, from outgassing of additives is found in an earlier hot pressed Al_2O_3 sample,
(Fig. 4). Agglomerates of poorly mixed MgO, added by milling for
grain growth control, continued gas producing impurities (e.g.,
$MgCO_3$) which resulted in the association of pores with these
agglomerates. EDAX analysis typically showed a Mg concentration
around such pores. Many of the pores in these hot pressed billets
showed a laminar character, presumably because of laminations in the
powder compaction process. Similarly, pores in cold pressed PZT
transducers often have a highly laminar character giving a 20-40%
anisotropy in strength.[9] Such pore lamination or alignment is fairly
frequent in cold pressing, and can occur in slip casting, extrusion,
etc., as well as hot pressing. Development of hollow agglomerates
in some spray dried powders, e.g., Al_2O_3, can also be a source of
large pores.

Agglomerates

Agglomerates per se can also cause failure, most commonly as
a source of pores, which are typically located around the periphery
of the agglomerate. An example of this is shown in Fig. 5 (see also
Fig. 4 of Ref. 2). The release of the cold pressing pressure can
result in more elastic relaxation in a less dense matrix than in a
dense agglomerate, leading to some porosity between the agglomerate
and the matrix. Also a denser agglomerate may densify more rapidly
during sintering because of (1) greater density or (2) impurities
pulling away from the matrix leaving a complete or a partial
peripheral pore or series of pores.

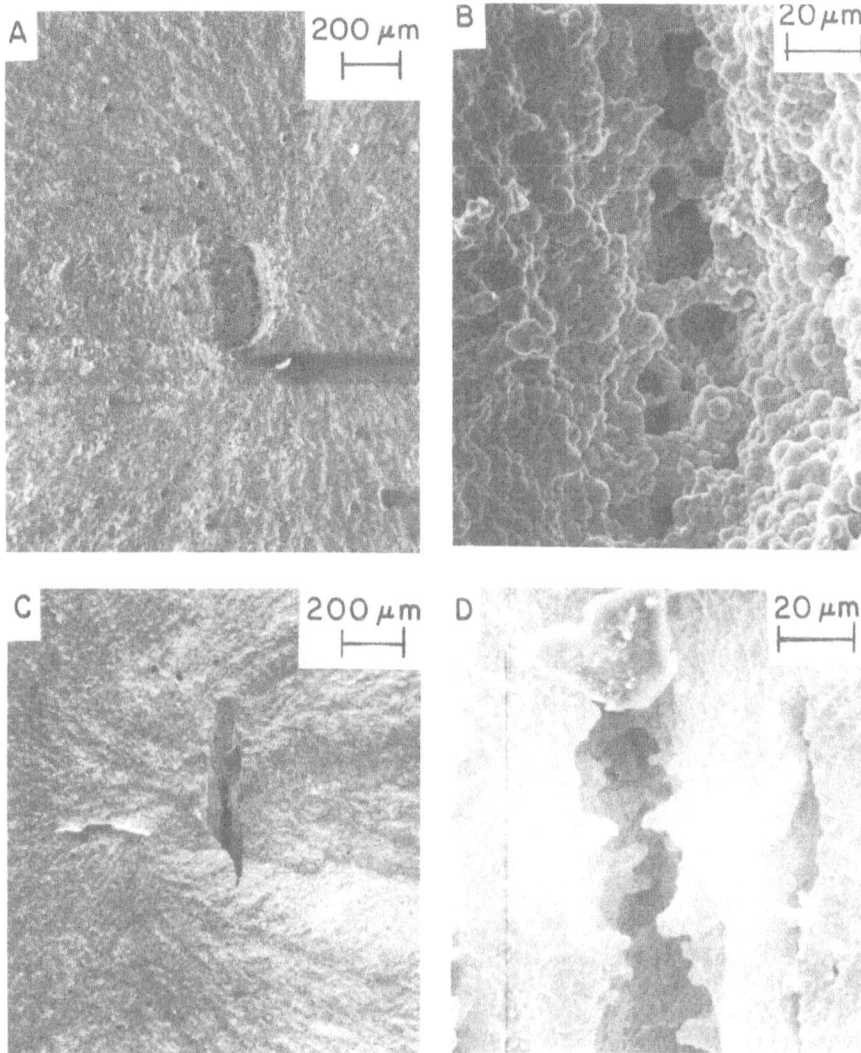

Figure 3. Hoop tension fractures of commercial PZT sonar transducer
           rings from large pores. The samples in A) and C), with
           respective higher magnifications of the failure-causing
           pores in B) and D), had σ= ~70 MPa (~10 ksi) and ~55
           MPa (~8 ksi), respectively. These pores, as discussed
           in the text, are tentatively attributed to binder
           agglomerates.

Second Phase Particles

     Particles consisting partly or wholly of a foreign, additive
or constituent phase different from the matrix may often be sources

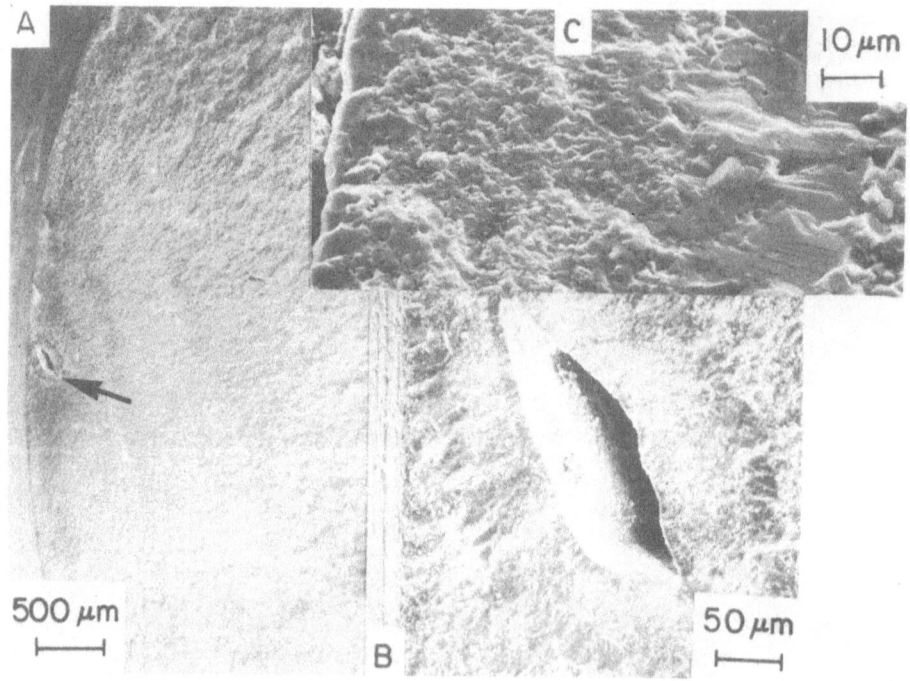

Figure 4. Failure of a hot pressed Al_2O_3 tensile test specimen from
           a large isolated pore (arrow in A); shown in higher magni-
           fication, in B. C shows the microstructure adjacent to
           the pore. Note large grains along the periphery of the
           pore at the left, then a finer grain structure and a ring
           of larger, somewhat columnar grains. As discussed in the
           text, EDAX analysis shows a high concentration of the MgO
           grain growth inhibitor associated with many such pores.
           σ_f ~190 MPa (~27 ksi).

of fracture with limited to substantial effects on strength, depending
on their size, number, and differences in properties from and bonding
to the matrix. Such particles can be introduced in various stages
of processing. For example, the fracture origins in Si_3N_4 (Fig. 6)
reflect (1) contamination from milling or the starting powder and
(2) poor mixing of the densification additive. Also, a few hot
pressed Al_2O_3 bodies similar to that shown in Figure 4 failed from
Mg-Si-rich inclusions, apparently introduced as impure agglomerates
from the MgO additive. In studies of fracture origins of commercial
PZT sonar transducer rings, some were observed to occur from Al_2O_3
particles believed to be chips from the Al_2O_3 milling media,[10] e.g.,
Figure 7.

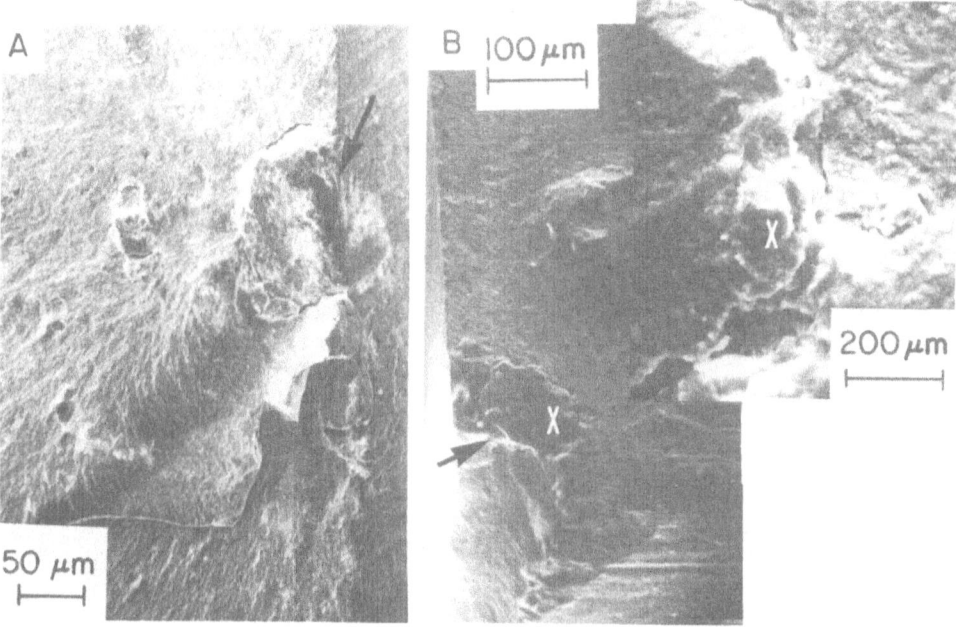

Figure 5.   Failure of high strength partially stabilized ZrO$_2$ from
agglomerates. A) Failure of a flexure bar from large
agglomerate (arrow); σ_f ~530 MPa (~77 ksi), tensile
surface at right edge of photo. Note another smaller
agglomerate below the main one and another still smaller
one to the left of the main agglomerate. B) A similar
sample (σ_f ~490 MPa 71 ksi) failing from an agglomerate
and associated spalled corner (arrow). Note the peripheral
voids largely surrounding these agglomerates, e.g., insert
of B). Failure of this very fine grained (0.4 μm) ZrO$_2$
body (partially stabilized with Y$_2$O$_3$) has occurred in the
absence of gross defects such as normal machining flaws.
The latter usually produces flexure strengths of 700 to
$\geq$ 1050 MPa (100 to $\geq$ 150 ksi). On the other hand, Curtis
Scott reports (private communication based on his Master's
thesis work at Alfred University) that fabrication of test
bars from highly agglomerated materials, e.g., as obtained
by centrifuging, though showing limited differences in
sintering results, gives σ_f ~210 - 280 MPa (30-40 ksi).
However, processing of initially agglomerate-free material
by repressing after incomplete breakdown of agglomerates
from the first pressing lowers strength 20-30%.

        A recent study of IR window quality, hot pressed MgF$_2$ illustrates
that foreign particles can be a problem even in what are considered
high quality, high technology ceramics. Biaxial testing of hot
pressed MgF$_2$ discs showed that between 60 and 80% of them failed from

single or clustered particles or grains ~2 or more orders of magnitude
larger than the average grain size.[18] Although the majority of the
fracture origins were from such particles, the latter were limited
in number (e.g., they were not commonly seen in laboratory test bars)
and did not result in a large strength reduction. EDAX and electron
probe analysis of these particles showed the great majority to be
SiO_2, Fig. 8. Consultation with the manufacturer revealed that
powder treatment was carried out in SiO_2 crucibles, having a coarse
sand finish exterior. SiO_2 particles similar to those found in the
MgF_2 (Fig. 8B) could be readily abraded into the powder and hence be
retained as a contaminant and subsequent source of failure. In this
same study, one MgF_2 sample was also found to fail from a zinc-rich
particle.[18] Apparently a granule or particle of the zinc compound
pressed in the same facilities and occasionally even in the same
dies either stuck to a die component or accidentally fell into a die
and subsequently contaminated the sample. Similarly, graphite particles, presumably from the graphite dies, have been observed in
earlier hot pressed Al_2O_3 (Fig. 9).

Large Grains

Isolated large grains or groups of larger than average grains
are not necessarily intrinsic sources of weakness; i.e., some fractures in bodies containing isolated large grains occur from normal
machining flaws in the fine grained matrix. However, large grains
commonly have other associated flaws such as pores, impurity particles,
or cracks, e.g., from machining or thermal expansion anisotropy. In
such cases larger grains provide lower energy paths for initial crack
growth from such associated flaws and thus result in strengths which
may be substantially lower than in bodies lacking such large grains.
Large grains may result from exaggerated grain growth, or an excess
or deficiency of additives or impurities. Such large grain origins
are common in Al_2O_3 bodies, being a predominant source of fracture
in many hot pressed Al_2O_3 bodies without,[2,3,13-15] as well as with,[17]
additives (Fig. 10). They are also often an important factor in limiting strengths of most commercial, higher strength Al_2O_3 bodies.

Large grains have been indicated as sources of failure in
$\beta-Al_2O_3$[19] and WC-Co.[20] They have been observed by this author in
commercial hot pressed B_4C; a few examples have been noted in commercial hot pressed MgF_2,[18] and one extreme case has been reported in
high optical quality CVD ZnSe.[6,21]

Machining and Handling Induced Flaws

Extensive studies of diamond grinding of glasses,[22] single
crystals and polycrystalline[23] samples show that typically two sets of
flaws are introduced in this common post-firing operation. The first

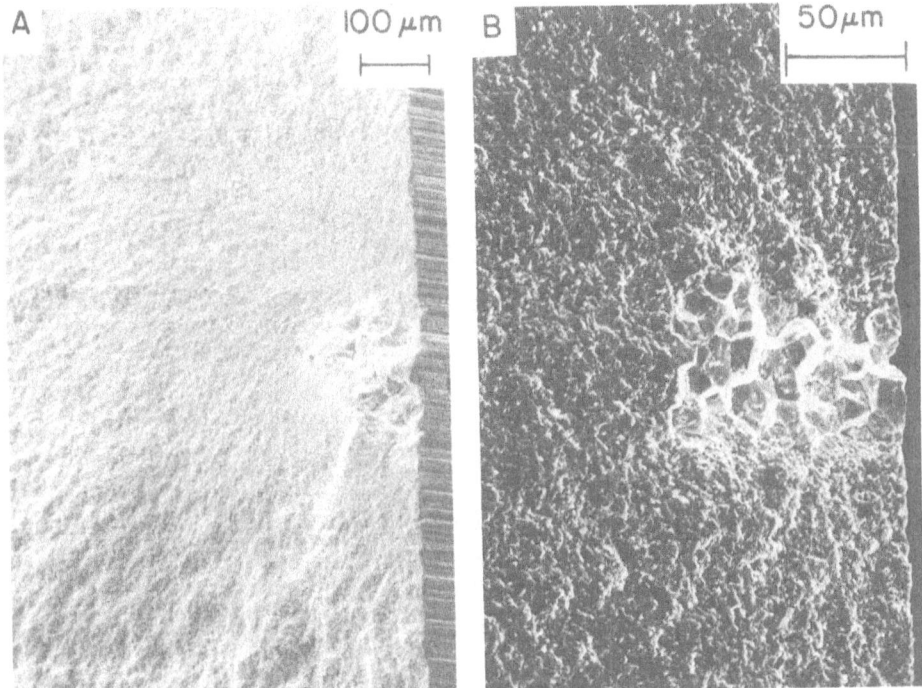

Figure 6. Failure of hot pressed Si_3N_4 flexure bars from second
           phase regions. Failure origins are from the obvious large
           grain regions at the intersection of the fracture surface
           and the tensile surface near the right hand edges of the
           photos. A) Commercial Si_3N_4 (HS-130) failing at
           σ_f ~430 MPa (~62 ksi) from a cluster of large grains
           which are primarily Fe and Si with some Mn, i.e., presum-
           ably an iron silicide from impurities introduced from
           processing. B) Experimental Si_3N_4 hot pressed with
           ~2% ZrO_2 additions, σ_f ~520 MPa (~75 ksi). The large
           grain fracture origin region is rich in Zr, indicating
           poor mixing of the ZrO_2 addition.

set of flaws generally form approximately perpendicular to the grinding
grooves left by the abrasive particles. These are apparently caused,
at least in part, by a stick-slip phenomena, and are sporadically
distributed along the grinding groove (Fig. 11A). The second set
forms underneath and along the bottom of the grooves made by the
abrasive particles, typically giving more elongated flaws either of
a continuous nature or as a result of partial or complete overlap of
shorter flaws (Fig. 11B). While machining parameters are important,
other factors can also enter in both the frequency and the orientation
of machining flaws, e.g., the orientation of preferred single crystal
cleavage or fracture planes relative to the grinding direction. The
typical orientation and size difference of the two sets of machining

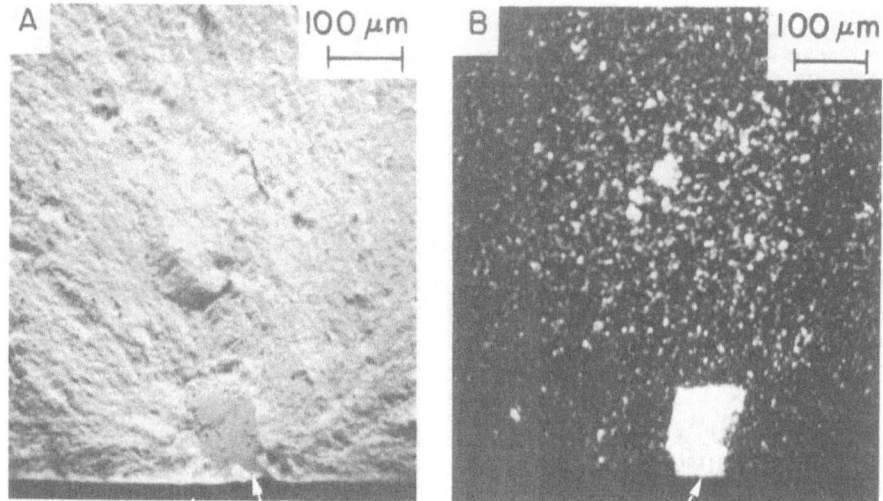

Figure 7.   Failure of commercial PZT sonar transducer materials from
            a large foreign particle. A) Shows the fracture surface
            with fracture originating from the foreign particle
            (arrow) at the junction of the tensile surface and the
            fracture surface at the bottom of the photo. B) Al x-ray
            fluorescence of the matching fracture half showing that
            the particle is Al-rich. Other analysis indicates this
            and the other Al-rich areas are Al_2O_3 particles which are
            believed to be chips from the milling media.

flaws relative to the grinding direction leads to substantial strength
anisotropy relative to the grinding direction (e.g., see Fig. 11).
Studies of polishing where samples are rotated randomly relative to
a rotating polishing disc typically results in failure from elongated
polishing flaws. This is attributed to the random direction of
abrasive particles over the surface which results in some elongated
flaws parallel with the direction of abrasive motion being perpendicular to the applied stress and hence dominating the failure process.

Handling defects have not been extensively studied, but Hertzian
cone cracks or flaws similar to those from machining are expected
(e.g., see Fig. 12).

Oxidation pits are an important example of another type of surface defect, recently observed as the source of failure in some hot
pressed Si_3N_4 bodies such as NC-132.[6,7,8,24,25] With increasing
oxidation, these pits become the source of weakness and failure.

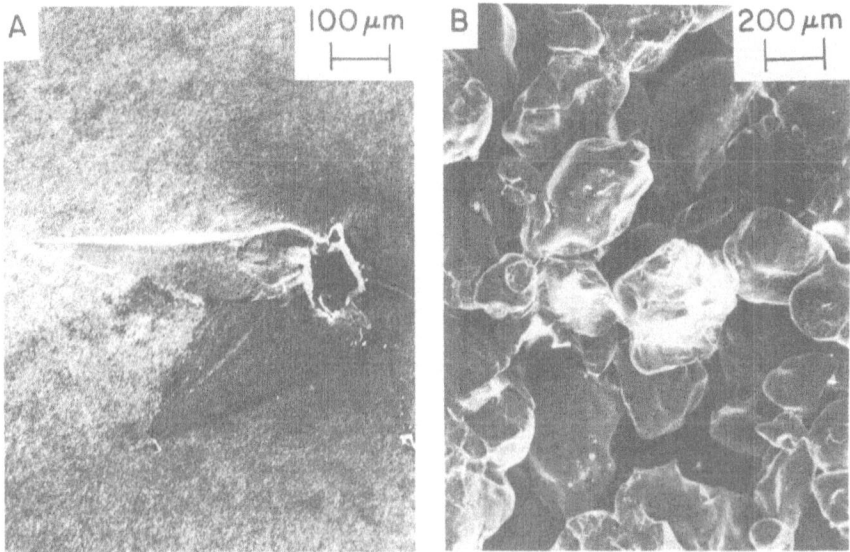

Figure 8. Failure of an IR quality hot pressed MgF_2 disk from SiO_2 particles. A) Fracture surface of a disk which failed under biaxial flexure, σ_f ~76 MPa (~11 ksi), from a particle near the right hand edge of the photo. The particle was shown by electron probe analysis to be SiO_2. B) SEM photo of the exterior surface of the SiO_2 crucibles used for powder processing. Note the SiO_2 grains of similar size that can be abraded off to provide impurity particles such as that shown in A).

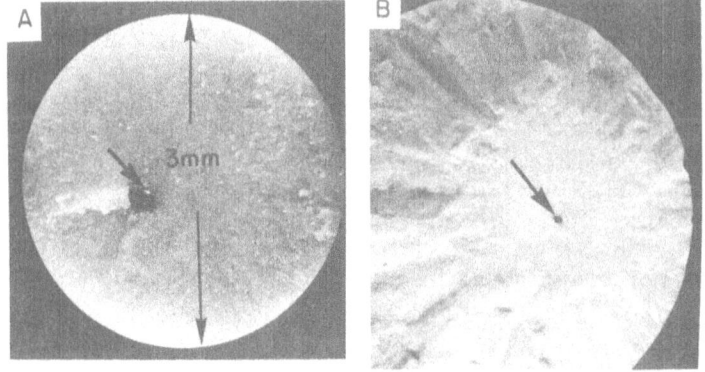

Figure 9. Examples of tensile tested hot pressed Al_2O_3 which failed from graphite particle inclusions (arrows). A) σ_f ~240 MPa (~35 ksi); B) 15 mm dia., σ_f ~220 MPa (~32 ksi).

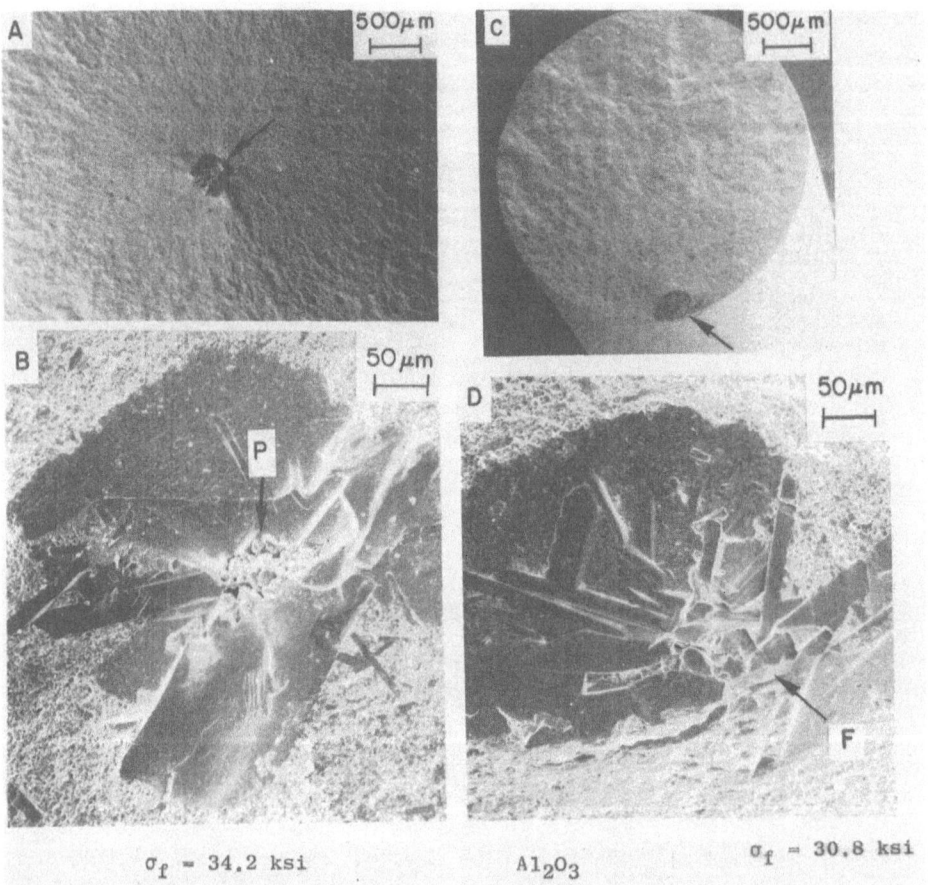

Figure 10.   Failure of tensile tested, hot pressed Al_2O_3 from large
             grains. A) Tensile sample failed at σ_f ~230 MPa (~34 ksi)
             from an internal cluster of large grains (arrow), shown
             in higher magnification in B). Note the associated
             porous region (b). C) Another sample failing from a
             cluster of large grains (arrow) at the surface of the
             tensile rod. D) Higher magnification of this large grain
             cluster; note the chipping indicative of machining flaws
             (F).

GENERAL AND SUMMARY COMMENTS

     Various processing defects can be interactive. For example
combinations of larger pores and grains are common sources of failure
in some commercial Al_2O_3 bodies, which may also be vulnerable to
origins stemming from smaller pores, porous regions, or machining
flaws with large grains, as noted earlier. Larger machining flaws

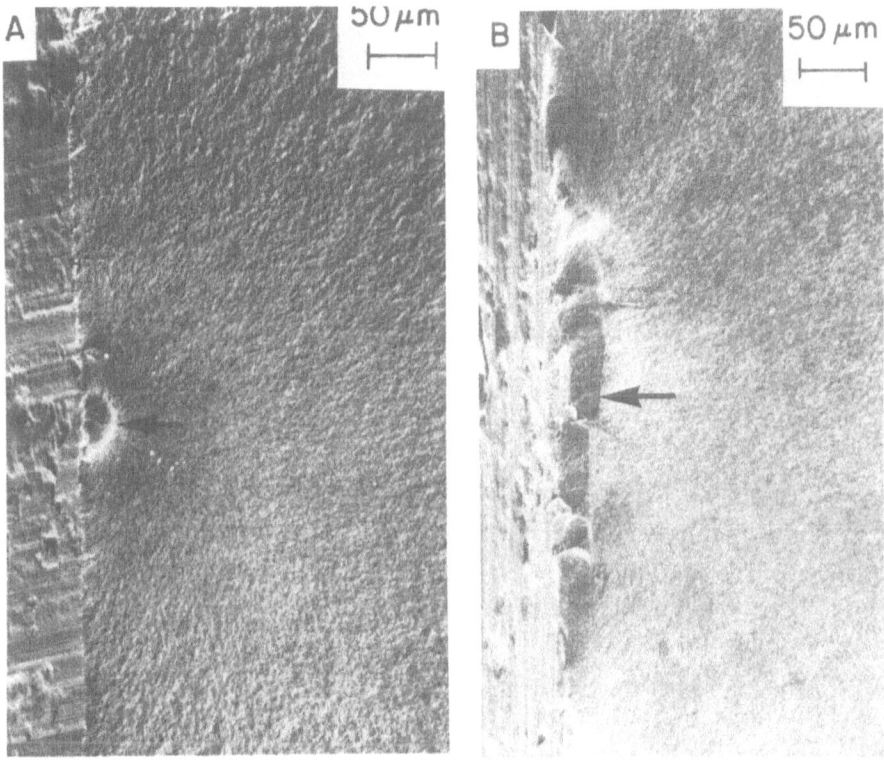

Figure 11. Characteristic machining flaws causing fracture initiation
due to machining: A) parallel with and B) perpendicular
to the tensile axis. Note the distinct difference in
the machining flaws (arrows) extending in from the surface
for these two different directions of machining relative
to the length of the commercial hot pressed MgF_2 flexure
bars. This difference in flaw shape, primarily length
as opposed to depth, and the associated strength
difference (σ_f ~41 vs ~83 MPa, i.e., ~12 ksi vs ~6 ksi,
respectively) is characteristic of a wide variety of
glasses, single crystals and polycrystalline bodies
tested.

may also preferentially form in more porous, impure, etc. regions or
join with them.[6]

The relative roles of processing defects depend not only on the
material and its processing, but also on the size and shape of the
component. Most, if not all, types of defects will increase their
extremes of size as the complexity, and especially the size, of the
component increases. As an example, it is instructive to compare
flaws in the flexural bars with those of the larger hoop tension

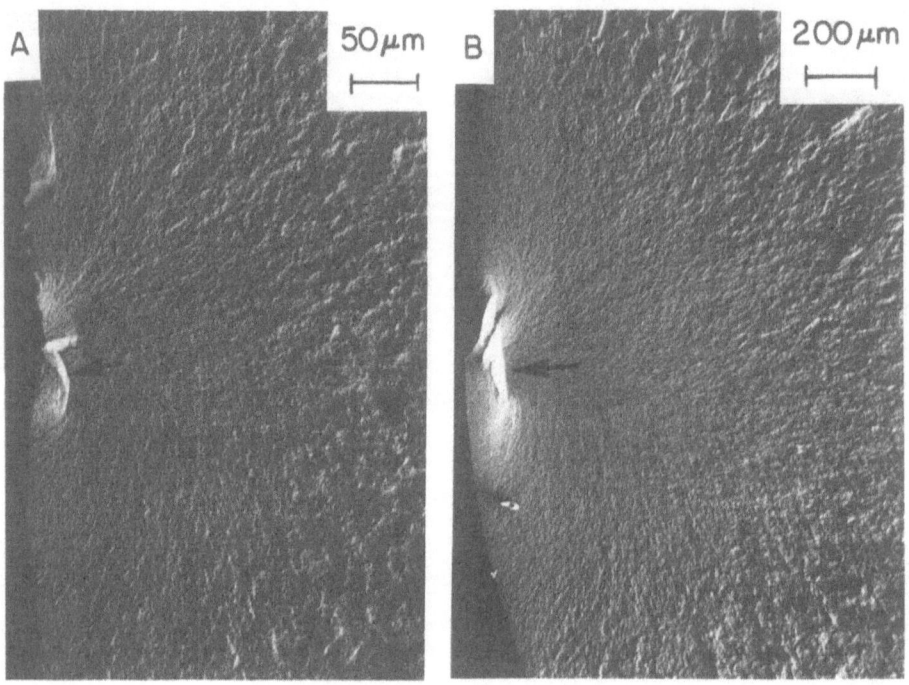

Figure 12.   Examples of fracture origins due to handling. Crystallized
              glass flexure rods failing at: A)σ_f ~260 MPa (~38 ksi)
              and B)σ_f ~130 MPa (~19 ksi) from impact-induced flaws
              (arrows) resulting from tumbling them with other rods.

rings and tensile test specimens. Of the process-related flaws, pores
are most likely to increase with specimen size, and large grains and
machining flaws least likely.

    Significant reduction of some processing induced defects is often
relatively easy, e.g., agglomerates, or foreign particles can often
be eliminated by screening or centrifuging. In highly competitive,
lower cost products even this simple a step may not be easily done
because of economic constraints. Also one problem may be exchanged
for another less frequent one, e.g., screening out agglomerates or
foreign particles will introduce screen fragments as the screens
wear. Further, there may be no benefit in removing one defect if the
distribution of others nearly matches it in terms of strength or other
required properties. It must be kept in mind that we will almost never
be dealing with defect-free bodies, so the challenge is to find the
critical defects which limit the required yield, acceptance, or performance levels. Fractography is a most effective tool in finding
such critical defects as illustrated in this paper.

ACKNOWLEDGMENT

The author wishes to acknowledge the skillful scanning electron microscopy of various colleagues, especially Mrs. Sarah Morey, and Dr. Paul Becher. The aid of Mr. G. Atkins of AFML and Dr. Stuart Starrett of Southern Research Institute in providing the tensile tested, hot pressed Al_2O_3 rods for examination was also most helpful.

REFERENCES

1. R. W. Rice, pp. 439-472 in Surfaces and Interfaces of Glass and Ceramics, ed. by V. D. Frechette, W. C. LaCourse and V. L. Burdick, Plenum Press, N. Y., 1974.

2. R. W. Rice, pp. 323-345 in Fracture Mechanics of Ceramics, ed. by R. C. Bradt, D. P. H. Hasselman, and F. F. Lange, Plenum Press, N. Y., 1974.

3. R. W. Rice, pp. 287-343 in Ceramics for High-Performance Applications, ed., by J. J. Burke, A. E. Gorum, and R. N. Katz, Brook Hill Publishing Co., 1974.

4. R. W. Rice and W. J. McDonough, pp. 394-403 in Mechanical Behavior of Materials, Vol. IV, The Soc. of Mat. Sci., Japan, 1972.

5. R. W. Rice and W. J. McDonough, ibid, pp. 422-431.

6. R. W. Rice, J. J. Mecholski, S. W. Freiman, and S. M. Morey, to be published in "Proceedings in Quantative NDE" (Conference held at Cornell University, June 1977) in Air Force Report #AFML-TR-77.

7. R. W. Rice, S. W. Freiman, J. J. Mecholsky, Robert Ruh, and Yoshiro Harada, to be published in Ceramics for High Performance Applications - II, Proceedings of Fifth Army Materials Technology Conference held in Newport, R. I., 1977.

8. R. W. Rice, S. W. Freiman, J. J. Mecholsky, and R. Ruh, to be published in the Proceedings of the DARPS/NAVSEA-Garrett/Airesearch Ceramic Gas Turbine Engine Demonstration Program Review, held in Castine, Maine, 1977.

9. B. K. Molnar and R. W. Rice, Am. Ceram. Soc. Bull., 52, 6, pp. 505-509, 1973.

10. R. C. Pohanka, R. W. Rice, J. Pasternak, P. L. Smith, and B. E. Walker, pp. 205-236 in Proceedings of the Workshop on Sonar Transducer Materials, ed. by P. L. Smith and R. C. Pohanka, February 1976.

11. W. J. McDonough, D. R. Flinn, K. H. Stern, and R. W. Rice, to be published in J. Mat. Sci.

12. S. R. Tan and G. J. May, J. Mat. Sci., 12, pp. 1058-1061, 1977.

13. H. P. Kirchner, R. M. Gruver, and W. A. Sotter, Mat. Sci. and Eng., 22, pp. 147-156, 1976.

14. R. M. Gruver, W. A. Sotter and H. P. Kirchner, Am. Ceram. Soc., 55 (2), pp. 198-204, 1976.

15. G. K. Bansal, W. H. Duckworth, and D. E. Niesz, J. Am. Ceram. Soc., 59 (11-12), 1976.

16. G. K. Bansal, W. Duckworth, and D. E. Niesz, Am. Ceram. Soc.
     Bull., 55 (3) pp. 289-307, 1976.
17. W. H. Rhodes, P. L. Berneburg, R. M. Cannon and W. C. Steele,
     Microstructure Studies of Polycrystalline Refractory Oxides,
     Avco Corp., Lowell, MA., Summary Report for Naval Air Systems
     Command, Contract N00019-72-C-0298, April 1973.
18. R. W. Rice, S. M. Morey, and F. W. Fraser, to be published in
     Fracture Initiation in Irdome Grade MgF$_2$, NRL Memorandum
     Report 3685, January 1978.
19. A. V. Virkar and R. S. Gordon, Am. Ceram. Soc., 60 (1-2)
     pp. 58-61, 1976.
20. H. Suzuki and K. Hayashi, Planseeberickte fur Pulvermetallurgie,
     Bd, 23, pp. 24-37, 1975.
21. S. W. Freiman, J. J. Mecholsky, J. C. Wurst, and R. W. Rice,
     Am. Ceram. Soc., 58 (9-10) pp. 406-409, 1975.
22. J. J. Mecholsky, S. W. Freiman, and R. W. Rice, Am. Ceram. Soc.,
     60 (3-4) pp. 114-117, 1977.
23. R. W. Rice, P. F. Becher, and J. J. Mecholsky, to be published.
24. S. W. Freiman, A. Williams, J. J. Mecholsky, and R. W. Rice,
     pp. 824-834 in Ceramic Microstructures '76, ed. by R. M.
     Fulrath and J. A. Pask, Westview Press, 1977.
25. S. W. Freiman, C. Cm. Wu, K. R. McKinney, and W. J. McDonough,
     to be published in the Proceedings of the DARPA/NAVSEA-
     Garrett/Airesearch Ceramic Gas Turbine Engine Demonstration
     Program Review, held at Castine, Maine, 1977.

DISCUSSION

Peter J. Gielisse (University of Rhode Island): Is it possible
from your experience to rate the "severity" of the strength degradation for the five centers for failure - pores, large grains,
foreign inclusions, machining damage, handling damage - as identified by your micrographic work. Can they furthermore be quanti-
fied relative to each other?

Author: It is difficult to rate the severity of the different
types of flaws, especially to do so quantitatively, since there
can be so many variables. The size and shape of the part, the
nature of the processing, and the body composition can all be interactive factors. Thus, bodies that are subject to exaggerated
grain growth are more likely to exhibit large grain origins than
bodies that show normal grain growth. Isopressing may often result in less severe flaws than cold pressing, extrusion, or slip
casting. Many flaws tend to increase with body size and shape
though impurity particles and especially voids can generally continue to increase more than large grains and particularly machining flaws with body size or shape. Thus, for example, laminating

in cold pressing, extrusions, or castings can often be quite sizeable; however, some service flaws may not vary much with size,
e.g., oxidation defects. On the other hand, handling damage may
depend on size, e.g., if it involves impact of the bodies with one
another or with another object.

William Rhodes (GTE Laboratories, Inc.): Impurity related defects
arise not only from the normal processing related sources but may
be present in the raw material powders as purchased. In an examination of 5 commercial 99.9+% pure Al_2O_3 powders we found particulate impurities in sufficient concentration to result in an impurity related defect every 3-5 mm in the consolidated body. It
would appear that this source of defects must be considered, and
further that full control of defects may require the processor to
synthesize his own powder.
Author: This amplification of Dr. Rhode's results noted earlier
further emphasizes the problem of foreign particulates and the
need for closer control of the whole process.

PART IV

DIELECTRIC AND MAGNETIC CERAMICS

PROCESSING OF HIGH DENSITY PIEZOELECTRIC CERAMIC COMPOSITIONS

Y. S. Kim and R. J. Hart

Bell Telephone Laboratories, Inc.

Allentown, Pennsylvania 18103

INTRODUCTION

Lead zirconate-titanate (PZT) ceramic compositions near the morphotropic phase transformation[1] exhibit a high dielectric constant and high radial coupling coefficient. Compositions found near the transformation are illustrated in Figure 1.

A composition, Pb $(Zr_{0.520} Ti_{0.465} Nb_{0.015}) O_{3.0}$, which lies in the transformation region, is of particular commercial interest for applications which require a piezoelectric transducer with high coupling coefficient.

One application is a ceramic transmitter for microphones which requires a specific PZT composition with maximum density and uniform microstructure that exhibits also good mechanical integrity when fashioned as a thin (75 μm) disc.

In preparing such a ceramic composition, it is difficult to maintain the material's stoichiometry during thermal processing steps such as calcining and sintering, because of the volatility of lead (Pb) or lead oxide (PbO). As can be seen in Figure 1, there is a very narrow band of compositions which exhibit useful piezoelectric properties.(*) Any deviation in the composition caused by the volatilization of components such as PbO or Pb, and contamination during processing adversely affects the densification of the PZT compounds. This results in degradation of the piezoelectric properties.

(*)PTZ-5A is a commercial product in this region.

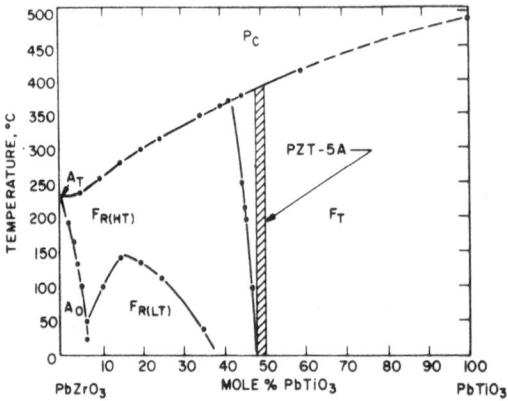

Figure 1.   Phase diagram of PbTiO₃-PbZrO₃ system.

Much work has been done in an effort to obtain a high density uniform PZT ceramic structure using advanced ceramic processing techniques[2] such as hot pressing and continuous hot pressing. Tape casting[3] and the conventional ceramic processing[4] of PZT composition have also been studied extensively. However, PZT with the maximum theoretical density and uniform microstructure has only recently been prepared by using usual ceramic techniques.[5]

This paper describes the development of a material, prepared by conventional ceramic processing, which yielded a reproducible piezoelectric compound with near theoretical density and superior electromechanical properties.

TABLE I

Chemical Composition Of PZT (Wt. %)

	PbO	ZrO₂	TiO₂	Nb₂O₅
PA-117	68.0	19.6	11.5	0.9
PA-118	68.2	19.6	11.5	0.6
PA-119	67.6	19.5	11.4	0.4
PA-120	67.0	20.0	11.1	1.15

SiO₂-0.01% and Al₂O₃-0.008% in all compositions.

EXPERIMENTS

The compositions were prepared using appropriate amounts of lead oxide (PbO), zirconia (ZrO_2), titania (TiO_2) and niobium pentoxide (Nb_2O_5) as listed in Table I. Four compositions were processed.

1200 gram batches were wet milled in a high density alumina ball jar. The dried powders were calcined at temperatures ranging from 600°C to 1100°C and then wet ball-milled again. During calcining, weight loss was determined on 1200 grams samples (including thermal gravimetric analysis using 400 µg sample). After calcination, the dried ball-milled powders were granulated and screened through sieves for isostatic pressing. Various sizes of samples (2.5 cm dia and up to 15 cm long) were pressed at 138 MPa (20,000 psi). The rod samples were sintered in a platinum lined container at temperatures from 1200°C to 1350°C, in air and oxygen atmospheres, respectively, using a tube furnace. The samples were covered with the same composition of PZT powder. The sintered rods were machined into discs to evaluate physical and electrical properties.

RESULTS AND DISCUSSION

PZT COMPOUND FORMATION DURING CALCINATION

Calcining the ball-milled mixture of oxides at temperatures greater than 600°C produced a material which was a single phase (tetragonal) PZT within detection limits of X-ray diffraction. Although the X-ray pattern of material calcined at 600°C shows only a single phase, the peaks are broad. Calcining at successively higher temperatures produces material with progressively sharper peaks as shown in Figure 2.

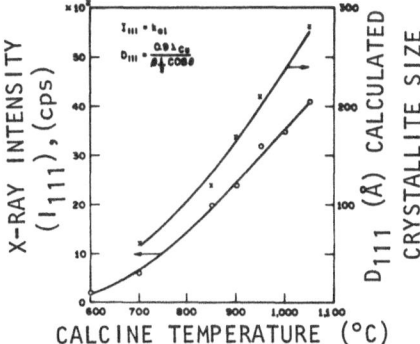

Figure 2. X-ray intensity (I_{111}) and Crystallite size (D_{111}) as a function of calcining temperature.

This also shows that the degree of crystallinity and the
particle size of the PZT increases with the calcining temperature. Since both these parameters are, in general, inversely
related to the reactivity of the calcined powder, the calcining
temperature is very important. This is consistent with previous
work.[4]

During the calcining treatment the powder loses weight. If
the final PZT composition is to be controlled, this loss must be
minimized. Figure 3 shows that: a) the weight loss depends on
and b) above the critical temperature the rate of weight loss increases sharply. It is preferable then to maintain the calcining
temperature below that point. From Figure 3 it is apparent that
for a 1200 gram batch the calcining temperature should be kept
below 1000°C where rate of lead loss is low. Therefore, 950°C $\pm$
50°C appears as an optimum calcining temperature.

This precise effect that calcining temperature has on sinterability is not fully understood. However, it is felt that the
particles will adsorb moisture and other atmospheric constituents
which affect the ability of the powder to flow and compact during
piece part formation. This is consistent with the observed difficulty in screening the powders calcined at temperatures below
900°C. Voids formed during the compaction of the powder calcined
at temperatures less than 900°C have a detrimental effect on the
sinterability. On the other hand, powders calcined above 1100°C
exhibit poorer properties, because of the above mentioned difficulty in controlling the composition.

IMPURITY CONTAMINATION DURING PROCESSING

The Si and Al picked up from the Burundum balls (96% Al_2O_3)
during the milling operation had a most detrimental effect on
the sinterability of the PZT powder as illustrated in Figure 4.
This effect has also been observed by others.[6]   The trend of
decreased sinterability as the Al and Si content of the powder

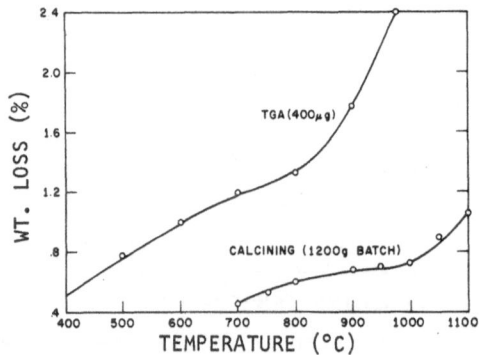

Figure 3.   Lead loss from PZT composition as a function of calcining
temperature.

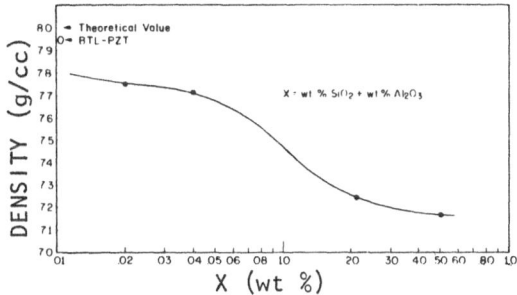

Figure 4. Effects of impurities on densification of PZT ceramics.
TAM-PZT powder sintered at 1205°C/2 h in air.

increases is of significance in manufacture. As it is important
that the ambient impurities SiO_2 and Al_2O_3 be less than 0.02 wt.
% in the milled PZT powder it follows that the raw materials must
be pure enough and processed under strict control to keep these
impurities below this critical limit.

PRESSING POWDER CHARACTERISTICS

It is felt that powder preparation prior to fabrication of
parts is one of the most critical steps in the process. Particle
size and size distribution as well as agglomerate size and agglomerate size distribution are important variables which affect the
ability of the powder to flow during compaction. An additional
variable, particle surface energy, also has a strong influence
on powder rheology and hardness of agglomerates. The pressing
powder prepared by the above techniques has an average particle
size of 4 μm and a typical particle size distribution as shown
in Figure 5.

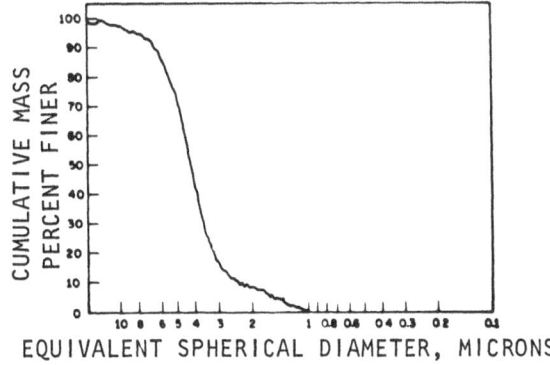

Figure 5. Particle size and distribution of PZT-BTL pressing powder.

SINTERING

From a study of various sintering parameters, it was found that the temperature had a strong influence on the sintering behavior of the four compositions mentioned in Table I (see Figure 6).

Our compositions sintered to a maximum density at different sintering temperatures which differed by as much as 70°C. The effect of sintering atmosphere on densification was also found to be significant as demonstrated in Figure 7. The sintering temperatures required for maximum density were increased by 50°C in an oxygen atmosphere as compared to an air atmosphere. Only in the oxygen atmosphere were the four compositions densified to near theoretical value. The sintering of PA-117 samples in oxygen yielded theoretically dense bodies; whereas in air parts from the same batches could only be sintered to 97% of theoretical density. With composition PA-117 it was demonstrated (Figure 6) that parts can be sintered to optimum density over a wide range of temperatures, especially 1275°C to 1325°C, in an oxygen atmosphere. It is believed that this is the first report of a PZT composition that has been densified to its theoretical value without using any hot pressing techniques.

Berlincourt[7] reported that a PZT composition fired in 100 psi of oxygen sintered to a higher density (97.7% theoretical

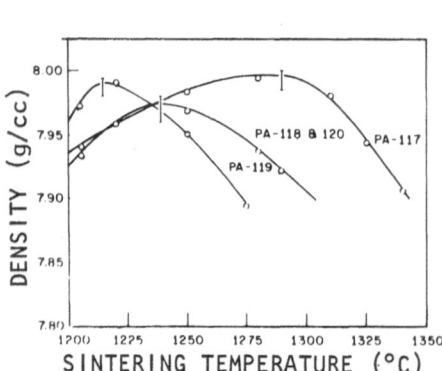

Figure 6. Density of PZT-BTL as a function of sintering temperature.

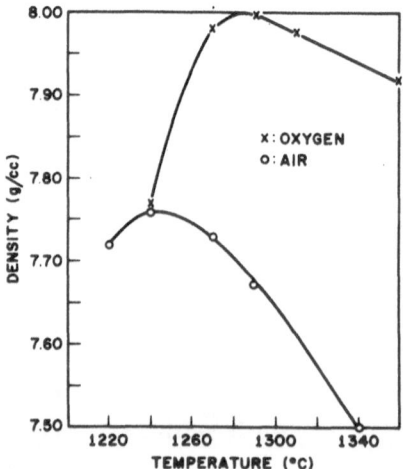

Figure 7. Effect of sintering atmosphere on densification of PZT.

value) than the same composition sintered in air. Murray and
Dungan[8] also found that densification of a similar PZT composition was enhanced when sintered in an oxygen atmosphere.
However, in neither case has a density approaching the theoretical value been achieved. Webster and others[9] obtained PZT
with 99.5% relative density from sintering in an oxygen atmosphere, but showed only a slight improvement in properties over
that of air sintered PZT. The enhancement of sintering with the
use of an oxygen atmosphere has been explained as follows:[10]
the oxygen replaces other gases, mainly nitrogen, in the open
pores of the powder compact during the early stages of sintering. This oxygen in residual pores can diffuse through the
lattice and along the grain boundaries, and will lead to eventual elimination of the pores during the final densification
stage.

DENSITY AND MICROSTRUCTURE

Examination of the microstructure of the four compositions
showed that all were highly dense materials with no apparent
second phase. All of these materials were virtually pore free
and exhibited a uniform grain size. A photograph of a typical
microstructure of PA-117 is shown in Figure 8, and a SEM photograph of the same material in Figure 9. The microstructure by
SEM revealed a dense structure with a relatively uniform grain
size in which a domain pattern can be seen.

PA-120 when sintered to a maximum density showed a significantly smaller grain size than the other three samples. The
fine grain size in PA-120 may be due to its higher niobium content. Others[11] have also observed that the higher the niobium
content in PZT composition, the smaller the grain size.

The microstructure itself has only a secondary effect on
the electrical properties of PZT as long as the grains are larger
than 5 μm.[12] The most critical parameter is density. It is
not necessary to achieve absolute control of grain size and grain

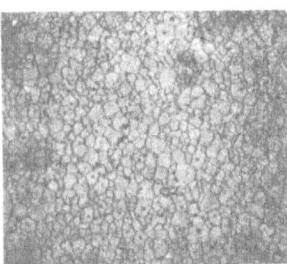

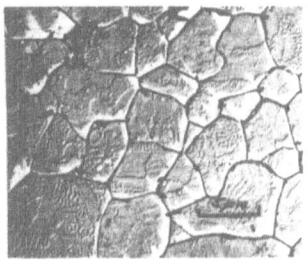

FIGURE 8   FIGURE 9

size distribution for the telephone transducer application. However, low porosity is essential not only for mechanical integrity but also to insure that the material is not sensitive to ambient humidity and that it has an optimum dielectric constant and breakdown strength. Gerson and Marshall[13] reported that voltage breakdown field decreases logarithmically with increasing porosity.

PHYSICAL AND ELECTROMECHANICAL PROPERTIES

Density, grain size, and electromechanical properties of PA-117 which exhibits the best overall characteristics are summarized in Table II. These properties are compared with other similar PZT ceramics.

CONCLUSIONS

A process has been developed to synthesize PZT ceramics with near theoretical maximum density and superior electromechanical properties. The following steps in the process sequence are especially critical:

1. Chemical Composition - An optimum composition was determined to be $Pb(Zr_{.520} Ti_{.465} Nb_{.015})O_{3.0}$.

2. Impurities - The combined Al_2O_3 and SiO_2 content must be less than 0.02 wt. % in the PZT powder used to fabricate final piece parts.

TABLE II

Physical And Electrical Properties

	ρ g/cc	Grain Size (μm)	$\varepsilon^T_{33}/\varepsilon_o$	$-k_p$	Qm
Powder A[+]	7.77	4.0	1555	.55	90
Hot Pressed[+]	7.98	2.0	1640	.58	105
G-1500[++]	>7.6	2.0	1700	.58	80
PZT-5[++]	7.75	4.0	1730	.60	75
BTL Hi-Density PZT	7.99	4.0	1850-	.62-	70-
		10	2000	.70	75

[+]Commercial Powder.
[++]Commercial Product (Literature Data).

3. <u>Calcining</u> – The optimum temperature is 950°C. Below this temperature the powder easily agglomerates and does not flow well during pressing. At higher temperatures lead loss becomes excessive.

4. <u>Pressing Powder Preparation</u> – The calcined and ball milled powder must be screened through sieves prior to pressing to remove hard agglomerates which can lead to a high concentration of low density areas in a sintered piece part.

5. <u>Sintering</u> – Pure oxygen in the sintering atmosphere enhances densification of PZT parts. During sintering the parts are covered with a PZT powder of the same composition as the parts.

6. <u>Electromechanical Properties</u> – Theoretically dense PZT ceramics exhibited a dielectric constant (ε) approaching 2000 and a radial coupling coefficient (k_p) greater than 60%. These properties have been reproduced from over 150 lots within a 2.7% standard deviation.

ACKNOWLEDGMENT

The authors gratefully acknowledge the continuous support of this work by H. M. Cohen.

REFERENCES

1. (a) E. Sawaguchi, J. Phys. Soc. Japan 8, 615 (1953).
   (b) S. Fushimi and T. Ikeda, J. Amer. Ceram. Soc. 50, (3), 129-132, (1967).
2. (a) G. H. Haertling, Amer. Ceram. Soc. Bull. 43, (12), 875-879, (1964).
   (b) F. J. Schnettler, 30-E-73, 75th Meeting Amer. Ceram. Soc., April, 1973.
3. (a) R. B. Runk, R & D Tech. Report, CC 4387, WE Eng. Res. Center, May 24, 1971.
   (b) J. J. Wentzel, U.S. Patent 3,517,093, June 23, 1970.
4. A. H. Webster, T. B. Weston and R. R. Craig, J. Can. Ceram. Soc. 34, 121-129, (1965).
5. Y. S. Kim, 4E73, 75th Meeting Amer. Ceram. Soc., April, 1973.
6. D. A. Buckner and P. D. Wilcox, Ceramic Bull. 51, (3), 218-222, (1972).
7. D. Berlincourt, Clevite Res. Report (Sandia Corp. Contract No. 51-4232), Project No. 30146, April 27, 1959.

8.   T. F. Murray and R. H. Dungan, Sandia Corp. (SC-R-65-874), March, 1965.

9.   A. H. Webster, T. B. Weston, and V. M. McNamara, J. Can. Ceram. Soc. 35, 61-68, (1966).

10. (a) A. H. Webster, et al., ibid.
     (b) R. Atkins, D. Eng. Thesis, Univ. California, Berkely, September, 1970.

11. F. Kulczar, J. Amer. Ceram. Soc. 42, (7), 343-349, (1959).

12. (a) A. H. Webster and T. B. Weston, J. Can. Ceram. Soc. 37, 111-114, (1968).
     (b) H. J. Gesenmann, Report on the 1st Int. Building Materials and Silicate Conference at Weiman, 1964.

13. R. Gerson and T. Marshall, JAP 30, (11), 1650-1653, (1959).

DISCUSSION

Robert L. Holman (Xerox): You've indicated that fully densified PtT specimens of the "optimum composition" exhibit optical transparency in thin slices; a new result for polycrystalline PZT ceramics. How would you compare its optical and electrooptical properties with that of the well known PLZT materials, and do you foresee any advantages inherent to your new material that would improve the likelihood of optical application?

Author: I thank you for your complimentary remark on my work. The optical and electro-optical properties of the PZT material are under investigation. The advantages will not be known until the results from the above mentioned study are available.

Frank G. Recny (General Electric Company): 1) In summary, to what do you attribute your success in achieving the high densities and high coupling coefficients? 2) What were the grades and who were the manufacturers of the raw materials used?

Author: 1) As indicated in the conclusions, it is attributed to the synergistic effects obtained from, e.g., control of composition, optimum calcination, pressing particle size distribution and sintering in an oxygen atmosphere. 2) These chemical grade raw materials were used: PbO-Hammond Lead Products, Inc.; ZrO_2-Tison Zirconium Chemical Corporation; TiO_2-Titanium Metals Corporation; and Nb_2O_5-Shieldalloy Corporation.

F. F. Lange (Rockwell International): Should not jet milling with plastic liners eliminate the contamination picked up during conventional ball milling?

Author: This is one of the well-known milling processes, and one may beneficially use jet milling with plastic liners.

THE ROLE OF ZrO_2 POWDERS IN MICROSTRUCTURAL DEVELOPMENT OF PZT CERAMICS

J.V. Biggers, D.L. Hankey, L. Tarhay

Materials Research Laboratory
The Pennsylvania State University
University Park, Pennsylvania 16802

Processing electrical ceramics to obtain reproducible properties is not an easy task. Seemingly inconsequential changes in raw materials or processing steps can produce ceramics with quite different properties.

The lead zirconate-lead titanate solid solutions used in manufacture of piezoelectric devices are a good example of processing sensitive materials. The major processing variables are generally considered to be: powder characteristics (composition and particle size distribution and morphology), material mixing, calcining and control of furnace atmosphere during firing.

The calcining step is usually limited to temperatures of about 900°C. A recent paper by Buckner and Wilcox[1] shows the strong influence of calcining temperature on the properties of the fired ceramics. These differences are certainly in large part related to the reactivities of the powders which in turn depend to a large extent on particle composition and morphology. Studies of reactions occurring during calcining[2,3] have shown that commercial ZrO_2 is the least reactive of the powders. Matsuo and Sasaki[2] suggest that the reaction sequence of the mixed oxide powders used to make PZT ceramics involves first the formation of $PbTiO_3$ and then reaction of this with ZrO_2 to form PZT. There is usually a mixture of several phases present after precessing, e.g. PZT, $PbTiO_3$ and ZrO_2. Rosolowski et al.[4] have determined that commercial calcines contain several weight percent of unreacted ZrO_2. They note also that the ceramics after final firing can contain 1-2 w/o ZrO_2.

While there is ample evidence to suggest that reactivity of the ZrO_2 is an important factor, it is extremely difficult to relate

observed differences in powder composition and morphology to the
reactivity. Recent work in this laboratory has shown that the only
noticeable difference in lots of powders judged good or bad (according to the ceramic produced) was the state of agglomeration[5].

Quantitative measurements of the agglomeration in powders are
very difficult. In order to test the hypothesis that the state of
agglomeration affects reactivity, a series of experiments was undertaken which compared ceramics produced under identical conditions
but with different ZrO_2 starting powders.

The ZrO_2 lots chosen were judged from a qualitative basis to
have differing agglomerate character. In addition, portions of
each lot were subjected to different high energy milling pre-
treatments in an attempt to change the particle and agglomerate
size distribution.

The different ZrO_2 powders were used to produce ceramics with
the composition $PbZr_{0.55}Ti_{0.45}O_3$. This composition was chosen as
being representative of many commercial PZT ceramics and it was far
enough from the morphotropic boundary to preclude anomalies associated with the phase transition[6].

EXPERIMENTAL

Three different lots of commercial ZrO_2 powders were selected
for this work--Tizon lot 367m and Harshaw lots 5-76 and 1-75. The
initial characterization of these lots using optical emission spectroscopy for impurities and wet chemical analysis for major elements
has been reported earlier[5].

Particle size and shape were measured using a beam controlled
SEM technique[7] and surface area was determined by a BET method.

The as-received ZrO_2 powders were then milled using two types
of equipment--a fluid energy mill and an attrition mill. Construction details and operation of both types of mills are described
by Wang[8].

The fluid energy mill (Fluid Energy Products, Hatfield, Pa.)
uses high-pressure air which accelerates the individual particles
and impacts them in a rubber-lined chamber. Operating air pressures
were 600 K Pa and the residence time for particulates in the mill
was about 10 minutes.

The attrition mill, built at this laboratory, used a polyurethane lined jar mill, a stainless steel impellor covered with tygon
tubing and a charge of 0.011 m diameter zirconia balls. The milling
was done for 20 minutes at 200 rpm impellor speed with equal volumes

of powder, ethyl alcohol and ball charge.

Figure 1 shows a processing flow chart used to produce ceramic discs. The powder mixing steps were carried out in rubber-lined jar mills. Pan drying was done at 100°C for 24 hours. The dried cake was broken up in an alumina mortar. This powder was calcined in closed 99.9 w/o Al_2O_3 crucibles for 2 hours at 900°C in an electrical resistance furnace.

The calcined powders were pressed into about 0.015 m diameter discs using a Stokes rotary press. Green densities were carefully controlled at 65 ± 1.0% of theoretical for all pellets. The green thickness of the pellets was about 0.0012 m. Firing of the pellets was carried out in closed alumina crucibles using a technique described by Klicker and Biggers[5].

Samples of the different pellets were distributed randomly on the setters inside the closed alumina crucibles to minimize effects of thermal and PbO atmosphere gradients.

RESULTS AND DISCUSSION

Table 1 shows the particle size distribution and surface area data for the as-received and milled powders. Table 2 shows in more detail the particle size distribution of the Harshaw lot 1-75 powders. The particle size changes after milling were not significant. Jet milling increased slightly the surface area of the three lots used while attritor milling appeared to slightly decrease the surface areas for the Harshaw lots and increase the area for the Tizon material.

As can be seen from Table 2, the milling did tend to remove large particles and shift the distribution slightly toward a lower mean diameter. The results were the same for the other lots.

The shape of the particles given by the min/max particle diameter ratios in Table 1 remains essentially constant.

Emission spectroscopy was used to check for impurity pick-up during milling and no contamination was detected. After calcining the phases present were checked using x-ray diffraction and a wet chemical method for PbO[9]. With all as-received lots the x-ray diffraction showed PZT and $PbTiO_3$. In all milled powders only PZT was detected. The results of the PbO analyses are shown in Table 3.

There was a significant difference in unreacted PbO between the Tizon and Harshaw as-received lots. Milling of the ZrO_2 powders did not, however, alter the amount of PbO in the calcine.

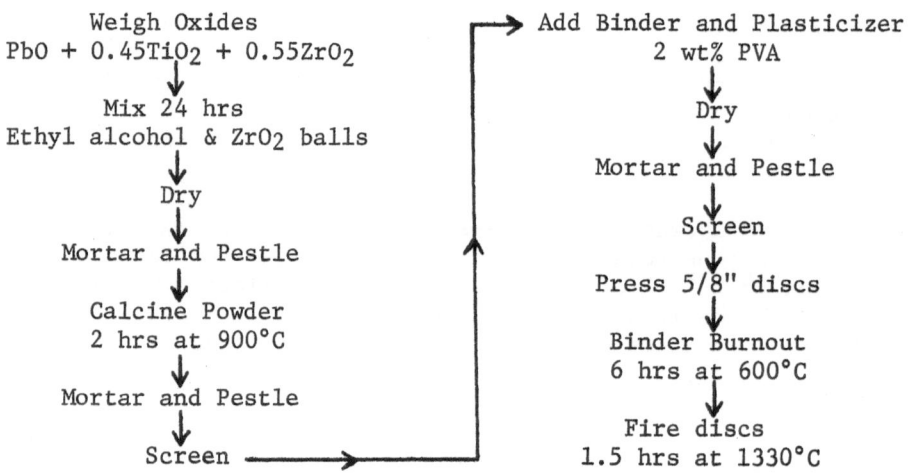

Figure 1.   Processing Flow Chart for ZrO_2 Study.

Table 1.   Data on ZrO_2 Powders from Automated SEM Image Analysis and Surface Area Analysis.

ZrO_2 Sample	Average Diameter		Min/Max Dia. Ratio		Surface Area (BET)
	Mean (μm)	σ	Mean	σ	$m^2/kg(x10^{-3})$
Tizon Lot 367m					
As received	0.69	0.5	0.47	0.17	18.4
3 jet mills	0.59	0.5	0.50	0.18	18.7
Attritor milled	0.68	0.5	0.50	0.18	19.4
Harshaw Lot 1-75					
As received	0.72	0.5	0.53	0.17	13.5
3 jet mills	0.62	0.3	0.49	0.17	14.6
Attritor milled	--	--	--	--	12.5
Harshaw Lot 5-76					
As received	0.89	0.5	0.44	0.18	14.6
3 jet mills	0.69	0.4	0.51	0.17	15.5
Attritor milled	0.68	0.4	0.50	0.17	12.8

Table 2.   Average Diameter Distribution for Harshaw Lot 1-75

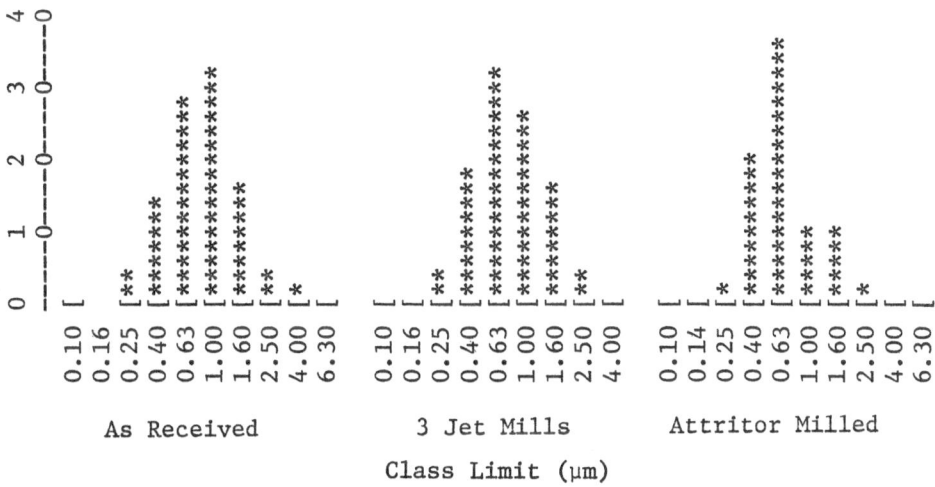

Class Limit (μm)

Table 3. Wet Chemical Analysis to Determine "Free" PbO

Specimen (ZrO₂ Specifications)	Weight % "Free" PbO After Calcining at 900°C For 2 Hours (σ=±0.1%)
Tizon-Lot 367m	
As received	1.1
3 jet mills	1.0
Harshaw-Lot 1-75	
As received	0.6
3 jet mills	0.4
Attritor milled	0.3
Harshaw-Lot 5-76	
As received	0.6
3 jet mills	0.4
Attritor milled	0.5

Table 4.   Grain Size Data for Fired PZTs

Fired PZT Spec.	Ave. Grain Size μm
Tizon 367m	
As received	8.9
Jet milled	4.2
Harshaw 1-75	
As received	8.26
Jet milled	3.90
Attritor milled	3.10
Harshaw 5-76	
As received	8.14
Jet milled	4.08
Attritor milled	3.25

Microstructures of the ceramics produced from the as-received and attrition milled powders from Harshaw Lot 5-76 are shown in Fig. 2.

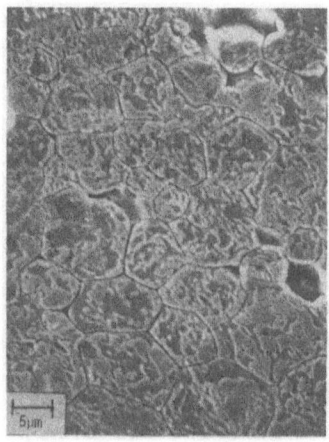 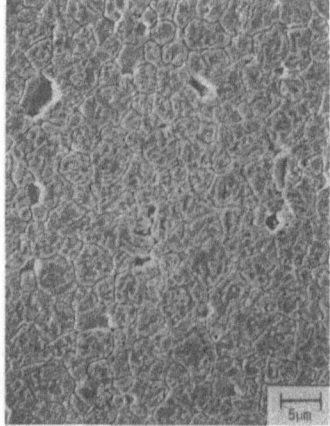

Fig. 2.   Harshaw 5-76 as-received; Harshaw 5-76 attritor milled.

Table 4 lists the average grain sizes determined by a linear intercept technique for the different ceramics. As can be seen, there is the same result in all cases--the as-received ceramics had a significantly larger grain size than those produced from milled powders. In each lot the attrition milled powders had the finest grain size. While it is only speculation at this point, it seems likely that the differences in grain size between the as-received and milled powders must be due to reactivity differences during calcining. It is presumed that the absence of the low melting PbTiO$_3$ phase and the presence of a PZT phase with higher ZrO$_2$ content would lead to a situation where less liquid phase was present during final sintering. This could result in a finer grained product.

Table 5 shows the fired density and dielectric properties of the different ceramics. The densities are essentially the same for all specimens. The weak field K and tan δ values measured using unpoled specimens are in general agreement with those for ceramics of the same composition reported by Berlincourt et al.[10]. There were, however, significant differences in K between as-received lots and as a function of milling. Some of the differences can be attributed to grain size effects, but different phase combinations present as a result of reactivity differences must account for some of the variations.

Table 5. Density and Electrical Data for PbZr$_{0.55}$Ti$_{0.45}$O$_3$

Specimen (ZrO$_2$ Spec.)	Ave. Density* of Fired Discs kg/m^3 (x10^{-3})	Ave. Dielectric[†] Constant (K)	Ave. Dielectric Loss (Tan δ)
Tizon Lot 367m			
As received	7.63	534 ± 16	.0026
3 jet mills	7.62	501 ± 8	.0036
Harshaw Lot 1-75			
As received	7.69	641 ± 16	.0033
3 jet mills	7.64	523 ± 13	.0040
Attritor mill	7.64	549 ± 13	.0043
Harshaw Lot 5-76			
As received	7.66	568 ± 13	.0031
3 jet mills	7.62	496 ± 15	.0041
Attritor mill	7.60	547 ± 11	.0048

*All density measurements are geometrical and were randomly checked using a mercury porosimeter.

[†]All electrical measurements were made at 22°C and 1 MHz.

CONCLUSIONS

We have shown that high-energy milling of different ZrO_2 powders used to produce PZT ceramics can have a significant effect on the ceramics' microstructure and electrical properties. Characterization of the powders showed that there was little difference in composition and particle morphology before and after milling. Qualitatively the milled powder appeared to contain fewer agglomerates.

The reactivity of the ZrO_2 powders with the other oxides during calcining was increased by milling. In all the calcines produced from milled powders, only PZT was detected.

The results of this work strongly suggest that the variability of the piezoelectric properties of commercial PZT ceramics is related to the reactivity of the ZrO_2 used in the process. More work is needed to determine the relation between the reactivity of the powders, the phases produced during calcining, and the resultant ceramic properties.

Acknowledgments

This work was carried out under Contract N0014-76-C-0515 under the joint sponsorship of the Office of Naval Research and the Defense Advanced Research Projects Agency.

References

1. D. A. Buckner and P. D. Wilcox, Am. Ceram. Soc. Bull. 51 [3] 218-222 (1977).
2. Y. Matsuo and H. Sasaki, J. Am. Ceram. Soc. 48 [6] 289-291 (1965).
3. V. M. McNamara, J. Can. Ceram. Soc. 33, 102-119, (1964).
4. J. H. Rosolowski, R. H. Arendt and J. W. Szymanszek, Annual Report, G. E., Schenectady, N. Y., Contract No. N00014-76-C-0659, Office of Naval Research (1977).
5. L. E. Cross, J. V. Biggers and R. E. Newnham, Semi-Annual Report, Contract No. N00014-76-C-0515, Office of Naval Research (1977).
6. B. Jaffe, W. R. Cook and H. Jaffe, Piezoelectric Ceramics, Academic Press, N. Y. (1971).
7. J. Lebiedzik, R. G. Burke, S. Troutman, G. G. Johnson and E. W. White. Scanning Electron Microscopy, Part 1, 26-33 (1973).
8. F. Y. Wang, Treatise on Materials Science and Technology, Vol. 9, Academic Press, N. Y. (1976).
9. A. E. Robinson and T. A. Joyce, Trans. Brit. Ceram. Soc. 61 [2] 85-93 (1962).
10. D. Berlincourt, C. Canolik and H. Jaffe. Proc. IRE, 48, 220-229 (1960).

NOVEL USES OF GRAVIMETRY IN THE PROCESSING OF CRYSTALLINE CERAMICS

Robert L. Holman

Webster Research Center
Xerox Corporation
Rochester, New York 14644

ABSTRACT

The recording microbalance has been utilized in studying a vapor phase reaction that occurs between two incongruently vaporizing ceramics at high temperature. The degree of nonstoichiometry and a model for solid state diffusion are established directly. Applications of the process are illustrated by determining crystal composition, by forming reproducibly tailored optical waveguides, and by reducing significantly the susceptibility to laser damage of $LiNbO_3$ single crystals.

INTRODUCTION

The recording electrobalance has proven to be a valuable tool in the study of high temperature chemical reactions. Its application to deposition, oxidation, and vaporization studies encompasses a wide range of inorganic, as well as organic materials. In this report, the use of the electrobalance is extended to the study of a new thermochemical process that allows direct characterization of nonstoichiometry and diffusion in such single crystalline ceramics as lithium niobate, $LiNbO_3$. The process, called vapor phase equilibration, is based in part on earlier efforts[1-3] to ascertain the phase boundary compositions of nonstoichiometric oxide powders. The process involves heating a $LiNbO_3$ crystal in close proximity with another solid of mixed lithium niobate phases whose mass considerably exceeds that of the crystal. The composition or nonstoichiometry of the $LiNbO_3$ crystal is then altered as a result of preferential lithium oxide transport that occurs between the two solids, dissimilar in their lithium oxide chemical potential. A recording elec-

trobalance is used to monitor the process and to provide process
kinetic information fundamental to characterizing nonstoichiometry
and to modeling the solid state diffusion.

The process is applied to adjust predictably and reproducibly
the nonstoichiometry of $LiNbO_3$ crystals, either homogeneously or
diffusively, from their surface into their bulk. As a result,
$LiNbO_3$ crystal composition, which cannot be assayed with sufficient
accuracy by standard analytical techniques, is related to the crystal's reaction weight-change. Crystal composition is also varied
to reveal any composition dependence displayed by an undesirable
optical property inherent to $LiNbO3$, known as optical damage. In
addition, the known relationship existent between refractive index
and crystal composition[4] is exploited to produce high quality
$LiNbO_3$ optical waveguiding layers, for the first time, without the
need of heating in the presence of a vacuum[5], a flowing gas[6], or a
diffusing impurity[7].

                          EXPERIMENTAL

$LiNbO_3$ crystals of the highest available purity (less than
20ppm Fe) were obtained from Crystal Technology Inc. and Union
Carbide Inc. as grown from the congruently melting composition
(48.6% mol. Li_2O; 51.4% mol. Nb_2O_5). These ferroelectric crystals
were first poled and then formed as optically polished 0.75" x 0.03"
discs, such that the polar c-axis was contained in the plane of the
disc.

Each disc was suspended from a recording electrobalance (Cahn
Inc. R-100), which allowed mass changes associated with any lithium
oxide diffusion into (or out of) the crystal to be monitored continuously. The suspended crystal was centered within one of two
special reaction crucibles, Fig. 1a, both formed from mixtures of
equilibrium lithium niobate phases. One curcible, made from the
lithium-rich mixture, M_H, Li_3NbO_4 + $LiNbO_3$, provides an infinite
source of lithium oxide for a $LiNbO_3$ crystal; the other crucible,
made from the lithium-deficient misture, M_L, $LiNbO_3$ + $LiNb_3O_8$, provides and infinite sink. The crucibles were formed by reacting isostatically pressed shaped, Fig. 1b, mistures of lithium carbonate and
niobium oxide at 1100°C until complete formation of the equilibrium
phases could be verified. A fugitive organic binder (napthalene)
was added to the mixture to assure sufficient porosity so that neither
lithium nor oxygen ion diffusivity within the crucible walls would
become rate limiting.

The selected reaction crucible, and independently suspended crystal, were positioned within the hot-zone of a platinum wound resistance furnace, heated (10°C/min.) to the desired reaction temperature (1050-1150°C), and maintained at temperature for 0.5-100

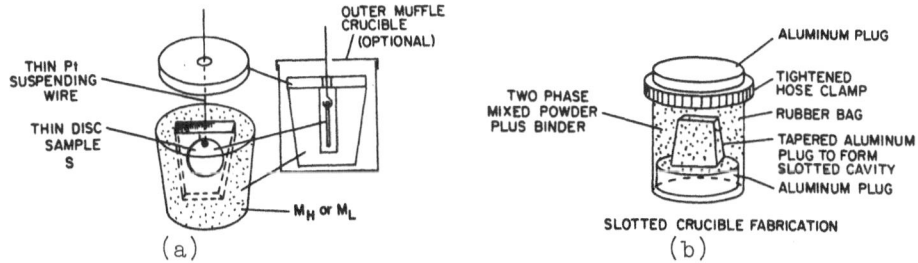

Figure 1.   a) Lithium niobate reaction crucible, b) Fabrication
   procedure.

hrs., depending upon the extent of reaction desired. A complete
description of a suitable thermogravimetric apparatus has appeared
elsewhere[1-3,8].

Crystal weight-changes were recorded continuously as a function
of time. Uniform $LiNbO_3$ crystals whose final composition corresponded to that of either the lithium oxide-rich or lithium oxide-
deficient phase boundary were prepared by completely equilibrating
(30-100 hrs.) crystals with crucibles M_H and M_L respectively at
1100°C. Uniform crystals exhibiting intermediate compositions were
prepared by heating for shorter time (0.5-10 hrs.) in the appropriate crucible, and then re-heating in air, at the same temperature,
until they became homogeneous by diffusion (50-100 hrs.). Low loss
optical waveguides were formed by partially equilibrating crystals
and lithium oxide deficient crucibles, M_L, (0.5-10 hrs.) without reheating.

Heat treatment caused no detectable change in crystal color,
surface quality, or degree of poling, although extreme care had to
be taken to cool the more nonstoichiometric crystals (those heated
in lithium deficient crucibles, M_L, rapidly enough (>20°C/min.) so
as to prevent nucleation of $LiNb_3O_8$. Microscopic examination with
partially crossed polarizers and side illumination was utilized to
reveal the presence of complex arrays of scattering centers associated with phase separation. In addition, x-ray diffraction was
used to establish the existence of $LiNb_3O_8$.

The optical damageability of bulk $LiNbO_3$ crystals was determined, before and after each high temperature heat treatment, by measurement of their laser induced birefringence[9]. This was accomplished by measuring the birefringence of each crystal through its
thickness (0.03") with a weak laser probe, before and after a strong,
optically damaging, standardized laser exposure (2 min., $50W/cm^2$,
441.6 nm).

The optical damageability of single domain $LiNbO_3$ waveguides
was assessed by means of a scanning pinhole detector which recorded

any laser induced changes in their transmitted output intensity profile as a function of time. A laser was coupled into (and out from) the crucible out-diffused and previously polished crystal surface with rutile prisms. The input-to-output prism separation was adjustable, allowing 2-15 mm optical path length in most crystals. Waveguide propagation was directed along either the x or y axes of $LiNbO_3$. The output beam was scanned perpendicularly to the crystallographic z-axis to reveal the waveguide's mode structure, and parallel to the z-axis to record any changes in beam profile due to laser interaction with the crystal.

RESULTS AND DISCUSSION

Nonstoichiometry and Phase Equilibria

A variety of commercially produced thin $LiNbO_3$ crystalline discs, as described previously, were heated alternately in lithium oxide-rich (M_H) and lithium oxide-deficient (M_L) reaction crucibles at 1050°C, 1100°C, 1125°C and 1140°C. The total reaction mass changes, $M(\infty)$ are summarized in Table 1. For example, $LiNbO_3$ crystal PE11, when reacted completely at 1100°C, first within a lithium

TABLE 1

Complete Vapor Phase Equilibration of $LiNbO_3$ Single Crystals

Crucible M_H: Li_3NbO_4 + $LiNbO_3$ Crucible M_L: $LiNbO_3$ + $LiNb_3O_8$

$$Li_{1-2x}\square_{2x}NbO_{3-x}\emptyset_x$$

$$(0 \leq x \leq x_b)$$

$$\nu = \text{Mole Fraction } Li_2O = (0.5-x)/(1.0-x)$$

SAMPLE		MASS-CHANGE M(∞) mgs				INITIAL		BOUNDARY		TEMPERATURE
No.	Weight(gm)	M_H	M_L	M_H	M_L	x_i(mol)	ν_i(%)	x_b(mol)	ν_b(%)	°C
PE1	0.3763^{12}	$+2.6^{04}$	-2.9^{89}	$+2.9^{77}$	-	0.0340^0	48.24^0	$0.0389^6 a$	47.97^3	1050
PE2	0.2978^{34}	-	-1.7^{91}	$+2.3^{56}$	-2.3^{17}	0.0297^0	48.46^9	$0.0389^0 b$	47.97^6	
PE10	0.5269^{11}	$+3.4^{10}$	-5.0^{50}	$+4.9^{97}$	-	0.0318^2	48.35^7	$0.0468^8 c$	47.54^1	1100
PE11	0.9967^{73}	$+5.7^{21}$	-9.4^{79}	$+9.5^{69}$	-	0.0282^4	48.54^7	$0.0470^1 d$	47.53^4	
V12	0.9998^{38}	$+5.4^{79}$	-9.3^{31}	$+9.6^{29}$	-	0.0269^3	48.61^6	$0.0466^6 e$	47.55^3	
V15	1.0126^{13}	$+6.0^{14}$	-	-	-	0.0292^2	48.49^5	-	-	
PE8.1	0.3553^{23}	$+1.7^{94}$	-3.7^{46}	$+3.7^{26}$	-	0.0248^6	48.72^5	$0.0517^7 f$	47.27^0	1125
PE8.2	0.3045^{11}	$+1.5^{46}$	-3.2^{12}	$+3.2^{38}$	-	0.0249^9	48.71^8	$0.0521^3 g$	47.25^0	
PE8.3	0.3402^{46}	$+1.7^{14}$	-3.5^{85}	$+3.5^{53}$	-	0.0248^0	48.72^8	$0.0516^5 h$	47.27^7	
PE9	0.4142^{19}	$+2.7^{76}$	-4.3^{68}	$+4.3^{82}$	-	0.0329^4	48.29^7	$0.0519^2 i$	47.26^2	
PE14	0.4756^{81}	$+3.5^{99}$	-5.7^{13}	$+5.6^{94}$	-	0.0371^6	48.07^0	$0.0588^4 j$	46.87^4	1140

NOTES: Standard Deviation $a \pm 0.00011$ $d \pm 0.00031$ $g \pm 0.00028$ $j \pm 0.00020$
$b \pm 0.00002$ $e \pm 0.00103$ $h \pm 0.00032$
$c \pm 0.00035$ $f \pm 0.00019$ $i \pm 0.00012$

oxide-rich crucible (M_H), was found to gain 5.721 mgs. The crystal, re-heated in a lithium oxide deficient crucible (M_L), was found to lose 9.479 mgs. When replaced in the original lithium oxide-rich crucible, the crystal was found to regain 9.569 mgs.

These reaction weight changes are indicative of compositional changes (with respect to lithium oxide) occurring within the crystal. The $LiNbO_3$ composition or nonstoichiometry is altered as a result of lithium oxide transport between the crucible and the crystal, each initially different in their lithium oxide chemical potential. Niobium oxides are not transported due to their substantially lower volatilities. The composition dependence of lithium oxide chemical potential at constant temperature, is shown schematically in Fig. 2. Consequently, $LiNbO_3$ crystal PE11, initially of composition S, reacts with the lithium oxide-rich crucible, M_H, by vapor transport from the crucible wall to the crystal's vapor exposed surface, followed sequentially by surface reaction and solid state diffusion within the crystal. As lithium oxide is incorporated within the crystal, its mass increases accordingly, and its composition changes from S to b (Fig. 2). In general, this $LiNbO_3$ nonstoichiometry is described, $Li_{1-2x}\square_{2x}NbO_{3-x}\emptyset_x$, $0 \ll x \ll x_b$, where x refers to the molar deviation from stoichiometry, x_b represents nonstoichiometry at the lithium oxide-deficient phase boundary, and $\square$, $\emptyset$ represent respectively, lithium and oxygen lattice vacancies. The lithium niobate composition at its lithium oxide-rich phase boundary has been established by the technique of mass-loss Knudsen effusion[10], as that of stoichiometric $LiNbO_3$, throughout the temperature range, 1050-1150°C[11]. Hence, the present reaction mass changes are related to the molar deviations from stoichiometry, by

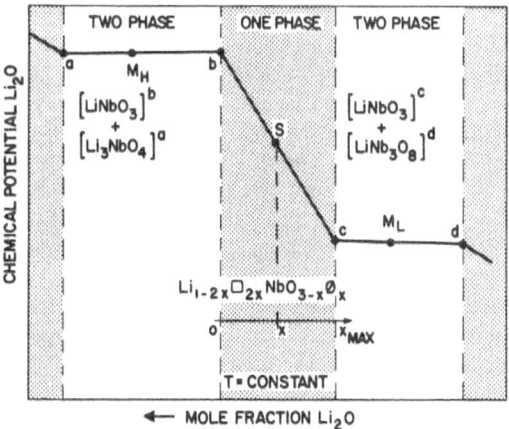

Figure 2. Isothermal composition dependence of lithium oxide chemical potential in the Li_2O-Nb_2O_5 system.

Eq. 1, which is used either to specify the initial composition of a
nonstoichiometric crystal, or to specify the compositional width of
the lithium niobate single phase region at a given temperature. The
molar deviation from stoichiometry, x, $(Li_{1-2x}\square_{2x}NbO_{3-x}\emptyset_x)$ is specified equivalently in terms of lithium oxide mole fraction, ν, by
means of Eq. 2. For example, $LiNbO_3$ crystal PE11 (Table 1) was
found to be initially nonstoichiometric, with x, as above, corresponding to a departure from stoichiometry of 2.82 moles Li_2O. In
addition, the lithium oxide-deficient phase boundary at 1100°C was
located at a composition corresponding to a 4.7 mole Li_2O departure.

$$x_{i,b} \; = \; \frac{M(\infty) \; M_{LiNbO_3}}{(M_i + M(\infty)) \; M_{Li_2O}} \; = \; \frac{M(\infty) \; M_{LiNbO_3}}{M_o \; M_{Li_2O}} \tag{1}$$

$$\nu_{i,b} \; = \; (0.5-x)/(1.0-x) \tag{2}$$

where $M(\infty)$ is the reaction weight change in grams, M_i is the initial weight in grams of a nonstoichiometric $LiNbO_3$ crystal, M_o is
weight in grams of a stoichiometric $LiNbO_3$ crystal, M_{LiNbO_3} and
M_{Li_2O} are molecular weights in grams/mole, x is the molar deviation
from stoichiometry (i:of an initially nonstoichiometric crystal; b:
at the lithium oxide-deficient phase boundary), and ν is the mole
fraction of Li_2O.

The initial composition (ν_i) or nonstoichiometry (x_i) of commercial crystals is referred in (Table 1) to the lithium oxide-rich
$LiNbO_3$ phase boundary which has been found to be temperature independent[11]. In contrast, the ultimate sensitivity of the gravimetric
technique is governed by the sensitivity of the weighing system, the
crystal mass, and the molecular weight of the vapor species, Li_2O.
Stoichiometric differences as small as 0.003 mol % (50 ppm) Li_2O can
be resolved by vapor phase equilibration of a 500 mg $LiNbO_3$ crystal
at 1100°C with the present apparatus. Hence, the crystal-to-crystal
variations apparent in Table 1 imply compositional variations inherent in the commercial crystal growth process. The mean initial crystal composition was found to be 48.42 mol % Li_2O, and the standard
deviation from this value was 0.20 mol % Li_2O. The compositional
difference between the most lithium oxide-rich crystal (PE8) and the
most lithium oxide-deficient crystal (PE14) was found to be 0.66 mol
% Li_2O. Compositional differences of this magnitude from crystal-
to-crystal would complicate many optical device applications. For
example, $LiNbO_3$ compositional variations as small as 0.2 and 0.6 mol
% Li_2O could result in refractive index variations of .003 and 0.01
respectively.

Compositional variations between $LiNbO_3$ crystals derived from
the same boule (i.e., PE1, 9, 10, and 14, or PE2, 8, 11, V12, and

V15) were found to be smaller than when crystals were assessed at random. Compositional variations between segments of the same $LiNbO_3$ crystal slice (PE8.1, 8.2, and 8.3) were found to be of the same order as the average experimental standard deviation as determined over all experiments.

The lithium oxide-deficient lithium niobate phase boundary compositions (x_b or v_b, Table 1) represent the maximum equilibrium $LiNbO_3$ departure from stoichiometry, and were found to be temperature dependent, decreasing in lithium oxide content as temperature increased. Fitting these results to the Arrhenius form, $x_b = x_{bo} \exp(-W_{V,i}/RT)$, the average energy of Schottky defect formation was found to be 14.7 kcal/mole. This value compares favorably with values reported for a range of inorganic crystalline materials[12].

All results (Table 1) are represented on a $Li_2O-Nb_2O_5$ equilibrium phase diagram in Fig. 3. The compositional variations observed in as-grown $LiNbO_3$ crystals are represented as isothermal bars. The predicted lithium oxide-deficient phase boundary locations are compared with previously reported lithium niobate phase equilibria (Raman[13], x-ray[14], NMR[4,15], Curie Temp.[4], and optical[16]). The phase boundary disagreement is attributed, for the most part, to experimental uncertainties of $LiNbO_3$ crystal composition determination, inherent in the earlier studies. Crystal composition was inferred indirectly, as the standard analytical techniques were insufficiently sensitive to resolve small compositional differences with respect to elements as light as lithium and oxygen.

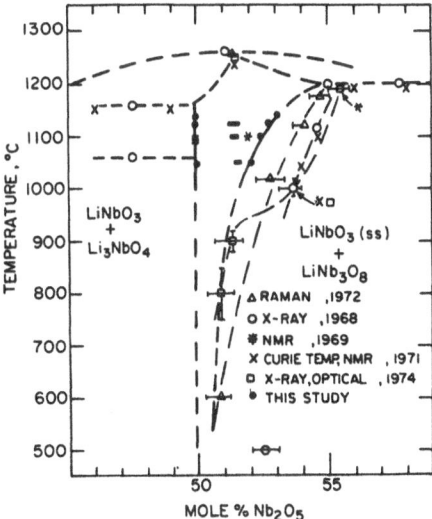

Figure 3. $Li_2O-Nb_2O_5$ equilibrium phase diagram.

Diffusion

Complete vapor phase equilibration has been applied to describe
the nonstoichiometry and phase equilibria of $LiNbO_3$. However, analysis of its kinetics offer a direct means to characterize the vapor/
solid equilibria and solid state diffusion. The reaction mass
changes occurring at $1100°C$ as several $LiNbO_3$ discs react with the
lithium oxide-rich crucible, M_H, are summarized in Fig. 4 as a function of time, t, normalized by the half thickness of the disc
squared, a^2. Similar results were obtained for crystal reactions
with lithium oxide-deficient crucibles. The general features of
the mass change kinetics were common to both the in- and out-diffusion of lithium oxide. Mass change was found to be approximately
a linear function of time, for short reaction times (t<1 hr.). Subsequent mass changes were found to proceed nonlinearly in the time
until the $LiNbO_3$ crystal and the crucible established a chemical
equilibrium.

Since the reaction crucibles were porous, two phased, and considerably larger in mass than the $LiNbO_3$ crystals (>100:1), they
served as infinite sources of, and provided infinite sinks for, the
vapor species lithium oxide. Consequently, the rate of crystal mass
change as indicated gravimetrically was limited only by the kinetics
of vapor/crystal reaction at the crystal's vapor exposed surfaces,
and by the kinetics of solid state diffusion within its bulk.

The boundary condition operative at such a vapor/crystal interface is derived by conserving the interfacial mass fluxes. When

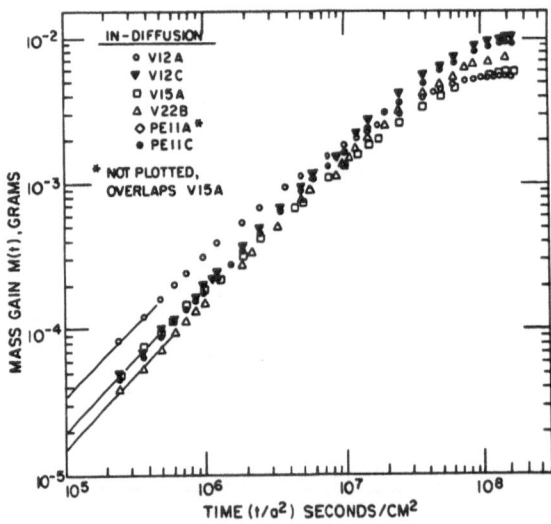

Figure 4. Experimental reaction mass change of $LiNbO_3$ crystals as
a function of time at $1100°C$.

the chemical potential of lithium oxide in the reaction crucible exceeds that in the $LiNbO_3$ crystal, the flux of lithium oxide molecules striking the crystal surface and reacting exceeds the flux rejected. Qualitatively, for a molecule to react with the crystal it must possess sufficient energy to exceed a surface-chemical barrier to such processes as molecular attachment, dissociation, and ionic incorporation in the crystal. Molecules able to surmount this barrier increase the lithium oxide concentration in the surface of the crystal relative to its bulk, thereby establishing a driving potential for the subsequent diffusion of lithium oxide (as lithium and oxygen ions) within the crystal. The flux of molecules striking the crystal surface minus the flux of molecules rejected equals the solid state diffusion flux. This boundary condition, applied to the present "plane sheet" geometry, is described by Eq. 3a and b, where the $LiNbO_3$ crystal is taken as initially uniform in composition, and as such, lithium oxide concentration changes, C^*, resulting from diffusion are considered in terms of an excess or deficit with respect to this initial condition.

$$-D\frac{\partial C^*(\overline{x},t)}{\partial \overline{x}}\Big|_{\overline{x}=a} = \alpha(C_e^* - C^*(\overline{x},t))\Big|_{\overline{x}=a} \qquad (3a)$$

$$-D\frac{\partial C^*(\overline{x},t)}{\partial \overline{x}}\Big|_{\overline{x}=-a} = \alpha(C_e^* - C^*(\overline{x},t))\Big|_{\overline{x}=-a} \qquad (3b)$$

where D is the lithium oxide diffusion coefficient in cm^2/s (assumed constant), α is a surface mass transfer coefficient in cm/s (assumed constant), C_e^* is the lithium oxide concentration change in $grams/cm^3$ effected when the vapor and the crystal are in equilibrium, and a is the half thickness of the $LiNbO_3$ crystal.

This boundary condition is directly analogous to a classic diffusion problem of water vapor in a porous solid[17], and to a more recent example, that of impurity diffusion at a semiconductor surface in a vacuum[18]. In either case, the rate of mass transfer at a vapor/solid interface was found directly proportional to the difference between the actual surface concentration at any time and the concentration that would be in equilibrium with the vapor remote from the surface.

Mathematical solutions to Fick's laws subject to these conditons have been treated extensively by Crank[17] and Carslaw and Jaeger[19] As applied to the present case, the lithium oxide diffusant concentration, C^*, at any time t, and depth $\overline{x}$, is given in Eq. 4, normalized to the lithium oxide diffusant concentration when the vapor and the crystal are in complete equilibrium, C_e^*.

$$\frac{C^*(\overline{x},t)}{C_e^{\ *}} \ = \ 1 - \sum_{n=1}^{\infty} \frac{2L\text{Cos}(\beta_n \frac{\overline{x}}{na})e^{-\beta_n^2 \lambda}}{(\beta_n^2 + L^2 + L)\text{Cos}\beta_n} \tag{4}$$

where $\lambda = Dt/a^2$, $L = \alpha a/D$, and β_n's are the positive roots of β tan $\beta = L$.

The total amount of lithium oxide diffusant entering or leaving the LiNbO$_3$ crystal, $M(t)$, up to a time, t, is obtained by integrating Eq. 4. This series solution has been solved numerically, and the log of the mass change, $M(t)$, normalized to the total mass change at full equilibration, $M(\infty)$, is plotted as a function of the log of normalized time, Dt/a^2, for a wide range of $L(\alpha a/D)$ values, in Fig. 5.

The model is characterized by three parameters, D, α, and $M(\infty)$. D represents the lithium oxide diffusion coefficient, and is assumed concentration independent. The mass transfer parameter, α, represents the rate at which the lithium oxide vapor and the LiNbO$_3$ crystal surface approach equilibrium, and is also assumed concentration independent. $M(\infty)$, which as before, represents the total mass change of the crystal in grams, is expressed in terms of concentration as $2 C_e^* a A$, where A is the surface area through which diffusion occurs. Since $M(\infty)$ is specified by the gravimetric experiment, the model is described by only two adjustable parameters, D and α.

The correlation shown between theory (Fig. 5) and experiment (Fig. 4) is excellent. For short reaction time, the rate of mass change displayed by both the model and the experiment, appears linear in reaction time. The conditions corresponding to a "best fit" are obtained by overlaying data upon theory, and performing a computer assisted graphical analysis. Model parameters $\alpha a/D$, $M(\infty)$ and D were specified as a direct result. The derived values of $M(\infty)$ were compared, in each case, with the value experimentally determined. A complete summary is provided in Table 2a (lithium oxide in-diffusion) for six LiNbO$_3$ single crystals, three as-grown (V12A, V15A, and PellA), two previously converted to the lithium oxide deficient phase boundary (V12C and PellC), and one of intermediate nonstoichiometry (V22B); similarly, in Table 2b (lithium oxide out diffusion) for three crystals, one as-grown (V13A) and for two with the lithium oxide-rich phase boundary composition (V12B and PEllB). The precision with which these parameters were able to be specified was restricted by the coarse incremental variation imposed by the semigraphical method. The present procedure allowed specification of $\alpha a/D$ within confidence limits of $\pm 30\%$. However, this rather broad uncertainty was not imposed upon D or $M(\infty)$, which were eval-

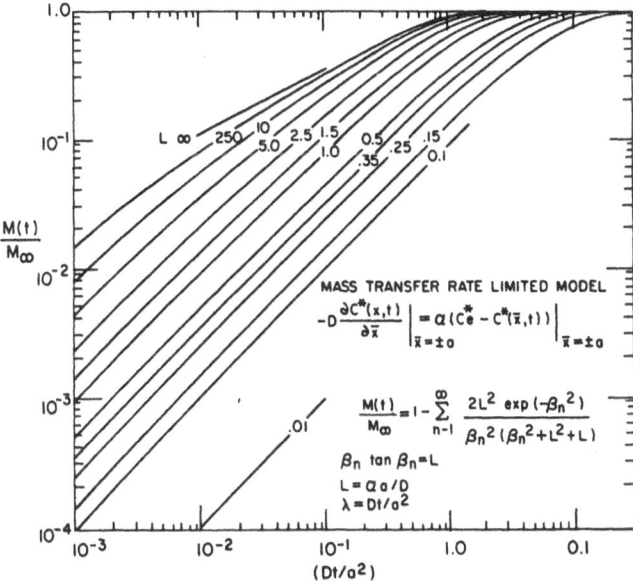

Figure 5. Theoretically predicted reaction mass changes of a plane sheet as a function of time.

uated within confidence limits of ±15% and ±5% respectively.

Even allowing these restrictions, the lithium oxide diffusion coefficient (Table 2a and b) is found to reflect a modest variation with LiNbO3 crystal composition, increasing as deviation from stoichiometry became larger, within the range 1 X 10⁻⁸ cm²/s to 8 x 10⁻⁸ cm²/s at 1100°C. Such behavior is not unexpected, considering that diffusion in complex oxide crystals occurs most often by a combination of vacancy and interstitial mechanisms, whereby ionic jump frequency (i.e., diffusivity) is increased as the number of vacant lattice sites are increased. However, while a completely rigorous process model must account for concentration dependent parameters, given relatively short reaction times and only small excursions in LiNbO3 composition, it is reasonable to describe the process, approximately, by an average diffusion coefficient.

The derived model parameters can be utilized in conjunction with a numerical solution to Eq. 4, to predict the shape of a lithium oxide diffusion profile for any set of process conditions. However, a detailed analysis of diffusion profiles in LiNbO3 is beyond the scope of the present report, and will be discussed separately.

TABLE 2

Summary of Graphical Correlation Between Diffusion Theory and
Gravimetric Experiment for $LiNbO_3$ Crystals at 1100°C

SAMPLE		MASS-GAIN, M(∞) mgs		MODEL PARAMETER	DIFFUSIVITY	RATE CONSTANT
NO.	Thickness	Theory	Experiment	L(=α a/D)	D($\times 10^8$)	α(=LD/a)$\times 10^6$ cm/s
a.						
V12A	0.0300"	5.6	5.4^{79}	5.	1.6^0	2.1
V15A	0.0302"	5.9	6.0^{14}	2.5	1.6^5	1.1
PE11A	0.0298"	5.8	5.7^{21}	2.5	1.6^5	1.1
V22B	0.0301"	7.4	7.1^{64}	0.35	6.7	0.61
PE11C	0.0298"	9.8	9.5^{69}	0.25	7.7	0.55
V12C	0.0300"	10.5	9.6^{29}	0.25	8.0	0.62
b.						
V12B	.0300"	9.5	9.3^{31}	2.5	1.2^0	0.78
PE11B	.0298"	9.6	9.4^{79}	2.5	1.3^5	0.89
V13A*	.0249"	3.1	2.8^{23}	1.0	4.0	1.26

* Half Disc, 0.422^{109} grams

Optical Properties

The process of vapor phase equilibration was applied to alter
the nonstoichiometry exhibited by commercially produced $LiNbO_3$ crystals, so as to assess its affect upon susceptibility to optical
damage. Some representative measurements for uniform $LiNbO_3$ crystals are given in Table 3. Increases in nonstoichiometry (due to
crucible out-diffusion) produced decreases in the amount of optical
damage, whereas decreases in nonstoichiometry (due to crucible in-
diffusion) produced increases in optical damage. For example, a
He-Cd laser (441.6 nm, 50 W/cm^2) induced 2.62 x 10^{-5} birefringence
in $LiNbO_3$ crystal 31 prior to crucible heat treatment. The amount
of laser induced birefringence diminished 13% to 2.27 x 10^{-5} as a
result of a one hour reaction within a lithium deficient crucible.
Additional reactions of 2 and 4.5 hours effected a 34% cumulative
reduction in the amount of laser induced birefringence. Each crystal was cooled equivalently and rapidly (>20°C/min.) to prevent

TABLE 3

Effect of Lithium Oxide In- and Out- Diffusion at 1100°C
on LiNbO$_3$ Laser Induced Birefringence at 441.6 nm

CRYSTAL[a]	PROCESS	t hrs	Δw mgs	Δn$_o$ x10^5	Δn$_f$ x10^5	Δ%[b] –
31	COD	1.0	.395	2.62	2.27	-13
		2.0	.700	2.27	2.04	-22
		4.5	1.19	2.04	1.73	-34
34	COD	8.0	2.48	3.28	2.01	-39
60	COD	7.0	1.95	2.41	1.51	-37
35	CID	0.7	0.77	4.01	5.35	+33
15	CID	80	6.01	1.88	4.71	+150

a - nominally congruent LiNbO$_3$ crystals
b - cumulative % change effected in the
    laser induced birefringence as a result
    of the indicated heating
Δn$_o$ - laser induced birefringence prior to
    the indicated heating
Δn$_f$ - laser induced birefringence after the
    indicated heating
Δw - crystal weight change as a result of
    indicated heating

COD - Crucible Out-Diffusion   CID - Crucible In-Diffusion

phase separation (Fig. 3). Whenever LiNb$_3$O$_8$ phase separation was
detected (by x-ray diffraction or optical microscopy), regardless
of the prior heat treatment, the amount of optical damage was found
to increase dramatically. Even apparently colorless transparent
crystals exhibited as much as 50-100% increased optical damage.
Crystals cooled more slowly developed opalescence and exhibited as
much as 1500% greater optical damage.

The process of vapor phase equilibration was applied to produce
graded lithium oxide concentration profiles (Eq. 4) in commercially
produced LiNbO$_3$ crystals, made more deficient at their surface than
in their bulk. The formation of LiNbO$_3$ optical waveguides by de-
pletion of lithium oxide in vacuum[5] or by an in-diffusing impurity[7],
is well known. Hence, LiNbO$_3$ crystals out-diffused for 2 to 4 hours
in lithium oxide deficient crucibles were found to serve as low loss
optical waveguides supporting one to three TE modes. For comparison,
equivalent vacuum out-diffused (VOD) and titanium in-diffused (TID)
LiNbO$_3$ waveguides were prepared as previously reported. The susceptibility to optical damage was assessed in each case.

All waveguides formed by the present process, single as well as
multi-mode, showed no detectable evidence of either beam broadening
or intensity change as a result of laser transmission, even at
relatively long optical path length (13 mm). This performance was
attained with 1 mW (500 W/cm^2) at 441.6 nm, Fig. 6a, and with 20 mW
(10kW/cm^2) at 488.0 nm in the waveguide. Conversely, both VOD and

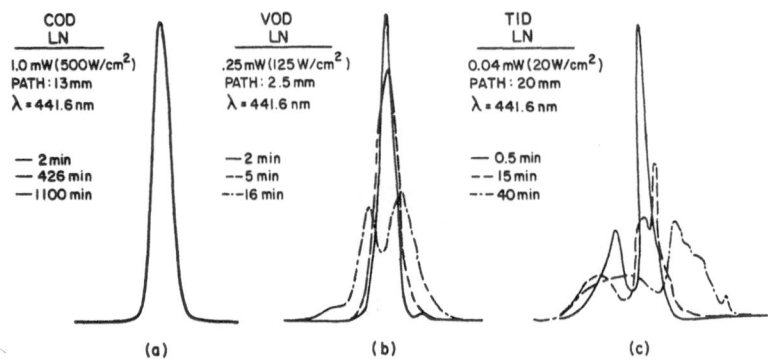

Figure 6. Output intensity profiles of LiNbO$_3$ waveguides along
their m-lines (z-axis) as a function of time (a) crucible out-
diffused (b) vacuum out-diffused (c) titanium in-diffused.

TID LiNbO$_3$ waveguides were found to be appreciably more susceptible
to the effects of optical damage. VOD waveguides were found to develop substantial output beam broadening during 0.25 mW (125 W/cm^2)
transmission at 441.6 nm, as shown in Fig. 6b. The output of TID
waveguides was found to broaden even more quickly and severely,
Fig. 6c, under similar conditions.

Optical damage is believed to be due to photogenerated electrons
from such substitutional impurities as iron, which are displaced
(along the crystal's positive c-axis) by an effective electric field
and trapped[20],[9].   This charge separation, persistent at room temperature, has been verified electrostatically[21], and is responsible,
via the Pockels effect, for the induced index inhomogeneity. Previous studies of optical damage in LiNbO$_3$ have correlated its occurrence with the number of photoionizable impurities found in the crystal[9], and with the relative ease they can be ionized[22]. Consequently, neither vacuum heating[5] (which inherently lowers the ionization energy of impurities) nor impurity diffusion[7] (which increases the number of ionizable impurities) appear to be practical
methods for forming high quality optical waveguides resistent to
optical damage.

The improved resistance to optical damage exhibited by both
LiNbO$_3$ crystals and optical waveguiding layers, as affected by the
present process, may be due to the development of charge trapping
defects, such as lattice vacancies introduced during the process.
Present in low concentration, charge trapping has been presumed
responsible for the observed persistence of optical damage[9].

Present in higher concentration, charge trapping may become competitive with the photogeneration process, and actually reduce the amount of optical damage. Further study is required, however, to identify the mechanism responsible unambiguously. Changes in stoichiometry may influence the photogeneration process itself by altering either $LiNbO_3$ band structure or the ionization energy of crystalline impurities.

CONCLUSIONS

Gravimetry is recognized as a useful tool for characterizing the high temperature processing of such crystalline ceramics as $LiNbO_3$. One such process, vapor phase equilibration, allowed direct determination of crystal composition and modeling of solid state diffusion kinetics. Use of the process was illustrated by forming tailored high quality optical waveguides. The process was further utilized to significantly reduce lithium niobate's inherent susceptibility to laser damage, both in crystals and in optical waveguiding layers. Significantly improved low loss optical waveguides formed by the process, supported between one and three TE modes, and tolerated more than 10 kW/cm^2 laser transmission at 488.0 nm without inducing any measurable distortion. This performance combined with high process reproducibility, makes feasible for the first time, the full exploitation of electrooptic $LiNbO_3$ at high laser power densities and low laser wavelengths.

REFERENCES

1. R.L. Holman and R.M. Fulrath, J. Am. Ceram. Soc. 55, 192-5 (1972).
2. R.L. Holman and R.M. Fulrath, J. Appl. Phys. 44, 5227 (1973).
3. R.L. Holman, Ferroelectrics 10, 185-90 (1976).
4. J.R. Carruthers, G.E. Peterson, M. Grasso, and P.M. Bridenbaugh, J. Appl. Phys. 42, 1846 (1971).
5. I.P. Kaminow and J.R. Carruthers, Appl. Phys. Lett. 22, 326 (1973).
6. V.E. Wood, N.F. Hartman, and C.M. Verber, J. Appl. Phys. 45, 1449 (1974).
7. R.V. Schmidt and I.P. Kaminow, Appl. Phys. Lett. 25, 458 (1974); J.M. Hammer and W. Phillips, Appl. Phys. Lett. 24, 545 (1974).
8. R.L. Holman, Ph.D. Thesis, University of California, Berkeley, LBL-880, 1972.
9. F.S. Chen, J. Appl. Phys, 40, 3389 (1969).
10. R.L. Holman, J. Vac. Sci. Technol. 11, 434 (1974).
11. R.L. Holman, unpublished.
12. N.N. Greenwood, Ionic Crystals, Lattice Defects and Nonstoichiometry, Butterworths, London, 1968.
13. B.A. Scott and G. Burns, J. Am. Ceram. Soc. 55, 225 (1972).

14. P. Lerner, C. Legras, and J.P. Dumas, J. Cryst. Growth 3, 231 (1968).

15. G.E. Peterson and J.R. Carruthers, J. Solid State Chem. 1, 98 (1969).

16. L.O. Svaasand, M. Eriksrud, G. Nakken, and A.P. Grande, J. Cryst. Growth 22, 230 (1974).

17. J. Crank, The Mathematics of Diffusion, Oxford Univ. Press, 1970.

18. F.M. Smits and R.C. Miller, Phys. Rev. 104, 1242 (1956).

19. H.S. Carslaw and J.C. Jaeger, Conduction of Heat in Solids, Oxford Univ. Press, 1947.

20. A.M. Glass, G.E. Peterson, and T.J. Negran, Nat. Bur. Stand. Spec. Publ. 372, 15 (1972).

21. L.B. Schein, P.J. Cressman, and F.M. Teshi, J. Appl. Phys. 48, 4844 (1977).

22. G.E. Peterson, A.M. Glass, and T.J. Negran, Appl. Phys. Lett. 19, 130 (1971).

DEFORMATION PROCESSING OF MAGNETIC HEXAFERRITES FOR

H_c MAXIMIZATION THROUGH GRAIN GROWTH CONTROL

T. J. Curci*, W. R. Bitler, and R. C. Bradt

Department of Materials Science and Engineering

Penn State University, University Park, PA 16802

ABSTRACT

Variations of compressive hot forging were investigated as a means of improving the magnetic properties of commercial barium and strontium hexaferrites. By selecting the proper combination of hot working conditions, an intense texture can be attained while minimizing grain growth. Imrpoved magnetic properties are realized by increased remanence, minimization of the decrease of coercivity, and an increased energy product.

Introduction

Single crystal, single domain magnetic properties are aniso-tropic in that the spontaneous direction of magnetization occurs in a specific crystallographic direction, termed the "easy dir-ection" of magnetization. This direction is the [00·1] in the magnetoplumbite hexaferrites, $MO \cdot 6Fe_2O_3$, where M is Ba, Sr, or Pb. The magnetic hardness of the magnetoplumbite hexaferrites arises from their large magnetocrystalline anisotropy. A randomly oriented polycrystalline magnetoplumbite hexaferrite will be isotropic thus reducing the contribution of the magnetic aniso-tropy. Reduced remanence (B_r) results. Grain sizes larger than single domains can result in decreased coercivity by allowing demagnetization to occur by domain wall nucleation and movement. The effects of randomization and grain size are reflected in the magnetic energy product, $(BH)_{max}$. In the case of $BaO \cdot 6Fe_2O_3$,

---

*Currently with the United States Air Force, Air Force Materials Laboratory, Wright Patterson AFB, OH 45433

a room temperature value of 7.16×10^3 T-A/m (0.9×10^6 gauss-oersteds) (1) is typical for a randomly oriented, multidomain, polycrystalline sample. This value is only about 15% of the theoretical maximum for a single domain crystal.

Attempts to achieve texture and thereby improve the magnetic properties of polycrystalline magnetoplumbite hexaferrites began with Stuijts, et al.[1] in the 1950's with the pressing of a powder slurry in the presence of a magnetic field. Other processes have included hot rolling loose powders in a steel tube,[2] hot forging sintered bodies[3,4,5,6,7,8] and forging while hot pressing.[3,4,5,9] All of these have resulted in improved texture, but are frequently accompanied by coercivity-degrading grain growth. This paper reports a means of improving the magnetic energy product of Ba- and Sr-magnetoplumbite hexaferrites by high temperature deformation grain growth, a process not unlike the "rate-controlled" sintering discussed in another paper at this conference.[10]

Experimental Procedure

Commercial, non-stoichiometric Ba- and Sr- hexaferrites* were studied. Molecular ratios were between 5.5:1 and the 6:1 true stoichiometry. Commercial grain growth inhibitors were present. These starting materials had bulk densities of approximately 93% theoretical and average grain sizes of about 4 μm. Magnetic, properties are listed in Table I. Specimens were cylindrical with axial ratios of 1.14:1 and 1.33:1 for the Ba- and Sr- respectively.

Table I. Magnetic Properties

Material	Remanence B_r (X10^{-2}T)	Coecivity H_c (X10^3A/M)	Intrinsic Coercivity H_{ci} (X10^3A/M)	Magnetic Energy Product $(BH)_{max}$ (X10^3T-Am)
"BaO·6Fe$_2$O$_3$"	24.5	153.2	245.1	9.79
"SrO·6Fe$_2$O$_3$"	36.3	239.5	259.0	24.99

Hot forging was accomplished by uniaxial compression is a conventional creep apparatus. Two deformation procedures were investigated, both involved heating the specimen to 1473K, allowing equilibration, then initiating deformation. In Procedure I, upon application of a stress of approximately 500 psi (3.95 MPa). the temperature was immediately increased at a constant rate, be-

*Stackpole Carbon Co., St. Marys, PA.

tween 2.5K/m and 4.4K/m, to a fixed final temperature, between
1498K and 1673K at increments of 25K, and held at that temperature
until noticeably rapid deformation ceased. Procedure II differed
from Procedure I in that stresses between 1000 psi (6.89 MPa) and
2000 psi (13.79 MPa) were applied and the temperature was increased
at a constant rate, between 1.5K/m and 3.5K/m, during the entire
interval of rapid deformation. Specimens were allowed to deform
during Procedure II until a true compressive strain about -1.2
was attained.

Other specimens were annealed for two hours at various temperatures in the range 1473K to 1673K, coinciding with the temperature range of deformation to independently determine the effect
of temperature on grain growth. Textures of deformed specimens
were determined from X-ray diffraction patterns from mid-sections
with a surface normal to the direction of deformation, as by
Hodge.[8] The degree of texture or preferred orientation was described
by Taft's "R" factor,[11] also employed by Haag[4] and by Hodge,[8] where:

$$"R" = \frac{(\Sigma A_{00\cdot 1}/\Sigma A_{hk\cdot 1})_{deformed}}{(\Sigma A_{00\cdot 1}/\Sigma A_{hk\cdot 1})_{non-textured}} \,. \tag{1}$$

"R" is the ratio of the quotients of the integrated areas of basal
diffraction peaks to the area of all the peaks of deformed and
non-textured specimens. The ratio for non-textured specimens was
taken to be the value obtained from a powder diffraction pattern
of isotropic stoichiometric materials.

Archimede's technique was used to determine bulk densities.
Microstructural examination included both reflected light microscopy and scanning electron microscopy. Grain growth is anisotropic
occurring preferentially in the basal plane. Grain sizes were
determined on the basal plane only, for the width dimension (basal
plane) of the flat platelike grains is the most important in
determining if domain walls will nucleate. For textured samples,
the basal grain dimension was estimated by the Heyn[12] linear
intercept technique on a plane of the microstructure which contained predominately basal planes. Since this technique may give
slightly low values for non-textured samples, the grain size of
non-textured samples was estimated by measuring the basal plane
dimensions of random grains.

Magnetic measurements were made on rectangular plates machined from the center of deformed specimens and oriented perpendicularly to the direction of deformation. A modified ballistic
galvanometer was employed for these measurements*. The magnetic

*Stackpole Carbon Co., St. Marys, PA.

field (H) was determined by a Hall probe. An integrator determined the induced flux density (B). The hysteresis loop was
plotted by an automatic flux recorder.

Results and Discussion

Deformation of the specimens proceeded with only minor edge
cracking to true compressive strains as high as -1.4. As the
forging progressed the rate decreased with the accompanying areal
increase, some grain growth, and slight densification of the specimens. An exception occurred when the deformation exceeded about
-0.95 where a density maximum of about 98% of theoretical was
observed. Further deformation produced slight decreases in overall density, although final densities always exceeded 93%. This
decrease was due to pore development on the specimen edges.
Figure 1 illustrates specimens in the before and after conditions.

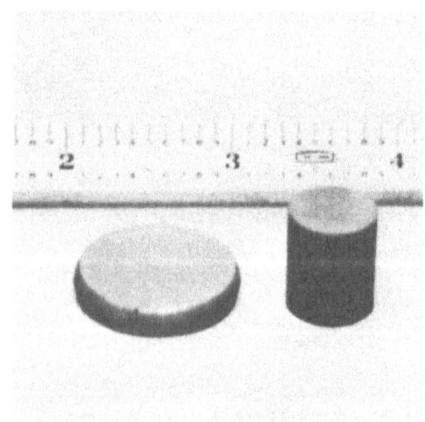

Figure 1. A commercial $SrO-Fe_2O_3$
specimen before (right)
and after (left) hot
working in compression
to a true strain of
-1.24.

Structurally, changes occurred in the specimens in terms of
the developed texture and the increase in grain size. The
development of texture was shown by Hodge[6] to be directly related to the amount of strain, to a first approximation. Figure
2 illustrates the variation of texture, as described by the
orientation factor "R" for the two materials hot worked by the
two procedures investigated in this study. Several points are
apparent, not the least of which is a variant of the gradual
development of increased textural intensity with increased strain.
The textural development does progress nearly directly with strain
for both materials deformed by Procedure I, holding at an elevated
temperature until sufficient deformation occurred; however. Procedure II, that of continuously heating while deforming to about
-1.2 yielded variable textures at a given strain. This is particularly evident in the $SrO-6Fe_2O_3$'s, ■, whose orientation factors
range from about 12-to 18 at the same level of strain, (about -1.2).

Apparently there are structural differences evident during the two procedures of hot working.

The differences in textural development as reflected in grain growth differences are evident in Figure 3. Again, the two procedures appear to yield different results. The constant temperature forgings, Procedure I, show increasing grain sizes with increasing degrees of texture, as described by the "R" factor. As suggested by Hodge, et al.[8] this substantiates that the development of texture during the deformation process may be intimately related to the grain growth process, at least for certain time/temperature/deformation schedules. The Procedure II results, as plotted in Figure 3, don't completely negate the role of grain growth in texture development, but clearly reveal that substantial orientation of individual grains is possible with only minimal grain growth, for any form of an "R" vs. grain size line would be nearly vertical. In retrospect, one may suggest that a possible interpretation to the apparent fine and coarse grain size regimes of Figure 3 is that a deformation texture develops at fine grain sizes by the sliding of the platelet-like grains over one another, then the texture is further enhanced by the growth of favorably oriented grains at the expense of non-aligned ones. Figure 4 illustrates typical grain structures before and after hot working.

Magnetically, Figures 5 and 6 illustrate the structure-property trends in remanence and coercivity, respectively. The remanence generally appears to follow the trend of a steady increase with increasing textural development. Only the "$BaO \cdot 6Fe_2O_3$" specimens deformed by Procedure I do not follow this trend. This is most likely the consequence of increased grain growth yielding multi-domain grains which although oriented, experience reverse domain growth on field removal. In spite of a high degree of texture, remanences remain below the theoretical maximum value of about 47×10^2 T (4700 Gauss).

The coercivity trends in Figure 5 exhibit precisely the variation expected with increasing grain size, namely a decrease as the grain size increases. The likelihood of multidomains becomes a fact and demagnetization occurs by domain wall motion instead of the more difficult rotation process. Nearly all of the fine grain size, high coercivity specimens result from Procedure II processing.

Table II lists the magnetic properties of a number of the materials deformed by Procedure II. Although they all exhibit some decrease in coercivity, due to grain growth, substantial increases in the energy product are possible.

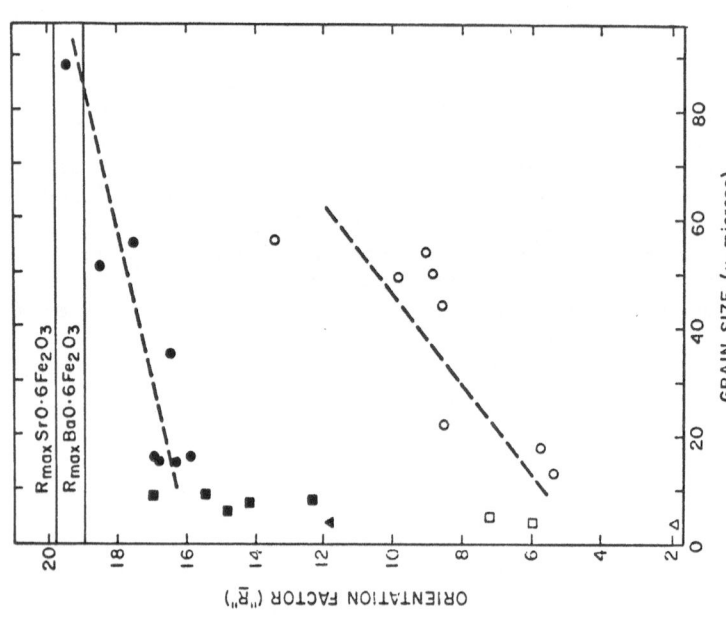

Figure 3. The grain growth of specimens as related to their orientation factor "R". Note the differences in Procedure I (o, ●) and Procedure II (□, ■) trends.

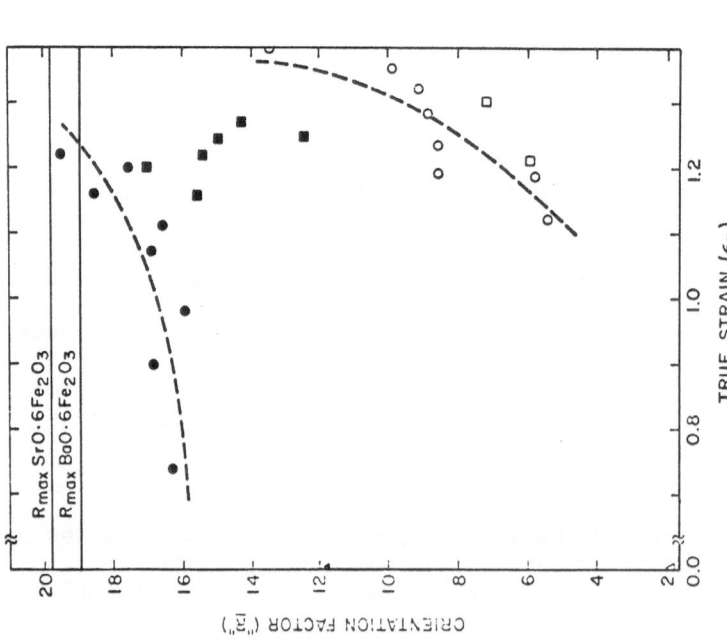

Figure 2. Textural development with strain. Key ▲Ba–as received, ▲Sr– as received, ◐Ba–Proc. I, □ Ba– Proc. II, ●Sr– Proc. I, ■Sr–Proc. II. Al~ though the strains are compressive, they are listed as positive for ease of representation.

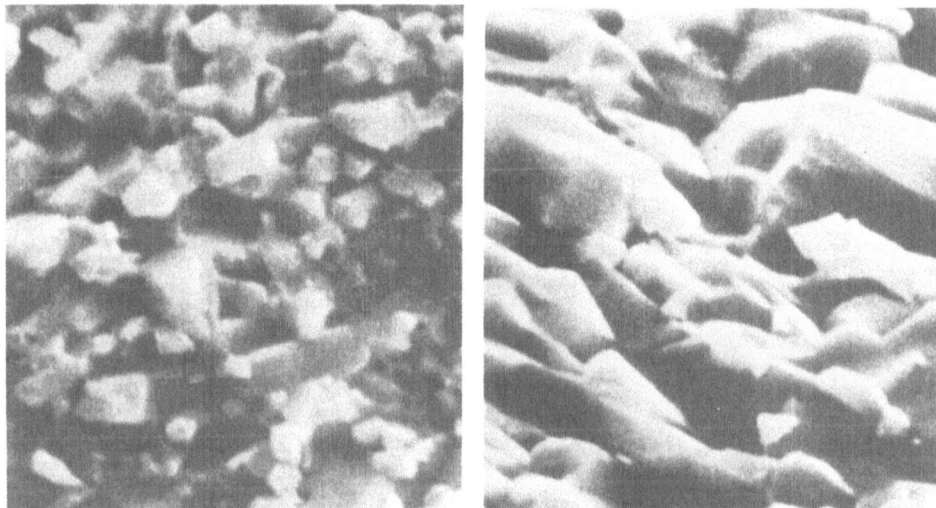

Figure 4. As-received (L) and deformation textured (R) "SrO-
6Fe$_2$O$_3$". The direction of forging is vertical, the
strain -1.2 and the "R" value, 17.9. The as-received
grain size is about 4 μm.

Table II. Magnetic Properties

Mat.	Def. Temp.	Strain	(B_r)*	(Hc)*	(Hc_i)*	(BH_{max})*
Ba-	as-received	------	24.5	153.2	245.1	9.79
Ba-	1551K	-1.21	29.7	144.0	164.9	14.40
Ba-	1517K	-1.30	31.3	160.3	173.9	16.23
Sr-	as-received	------	36.3	239.5	259.0	24.99
Sr-	1568K	-1.27	38.8	144.0	145.6	27.37
Sr-	1585K	-1.22	39.6	141.3	144.4	28.33
Sr-	1558K	-1.20	39.2	153.6	154.4	29.05
Sr-	1557K	-1.24	37.8	150.4	151.2	26.10
Sr-	1548K	-1.16	38.7	161.1	162.3	28.25
Sr-	1575K	-1.24	39.3	146.0	148.0	28.49

* Same units as in Table I.

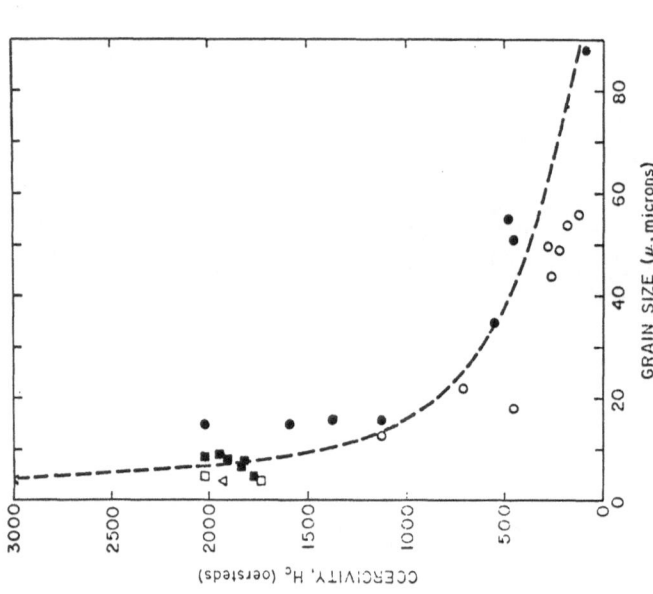

Figure 6.   The decrease in coercivity (H_c) with
             increasing grain size.

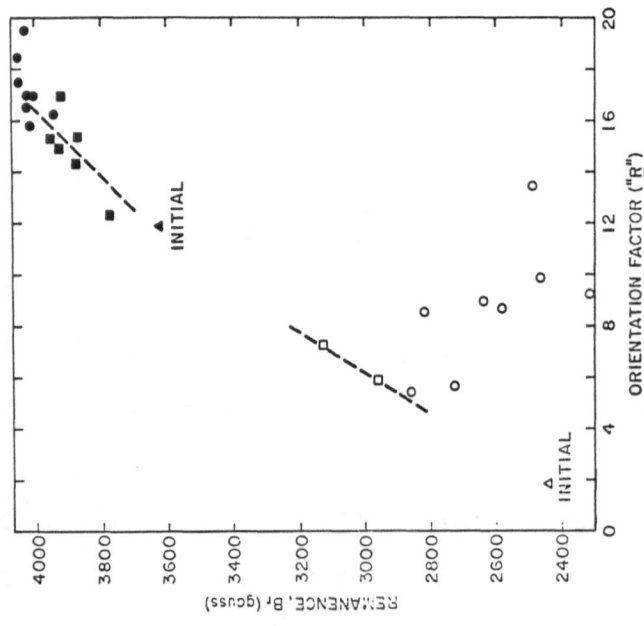

Figure 5.   The variation of remanence (B_r)
             with orientation.

Conclusions

By choosing the proper hot working conditions, hexaferrites
can be textured in such a manner as to minimize grain growth.
This results in increased texture and magnetic energy product
with only a moderate decrease in coercivity due to grain growth.

Acknowledgements

The authors acknowledge the financial support of the
National Science Foundation and the technical assistance of
J. Proske, T. Haberberger, and C. Dangelo of the Stackpole Carbon
Company.

References

1. A. L. Stuijts, G. W. Rathenau, and G. H. Weber, Philips Tech.
    Rev., 16, (5-6), 141-147 (1954).
2. A. L. Stuijts and H. P. J. Wijn, Philips Tech. Rev., 19, (7-8),
    209-217 (1957/58).
3. R. M. Haag, AVSD-0047-69-CR (1969).
4. R. M. Haag, "Magnetic-Crystallographic Orientation Produced in
    Ferrites by Hot Working," AVSD-0128-70-CR (1970).
5. R. M. Haag, "Magnetic-Crystallographic Oreintation Produced in
    Ferrites by Hot Working," AVSD-0249-71 (1971).
6. M. H. Hodge, Ph.D. Thesis, The Pennsylvania State University
    (1973).
7. T. Nishikawa, T. Nishida, K. Inoue, H. Inoue, and I Uei, Yogya-
    Kyokai-Shi, 82 (5), 241-247 (1974).
8. M. H. Hodge, W. R. Bitler, and R. C. Bradt, J. Amer. Cer. Soc.
    56 (10) 497-501 (1973).
9. N. Ichinose and Z. Tanno, J. Elec. Ceramics of Japan, 3, (9),
    57-61 (1972).
10. M. L. Huckabee, T. M. Hare and H. Palmour III, this volume.
11. D. R. Taft, RADC-TR-67-614 (1967).
12. E. E. Underwood; pp. 37-47 in Metals Handbook, Vol. 8, American
    Society for Metals, Metals Park, Ohio, 1973.

Discussion

David W. Johnson, Jr. (Bell Laboratories): How much is an increase in texture consistent with no grain growth? Is the orientation mechanism one of reorientation of existing grains or is it
one of growth of properly oriented grains at the expense of im-
properly oriented grains?
Authors: All our evidence for both the stoichiometric ferrites
and the commercial doped ferrites indicate that texture development occurs by the concurrent processes of preferential grain
growth and grain reorientation produced by grain boundary sliding.
It was the purpose of the current study to maximize the latter

process and minimize the former to avoid the deleterious effect associated with grain growth. The results for the most favorable case corresponded to an increase in the 'R' factor from 2-6 for barium ferrite with negligible grain growth, i.e., the increase in texture can be entirely ascribed to grain reorientation.

Max Paulus (CNRS): 1) I believe, in fact, that the orientation of the grains occurs mainly by hexagonal planes gliding as we observe during cold pressing. 2) From the curve H_c versus grain size, it seems that the smallest grains are 3 or 6 μm in diameter. This is much over the single domain size. I think you have to start with much smaller grains.

Authors: 1) SEM and TEM studies on our samples lead us to believe that the primary mechanism for grain reorientation is grain boundary sliding and not internal plastic deformation of grains by dislocation glide. TEM studies did show some twinning, but only rarely and never of a magnitude to account for the observed texture development. 2) We are aware that the grains are larger than single domain size, and our TEM studies of these materials have documented this. We were constrained to use samples with the reported initial grain sizes. The sample material was obtained from a commercial ferrite producer, Stackpole Carbon, because we wished to examine material containing the commercial proprietary dopants used to hinder grain growth. To obtain single domain starting materials, a process other than mechanical comminution would likely be required.

H. Palmour III (North Carolina State University): The contribution of plastic flow processes to texture development could probably better be detected by monitoring the stress-strain rate dependence, n, ($\dot{\epsilon} \propto \sigma^n$), since even if dislocations are involved in quite considerable numbers, dislocation climb processes are quite likely to anneal most of them out before the hot worked specimen can be cooled and prepared for TEM examination.

Authors: We have examined the stress exponents for both the stoichiometric ferrites and the commercially doped materials. The stoichiometric ferrites do in fact exhibit stress exponents in the range 3 to 4.5, indicative of flow by dislocation processes. However, the commercial ferrites described in this paper exhibit much lower stress exponents, $1 < n < 2$, suggestive of a much smaller (perhaps negligible) contribution from dislocation process.

PROCESSING AND MAGNETIC PROPERTIES OF LOW-LOSS AND HIGH-STABILITY Mn-Zn FERRITES

B. B. Ghate

Bell Telephone Laboratories, Inc.

Allentown, Pennsylvania   18103

INTRODUCTION

Telecommunications and entertainment electronics have remained the two principal areas of application for soft linear ferrites.[1] And low-loss high-stability Mn-Zn ferrites find the widest application as inductor core materials for use in high selectivity filters in the frequency range from a few kHz to approximately 1.5 MHZ.

The requirements imposed by the telephone industry are stringent, and a discussion based on material requirements for high quality inductors would also be relevant to other applications of Mn-Zn ferrites.

In this paper, I will review some of the work which has led to improved material properties, and describe methods used to control them in manufacturing. The reader may also refer to a number of recent reviews on this subject by Roess,[2] Sibille,[3] Hanke and Zenger,[4] Paulus,[5] and Stuijts.[6]

CURRENT STATUS

The inductor designer is primarily interested in (i) correct inductance, (ii) low electrical losses, and (iii) a high degree of stability over long periods of time. The inductor should be stable to detuning by thermal, mechanical, and electrical shocks that can be encountered in transport of the assembled filter unit, or during its use. Thermal stability requirements in particular are quite stringent since inductors can be exposed to temperatures of -40°C in Minot, ND and $+60^{\circ}$C in Galveston, TX.

Translated into ferrite material requirements, the ferrite should have a (i) high initial permeability, μ_i or simply μ, (ii) high quality factor, Q, over the range of[1] operating frequencies, (iii) linear and tightly controlled temperature factor, TF, and (iv) low disaccommodation factor, DF, which is a measure of stability with time.

The major problem in developing a low-loss and high-stability Mn-Zn ferrite lies not in optimizing the material properties but in striking a correct compromise between them. Peloschek and Kooy[7] exemplify this problem by their comment, "improvement in any one property is most likely to be paid for by a deterioration of one or more other properties."

In the last decade, the major improvements in material properties have occured by (i) a careful choice of composition, (ii) raw materials and processes which introduce a minimum of impurities, and yield a homogeneous product, (iii) incorporation of suitable additives, and (iv) a greatly improved understanding of the sintering process. In addition, availability of improved raw materials and advanced analytical tools have also contributed greatly towards achieving the material objectives.

Improvements in material properties, attained during the past 10 years, are shown in Table I, and the resulting inductors with improved performance and reduced size are shown in Figure 1. For

TABLE I. Typical Properties of High Quality Mn-Zn Ferrites for Inductor Applications

	1966	1970	1977		
	Neferrite[†]	Super Neferrite[†*]	Roess[2*]	N48[††]	H6H3[Δ]
Max. freq. MHZ	1.0	1.0	3.0?	0.1	0.8
μ nominal	2000	1100	500	2000	1300
μQ_{min} (X10^5)					
at 0.1 MHz	6.6	12.0	30	5.0	7.8
0.5 MHz	1.0	6.0			2.8
1.0 MHz		2.3			0.6
TF, ppm/°C from 5 to 55°C	1.0	0.5	0.3	0.8	0.6
DF, ppm/decade	3.0	1.5	2.5	2.0	5.0

*stabilized perminvar type; †NEC; ††Siemens; ΔTDK.

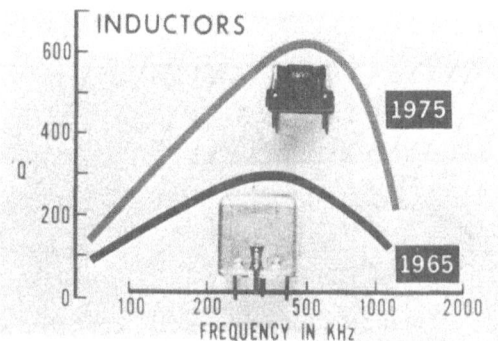

Figure 1.   Improvements in quality and size of a ferrite
            inductor from 1965 to 1975

nomenclature and a detailed discussion of the magnetic properties,
and the factors which control them, the reader may refer to standard texts, [8,9]

AREAS OF MAJOR IMPROVEMENT

Choice of Compositions

Mn-Zn ferrite has the cubic spinel structure, and it can exist
over a wide range of compositions. The stoichiometric composition
can be expressed as $Mn_x^{2+} Zn_{1-x-y}^{2+} Fe_y^{2+} Fe_2^{3+} O_4$, x and y being determined
by the properties desired. In addition, oxidation-reduction
kinetics which occur during sintering influence the vacancy content and the stoichiometry of the sample.

An enlarged view of the portion of a composition diagram[2]   is
shown in Figure 2. Marked on it are regions where a given property
such as μ, μQ, or DF can be optimized. Low Fe_2O_3 favors low DF
whereas high μQ is attained with high Fe_2O_3. However, the range
of compositions is narrowed by the requirement[2]   that the crystal
anisotropy constant, K_1, and the magnetostriction constant, λ_s,
should be close to zero. Lines corresponding to K_1 = 0, and λ_s = 0
are marked on the diagram.[10]   Also shown on the diagram are regions which produce either Perminvar loop or Isoperm loop in the
ferrite. Mn-Zn ferrites of the Perminvar type, such as the Super
Neferrite and Roess's material as shown in Table I, have excellent
low-loss properties. However, under the influence of high field
or stress they revert to the Isoperm type loosing the initial advantage. Hence, it is desirable to choose a composition, and process it in a way to achieve the best Isoperm loop.[11]

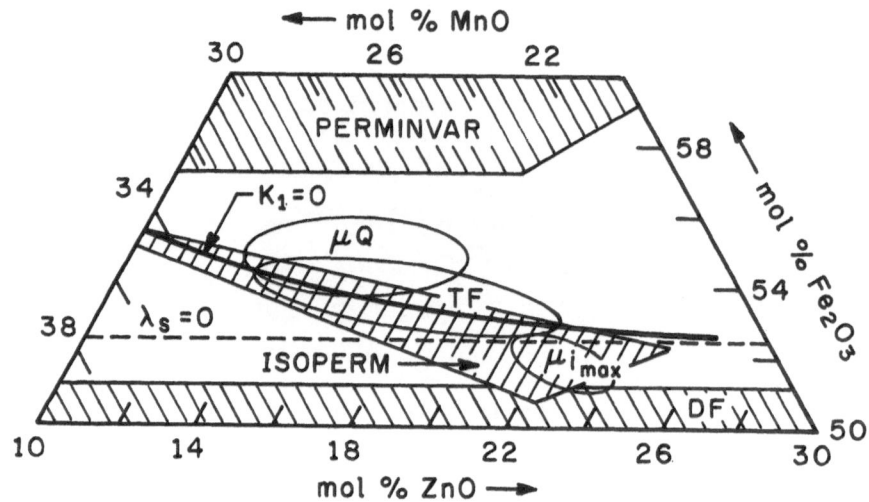

Figure 2.   Portion of the MnO-ZnO-Fe$_2$O$_3$ composition diagram (after
            Roess.)[2]   Also shown are lines corresponding to K$_1$ = 0
            and λ_s = 0.

Selection of Raw Materials

The conventional raw materials used to make high-stability
low-loss ferrites are Fe$_2$O$_3$, MnCO$_3$, and ZnO. Lately, ferrite-grade
raw materials have been developed which have low impurities, and
have well defined particle characteristics. In spite of this, the
conventional processing has two important drawbacks: (i) uncertainty in achieving the exact bulk composition, and (ii) insufficient
chemical homogeneity.

Using neutron activation analysis on a ferrite sample,
Lobanev[12]   showed that the mean square scatter of Zn and Mn concentration was 3.8 to 5.0% and 3.8 to 4.6% resp. The maximum scatter for Zn was 16.2% and for Mn 12.8%. Inhomogeneity in the mix
adversely affects the variability of TF.

Coprecipitation of raw materials and their conversion to spinel,
prior to fabrication of parts, offers distinct advantages over conventionally prepared powders. Coprecipitation, or similar techniques of chemical preparation, permit vastly improved control of
the nominal composition, a further lowering of impurities, improved
chemical homogeneity, and a reactive powder. Via coprecipitation,
it is also possible to have a narrow distribution of particle
size.[13,14]

A distinct advantage in using coprecipitated powders is seen in reduced variability of TF of the ferrite. Akashi[15] observed that for such a ferrite, with a nominal TF of 0.5 ppm/$^\circ$C, the variability decreased from ± 0.25 to ± 0.05. Another benefit of coprecipitated powders is the Improvement in µQ as seen by Goldman.[16] He doubled the µQ product from 5 x 10^5 for conventional material to 10^6 for a ferrite made from coprecipitated powder.

Newer techniques[16, 17] of coprecipitation have resulted in further control of impurities such as Na$^+$ and K$^+$. They are simple, economical, and are becoming increasingly available. However, there are two unanswered questions: (1) what degree of chemical homogeneity is desired to attain the "minimum variability"?; and (2) is the cost of using coprecipitated materials justifiable? These questions should be answered soon.

Role of Additives

The initial impetus to the study of additives came from an observation by Guillaud[18] that CaO, present as an impurity, lowered the eddy current losses significantly. It was observed that Ca segregated at the grain boundaries creating an electrically insulating layer between the grains. Since then a variety of additives have been examined and are being used to control all the important magnetic properties. The additives can be conveniently divided into three groups.[19]

The first group yields a liquid phase during sintering and enhances densification. Bradley[20] showed that 2 to 5 wt % Bi$_2$O$_3$ additions to Mn-Zn ferrite can lower the sintering temperature by nearly 250°C, and increase the relative density from 0.85 to 0.95. The limitation of using such additives is that not enough time is available for reaching equilibrium with the atmosphere. And therefore, they can be used only with ferrite powders which are fully reacted.

The second group normally segregates at the grain boundaries, and thereby reduces losses. These additives are needed to the extent of only a few hundredths to a few tenths of a percent to yield the desired results. Excess quantities tend to promote discontinuous grain growth. Examples of this group are CaO, CaO-SiO$_2$, CaO-B$_2$O$_3$, and BaO-B$_2$O$_3$.[21,22] The maximum benefit is derived from these additives by adding them to the precalcined powder during the ball milling stage prior to spray drying.[23]

The third group of additives alter the magnetic properties of the host lattice. They consist of cations having ionic radii between 0.06 to 0.09 nm (0.6 Å to 0.9 Å). Generally they substitute for host ions on the octahedral sites. Examples of this group of additives are TiO_2,[24] SnO_2,[25,26] Li_2O,[26] Sc_2O_3[27] and In_2O_3.[28]

A common feature of all the third group of additives is that they alter the Fe^{2+} content of the host lattice. The mono and divalent substituents decrease the divalent iron content, and the tri- and quadrivalent ions increase the Fe^{2+} content. The vacancy content is also affected, but the effects which are due to change in Fe^{2+} content are of great technological importance.

An important concept in controlling the secondary maximum in permeability (SMP) is that of anistropy compensation.[10,29] Adjusting the temperature of SMP on the $\mu(T)$ curve permits adjustment of TF. A schematic representation of the $K_1(T)$ relation for the host lattice and for the Fe^{2+} ions of the host lattice is shown in Figure 3. At the temperature at which K_1 becomes zero, the permeability rises sharply giving rise to the SMP. Thus, with increasing Fe^{2+}, the SMP can be shifted to a lower temperature, within certain limits. This can be achieved either by increasing the Fe_2O_3 content, incorporating the additives which increase the Fe^{2+} of the host lattice, or by lowering the oxygen pressure in the sintering atmosphere. Increasing the Fe_2O_3 content of the mix invariably raises DF, and in some cases will also lower the resistivity of the ferrite; hence, this method is undesirable. Use of additives such as SnO_2 and TiO_2 offer a simple way to increase the Fe^{2+} content.[24,30] Figure 4 illustrates this effect for TiO_2.

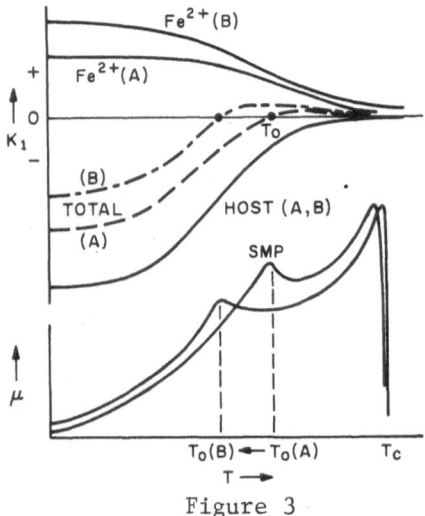

Figure 3

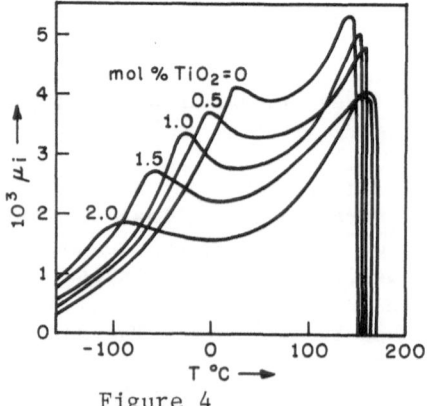

Figure 4

A schematic of anistropy compensation (after Stijntjes et al.[24]) and shift in SMP with change in Fe^{2+} content.

Variation of permeability with temperature for a composition $Mn_{0.569}Zn_{0.369}Fe_{2.062}O_4$ with varying TiO_2 (after Roess).[30]

Additions of ions with incompletely filled $3d^0$ and $4s^0$ shells, such as Sc^{3+}, Ti^{4+}, and V^{5+}, are of further interest. In addition to increasing the Fe^{2+} content, they act as "electrostatic traps" for exchange of electrons passing between Fe^{2+} and Fe^{3+} ions resulting in localizing of the Fe^{2+}. This is thought to cause increased resistivity, and hence improved loss characteristics in the substituted ferrite. Ti^{4+} is especially noted for this effect. Roess[11] increased the μQ of a Ti^{4+} substituted ferrite to 10^6 at 100 kHz and to 2×10^5 at 1 MHz.

Another interesting effect of the third group of additives is their influence on the disaccommodation temperature spectrum. Once again, Ti^{4+} and Sb^{5+} show rather peculiar effects.[25] The room temperature peak observed in the disaccommodation spectrum shifts to lower temperatures, and the high temperature peak nearly vanishes. Precise explanations of this behavior are not yet available; however this effect provides for good stability of permeability over long periods of time. If we omit these special cases, it is generally observed that ions which effectively raise the Fe^{2+} content of the host lattice also increase the DF at room temperature, and those which lower the Fe^{2+} content also lower the DF. Illustrations may be seen in Figure 5, where the Fe^{2+} content of the basic composition, $MnO:ZnO:Fe_2O_3=28:20:52$ and marked b.c., was varied by adding several additives. The curve marked Fe^{2+} corresponded to a composition, $MnO:ZnO:Fe_2O_3=28:17:55$. The compositions are quoted in mol%. y corresponds to calculated maximum Fe^{2+} in the sample.[19]

Additives should be used with caution. Yan[31] showed that a number of additives can produce a small amount of liquid phase which will initiate discontinuous grain growth. TiO_2 is one such additive. Any coarsening of the grain contributes to hysteresis losses. To circumvent such a situation, Roess[11] suggested other additives such as lanthanum and niobium oxides. They will increase

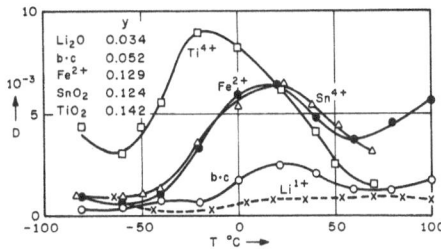

Figure 5

Disaccommodation $D = [\mu(t_1)-\mu(t_2)]/\mu(t_1)$ as a function of temperature T. The permeabilities were measured at $t_1=10s$ and $t_2=100s$ after demagnetization (after Koenig).[19]

the resistivity by a factor of 10 to 20 at concentrations of a mere
0.025% per each constituent.

Sintering and Microstructure Control

Two aspects of sintering may be cited, which are special to
the sintering of iron-rich Mn-Zn ferrites. They are the attainment
of (i) correct Fe^{2+}/Fe^{3+} ratio, with a minimum vacancy content, and
(ii) a dense, homogeneous structure consisting of pore-free grains
of 4 to 6 μm grain size.[4]

The oxygen pressure used during sintering and cooling affects
the Fe^{2+} and the vacancy content. Also, inadequate pO_2 control
can cause precipitation of a second phase. Hence, use is made of
special equilibrium cooling curves of pO_2 versus T. Morineau and
Paulus[32]   have presented such curves for a broad range of Mn-Zn
ferrites. Also, one may refer to the earlier works by Slick and
Blank mentioned in Ref. 32.

Reijnen[33]   showed that a good sintered product, but with an
exaggerated pore growth can be obtained by maintaining sufficient
pore-mobility along the grain boundaries. Therefore, in order to
attain high density and yet maintain small grain size, long sintering times at moderate temperatures are needed, which are unattractive from the manufacturing point of view. Further information on sintering and grain growth of ferrites may be found in
Stuijts,[6] Yan,[31] and Carpay.[34] The explanation for the grain
size requirement is given, recently, by Knowles.[35] He noted that
6 μm grain size is optimum for attaining maximum rotational permeability, and minimum hysteresis and residual loss factors. The
reduction in losses is attributed to the disappearance of domain
walls.

The sintering of ferrites in tunnel kilns is handicapped by
inadequate atmosphere control, difficulty in maintaining equilibrium
cooling, and economic necessity for fast through-puts. Fluctuations
in oxygen pressure result in shifts in the SMP and increased variability of TF. Inadequate time for equilibration of cation vacancies, and cation short range order[36]   result in low quality product. Oxidation during cooling may give rise to higher DF and TF.

In recent years, improved kiln designs have been introduced
to remedy these problems.[37] These designs offer (i) improved
temperature uniformity, (ii) special diffusion shields to minimize
interplay in O_2 levels in adjoining zones, and gas seepage through
the walls, and (iii) drop arches and narrow clearances over saggars
to achieve better control and uniformity of O_2 pressure across the
width of a saggar. Also, cooling zones have been designed to simulate equilibrium cooling.

SUMMARY

Using the ferrite inductor as an example, I have tried to focus on those areas of materials processing which have led to improved properties in Mn-Zn ferrites. They are: (i) choice of a suitable composition, (ii) use of pure and chemically homogeneous raw materials, (iii) control of the ferrous ion content by suitable choice of additives, and (iv) an improved understanding of the special problems of sintering iron-rich Mn-Zn ferrites.

For those of us in ferrite development, many current problems are related to transferring the lab know-how to real-life manufacturing. Ultimately, the economics will decide whether a given product should be made or not. There are encouraging signs in terms of growth potential for high-quality, high-stability ferrites. With improved raw materials becoming readily available, and with improved processing, it is reasonable to predict that these ferrites will extend their use up to a maximum frequency of 3 MHz.[38]

ACKNOWLEDGMENTS

Special thanks are due to Dr. M. Dixon for many helpful discussions, suggestions and a critical review of the paper. The interest and counsel of Drs. A. Olsen, P. I. Slick, and T. S. Stakelon are equally appreciated. Thanks are also due to Messrs. R. J. Holmes and C. E. Pass for general assistance.

REFERENCES

1. E. C. Snelling, IEEE Spectrum, 9(1) 42-51, & 9(2) 26-32 (1972).
2. E. Roess, J. Magn. & Magnetic Materials, 4(1-4) 86-94 (1977).
3. R. Sibille, Rev. Phys. Appl., 9(5) 837-845 (1974).
4. I. Hanke and M. Zenger, ibid Ref. 2, pages 120-128.
5. M. Paulus, Sintering and Related Phenomena, ed. G. C. Kuczynski, 220-245, Mater. Sc. Res. V 6, Plenum Press, (1973).
6. A. L. Stuijts, ibid, 331-350; Ceramurgia, 5(4) 189-195 (1975).
7. H. P. Peloschek and C. Kooy, 1965 Proc. of Magnetic Core Conference, Magnetic Powder Core Association, New York.
8. C. Heck, Magnetic Materials and Their Applications, translated into English by S. S. Hill, Crane Russak & Co. Inc. N.Y. (1974).
9. D. J. Craik, Magnetic Oxides, Pt. I, Wiley & Sons, N.Y. (1975).
10. K. Ohta, J. Phys. Soc. Japan, 18(5) 685-690 (1963).
11. E. Roess, Proceedings Int. Conf. FERRITES, Kyoto 1970, ed. Y. Hoshino et al, 187-190, University Park Press, (1971).
12. E. M. Lobanev and E. I. Girei, Dokl. Akad. Nauk. BSSR, 20(12) 1072-1073 (1976).
13. T. Takada and Y. Kiyama, ibid Ref. 11, 69-71.
14. I. Kiwase et al., Brit. Pat. #1,142,215 (1969).
15. T. Akashi, ibid Ref. 11, 183-186.

16. A. Goldman and A. M. Laing, Int. Conf. Ferrites 2, Paris 1976, J. de Phys., 38(4) C1-297-301, Colloque 1, C-1 (1977).
17. I. Makoto, S. Kazuo, O. Etsuo, Jap. Pat., #75 15997 (1975).
18. C. Guillaud, Proc. IEE, 104B, Sup. 5, 165-173 (1957).
19. U. Koenig, Appl. Phys., 4(3) 237-242 (1974).
20. F. N. Bradley, J. Australian Cer. Soc., 5(5) 9-15 (1969).
21. T. G. W. Stijntjes et al, ibid Ref. 11, 194-198.
22. E. Takama, S. Mishima, and S. Kawahara, ibid Ref. 16, C1-349-352.
23. K. Muramori et al, U.S. Pat. #3,375,196 (1968).
24. T. G. W. Stijntjes et al, Philips Res. Repts., 25 95-107 (1970).
25. J. E. Knowles, ibid, 29 93-118 (1974).
26. I. Sugano et al, IEEE Trans. Magn., 8(3) 708-710 (1972).
27. L. A. Vladimirtseva et al, Inorg. Mat'ls., 10(5) 747-750 (1974).
28. Y. Shichijo et al, Jap. Pat. #7,653,299 (1976).
29. U. Enz, Proc. IEE, 109B, Sup. 21, 246-247 (1962).
30. E. Roess and I. Hanke, Phys. stat. Soli., (a) 2(3) K185-K187 (1970).
31. M. Yan and D. W. Johnson, Jr., this conference.
32. R. Morineau and M. Paulus, IEEE Trans. Magn., Mag-11(5) 1312-1314 (1975).
33. P. J. L. Reijnen, Problems of Nonstoichiometry, ed. A. Rabineau, 219-238, North Holland Publ. (1970).
34. F. M. A. Carpay, Ceramic Microstructures '76, ed. R. M. Fulrath and J. A. Pask, 261-275, Westview Press (1977).
35. J. E. Knowles, ibid Ref. 16, C1-27-30.
36. S. Iida, ibid Ref. 11, 17-23.
37. F. Petzi, talk presented at the Annu. Mtg. of the Am. Ceram. Soc., 4-E-77, April 1977, Chicago; also see Keram. Z., 26(3) 144-153 (1974).
38. A. Olsen, Transmission Div., BTL-Merrimack Valley, Personal communication.

DISCUSSION

Jim Reed (Alfred University): What is the coordination of Ti^{4+} in the spinel phase? How does this affect the valences of Fe?
Author: It is generally assumed that Ti^{4+} is in octahedral coordination. I do not have data on site preference energies or other evidence to say that some Ti may not be found on the tetrahedral sites. Chemical analysis of ferrites containing titanium show (see Reference 30) a Fe^{2+} concentration corresponding to the formula $Ti^{4+}Fe_2^{2+}O_4$; thus, each mole of TiO_2 induces the formation of two moles of FeO, provided proper firing conditions are used. You may wish to refer to References 19 and 24 for details on how TiO_2 affects the valence of Fe.

D. R. Clarke (Rockwell Science Center): You mentioned that the
CaO goes to the grain boundaries to influence the eddy current
losses; does this form a glassy layer and, if so, is there any direct evidence that it forms a continuous layer surrounding each
grain?

Author: There is no direct evidence that the phase at the grain
boundary is continuous or that it is glassy. However, results of
autoradiography, electron probe analysis, electrical resistivity
measurements, and Auger analysis suggest that most of CaO and SiO_2
concentrate at the boundary resulting in an insulating phase. The
amount of grain boundary phase in relation to the ferrite phase is
so small that sophisticated experimentation has yet to be carried
out to give a definitive answer whether that boundary phase is
really glassy. For further information, you may refer to References 18 and 21 and also an article by M. Paulus in The Role of
Grain Boundaries, Materials Sc. Res. Vol. 3, p. 31-47, eds.
W. Wurth Kriegel and Hayne Palmour III, Plenum Press, 1966.

RELATIONSHIP BETWEEN PROCESSING CONDITIONS, OXYGEN STOICHIOMETRY AND

STRENGTH OF MnZn FERRITES

David W. Johnson, Jr.

Bell Laboratories

Murray Hill, New Jersey   07974

INTRODUCTION

Magnetic ceramics, better known as "ferrites", have become technologically important in telecommunications systems, as well as in many other electronic industries, as inductor and transformer materials. In general, the processing variables which can be adjusted for the manufacture of ferrites, are set to optimize the desired magnetic properties of the product. The mechanical properties of ferrites, on the other hand, are often taken for granted during the manufacture and testing. Despite this, the strength of ferrites can be important particularly for parts which are tightly clamped during their assembly and use to assure minimum changes in device performance with mechanical shock, temperature, vibration, etc.

In a general experimental review of the strengths of some MnZn ferrites, it was found that the strength of some ferrites was severely dependent on small changes in firing conditions which apparently affect the magnetic properties only slightly. This current work serves mainly to discuss the probable mechanisms relating strength to firing conditions for these ferrites in light of the preliminary experimental evidence available at this time.

The strength, its statistical nature and models for brittle solids have been discussed.[1-4] These, in general, relate strength to the microstructural parameters, porosity and grain size. Some data on ferrites have served to substantiate some of the models.[5]

There are of course other less well understood parameters which can affect the strength of ceramics through processing. For example,

Ishino et al.[6] have correlated the bending strength of MnZn ferrite
with the oxygen content of the atmosphere used during cooling after
sintering. They found an increase in strength as the surface of the
specimen was oxidized and then a decrease in strength when further
oxidation of the surface led to the formation of hematite. Tanaka[7]
has also observed an increase in strength with oxidation up to a
point and then a decrease in strength with further oxidation.

EXPERIMENTAL

A MnZn ferrite dry pressing powder was produced at a Western
Electric ferrite manufacturing facility by the conventional process
of mixing oxides or carbonates, calcining, milling and spray drying.
The approximate composition is 52.3 Fe_2O_3, 26.4 MnO and 21.3 ZnO
(mole %) with a minor addition of TiO_2, CaO and SiO_2.

For this preliminary study ten bars were pressed from the spray
dried powder at a load of 67,000 N (15,000 lb.) on a bar die whose
cross sectional dimensions are 5.13 × 0.59 cm. The green density
was measured using dimensions and weight.

The removal of organic binders and sintering of these test
specimens was done with continuous furnaces used for production
firing of this ferrite composition. The specimens were grouped into
three sublots and each was fired in a different furnace (A, B, and
C), all producing the same components.

After firing, the density was again measured and the bars broken
in a 3 point bend test using an Instron testing apparatus. A span
of 3.8 cm. and a cross-head speed of 0.13 cm. per min. were used.

Other characterization included reflected light microscopy on
polished and etched specimens, scanning electron microscopy of
fracture surfaces, electron microprobe analyses, indentation hardness using a Vickers Pyramid, and divalent iron analyses using a
coulometric titration technique.[8]

RESULTS AND DISCUSSION

Table I shows the densities and strengths of the test specimens
along with the furnace used for firing. Note the large scatter in
the strength data which is not typical of other data generated for
larger lots of bend test specimens all fired in a single furnace.

Examination of the data in Table I shows some correlation

TABLE I

Ferrite Properties

Bar #	Green Density kg/m³	Fired Density kg/m³	Modulus of Rupture MPa (psi)	Furnace #
1	3060	4350	94.5 (13,700)	C
2	3060	4320	59.6 (8,640)	C
3	3070	4360	88.9 (12,900)	B
4	3070	4370	92.4 (13,400)	B
5	3060	4340	82.0 (11,900)	B
6	3060	4330	73.8 (10,700)	C
7	3060	4330	85.5 (12,400)	C
8	3080	4370	38.9 (5,640)	A
9	3060	4360	30.1 (4,360)	A
10	3060	4370	58.7 (8,510)	A
Mean	3070	4350	70.4 (10,200)	
Std. Dev.	10	20	22.7 (3,300)	

between the strengths and the furnace used for firing. In particular, the samples from furnace A are the weakest samples. The mean
strengths for samples from each furnace are summarized in Table II.
It can be seen that the mean strength of bars from Furnace A is
about half of those fired in the other furnaces.

TABLE II

Average Density and Strength of Ferrites
According to Firing Furnace

Furnace #	Fired Density kg/m²	Modulus of Rupture MPa (psi)
C	4330	78.3 (11,400)
B	4360	87.8 (12,700)
A	4370	42.5 (6,170)

Another difference between samples fired in Furnace A and the
others involves the nature of the fracture surface. A reflective

layer near the edge of the weak samples was first noted by visual
inspection of the fracture surface and is shown at low magnification in Fig. 1.   While the reflectivity of the Bar #1 (Furnace C)
is constant across the fracture surface, a more reflective layer
about 300 μm thick is clearly visible on Bar #9 (Furnace A).

By comparing the scanning electron micrographs of these fracture
surfaces near the edge and near the center of the samples, Figs. 2
and 3, it is obvious that the shiny surface layer is due to changes
in the fracture surface.   In Bar #1, Fig. 2, the fracture surface is
similar in the center and edge with a predominantly grain boundary
fracture mode.   In contrast, Bar #9, Fig. 3, shows the normal grain
boundary fracture near the center but a predominantly transgranular
fracture mode near the edge.   It is the reflective nature of this
transgranular fracture layer which gives the bright layer shown in
Fig. 1.

Two effects, lower strength and transgranular fracture near the
surface, have thus been identified for samples fired in Furnace A.
These two effects are presumably related suggesting that lower energy
fracture paths through the grains exist near the surface of the
furnace #A bars which results in lower strengths.   The rest of this
report will cover the possible causes of this observed behavior and
a tentative mechanism will be proposed.

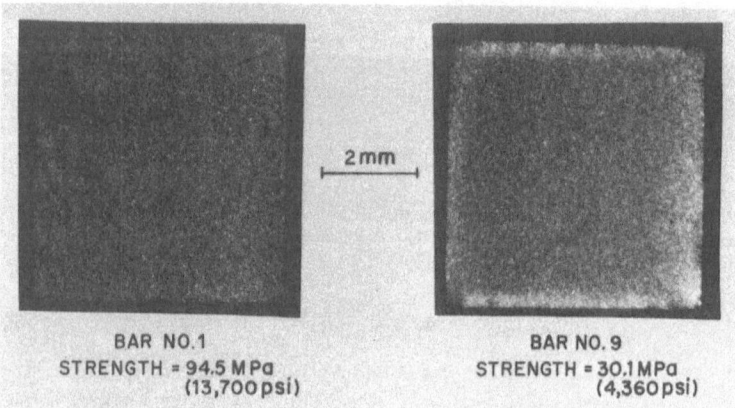

Fig. 1

Fracture surfaces of MnZn ferrites showing a reflective surface
layer for weak Bar #9.

BAR NO. 1-94.5 MPa
(13,700 psi)

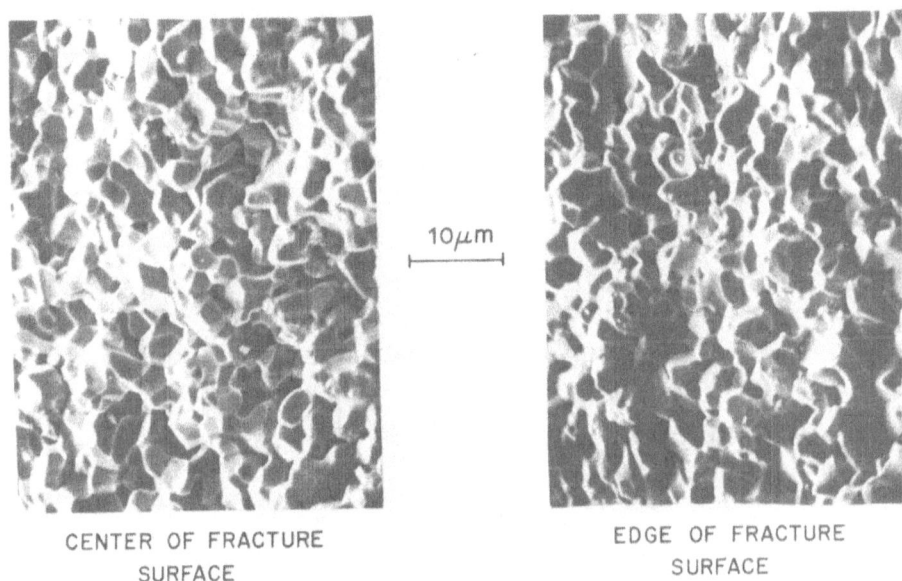

10 μm

CENTER OF FRACTURE
SURFACE

EDGE OF FRACTURE
SURFACE

Fig. 2

Fractographs of high strength sample showing grain boundary fracture
throughout the sample.

1. <u>Porosity</u> - The effect of porosity on strength can, of course,
be great.[1,4] However, Table II shows no significant change in
density for samples fired in each furnace eliminating this as the
strength controlling mechanism.

2. <u>Grain Size</u> - Figure 4 shows polished and etched surfaces of
the bar samples #1 and #9. The grain sizes measured by the intercept method were 5.9 μm for the strong sample #1 and 4.4 μm for the
weak sample #9. With the accepted models,[2-4] the strength is
expected to increase with decreasing grain size but just the
opposite behavior was observed and, thus, grain size effects cannot
account for the observed strength differences. Also, observations
of the microstructure through the cross-section of the bars showed
no significant grain size or porosity differences from the volume
near the surface to that near the center.

3. <u>Chemistry</u> - It can be hypothesized that the transgranular
fracture area seen near the surface of the weak bars fired in
Furnace A is due to a difference in chemistry caused by the furnace.
Here the term chemistry refers only to the overall cation content.
Two possible mechanisms would be: (1) volatilization of Zn into the

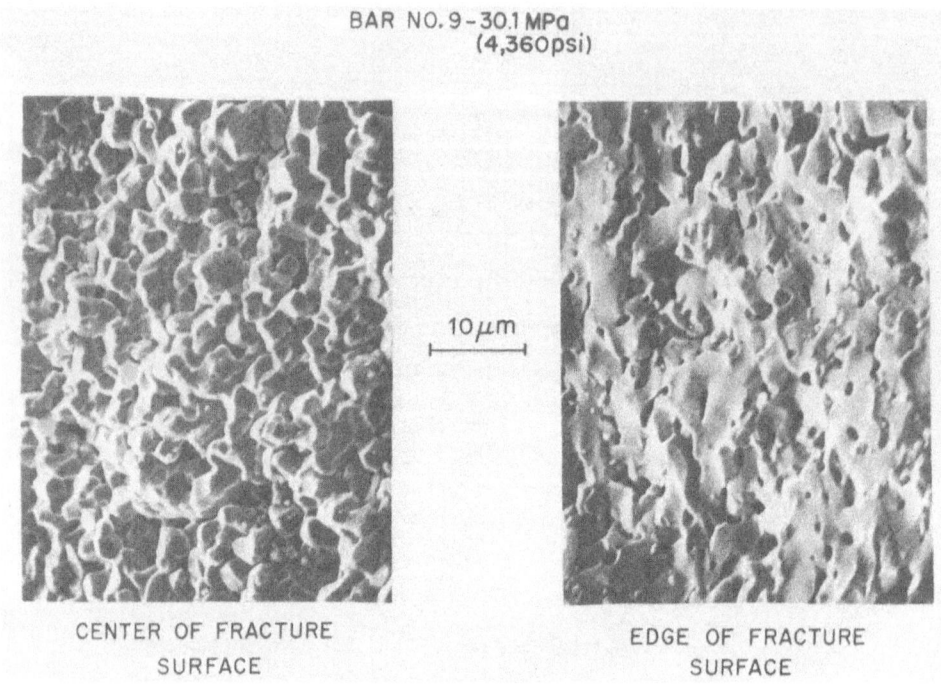

Fig. 3

Fractographs of low strength sample showing transgranular fracture
layer near surface.

furnace during the firing cycle; and (2) diffusion of some volatile
cationic impurity into the sample from the furnace heating elements
or refractories. Neither of these could be confirmed using a non-
dispersive X-ray analyzer on a scanning electron microscope.
Furthermore, if there were a difference in the chemistry which would
serve to weaken the sample, the effects should also be observed in
an indentation hardness experiment. The strong Bar #1 had a Vickers
Pyramid Number of 377 MPa and the weak #9 bar showed 452 MPa.
Assuming that these values in the absence of other effects, are
proportional to the inverse square root of grain size, then the 377
MPa for the 5.9 μm sample #1 (Fig. 4) would translate to 437 MPa
if it had a grain size of 4.4 μm as in Sample #9. However, this is
still slightly less than the observed 452 MPa and this further
refutes the argument that such chemical differences cause the
strength variations.

   4. <u>Microcracks</u> - There is no doubt that because the overall
strength of these brittle materials is controlled by flaw sizes, any
cracks near the surface could severely affect the strength. It is,
however, difficult to determine how the microcracking would change

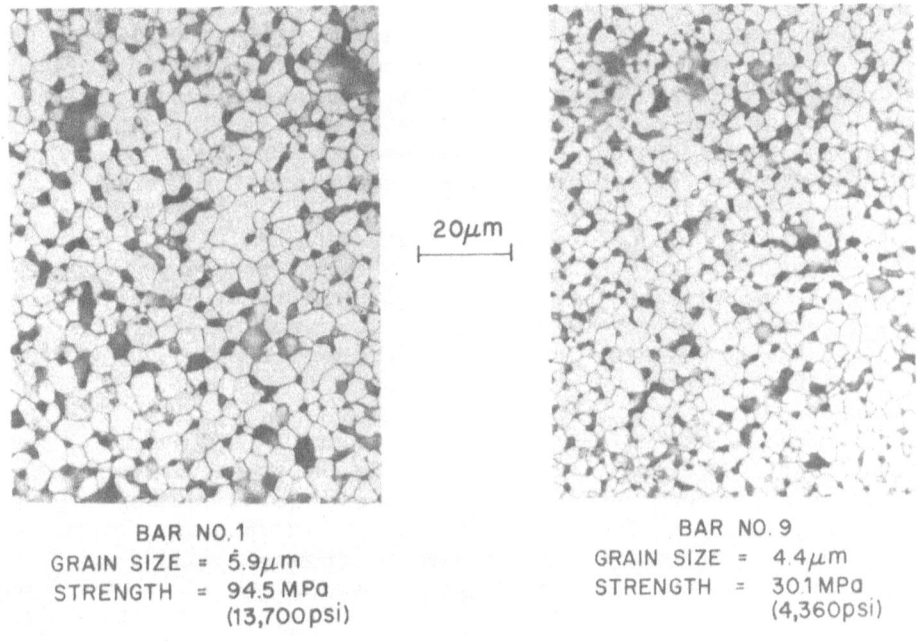

BAR NO. 1
GRAIN SIZE = 5.9μm
STRENGTH = 94.5 MPa
(13,700 psi)

BAR NO. 9
GRAIN SIZE = 4.4μm
STRENGTH = 301 MPa
(4,360 psi)

Fig. 4

Typical microstructures of strong and weak bars. Microstructure is
independent of position on the cross-section of the bars.

the mode of fracture to a depth of 300 μm on all four sides of the
sample. The potential mechanisms for causing microcracks are:
(1) Zn loss which has already been discussed as being not
measurable; (2) thermal shock from a faster cooling in one furnace
than another; and (3) second phase precipitation which will be discussed in more detail later. Careful examination of the as-fired
surfaces revealed no evidence of microcracks in either the strong
or weak samples. There were in all samples, however, rather large
grooves between the relics of the spray dried spheres as they were
arranged during dry pressing.

  5. Oxygen stoichiometry - It is well recognized that oxygen
stoichiometry is important in controlling magnetic properties of
ferrites and that hematite can precipitate from MnZn ferrite if
the conditions are sufficiently oxidizing at the temperature of a
heat treatment. In fact, several investigators[9,10] have constructed
the phase diagrams for MnZn ferrites showing the position of this
phase boundary. If hematite does precipitate, then it has been
shown[6-7] that strength drops because of the cracks that form. In
this study, however, no cracks have been seen and no evidence of
Fe_2O_3 can be found in the samples using reflected light microscopy.

Nonetheless, there is some compelling evidence which links this fracture behavior to oxygen stoichiometry. Slabs of 0.5 mm thick ferrite were cut parallel to the surface from Bars #1 and 9 such as to include the region of transgranular fracture in Bar #9. Samples were also cut from the interior volume of each sample. These were analyzed for Fe^{2+} content as a measure of the degree of oxidation of the surface vs. the center. The results are shown in Table III. The ferrous content of both was quite similar and the exterior sections of both are more oxidized (less ferrous) than the center indicating that, during the firing both furnaces were probably at nearly the same temperature and partial pressure of oxygen, but during cooling, the conditions led to the formation of a more oxidized skin. However, this skin of the weak sample (#9) was more oxidized than that of the strong sample (#1).

It was also found that if the weak samples from Furnace A were annealed at 1100°C in N_2 and cooled in N_2 the transgranular surface layer was no longer present after fracture. Because this heat treatment would not affect the microstructure, this also suggests that the observed behavior is related to an oxidized surface. Furthermore, analysis of the gas flow data through the furnaces used in firing these samples shows considerably slower flow rates of N_2 into cooling zones of Furnace A than into Furnace B or C. Since these continuous kilns are by no means air tight, slower flow rates of gases into the cooling zone would allow a greater chance of diffusion of air into the zones allowing oxidation during cooling.

While no hematite was observed, the precipitate and associated flaws may be in a form too small to observe. Also, it has been recently observed[11] that the magnetic properties of MnZn ferrite can be drastically affected by firing conditions which are close to the phase boundary but yet within the reported single phase spinel region[9,10] and there too no second phase was observed. It was proposed[11] that an incipient second phase formation or a clustering phenomenon may account for the magnetic properties through the generation of regions of high stress. Similarly, these regions of high stress may contribute to the observed strength and fracture mode behavior in the oxidized samples.

TABLE III

Wt.% Fe^{2+} of Exterior and Interior Sections of Bend Test Specimens

Sample #	Outside 0.5 mm	Inside Volume
1 (strong)	1.81	2.13
9 (weak)	1.68	2.15

It has been shown here that the strength of these MnZn ferrites can be altered substantially by the difference in firing conditions from furnace to furnace. However, the magnetic properties of the ferrites fired in these same furnaces have fallen within the specifications set for the ferrite application. Thus, understanding the conditions which produce both high strength and acceptable magnetic properties would allow the inductor designer a greater latitude of hardware designs which may stress the ferrite parts. More detailed studies are currently underway in this area.

SUMMARY

Samples of MnZn inductor ferrite fired in different furnaces were found to have similar microstructures but differing strengths and fracture modes. These strength and fracture mode differences have been related to an oxidized layer found on the surface of the weak samples. Since the magnetic properties of the ferrite parts fired in each of the furnaces are acceptable, this study raises the possibility that the processing conditions can be adjusted to give the proper magnetic properties and, at the same time, avoid those conditions which result in low mechanical strength, thereby allowing the inductor designer greater leeway in designs which may rely on the mechanical strength of the ferrite. Further work is currently under way to correlate fracture toughness with oxygen content.

ACKNOWLEDGMENTS

The author wishes to thank H. Hayes for assistance in processing the samples, P. K. Gallagher for providing the Fe^{2+} analyses, C. M. Preece for the microhardness measurements, S. Vaidya for some of the scanning electron microscopy, and R. Hart for aid in setting up the bend strength testing apparatus. The author also thanks G. Y. Chin, P. K. Gallagher, C. M. Preece, S. Vaidya, N. Sharma and P. Slick for helpful discussions.

REFERENCES

1. E. Ryshkewitsch, J. Am. Ceram. Soc., 36 (2) 65-68 (1953).
2. E. Orowan, Z. Physik, 86, 195-213 (1933).
3. N. J. Petch, J. Iron Steel Inst. (London), 174 (1), 25-28 (1953).
4. F. P. Knudsen, J. Am. Ceram. Soc., 42 (8), 376-387 (1959).
5. C. P. Chen and R. S. Weisz, Am. Ceram. Soc. Bull., 51 (6) 532-538 (1972).
6. K. Ishino, M. Makino, and E. Idehata, pp. 252-254 in FERRITES, Proceedings of the International Conference, 1970, eds. Y. Hoshino, S. Iida, and M. Sugamota, University Park Press, Baltimore, 1971.

7.   T. Tanaka, Jap. J. Applied Physics, 14 (12) 1897-1901 (1975).

8.   P. K. Gallagher, to be published by the Am. Ceram. Soc.

9.   P. I. Slick, pp. 81-83 in FERRITES, Proceedings of the International Conference, 1970, eds. Y. Hoshino, S. Iida, and M. Sugamoto, University Park Press, Baltimore, 1971.

10. R. Morineau, and M. Paulus, Comp. Rend. Séances Acad. Sci. Paris, Séric C, 277 (12), 437-40 (1973).

11. P. K. Gallagher, E. M. Gyorgy, and D. W. Johnson, Jr., to be published.

DISCUSSION

Frans Carpay (Philips Research Laboratories): Was there a difference in grain size around the edges of the fracture surfaces when comparing the strong and the weak bars? I suppose the structures you have shown were taken out of the center sections. Maybe the areas with intragranular fracture show discontinuous grain growth resulting from different pore mobility caused by different Fe^{2+} concentrations.

Author: The microstructures shown are representative of the grain sizes across the entire cross section, and no evidence of larger grains near the surface of the weak bars could be found.

Max Paulus (CNRS): Concerning the effect of atmosphere on fracture, we have observed that grain boundary energy changes with partial pressure of oxygen. We deduced these results from study of the grain growth rate. In slightly "oxidizing" atmosphere the grain boundary energy is greater than in a slightly "reducing" one. And we correlated this with fracture energy. But you also have to take into account the very common superficial oxidation of the surface which comes about because equilibrium cooling is very difficult to achieve. This oxidation induces a decrease of the lattice parameter and puts the superficial layer in tension and reduces the fracture strength.

Author: I agree that the observed strength decrease with surface oxidation may be due to a tensile layer. This mechanism necessitates the formation of some new ferrite unit cells at the surface using cations which have diffused from a subsurface layer of the bulk.

Arno Gatti (General Electric Company): You describe what appears
to be a surface-related strength dependence. Have you tried to
mechanically remove surface layers and study this effect?
Author: No, we have taken the approach of changing the surface
layer through proper oxygen anneals. Grinding could be incorporated into the ongoing studies, however.

F. F. Lange (Rockwell International): Following Prof. Paulus'
comment, the stress distribution caused by the lattice contraction
with loss of oxygen can be calculated. The stress is highest for
short periods, i.e., when the diffusion layer is thin, and lowest
when the layer is thick.*
Author: Thank you.

---

*Richmond, Leslie and Wriedt, Trans ASM 57, 294 (1964).

SINTERING OF HIGH DENSITY FERRITES

M. F. Yan and D. W. Johnson, Jr.

Bell Laboratories

Murray Hill, New Jersey   07974

ABSTRACT

Low melting eutectic compositions in MnZnTi ferrites, in the amount of 0.2-10 wt %, are used to promote sintering at low temperatures. The unique aspects of this approach are the use of low melting point eutectics per se and that the source of the low melting composition is derived from components which are normally part of the ferrite composition. During oxygen sintering, a second phase of the hematite structure is stablized by the TiO_2 content in ferrites. After the hematite phase is reduced by appropriate nitrogen anneal, dense-single phase ferrites with uniformly large grains can be formed at temperatures as low as 1250°C.

Introduction

In a previous study on the effects of TiO_2 and other dopants on the exaggerated grain growth in MnZn ferrite,[1] a liquid phase was observed in the ferrite-TiO_2 interface after a short sintering time. Dissolution of this liquid phase and subsequent diffusion of Ti into the ferrite matrix induce the exaggerated grain growth in the ferrite. The purposes of this paper are (a) to identify the origins of the transient liquid phase; (b) to explore the beneficial effect of this liquid phase on ferrite sintering and (c) to investigate the proper heat treatments to give dense, single phase ferrite with uniformly large grains.

A phase diagram of the system, Ti-Mn-Zn-Fe-O, has not been reported in the literature, except for some concerned ternary systems e.g. TiO_2-FeO,[2][3] TiO_2-Fe_3O_4,[4] TiO_2-MnO[5] and TiO_2-ZnO.[6] Table 1 summarizes the eutectic temperatures and compositions in

these systems. These data show that certain compositions in the
TiO_2-MnO and TiO_2-FeO systems have low eutectic points within the
range of usual sintering temperatures (1200 - 1400°C) for MnZn ferrites. Furthermore, some of these low melting compositions are
rather dilute in TiO_2 content (5-35 wt %).

A liquid phase, preferably between grains, provides a relatively
fast transport medium for the host ions and thus promotes sintering.
Consequently, some of the listed low melting eutectic compositions
may promote ferrite sintering and grain growth at a lower temperature. A lower sintering temperature effectively reduces the zinc
loss which has deleterious effects on many magnetic properties.

Furthermore, Ti is a commonly added component in ferrites to
achieve a low temperature coefficient in permeability. However, for
high permeability applications, high purity MnZn ferrite is usually
used. It has been suggested that alien ions with ionic misfit within
the ferrite matrix constrain the domain wall movement and thus
compromise the high permeability magnetic properties. Nevertheless,
with improved microstructures and reduced zinc loss, the high permeability qualities of the ferrite may be improved after proper adjustment in the ferrite composition to accommodate the Ti addition.

Experimental

Reacted mixtures of MnO-TiO_2 with three nominal compositions,
namely 66.7, 50 and 33.3 wt % TiO_2, were used as sintering aids for
ferrites. Reagent grade TiO_2 and $MnCO_3$ were used to prepare the low
melting compositions. The components were homogenized in a stainless steel blender for 40 minutes in de-ionized water. After
filtering, drying and screening through a 20 mesh stainless steel
sieve, the powder was placed in alumina boats and calcined at 1010°C
for 21 hours in nitrogen with an oxygen partial pressure less than
10^{-3} atm. X-ray diffraction studies and differential thermal
analyses were performed on the as-calcined powder.

Ferrite powder used in this study had a composition of $Mn_{0.51}$
$Zn_{0.43}Fe_{2.06}O_4$* and was prepared by mixing, calcining Fe_2O_3, $MnCO_3$
and ZnO and milling. The ferrite powder had a surface area of
~ 2.8 m^2/gm as measured by the BET technique.

The ferrite powder was mixed with 10 wt % Halowax binder and a
measured amount of the pre-reacted TiO_2 and $MnCO_3$ powder. An
appropriate amount of ferrite and the sintering aid along with the
Halowax, steel balls and CCl_4 were milled in a steel jar for 16 hours;
beaten dry using a planetary mixer and pressed through a 20 mesh
screen. Discs of this powder were uniaxially pressed at a pressure
of 2.52×10^8 N/m^2 (36,500 psi).

---

* Chemical analyses performed by F. Schrey, Bell Laboratories.

TABLE 1

Eutectic Temperatures and Compositions in Ti-Fe-Mn-Zn Oxide
Systems.

System	Eutectic Temp., °C	TiO_2 Content, wt %
(1a) TiO_2-FeO[2]	1305	5
	1320	42
	1330	68
(1b) TiO_2-FeO[3]	1312	10
	1363	47.1
	1390	57.9
	1430	82.2
(2) TiO_2-Fe_3O_4[4]	1493	26.4
	1515	57.2
(3) TiO_2-MnO[5]	1330	34.4
	1290	64.4
(4) TiO_2-ZnO[6]	1537	31.6
	1418	57.6

Doped ferrite samples were sintered in oxygen at ambient
pressure with a heating rate of 200°C/hr and cooled at 320°C/hr in
nitrogen with an oxygen partial pressure less than 10^{-5} atm. Some
as-sintered samples were nitrogen annealed below the sintering
temperature before cooling down to room temperature. Temperatures
were controlled within 1°C of the set point, and the gas flow
velocity was maintained at 150 cm/sec in a mullite tube with a
cross-sectional area of 11 cm^2.

X-ray diffraction was used to identify the phase composition
in sintered samples. Cross sections of specimens were polished,
etched with an acid solution of 20 vol. % HF and 80 vol. % HCl for
15-25 seconds at room temperature, and examined by optical microscopy. The volume fraction of the second phase was measured from
optical micrographs by the reflectivity contrast of the two phases.

Results and Discussion

Three crystalline phases, namely, rutile (TiO_2), pyrophanite
($MnTiO_3$) and hausmannite (Mn_3O_4) were identified by x-ray diffraction in the calcination products of TiO_2 and $MnCO_3$ powder
mixtures after 21 hours at 1010°C in nitrogen with an oxygen
partial pressure less than 10^{-3} atm. Weight fractions of these
phases in each sintering aid composition were calculated from the
averaged areal integrals of x-ray diffraction peaks from {110} and

{101} planes of the rutile phase, {104} and {110} planes of the pyrophanite phase and {211} and {224} planes of the hausmannite phase, Table 2. A calibration curve prepared from the measurements of unreacted mixtures of rutile, pyrophanite and hausmannite in various weight ratios was used.

DTA analysis of the reacted low melting compositions were performed at a heating rate of ∿30°C/min in nitrogen and oxygen atmospheres. The DTA data show the eutectic temperatures are ∿1290 ± 10°C, Table 2, and have a qualitative agreement with the literature data.[5]

Low temperature sintering of MnZn ferrites mixed with pre-reacted TiO_2-MnO dopant yields samples with low porosity. Figure 1 shows the microstructures of a ferrite sample with 1.7 wt % TiO_2 and 0.8 wt % MnO dopant, sintered at 1200°C for 20 hrs. in oxygen.

TABLE 2

Crystalline Phases and Melting Points of the As-calcined* TiO_2-MnO Dopants

| | TiO_2, wt % | Crystalline Phases[‡] wt % | | | "Melting" Pt, °C |
		TiO_2	$MnTiO_3$	Mn_3O_4	N_2, O_2
(A)	66.7	22	78	0	1304, 1286
(B)	50.0	0	100	0	1399, 1364
(C)	33.3	0	76	24	1279, 1280

* Calcination conditions: 1010°C for 21 hrs in nitrogen.

‡ TiO_2, $MnTiO_3$ and Mn_3O_4 refer to Rutile, Pyrophanite and hausmanite phases respectively.

The porosity is less than 0.5% and the grain size is ∿11μm. The low porosity is due to the rapid sintering promoted by the TiO_2-MnO dopant. Figure 1 also shows the drastic contrast in both the porosities and grain sizes of the doped and undoped samples after the same sintering conditions.

However, at sufficiently low temperatures in oxygen these low melting compositions also stabilize a second phase of the hematite structure, Fig. 1, which has deleterious effects on the intended magnetic properties of the ferrites. From x-ray diffraction data, the lattice parameters for the hematite second phase are a=5.044Å and c=13.752Å. Microprobe studies in the hematite phase grains showed

that Fe is the major cation component, and Mn, Ti and Zn are less
than 20 cation %.[7] The hematite grains inhibit the grain growth due
to their sizes and the volume fraction. In Fig. 2, the grain sizes
are correlated with the volume fraction of the hematite grains in
oxygen-sintered ferrite specimens with a net TiO_2 content ranging
from 0.7 wt % to 3.3 wt %. This correlation is independent of the
dopant content and composition, sintering temperature and time.
Figure 2 also shows that when the volume fraction of the hematite
grains decreases below ∿4%, the hematite phase can no longer prevent
the discontinuous grain growth.

     Before the exaggerated grain growth occurs, the size and shape
of the hematite grains are similar to those of the spinel grains.
Occasionally, some hematite grains assumed a spherical shape and
were embedded within spinel grains, Fig. (1). This suggests an isotropic interface energy and a high contact angle between these two
phases. When the discontinuous grain growth occurs, some second
phase grains are trapped within the exaggerated grains. The phase
boundaries between the trapped hematite grains and the spinel matrix
are usually straight in contrast to the spherical interface observed
in samples sintered at low temperatures and without discontinuous
grain growth. The straightness of the phase boundaries was probably
imposed by the grain growth process, during which the hematite grains
were stretched, pushed and cut through by the migrating grain
boundaries.

     The motion of the hematite grains is controlled by either the
interface kinetics of the oxidation and reduction process at the
phase boundaries or the diffusion process of the slowest moving ionic
species. If oxygen diffusion along boundaries is the rate controlling step, the maximum allowable velocity V_m, can be given as:

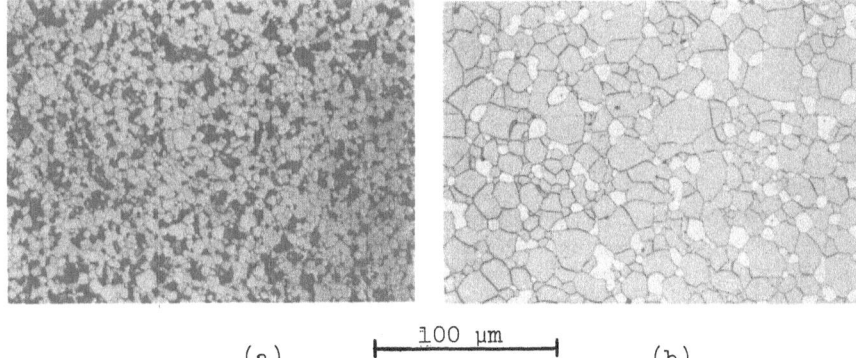

| 100 μm |
| (a)   (b) |

Fig. 1. (a) Undoped MnZn ferrite; (b) MnZn ferrite mixed with 2.5 wt
        % low melting composition, (A) in Table 2. Both were sinter-
        ed at 1200°C for 20 hrs in oxygen. Note the circular shape
        of the second phase embedded within spinel grains, (b).

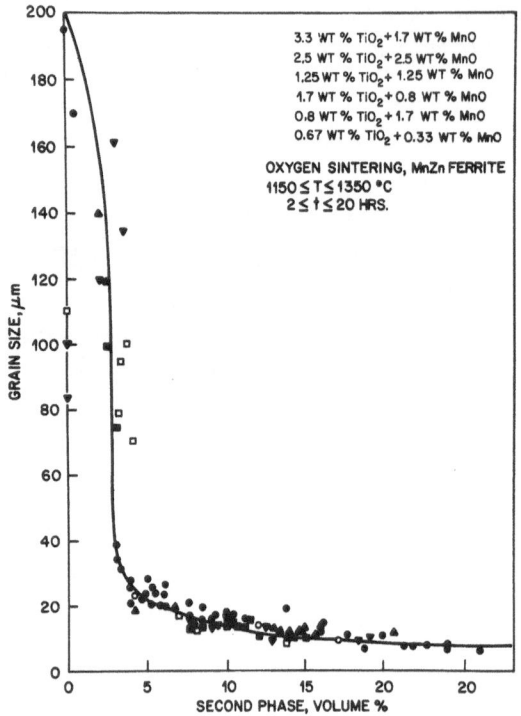

Fig. 2. Ferrite grain sizes are correlated with the volume fraction of the second phase. The correlation, derived from ferrite specimens after oxygen sintering, is independent of dopant content and composition, sintering temperature and time. Discontinuous grain growth occurs if second phase grains are less than ~4 vol. %.

$$V_m = \frac{\gamma(\delta D_b)\Omega}{R^3 kT} \qquad (1)$$

where γ is the grain boundary energy which is assumed to be 300 dyne/cm for ferrite grain boundaries, δ is the width of the transport path, D_b is the diffusivity along the interface boundaries, Ω is the resultant volume transported by each ion of the rate controlling species and R is the radius of the hematite grains. If we assume that the diffusivity at the phase boundary region is same as the grain boundary diffusivity and that $\delta D_b = 5.8\times10^{-11}$ cm^3/sec at 1197°C as interpreted from the diffusion data in nickel ferrite,[8] the maximum velocity, V_m, is calculated to be 0.48μm/hr. at ~1200°C. This velocity is only slightly higher than the average grain growth rate and can still be exceeded by some fast moving grains; thus leading to entrapment of hematite particles within spinel grains. Furthermore, following the analyses of Brook[9] and Carpay,[10] it can be shown that in the presence of hematite particles of 10μm size and an areal density of ~5×10^4 cm^{-2}, discontinuous grain growth can occur if the grain sizes are within the 32-78 μm range.[11] Given the qualifying assumptions in this calculation, there exist qualitative agreements with the microstructural observations.

In general, the relative amount of the hematite phase is a function of the TiO_2 dopant content, the sintering temperature and atmosphere. After sintering in oxygen, the volume fraction of the hematite phase increases with an increase in the net TiO_2 content and a decrease in the sintering temperature. For ferrites sintered in oxygen with a given TiO_2 dopant content, there exists a critical temperature above which the samples remain in the spinel single phase. This critical temperature increases with the dopant content of TiO_2, which stabilized the hexagonal structure of the hematite phase, Fig.3. These critical temperatures may correspond to the spinel single phase boundary. In fact, the observed phase boundary in Fig. 3 is the same as that derived from water-quenched specimens with a cooling rate $\sim$-$10^6°C$/hr. Thus, during the slow cooling process (-$400°C$/hr) in nitrogen a negligible amount of the hematite phase was reduced to the spinel phase. Similar data on the stabilizing influence of TiO_2 on the hematite structure relative to the spinel structure and the single phase boundaries have been reported for the Fe-Ti oxides system at other oxygen partial pressures.[4]

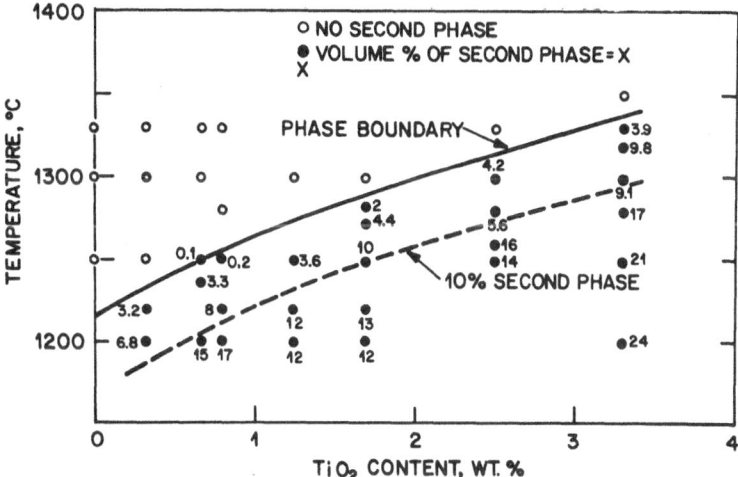

Fig. 3. Phase equilibria diagram for MnZn ferrite with TiO_2 addition. Above the phase boundary (solid line) spinel phase is stable, otherwise hematite and spinel phases co-exist.

The hematite second phase in these sintered ferrite samples can be reduced by annealing in nitrogen. Since the sintered specimens have little open porosity, the reduction of the hematite phase is rate limited by solid state diffusion. Thus, the hematite phase may have a spatial distribution in annealed specimens; and the distribution will be a function of the specimen thickness, the annealing time and temperature. After a short time nitrogen anneal, it has generally been observed that most of the hematite second phase was

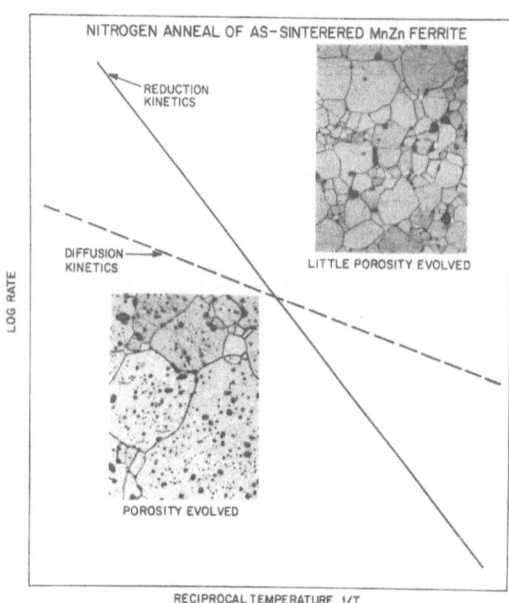

Fig 4.   Schematic diagram showing the
kinetic processes relating to the pore
evolution during nitrogen anneal.

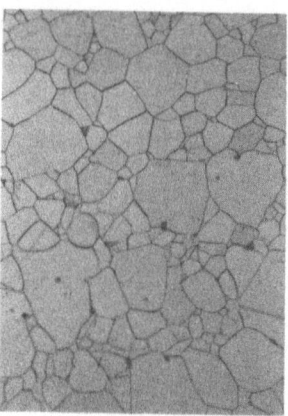

100 μm

Fig 5. MnZn ferrite with
2.5wt% dopant composition
A, Table 2. After sintering
at 1200°C, the sample was
heated slowly to 1250°C
for grain growth. Subse-
quent nitrogen anneal at
1150°C removes the remaining hematite phase.

reduced to the spinel phase within a surface layer of a few hundred
microns thick while the hematite phase in the bulk of the specimen
remained unchanged. The non-uniform distribution of the hematite
phase induces a stress gradient within the specimen and has a
deleterious effect on the high permeability magnetic properites.

Furthermore, if the ferrites are annealed at a sufficiently
high temperature, significant grain growth occurs in regions where
the hematite grains are rapidly reduced to the spinel phase. Since
the reduction of the hematite phase starts from specimen surfaces
and propagate into the bulk, columnar exaggerated grains normal to
specimen surfaces have been observed after post-sintering heat treatments in nitrogen. Consequently, it is desirable to anneal the
specimens at relatively low temperatures to avoid the discontinuous
grain growth and for a long anneal time to have a uniform removal of
the second phase.

The conversion of the hematite phase to the spinel phase can be
given as

$$3Fe_2O_3 \text{ (hematite)} \rightleftharpoons 2Fe_3O_4 \text{ (spinel)} + 1/2 \ O_2 \text{ (gas)}   (2)$$

where oxygen gas is generated. When in excess of the solubility
limit in the ferrite matrix, oxygen ions either diffuse out of the
specimen or accumulate to form voids within the specimen. The void

formation is the balance of two competitive processes: hematite
reduction to generate oxygen and diffusion kinetics to remove the
excess oxygen. If oxygen generated in Eq. (2) is allowed to form
voids, significant porosity will be developed. For example, when a
ferrite sample with a 1 vol.% hematite phase is reduced to the spinel
phase, porosity will increase by about 10 vol.%, assuming that the
porosity is in the form of ∿1μm pores and under an effective pressure
of ∿13 atmospheres due to the surface energy of ∿300 dyne/cm.

However, if all oxygen generated in Eq. (2) diffuses out of the
sample, the loss of oxygen introduces a negligible amount of porosity.
For example, when a specimen with 1 vol. % hematite is reduced to
spinel, the loss of oxygen results in only 0.02 vol. % porosity, or
500 times less than the porosity when oxygen accumulation occurs.

Thus, it is important to minimize the accumulation and coalescence of oxygen evolved in Eq. (2) to form pores by increasing the
diffusivity for oxygen removal and decreasing the rate of hematite
reduction. Both the reduction and diffusion processes are thermally
activated but with different activation energies as illustrated in
the schematic diagram, Fig. 4. The choice of a higher activation
energy for the hematite reduction process was suggested by the microstructural observations after nitrogen anneals showing more porosity
was evolved at a higher annealing temperature.

Furthermore, the reduction rate of the hematite grains can be
controlled by the heating rate. During a slow heating process, the
hematite grains dissociate in a controlled rate while the evolved
oxygen is removed from the specimen at an increasing rate due to
the higher diffusivity at an elevated temperature. Due to the disappearance of the hematite second phase and the higher temperatures,
grain sizes increase gradually with little or no discontinuous grain
growth. The resultant microstructures of doped ferrite after these
heat treatments show dense, large grain and single phase bodies,
Fig.5.

Conclusions

1. Low melting compositions near the eutectics of the $MnO-TiO_2$
   systems are used to promote MnZn ferrite sintering at relatively low temperatures (1100-1300°C) in oxygen.
2. A second phase of the hematite structure is stablized by the
   TiO_2 content in the dopant.
3. Grain sizes in oxygen-sintered ferrite specimens are correlated
   with the volume fraction of the second phase. Exaggerated grain
   growth is inhibited by the second phase in excess of ∿4 vol %.
4. Anneal in nitrogen can reduce the hematite second phase to the
   spinel phase. However improper anneal schedules at high temperatures introduce porosity into the originally dense specimens.

5.   Controlled heating rates and annealing atmospheres give ferrite
      specimens with 30-40 μm grain sizes and less than 0.5 vol %
      porosity.

Acknowledgements

        The authors wish to thank G. Y. Chin for useful suggestions
and discussion during this study; F. Schrey for performing chemical
analyses; W. W. Rhodes for experimental assistance and H. Schreiber
for providing yet unpublished data.

References

1. M. F. Yan and D. W. Johnson, Jr., J. Am. Ceram. Soc., submitted
        for publication (1977).
2. J. Grieve and J. White, J. Roy. Tech. Coll. (Glasgow), 4,
        p. 441-448 (1939).
3. J. B. MacChesney and A. Muan, Am. Mineralogist, 46, p. 572-582
        (1962).
4. J. B. MacChesney and A. Muan, ibid., 44, p. 926-945 (1959).
5. J. Grieve and J. White, J. Roy. Tech. Coll. (Glasgow), 4, 661
        (1940).
6. F. H. Dulin and D. E. Rase, J. Am. Ceram. Soc., 43, p. 125-131
        (1960).
7. H. Schreiber, unpublished research, Bell Laboratories (1977).
8. H. M. O'Bryan and F. V. DiMarcello, J. Am. Ceram. Soc., 53 [7]
        p. 413-416 (1970).
9. R. J. Brook, ibid., 52 [1] p. 56-57 (1969).
10. F. M. A. Carpay, p. 261-275, "Ceramic Microstructures '76", Eds.
        R. M. Fulrath, J. Pask, Westview Press (1977).
11. M. F. Yan, unpublished research, Bell Laboratories (1977).

PART V

ENERGY RELATED CERAMICS I:

Fast Ion Conductors; MHD,
Nuclear and Refractory Ceramics

PROCESSING AND CHARACTERIZATION OF POLYCRYSTALLINE β''-ALUMINA

CERAMIC ELECTROLYTES

R. S. Gordon, B. J. McEntire, M. L. Miller, and
A. V. Virkar

Department of Materials Science and Engineering
University of Utah, Salt Lake City, Utah 84112

I.   INTRODUCTION

In this paper recent developments in the processing and
characterization of polycrystalline β''-alumina ceramic electrolytes
will be discussed. β''-alumina ceramics have potential uses in
diverse devices such as the electrolyte or sodium-ion separator in
(1) the sodium-sulfur energy storage battery [1] designed for either
electric utility load leveling or automotive propulsion, (2) a high
temperature sodium heat engine or thermoelectric device.[2] (3) a high
electrowinning of sodium metal from molten sodium salts, (4) the
purification of sodium metal and (5) various electrochemical devices
for measuring sodium and perhaps even oxygen activities.

The sodium-sulfur cell is an electrochemical device of unique
design in that the electrodes are liquid sodium and sulfur (or
sodium polysulfide) while the electrolyte is a solid ceramic which
is permeable to sodium-ions and essentially impermeable to electronic
charge carriers. The elctrolyte is usually in the form of a thin
walled cylindrical tube and consists of β or β''-alumina in the
polycrystalline state. Two structural forms of beta-type alumina
exist.[3] One, β-alumina, is commonly referred to as the two-block
spinel structure with an alumina-soda mole ratio between approximately 9 and 11. The second is β''-alumina which is composed of
three Al-0 spinel blocks per unit cell with a molal ratio of alumina
to soda of approximately 5. The three block structure is believed
to be unstable at elevated temperatures (>1500°C) where sintering
operations are normally conducted unless small amounts (up to
1-4% by weight) of Li_2O or MgO or both are present as stabilizing
agents. To insure the presence of reversible liquid electrodes

and sufficient sodium ion conductance in the electrolyte, the
projected operating temperature of the Na-S cell is between 300
and 400°C. Operation of the cell at the high end of this temperature range is desirable for the achievement of high battery
efficiencies and power densities.

Two major applications are foreseen for the sodium-sulfur
battery: (1) electric vehicle propulsion and (2) energy storage
systems for load leveling in electric power generating plants.
Both of these requirements at full scale production will require
up to 10^7–10^9 ceramic electrolyte tubes on an annual basis. It
is clear from the number of cells which are potentially required
that the unit cost of each sodium-sulfur cell including the ceramic
electrolyte must be kept to a minimum. Furthermore, for a low
capital investment the number of cells in a load leveling installation must be kept as small as possible. With respect to the number
of required cells, perhaps the most important property is the
sodium ion resistivity of the electrolyte. To achieve a high
power density for vehicular propulsion, a low resistivity ($\leq$ 5
ohm-cm at 300°C) is essential. Although higher resistivities
(10-15 ohm-cm) may be acceptable for load leveling storage batteries,
both the battery efficiency and the initial battery cost are related to the resistivity of the electrolyte. A given performance
can be obtained with much fewer cells and correspondingly lower
capital costs by keeping the resistivity of the electrolyte small.

II. PHYSICAL PROPERTIES

Over the last several years both polycrystalline β and β''-
alumina electrolytes have been fabricated into cylindrical tubes
by a number of techniques including isostatic pressing, extrusion,
electrophoretic deposition and centrifugal casting. These electrolytes have attractive resistivities (3-15 ohm-cm at 300°C) and
reasonably high strengths (170-240 MN/m^2). They have exhibited
encouraging performance characteristics in operational Na-Na*
and Na-S** cells.[4,5,6]
Due to its intrinsically low resistivity,*** work in our
laboratories has been concentrated on lithia-stabilized β''-alumina

---

* 3750 Ah/cm^2 (one direction) at 1.25A/cm^2.[6]

** 3500 Ah/cm^2 (both directions) in 12,000 cycles.[6]

*** β''-alumina ceramics stabilized with Li_2O, MgO or both exhibit
lower resistivities (3-5 ohm-cm) at 300°C than β-alumina compositions ($\sim$ 10-20 ohm-cm).[3,4,5]

with a composition in the range: 8.7–9.0% Na_2O, 0.7–0.8% Li_2O, and 90.2–90.6% Al_2O_3 (by weight).

A. Sodium–Ion Resistivity

Most β or β''-alumina electrolytes are fabricated by starting with α-alumina and various sources of Na_2O (e.g., Na_2CO_3, NaOH, $NaAlO_2$), Li_2O (e.g., $Li_2C_2O_4$ or $LiNO_3$) and/or MgO. These materials are reacted together to form the β and β'' phases either in a pre-sinter calcination step[7] or during the sintering operation itself.[5] Consequently during reactive phase sintering three processes are normally proceeding in parallel: (1) phase conversion to β or β'' – alumina (2) densification presumably by liquid phase sintering and (3) microstructure development with the possibility for exaggerated idiomorphic grain growth always present. All three of these processes are mutually dependent and exert influence to some degree on the final resistivity which can be achieved.

Early in our program hot-pressing was used to prepare dense, fully converted, and fine-grained polycrystalline β''-alumina.[8] The objective of these studies was to determine if both a fine-grained and conductive material possessing a high fracture strength could be produced in the lithia-stabilized system. Previous work on isostatically pressed and sintered material indicated that conductive material (≤5 ohm–cm at 300°C) could be achieved only with a coarse-grained ($\sim100\mu m$) microstructure.[9]

In Table 1, a summary is given for typical hot-pressing conditions which are required to achieve complete densification* in polycrystalline lithia-stabilized β''-alumina.[5,8] Hot-pressing times varied between about 15 and 60 minutes. Pressing pressures ranged between 30 and 40 MN/m^2. The microstructure after hot-pressing consisted of grains under 3–5 μm in size. The fracture strengths varied between 207 and 276 MN/m^2. The resistivities of hot-pressed ceramic varied between 6 and 11 ohm–cm for partially converted material.** Annealing treatments at 1300°C resulted in further conversion to β''-alumina with a drop in resistivity to around 5–6 ohm–cm with essentially no grain growth. Annealing at higher temperatures ($\sim 1450^{\circ}$C) resulted in complete phase conversion to β''-alumina, a further drop in resistivity to under 3 ohm–cm, and in grain growth (up to 250μm in some cases) with a corresponding drop in fracture strength to around 104 MN/m^2.

---

*All densities $\geq$ 3.25 g/cm^3 (3.27 g/cm^3 was the highest density measured).

**All resistivities in the paper are quoted at 300°C unless stated otherwise.

TABLE I

PROPERTIES OF HOT-PRESSED AND ANNEALED LITHIA-STABILIZED β"-ALUMINA

Composition		Temp.	Resistivity (ohm-cm)		
%Na$_2$O	%Li$_2$O	(°C)	As Pressed	+ 1300°C (22h)	+ 1450°C (20h)
8.8	0.75	1440	7.8	5.3	2.5
8.8	0.75	1417	7.2	6.7	2.6
8.8	0.75	1375	7.0	6.1	3.4
			As Pressed	+ 1350°C (20h)	+ 1380°C (20h)
8.8	0.75	1420	6.4	5.5	4.9
8.8	0.75	1390	7.8	5.7	5.1
8.8	0.75	1350	8.7	4.8	5.2
8.7	0.70	1500	3.4*	-	-
8.7	0.70	1450	3.2*	-	-
10.0	0.80	1370	10.2	6.6	5.1
10.0	0.80	1330	10.6	5.9	5.1

*100% β"-alumina before hot-pressing; ∿10μm grain size after pressing with a 245 MN/m^2 fracture strength. (At all other compositions a calcined (1250°C) mixture of α-Al$_2$O$_3$, Na$_2$CO$_3$ and LiNO$_3$ was used as the starting powder). The fully converted β"-alumina powder was prepared by grinding sintered and thoroughly annealed specimens.

Our studies on hot-pressing and sintering have revealed that the lack of complete conversion to β"-alumina can have the most pronounced effect in increasing the resistivity (up to factors of 4 to 5) after firing.[7,10] Grain-size effects are much smaller.[10] For example, an increase in grain-size from approximately 2 to 100μm leads to a decrease in the resistivity by about a factor of 1.6 (4.5 to 2.8 ohm-cm) in fully converted lithia-stabilized β"-alumina. Furthermore, Virkar, et.al.,[11] have shown that the grain-size effect on the resistivity becomes relatively insignificant for grain-sizes over about 5μm. Consequently, to achieve high fracture strengths, fine-grained (≤ 10μm) ceramics can be fabricated with resistivities under 5 ohm-cm.

The degree of phase conversion and grain-size also affect the activation energy for sodium ion conduction.[10] Both the resistivity and activation energy decrease with increasing conversion to β"-alumina. In fully converted material the activation energy decreases both with decreasing temperature and grain-size due to the influence of grain boundary conduction which becomes predominant at small grain-sizes and low temperatures.*

Since sodium ion conduction in β- or β"-alumina is two
dimensional, one might expect some conduction anisotropy in polycrystalline material. To check this possibility, the effects of
texture on sodium ion conduction were measured on fully converted
hot-pressed material.[12] Hot-pressed polycrystalline β"-alumina
(lithia-stabilized) possesses an (00ℓ) texture which is perpendicular to the pressing direction. The sodium ion resistivity,
measured in a direction parallel to the hot-pressing direction, was
a factor of 1.4 to 1.8 larger than the resistivity measured in the
perpendicular direction for small (∼ 2μm) to large (∼ 100μm) grained
microstructures, respectively. The lowest resistivites (1.8 ohm-
cm) and activation energies (3.1 kcal/mole) were observed perpendicular to the pressing direction for the large-grained material.
The highest resistivities (6.8 ohm-cm) and activation energies
(5.1 kcal/mole) were observed parallel to the pressing direction
for small grain size material. These results on conduction
anisotropy were then extended to explain qualitatively why the
radial resistivity for an isostatically pressed and sintered β"-
alumina tube is typically 25% higher than the axial resistivity.

B. Mechanical Properties

For the sodium-sulfur battery to be economically attractive,
it must have a reasonably long life (i.e., 5 years or 2000 charge-
discharge cycles in automotive devices and 10-20 years or up to
8000 cycles in load leveling applications). Hence the ceramic
electrolyte must be able to withstand many charge and discharge
cycles at current densities up to 0.25 A/cm^2 without any mechanical
degradation.   Furthermore the ceramic must be capable of withstanding stresses which arise in sealing during cell construction because
of thermal expansion mismatch. These requirements suggest that the
fracture strength of polycrystalline β"-alumina should be as high as
possible, consistent with a sodium ion resistivity under 5 ohm-cm.

To determine the optimum mechanical properties, fracture
studies were conducted on hot-pressed and sintered β"-alumina
which was dense and fully converted.[13] Various annealing temperatures and times were employed to achieve microstructures with grain
sizes between 4 and 400μm. Two regimes of fracture were identified.
In the small grain-size limit (≤ 120μm) the strength (four-point-bend)

---

*
 Typical activation energies for sodium ion grain boundary conduction in lithia-stabilized β"-alumina lie between 6.6 and 8.0 kcal/
mole.[11] For intra-granular conduction, the activation energy is
approximately 3 kcal/mole.[12]

was high (>170 MN/m^2) and weakly dependent on the grain-size
(D). For larger grain sizes, the strength decreased rapidly
($\propto$ D$^{-\frac{1}{2}}$) with increasing grain-size. The strengths of sintered
β"-alumina were comparable in most cases to values achieved with
hot-pressed material, particularly in the limit of small grain
sizes ($\leq$ 40μm).

While the fracture strengths can be a strong function of the
grain-size (>120μm) the critical stress intensity factor K_{IC},
(2.6-3.8 x 10^6 N/m$^{3/2}$) was found to be nearly independent of
grain-size. However both K_{IC} and the strength were found to be
orientation dependent in hot-pressed material as might be expected
for non-cubic β"-alumina.[13]

The elastic modulus of sintered polycrystalline β"-alumina
has been determined over a range of grain sizes (2-100μm) and
porosities (0.5 to 5%). The modulus (E) is independent of grain-
size (up to 100μm) and varies with the volume fracture porosity (P)
according to the relation, $E = E_0 exp(-bP)$ with b = 0.052. The
modulus at zero porosity E_0 was found to be 2.11 x 10^{11}N/m^2
(30.4 x 10^6 psi) and in good agreement with that measured for 99%
dense hot-pressed β"-alumina (2.02 x 10^{11}N/m^2). The fact that the
modulus was independent of grain-size up to about 100μm indicated
the absence of any significant microcracking in the microstructure.
Microcracks can be expected to be present in material with duplex
microstructures containing very large grains because of the
anisotropy in the coefficient of thermal expansion. In hot-pressed
β"-alumina a thermal expansion anisotropy of approximately 10% has
been measured with the thermal expansion being larger in the
direction perpendicular to that of hot-pressing.

C. Electrolytic Degradation

In addition to satisfactory out-of-cell electrical and
mechanical properties, the β"-alumina electrolyte must be resistant
to electrolytic degradation during cell operation. Previous
work [14],[15],[16] has shown that the ceramic will degrade mechanically
after it is subjected to a supercritical charging current density
(i.e., normally in excess of 1A/cm^2 for a properly prepared
electrolyte with the optimum composition). Cracks are found to
form on the ion neutralization surface (Na$^+$ + e$^-$ → Na). Early
theories, which have been used to account for this type of electrolytic degradation, are based on a stress corrosion model for sub-
critical crack growth. Crack growth is assumed to take place by
the stress assisted dissolution of the ceramic in liquid sodium.

To check on the viability of a stress corrosion mechanism, sub-critical crack growth experiments were conducted in liquid sodium using DCB specimens in a load relaxation technique.[17] Two compositions were studied: (1) 8.8% Na_2O-0.75% Li_2O (2) 10% Na_2O - 1.2% Li_2O. Tests were conducted at $300^\circ C$ on dense, fully converted, sintered specimens. Exceptionally large exponents N (i.e., $V = AK_I^N$ in which V is the sub-critical crack velocity and K_I is the stress intensity factor) ranging between 355 and 562 discounted any serious stress corrosion in the 8.8 - 0.75 ceramic. The N values for the 10 - 1.2 ceramic were lower (35 - 120) indicating an increased tendency for stress corrosion. However, even in this case, the relatively large N values indicated a small stress corrosion component for sub-critical crack growth. It is of interest to note that the Richman-Tennenhouse stress corrosion model[19] predicts N = 2 for the situation when the stress induced dissolution of the ceramic is dominant.

Having essentially eliminated the stress corrosion mechanism, it is currently believed that electrolytic degradation takes place by a critical fracture process.[17] In this case crack extension occurs when the stress intensity at the crack tip reaches K_{IC}. However, fracture initiated by an internal fluid (e.g. Na) pressure is not catastrophic. If the fluid amount is fixed, the crack arrests immediately. For the crack to grow, additional fluid must be brought continuously into the crack (i.e., the situation during charging). The rate of crack growth at any instant is determined by the rate of Na flow into the crack which depends upon the current density and the viscosity of liquid sodium. Experiments are underway to check the validity of this model. If the model is correct, the critical charging current density should increase with increasing temperature because of the decrease in the viscosity of liquid sodium.

D. Resistance to Attack by Water Vapor

Resistance to attack by water vapor is desired to avoid the requirement of storing finished electrolytes in an ultra dry environment. Some β''-alumina ceramics stabilized with Li_2O or Li_2O and MgO[18] have exhibited a high resistance to electrical and mechanical degradation by water vapor, while, on the other hand, ceramics with similar compositions have demonstrated very poor resistance to degradation. For example, partially converted lithia-stabilized β''-alumina with a resistivity of 7.2 ohm-cm cracked after storage at 75% relative humidity ($25^\circ C$) for two days. After 15 days the resistivity increased to 9.2 ohm-cm and after one year the electrolyte tube was completely covered with a dendritic growth containing a NaOH hydrate indicating leaching of the soda component. On the other hand a fully converted, hot-pressed,

and annealed β"-alumina ceramic which was stored at 25°C in
environments ranging between 20 and 100% relative humidity, did
not show any evidence of dendrite formation for periods up to
one year or any change in resistivity for periods up to 6 months
at 20% relative humidity and up to approximately 1 month at 75%
relative humidity.

Recently, humidity tests were conducted on sintered β"-alumina
(lithia-stabilized) which was processed using $Li_2O \cdot 5\ Al_2O_3$ as the
source of lithia. Bar specimens were sintered at 1587°C for 2
minutes and subjected to successive annealing treatments of 2 days
at 1250°C, 2 days at 1330°C, and 3 days at 1350°C to obtain specimens in which the conversion to β"-alumina was complete and yet in
which the grain-size was reasonably small ($\sim$ 50μm). After processing, the bars were stored at 100% relative humidity at 25°C for
various periods of time up to seven days. After the requisite
number of days in storage, the resistivities, densities, and
strengths (four-point-bend) of the bars were measured. These
measurements are summarized in Table 2. The resistivity and
density increased slightly as the storage time at 100% relative
humidity increased. However, storage in a humid environment had
no effect on the strength for a period of at least up to seven
days. After annealing at 650°C for 24 hours, both the density and
resistivity returned to their original values. Thus it is clear
that water can migrate into the β"-alumina structure, probably in
the conducting planes. However, this absorption of water is a
reversible process and leads to no apparent mechanical degradation.
These data clearly establish that fully converted polycrystalline
β"-alumina (lithia-stabilized) is very resistant to mechanical
degradation by water vapor. It is suspected that the processing
history, phase composition, and perhaps, the overall composition
are important factors in determining whether or not a given β"-
alumina ceramic will withstand electrical and mechanical degradation
in humid environments.

TABLE 2

EFFECT OF HUMIDITY ON MECHANICAL AND ELECTRICAL PROPERTIES OF
POLYCRYSTALLINE LITHIA-STABILIZED β"-ALUMINA

No. of Days at 100% Rel. Humidity (25°C)	Density (g/cm^3) After Humidity Treatment	Density (g/cm^3) After Drying	Resistivity (ohm-cm) After Humidity Treatment	Resistivity (ohm-cm) After Drying	Fracture Stress (MN/m^2)
0	3.23	3.23	4.58	4.58	199
1	3.28	3.24	4.51	4.34	228
2	3.33	3.24	4.14	4.39	205
3	3.26	3.22	4.65	4.39	181
4	3.31	3.22	6.04	4.70	207
7	3.29	3.23	5.41	4.56	205

III. PROCESSING

In addition to the requisite properties, β''-alumina electrolytes must be fabricated using inexpensive raw materials and by processes which are economically viable. The pre-pilot unit operations of isostatic pressing and sintering will be described in terms of their present stage of development and with respect to processes envisioned for large scale production. Currently β''-alumina electrolyte tubes (15-34 mm OD by 200-300 mm long) are fabricated in a pre-pilot facility capable of producing up to 15 (34 mm OD) units per day.[19]

A. Raw Materials

During the course of our studies on the fabrication of β''-alumina electrolyte tubes, several raw material sources for α-alumina have been evaluated.[19] These have ranged in cost between $1.00 and $50.00 per kilogram. As might be expected, the higher cost materials possess smaller concentrations of possibly undersirable impurities (Si, Ca) than do the lower cost materials. In Table 3 a summary of quantitative emission spectrographic analyses and the ultimate particle sizes of several candidate α-aluminas are presented. Electrolyte tubes with acceptable out-of-cell and in-cell properties have been fabricated with these raw materials. These tubes have tested successfully in Na-Na cells at 300°C (1000 Ah/cm^2 of charge at current densities over 1 A/cm^2) with no evidence of any mechanical degradation. Electrolyte tubing processed with lower cost ($\sim$ $1/kg) α-alumina powders containing higher levels of silicon and/or calcium showed more or less severe cracking of the ceramic during Na-Na cell testing at current densities over 1 A/cm^2 for charges between 200 and 1000 Ah/cm^2. It is of course possible that electrolyte tubes made with relatively impure powders could pass 1000 Ah/cm^2 of charge at much lower current densities (0.125-0.25 A/cm^2). In this paper emphasis will be placed on the processing and properties of tubes fabricated with B and D* alpha aluminas. The A material, although employed in the early fabrication studies, was discarded because of its high cost.

The minor constituents (Na_2O and Li_2O) are derived from Na_2CO_3 and $Li_2C_2O_4$ (or $LiNO_3$). Since the Li_2O is present in small amounts, it is difficult to achieve a uniform distribution of this component in the powder compact, which is essential for the complete conversion to the β'' phase after sintering. Thus instead of adding

---

* D is a much less expensive variant of C, containing about 0.12 wt% Na_2O.

Li_2O directly by calcination of a stoichiometric mixture containing $Li_2C_2O_4$, the Li_2O ($Li_2C_2O_4$) is prereacted with α-alumina
to form zeta lithium aluminate ($Li_2O \cdot 5\ Al_2O_3$). This larger amount
of zeta component is then used as the starting lithium-bearing raw
material. It is mixed with an appropriately calcined mixture of
α-alumina and Na_2CO_3 to form the so-called 'zeta process' powder.[7]

B. Isostatic Pressing

Closed end electrolyte tubes are fabricated by wet-bag
isostatic pressing. The most important factors in isostatic
pressing are, (1) powder flowability, (2) powder packing and
compaction, and (3) pressing pressure. For automated isostatic
pressing (dry or fixed bag), it is essential that flowable and
fillable powders are available so that rapid bag filling,
reasonably low compaction ratios, and high green densities can be
achieved.

Higher green densiities (up to 60% of theoretical) can be
achieved with the D-based powder than can be achieved with powder
B (up to 52%). This is in marked contrast to the effect of
pressing pressure on sintered density, particularly for the B-
based powder. In this case the sintered density is excellent

TABLE 3

EMMISSION SPECTROGRAPHIC ANALYSIS AND ULTIMATE PARTICLE
SIZE OF ALPHA ALUMINAS*

Element (ppm)	A	B	C	D
Si	142–179	59	106–111	130–146
Ca	12–16	<9	42–52	8–10
Fe	25–48	12	28–29	21–25
Ni	<2	<2	<2	<2
Particle Size (μm)	1	0.3	0.5	0.5
Cost ($/kg)	40–50	16–20	∿10	∿4

---

*A-Linde C, B-Meller (deagglomerated), C-Reynolds RC-HP-DBM,
D-Reynolds RC-HPS-DBM.

($\sim$98%) and is independent of pressing pressure between 34 and 379 MN/m^2. In the case of D-based powders, the sintered density is acceptable and independent of pressing pressure above 172 MN/m^2 (25 kpsi). Below this pressure the sintered density falls off dramatically.

While tubes produced in the pre-pilot plant are fabricated with powders which have not been specially designed to be flowable and fillable, a research effort has been conducted on the development of a slurry spray drying process. In this technique an appropriate (stabilized and low viscosity)* slurry (water-based) of calcined powder is passed through a spray dryer resulting in the formation of relatively uniform (60-100μm) sized spheres which render the powder fillable and flowable. Powders prepared in this fashion can fill an isostatic pressing mold rapidly ($\leq$ 30 seconds) to a high fill density.

Using specially designed laboratory spray dryers, preliminary pre-pilot studies have been conducted on the spray drying of slurries prepared from C-based powder (calcined). This powder was selected because of its high purity, relatively low cost, and ease of forming high-solids-content, low-viscosity aqueous dispersions. Depending on the sintering conditions, partially converted** and zeta-process powders, after slurry spray drying, may be sintered to dense, reasonably fine-grained material (<50μm) with a low sodium ion resistivity (3.0 - 6.5 ohm-cm). β''-alumina tubes fabricated from spray dried powders have successfully passed 1000 Ah/cm^2 at current densities over 1 A/cm^2 in Na-Na cell tests at 300°C with good electrical performance and no evidence of any mechanical degradation.

D. Sintering

Essentially all of the sintering of either β or β''-alumina ceramics is conducted at temperatures $\geq$ 1545°C in batch or continuous pass-through modes of operation in which the time at the maximum temperature is normally under 10 minutes. Most batch operations utilize some type of specimen encapsulation (platinum; or MgO, β-alumina or β''-alumina refractory containers) to prevent the loss of soda.[4,5,7] β-alumina ceramics are typically fired at temperatures in excess of 1700°C while β''-alumina compositions

---

*2.1% tartaric acid, 1.2% carbowax 200 polyethylene glycol and 2.8% Gelvatol 20-30 PVA, all based on the powder by weight.

**Mixing α-alumina, Na_2CO_3, and $LiNO_3$ prior to calcination at 1260°C without the use of $Li_2O \cdot 5\ Al_2O_3$.

can be sintered easily at temperatures between 1545 and 1600°C.
In the case of β"-alumina the sintering process is complicated
by the simultaneous phase conversion of alpha or beta alumina to
the final phase, β"-alumina.[7] In a fast sintering process, the
rate of densification usually proceeds at a much faster rate than
the rate of phase conversion. Hence after sintering, even though
the density is high and the grain size is small, the ionic
resistivity is also high (8-25 ohm-cm) because conversion to β"-
alumina is incomplete. By annealing the sintered specimen at a
temperature 50 to 200°C below the sintering temperature for various
periods of time (1-12 hours) conversion can be completed without
resulting in any excessive and undesirable grain growth. Thus by
a two step sintering and annealing process, a fine grained
(∿ 10-50μm), strong(200 MN/m^2), and conductive (3-5 ohm-cm) ceramic
electrolyte can be achieved.[7]

For annealing times ≤ 1 hour it is possible to combine the
firing and annealing schedules into a single pass-through
continuous firing operation. On the other hand if the annealing
cycle is much longer, then a separate batch annealing operation
is required. If the zone sintering operation is rapid, sintering
can be accomplished in open air presumably without encapsulation.[4]

Typically unconverted, partially converted, or zeta-process
powder mixtures are prepared prior to sintering and annealing.
Unconverted powders* consist of a mixture of α-alumina, $NaAlO_2$,
and $LiAlO_2$ with no β or β"-alumina. Partially converted powders
consist primarily of β and β"-alumina in equal amounts with no
α-alumina. After green forming the specimens are heated rapidly
to the sintering temperature; densification is rapid (≤ 5 minutes)
particularly at 1575-1600°C. The rate of phase conversion to β"-
alumina lags behind the rate of densification as evidenced by the
relatively high resistivity (8-15 ohm-cm) after short time sintering. A post-sintering anneal can lower the resistivity to values
under 4-7 ohm-cm. At 1415-1420°C annealing times on the order of
8-12 hours are required. At this temperature grain growth is
minimal. Shorter annealing times (1-2 hours) can be used at
higher temperatures (1490-1550°C) to promote the conversion to
β"-alumina; however, at these temperatures rapid grain growth
proceeds along with chemical conversion and in some cases duplex
microstructures with large grains (200-300μm) develop. At all the
annealing temperatures, the drop in resistivity coincides with the
conversion to β"-alumina.[7,10]

---

*Prepared by calcining α-alumina, Na_2CO_3, and $LiNO_3$ ($Li_2C_2O_4$) at
1000°C.[7]

An improved method of distributing Li_2O in the green compact
has been developed whereby the lithia component is introduced in
the form of $Li_2O \cdot 5\ Al_2O_3$ (zeta lithium aluminate spinel).[7,20]
Using $Li_2O \cdot 5\ Al_2O_3$ in powder mixtures of α-alumina and Na_2CO_3 precalcined at $1250°C$, rapid densification (2-10 min) can be achieved
at $1585-1600°C$. Post sintering anneals for one hour at $1475-1550°C$
lead to the development of strong ($200 \pm 27\ MN/m^2$) and conductive
($\sim$4 ohm-cm) ceramics with reasonably fine grain microstructures
($\sim$10-50μm). Thus the more uniform distribution of lithia allows
for a faster conversion to β''-alumina and at the same time delays
the onset of exaggerated grain growth.

Recently[21] a technique has been developed wherein seed crystals
of fully converted β''-alumina are added to an unconverted or
partially converted powder mixture of α-alumina, sodium aluminate
and lithium aluminate prior to sintering. These seeds catalyze
the conversion to β''-alumina such that it no longer lags significantly behind the densification process. Thus reasonably uniform
and fine grained ($\sim$50μm) microstructures can be produced in a
conductive ceramic without any post sintering anneal. This
technique along with a thorough discussion of microstructure and
its development in β''-alumina bodies will be discussed in a
companion paper in these proceedings.

Attempts have been made to lower the sintering temperature
(under $1600°C$) of β''-alumina via the introduction of transient
liquid phases. Even though a $100-150°C$ drop in the sintering
temperature has been achieved, the relatively long sintering
time ($\sim$ 3 h) required for conversion makes these processes somewhat unattractive.[22]

Some attempts have been made to synthesize completely converted
and very fine powders of β''-alumina by gel processing and
solution spray drying methods.[22] While a uniform distribution
of components has probably been achieved in these studies, low
temperature conversion to β''-alumina was never achieved. Other
low temperature phases (λ, β, etc.) were always favored over the
β'' phase. These powders could be sintered to density only at
temperatures over $1600°C$. Evidently some inhomogeneity in the
system is required to promote liquid phase sintering between 1450
and $1600°C$.

IV. PHYSICAL PROPERTIES AND DIMENSIONAL
CHARACTERISTICS OF ELECTROLYTE TUBING

In Table 4 typical properties of standard pre-pilot β''-alumina
tubing are summarized with standard deviations indicated for
density, sodium ion resistivity and diametral strength. The
microstructures of these tubes consist primarily of fine grains

(<10μm) with isolated grains up to 100μm is size. The fairly high strengths (> 207 MN/m^2) indicate that the microstructure is sufficently fine for the use of these tubes in Na-S cells.

Typical dimensions with standard deviations are presented in Table 5 for large and small diameter tubing fabricated from both α-alumina powder sources. As might be expected the degree of ovality, as measured by MAX OD/MIN OD at the open end, increases for the large diameter tubing when unsupported horizontal sintering is employed. To reduce the ovality in large diameter tubing caused by creep or sag during sintering, the tube can be supported on a ceramic V setter (e.g. H brick - Carborundum Monofrax H) during firing. Preliminary results indicate that the wall thickness can be reduced to about 2 mm without increasing significantly the degree of ovality from that which is typical of 15 mm OD tubing. In no case was the MAX OD/MIN OD ratio greater than 1.015 in supported sintering.

TABLE 4

PHYSICAL PROPERTIES OF "ZETA-PROCESS"
PRODUCTION β"-ALUMINA TUBING

Nominal OD (mm)	Number Tubes	Density (g/cm^3)	Resistivity 300°C (ohm-cm)	Diametral Strength(MN/m^2)
15	40	3.20 ± 0.01	4.12 ± 0.37	285 ± 48
33.4	15	3.20 ± 0.01	4.40 ± 0.28	--
33.4	17	3.21 ± 0.01	4.28 ± 0.27	--
33.4	25	3.21 ± 0.02	4.18 ± 0.40	--
33.4*	38	3.22 ± 0.02	4.22 ± 0.25	214 ± 40

*D-based raw material; all others B-based.

TABLE 5

DIMENSIONS OF "ZETA-PROCESS"
PRODUCTION β"-ALUMINA TUBING

| Number Tubes | Diameter - Open End | | Wall | |
	Average OD (mm)	MAX OD / MIN OD	Ave. (mm)	MAX WALL / MIN WALL
40	14.80 ± 0.09	1.01 ± 0.010	1.08 ± 0.05	1.05 ± 0.020
42	33.02 ± 0.69	1.034 ± 0.010	2.45 ± 0.08	1.02 ± 0.010
38*	34.03 ± 0.25	1.030 ± 0.009	2.55 ± 0.07	1.026 ± 0.016

*D-based raw material, all others B-based.

ACKNOWLEDGEMENTS

The National Science Foundation under Contract NSF (RANN) C-805 and the Energy Research and Development Administration under Contract EY-78-C-02-2566 are acknowledged for the support of this work. The assistance of D. K. Shetty, J. P. Singh, T. Ketcham, G. E. Youngblood and H. Beuhler in some experimental aspects of the work is acknowledged.

VI. REFERENCES

1. J. T. Kummer and N. Weber, Soc. Automotive Engs. Trans. 76 1003 (1968).

2. Neill Weber, Energy Conversion 14 1 (1974).

3. J. T. Kurmer, Prog. Solid State Chem. 7 141 (1972).

4. R. M. Dell, J. L. Sudworth, and I. Wynn Jones, Eleventh Inter-society Energy Conversion Engineering Conference, Proceedings, pp. 503-509, Stateline, Nevada, Sept. 12-17, 1976.

5. J. L. Sudworth, A. R. Tilley and J. M. Bird, Presentation Fall Meeting Electrochemical Society, Atlanta, 1977.

6. "Research on Electrodes and Electrolytes for the Ford Sodium-Sulfur Battery" Semiannual Report for period June 30, 1976 to December 31, 1977, January 1977, National Science Foundation, Contract #NSF C-805.

7. G. E. Youngblood, A. V. Virkar, W. R. Cannon and R. S. Gordon, Bull. Amer. Ceram. Soc. 56 (2) 206 (1977).

8. A. V. Virkar, G. J. Tennenhouse, and R. S. Gordon, J. Amer. Ceram. Soc. 57 (11) 508 (1974).

9. T. J. Whalen, G. J. Tennenhouse, and C. Meyer, J. Amer. Ceram. Soc. 57 (11) 497 (1974).

10. G. E. Youngblood, G. R. Miller, R. S. Gordon, J. Amer. Ceram. Soc. (in press).

11. A. V. Virkar, G. R. Miller, and R. S. Gordon, J. Amer. Ceram. Soc. (in press).

12. G. E. Youngblood and R. S. Gordon, Ceramurgia International (in press).

13. Anil V. Virkar and Ronald S. Gordon, J. Amer. Ceram. Soc. 60 (1-2) 58 (1977).

14. G. J. Tennenhouse, R. C. Ku, R. H. Richman, and T. J. Whalen, Bull. Amer. Ceram. Soc. 54 (5) 523 (1975).

15. R. H. Richman and G. J. Tennenhouse, J. Amer. Ceram. Soc. 58 (1-2) 63 (1975).

16. R. D. Armstrong, T. Dickenson and J. Turner, Electrochimica Acta 19 185 (1974).

17. D. K. Shetty, A. V. Virkar and R. S. Gordon, Proceedings of an International Symposium on Fracture Mechanics of Ceramics, July 27-29, 1977, University Park, Pennsylvania, Plenum Press (in press).

18. James H. Duncan and Brian K. Hick, U. S. Pat. 3,765,915,
       October 16, 1973.
19. M. L. Miller, B. J. McEntire and R. S. Grodon, Bull. Amer.
       Ceram. Soc. (in press).
20. Anil V. Virkar and Ronald S. Gordon, pp. 610-620, "Ceramic
       Microstructures, '76", edited by R. M. Fulrath and J. A. Pask,
       Westview Press, Boulder, Colorado, 1977.
21. Arun D. Jatkar, Ivan B. Cutler and Ronald S. Gordon, ibid,
       pp. 414-422.
22. Research on Electrodes and Electrolytes for the Ford Sodium-
       Sulfur Battery, Annual Report for the period June 30, 1975
       to June 29, 1976, NSF (RANN) Contract NSF C-805, July 1976.

DISCUSSION

L. De Jonghe (Cornell University): β-alumina seems to respond to
increased green forming pressure above 20 ksi. The increase in
final density is favorably affected by isostatic pressing at ∿100
ksi, especially if the sintering temperature is lowered. I wonder
if the insensitivity of the Meller material to increased green
forming pressure is due to adsorbed gases.
Author: We have not conducted isostatic pressing at pressures over 60 kpsi. It is possible that slightly lower sintering temperatures could be achieved by pressing at very high pressures. I
suspect, however, that factors such as the presence of liquid
phases and simultaneous grain growth play a greater role in determining the end point density than does the isostatic pressing
pressure.

I do not believe that absorbed gases play a significant role
in the sintering of Meller-based powders. All powders are calcined at $1260^{\circ}C$ prior to sintering to convert the α-alumina to a
mixture of β and β"-alumina. With the operations of calcination
and milling taking place prior to pressing, the original powder
loses most of its identity, other than, perhaps, some remnant relic structure.

R. F. Davis (North Carolina State University): Do you attempt to
closely control the particle size distribution in the starting
powder? Do you think such a distribution plays a major role in
the exaggerated grain growth?
Author: No, although we do measure the particle size distribution
of the calcined powder after milling* by the Sedigraph technique.
It is possible that the very large particles present in the distribution may act as nuclei for exaggerated grain growth provided
they are fully converted to lithia-stabilized β"-alumina.

---

*The milling time is selected to achieve a minimum firing
  shrinkage.

MICROSTRUCTURAL CONTROL DURING SINTERING OF β"-ALUMINA COMPOSITIONS

THROUGH CERAMIC PROCESSING MODIFICATION

Arun D. Jatkar, Ivan B. Cutler, Anil V. Virkar
and Ronald S. Gordon

Department of Materials Science and Engineering
University of Utah, Salt Lake City, Utah, 84112

INTRODUCTION

The exaggerated grain growth during high temperature (1585°-1600°C) sintering of Li_2O-stabilized β"-alumina compositions is essentially a process of very rapid growth of stabilized β" nuclei to large dimensions (>100 μm) while the matrix remains extremely fine-grained (< 3 μm). The final grain size is therefore inversely proportional to the cube root of the volume fraction originally occupied by these nuclei.[1] Such a relationship was reported by Cutler[2] in the case of seeded alpha alumina powders and more recently by Lacour and Paulus[3] in the case of barium-hexaferrite. The occurrence of an exaggerated grain growth during sintering of ceramics has been shown to depend on the flux-growth mechanism wherein a film or channel of liquid phase surrounds the growing nuclei. The growth of nuclei to large dimensions takes place by the solution of surrounding fine grains in the liquid and their precipitation on the growing crystal surface. While Cooy[4] and Stuijts[5] reported accidental impurities as the cause of liquid phase formation, Lacour and Paulus[3] showed that local inhomogenities can also give rise to a liquid phase responsible for the flux growth.

The estimated linear growth rates of stabilized β" nuclei (either inherently present or externally added) during sintering of Li_2O-stabilized β"-alumina compositions (8.7-9.0 w/o Na_2O; 0.7-0.8 w/o Li_2O; balance Al_2O_3) are around 20 μm/min,[6] which are too high to account for on the basis of solid state diffusion and therefore suggest a liquid-phase-assisted growth mechanism. In the present work, the postulate of liquid phase-assisted exaggerated

421

grain growth during sintering of Li_2O-stabilized β''-alumina compositions was experimentally verified[2].

EXPERIMENTAL RATIONALE AND PROCEDURE

In line with the current practice of fabricating Li_2O-stabilized β'' by sintering, polyphase powders obtained by calcining mixtures of sodium and lithium salts and alpha alumina were used as β''-precursors for pressing the green shapes. The various types of precursors used are shown in table-1.

Table 1. Li_2O-Stabilized β''-Alumina Precursors

Type	Method of Preparation (All calcines vibratory milled for four hours)	Phases Present
Partially converted powder	Na_2CO_3 + $LiNO_3$ + α-Al_2O_3* calcined at 1250°C for 2 hrs.	β, β'', $NaAlO_2$ and $LiAlO_2$
Zeta Process Powder **	'SZ-60' + 'L-60A'; 'SZ-60': Na_2CO_3 + α-Al_2O_3* calcined at 1250°C/2 hrs 'L-60A': $LiNO_3$ + α-Al_2O_3* calcined at 1250°C/2 hrs	β, β'', $NaAlO_2$ $Li_2O \cdot 6Al_2O_3$ (Zeta alumina)
Mixture of compositions '1' and '11' **	'1': Na_2CO_3 + $LiNO_3$ + α-Al_2O_3*, calcined at 1250°C for 2 hrs. '11': Na_2CO_3 + α-Al_2O_3*, calcined at 1250°C/ 2 hrs	$NaAlO_2$, $LiAlO_2$. β, β'', $NaAlO_2$
* Meller, deagglomorated; ** see Fig. 1		

The procedure for preparing stabilized β" seeds has been described elsewhere.[1]

The sandwich compacts '1/11/1' and 'SZ/L/SZ' represent a far larger extent of inhomogeneity than normally encountered in the current practice because the lithium-bearing component is highly localized (see Fig.1). The zeta process powder represents a more uniform distribution of lithium than the partially converted powder (better by a factor of about 5[6]). The pre-treatment (iso-thermal holding) at 1480°C was given to the specimens of partial-ly converted and of zeta process powders to make them more homo-geneous with respect to the phase distribution and to promote further conversion.

Rectangular bar-shaped specimens (0.5 x 0.5 x 2.5 cm) of partially converted powder, zeta process powder, seeded zeta process powder (zeta process powder + 2 w/o stabilized β" seeds, -400 mesh size) and the sandwich compacts were all isostatically pressed at 380 MN/M^2 and subsequently bisque-fired in open air at 800°C. Pre-treatments as well as isothermal sintering were carried out in an electrical resistance furnace with specimens placed in tightly closed platinum envelopes for preventing the soda loss by evaporation.

Grain size measurements on sintered specimens required the use of an optical microscope for grains larger that 5 μm and a SEM for finer grains. The volume fraction of coarse grains was estimated by a point count method and the composite grain size distribution histograms constructed using Spektor analysis.[7][8] For the sake of convenience, the linear intercepts on individual grains were grouped into various size ranges for Spektor analysis which assumes a spherical grain geometry. (Such an assumption is not strictly valid in the present case due to a pronounced plate-type grain morphology and therefore the histograms would be regarded as trend indicators and not a absolute size distributions).

For other measurements such a density and ionic resistivity and the results thereof, the reader is referred to Ref. 6 and 9.

Results

1. Sintering of sandwich compacts:

Fig. 1. shows the various compositions used in making sandwich compacts and the way they were fractured after sinter-ing. The central layer of '1/11/1' compacts was fully densified after sintering at 1500°C for one hour. A fracture through this densified layer revealed several shiny crystals (Fig. 2) of

β"-alumina as confirmed by x-ray diffraction patterns. In the
case of 'SZ/L/SZ' compacts, the 'SZ' layers densified preferentially, while the central 'L' layer remained porous even after sintering the compacts at 1600°C. The through-thickness sections of
both 'l/ll/l' and 'SZ/L/SZ' compacts showed a variation of grain
size, the largest grains occurring away from the l/ll or SZ/L
interface.

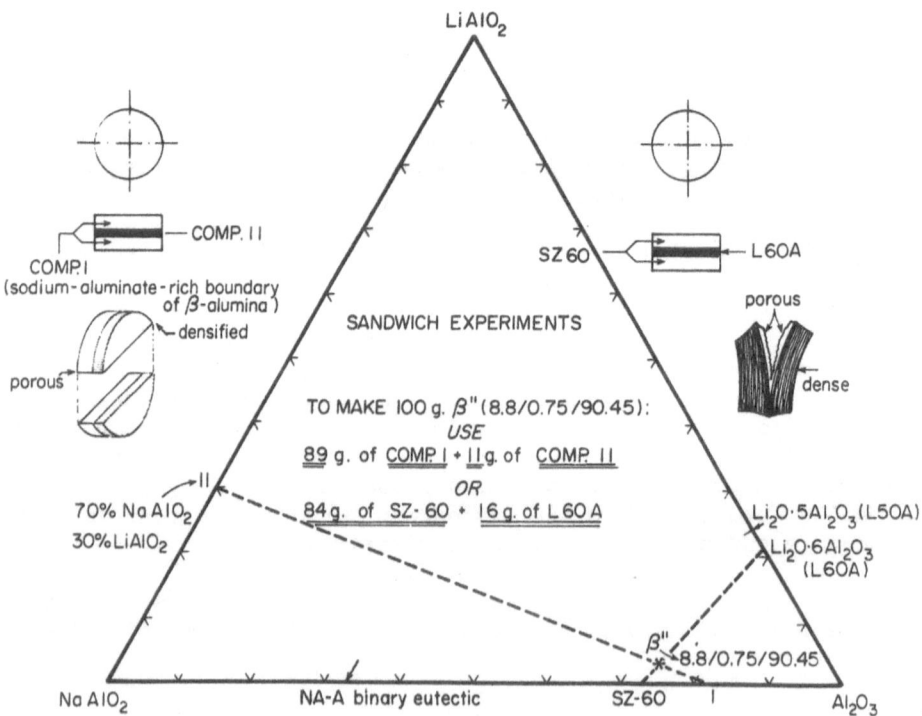

Fig. 1 Sandwich Compacts

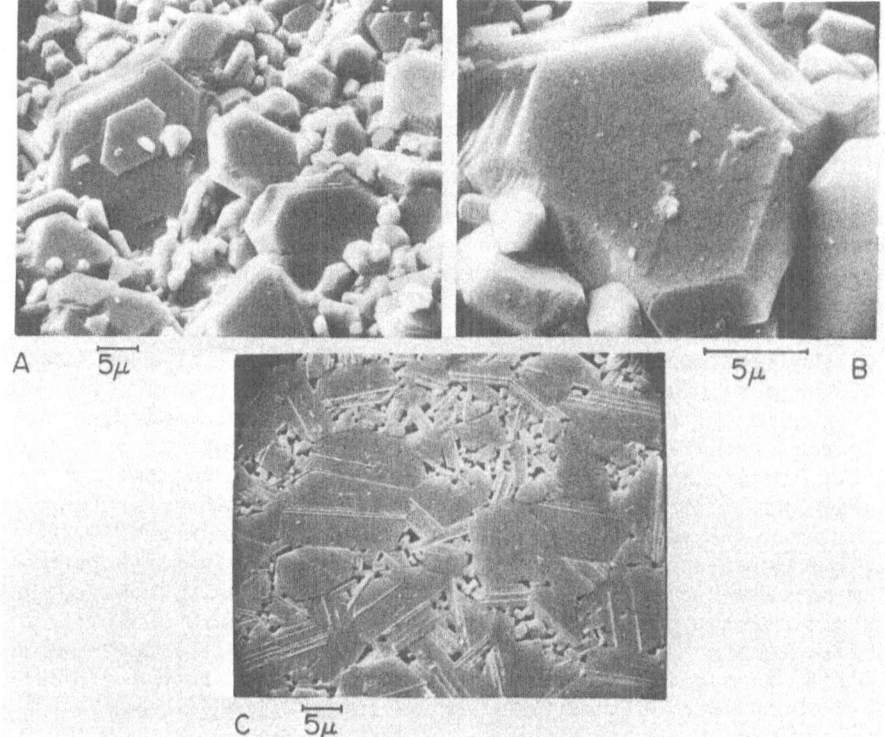

Fig. 2   Fracture Surface of '1/11/' Compact (A & B)
Same Surface After Polish and Etch (C).

2.   Sintering of Pre-Treated Specimens:

Rectangular bar-shaped specimens of partially converted
powders were pre-treated at 1480°C for 2 and 4 hours and those
of zeta process powders for 2 hours. Sintering of these pre-
treated specimens and of control specimens at 1600°C for 8,15,30
and 60 minutes produced a variety of grain-size distributions
as shown in Figs. 3-a and 3-b. The increase in average grain size
as a function of sintering time at 1600°C is plotted in Fig. 4.

3.   Sintering of Seeded Zeta Specimens:

Rectangular bar specimens of zeta porcess powders with
seeds (2 w/o, -400 mesh size) were sintered at 1590°C for 8, 15
and 30 minutes. The resulting microstructures are shown in Fig. 5.

DISCUSSION

The densification of the central portion of '1/11/1' compact at 1500°C while the peripheral regions remained porous is in agreement with the earlier findings[6] that the lowest melting point in the $NaAlO_2-LiAlO_2-Al_2O_3$ system is in the range 1460-1470°C and that is a ternary melt. The liquid phase thus first appeared at the '1/11' interface. A similar situation may exist in the 'SZ/L/SZ' compact at the 'SZ/L' interface. However, in the preferential shrinkage and densification of the 'SZ' layers, the 'L' layer remains porous, implying that the formation of a liquid phase at the 'SZ/L' interface requires a higher temperature (probably in the vicinity of 1585°C, the melting point of $NaAlO_2-\beta$ eutectic). It is thus obvious that if the zeta process powder is used for fabrication β'', the liquid phase formation is postponed until a higher temperature is reached as compared to the liquid phase appearance temperature in the partially converted powders. It must be borne in mind that not only are these liquids transient, but the entire system is in a non-equilibrium state except for those grains which have attained the stoichiometry of stabilized β'' that is thermodynamically an equilibrium phase for a given sintering temperature. Only such grains then represent the nuclei for exaggerated grain growth while the surrounding unstable grains dissolve in (or react with) any available liquid film to precipitate onto the growing crystal face (or attain the equilibrium stoichiometry by exhausting the liquid phase).

The amount of the liquid phase available at the sintering temperature is also governed by the extent of chemical heterogeneity; the larger the heterogeneity, the larger the amount of liquid phase available, provided that the specimen is heated rapidly to the sintering temperature. Thus, the isothermal hold at 1480°C (referred to as 'pre-treatment') must reduce the amount of liquid phase available during the isothermal sintering at 1600°C. As a consequence, fewer nuclei will be capable of exaggerated growth via a flux-growth mechanism and the resulting grains will be of larger dimensions. The results of Fig. 3-a bear this out. The termination of the flux-growth process by grain impingment is fastest in the case of seeded specimens of partially converted powders and requires the longest time for the pre-treated specimens of partially converted powders as seen in Fig. 4. An even more drastic effect of pre-treatment is seen in the zeta process specimens. While the control specimens of zeta process powder exhibit a considerable extent of exaggerated grain growth after one hour at 1600°C the process has barely begun in the pre-treated zeta specimens (see Fig. 3-b). It is noteworthy that the control specimens of the zeta process powders are fine-grained relative to the control specimens of the partially converted powders at all the stages; which makes the zeta process powders more desirable than the partially converted powders in the current production practice.

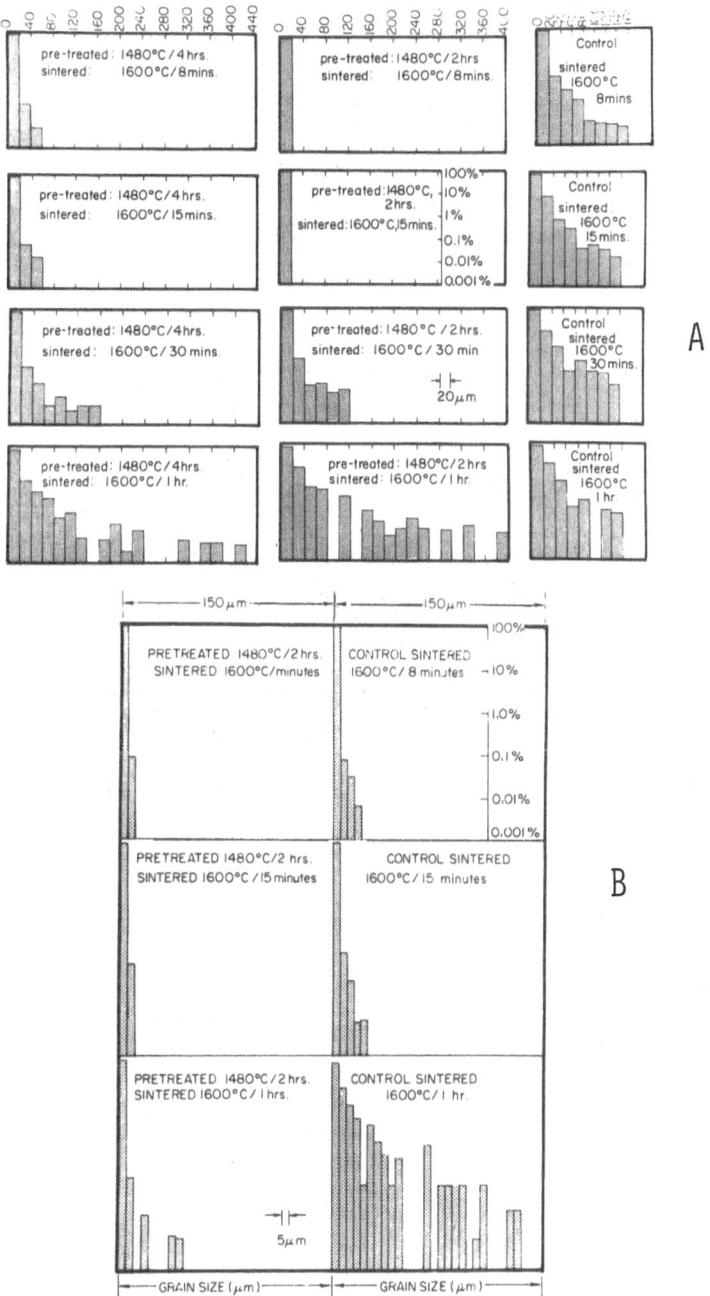

Fig. 3. Grain size distribution histograms for 1600°C-sintered control and pretreated specimens of partially converted powder (A) and zeta process powder (B).

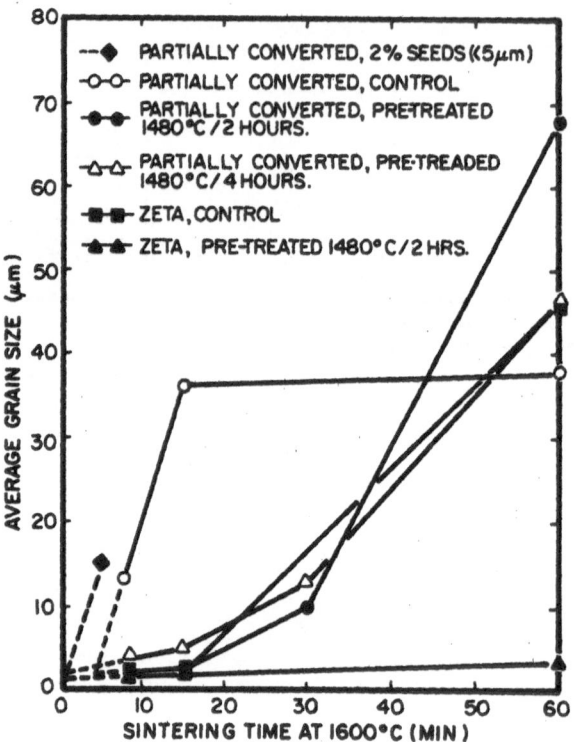

Fig. 4. Comparison of grain growth rates during sintering at 1600°C.

The results in Fig. 5 clearly indicate that the growth of externally added seeds is slowed down when placed in a zeta process powder
matrix wherein the liquid phase is not available readily and/or
in sufficient amount, and that it is even more hindered by a pre-
treatment that further reduces the amount of liquid.

    The results thus support the postulate of liquid phase-
assisted exaggerated grain growth during sintering of Li_2O-stabil-
ized β'' alumina compositions; however, the agreement between the
postulate and the results is only qualitative at present.

    Given the polyphase nature of the β'' precursors used in the
current production practice, the exaggerated grain growth during
the high temperature sintering is unavoidable; which makes the
procedure of short-time-sintering followed by low-temperature
annealing mandatory. While the external additions of fine seeds
(< 5 μm) to partially converted powders have shown the promise

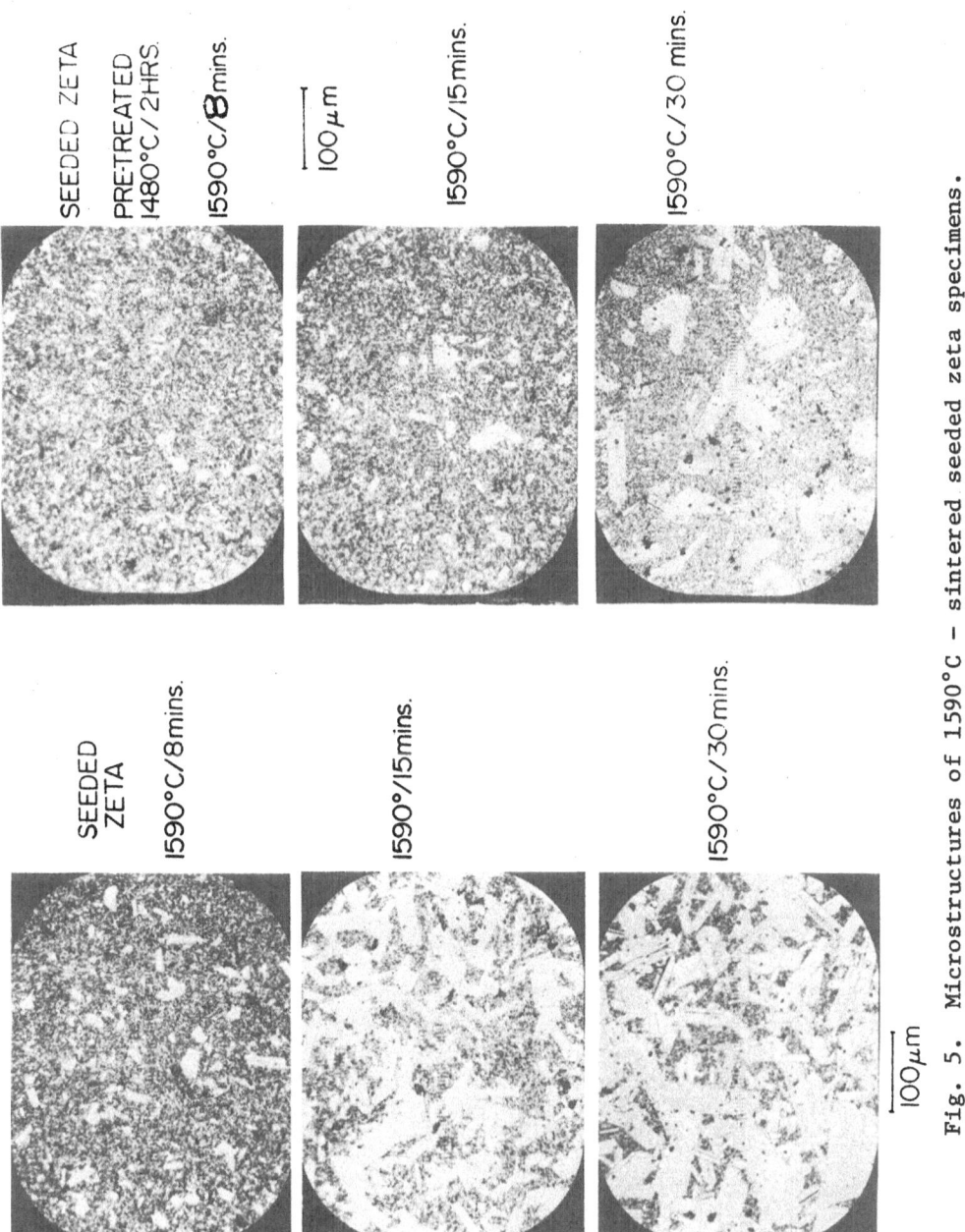

Fig. 5. Microstructures of 1590°C – sintered seeded zeta specimens.

as a viable alternative, its adoption must await the successful in-cell operation of β" electrolyte fabricated by the seeding process. If the sintering operation is to be carried out in a batch-mode, the use of zeta process powders, coupled with a pre-treatment offers a very fruitful way to retain a fine-grained microstructure. The resistivity of the pre-treated zeta specimens was 6 ohm-cm at 300°C after 8 minutes sintering at 1600°C and did not decrease any further upon further sintering;[6] nor did the microstructure coarsen appre-ciably, as seen in Figures 3-b and 4. For load-leveling operations, wherein this resistivity value is not objectionable, the combination of zeta process and pre-treatment merits consideration.

ACKNOWLEDGEMENTS

The support of this work by the National Science Foundation under Contract NSF-C-805 is acknowledged.

REFERENCES

1. Arun D. Jatkar, Ivan B. Cutler and Ronald S. Gordon, pp. 414-22 in Ceramic Microstructures '76. Edited by Richard M. Fulrath and Joseph A. Pask. Westview Press, Boulder, Colorado, 1977.
2. Ivan B. Cutler, pp. 120-27 in Kinetics of High Temperature Processes. Edited by W. D. Kingery. John Wiley and Sons, Inc., New York, 1959.
3. C. Lacour and M. Paulus, Phys. Stat. Solid. (a), 27 [2] 441-56 (1975).
4. C. Kooy, Science of Ceramics, 1, 21-34 (1962).
5. A. L. Stuijts, pp. 331-50 in Materials Science Research, Vol. 6, (Sintering and Related Phenomena). Edited by G. C. Kuczynski, Plenum Press, New York, 1973.
6. Arun D. Jatkar, Ph.D. Dissertation, University of Utah, Autumn 1977.
7. Ervin E. Underwood, pp. 126-39 in Quantitative Stereology. Addison-Wesley Publishing Co., Reading, 1970.
8. Anil V. Virkar and Ronald S. Gordon, pp. 610-20 in Ceramic Microstructures '76. Edited by Richard M. Fulrath and Joseph A. Pask. Westview Press, Boulder, Colorado, 1977.
9. Gerald E. Youngblood, Anvil V. Virkar, W. Roger Cannon and Ronald S. Gordon, Bull. Amer. Ceram. Soc., 56 [2] 206-12 (1977).

DISCUSSION

Frans Carpay (Philips Research Laboratories): Plotting of the distribution of grain sizes, like you have shown, is not very sensitive to detect discontinuous grain growth. Bimodel or duplex microstructures show up much earlier when you plot either the cumulative frequency on probability paper or make a Gaussian distribution plot.

TRANSIENT EUTECTICS IN SINTERING OF SODIUM BETA ALUMINA

Lutgard C. De Jonghe and Edward Goo

Department of Materials Science and Engineering

Cornell University, Ithaca, New York 14853

I. INTRODUCTION

Sodium beta alumina solid electrolytes for use in the sodium/ sulphur battery[1] have been prepared by a wide variety of methods. Two different powders can be used: component oxides or other precursors to give a reactive sintering[2-4], or prereacted powders[5] with or without additives[6]. Fast heat-up rates and zone sintering[2] seem now universally accepted as the preferred sintering schedule. Additional improvements in ionic conductivity and mechanical strength can be obtained by post-annealing treatments[4]. Generally, for prereacted β alumina powders, such as the Alcoa powder XB-2 "superground", heat-up rates are between 50-100°C min^{-1}, and time-at-temperature ($\sim$1750-1800°C) is between 5 to 10 minutes. The high sintering temperatures are not an absolute necessity, as was recently demonstrated by Cannon and Chowdry[6] who obtained 97% dense β-alumina from the submicron agglomerate fraction of preconverted XB-2 powder sintered at 1530°C. Sintering times then get prohibitively long for commercial application, however. The high rates of densification during zone sintering led to the postulate that a transient liquid phase might be active during the densification of sodium beta alumina[7]. Additionally, preconverted beta alumina (Alcoa XB-2 "superground") mixed with a small amount of metastable NaAlO$_2$-β alumina eutectic showed improved sinterability[8,9] lending further support to the hypothesis that transient liquid phases are present at one point during the zone sintering process. In this paper we explore further if indeed a transient eutectic liquid phase is responsible for the improved sinterability of preconverted sodium beta alumina powders when mixed with a small amount of metastable eutectic additive.

II. EXPERIMENTAL

The principle of the transient eutectic additive sintering is illustrated in Fig. 1. Eutectic material is added to the ceramic such that its composition after sintering remains within the useful phase field. The range of the single phase field for β-alumina[10] permits the addition of 5-10 wt.% of eutectic material.

Two types of experiments were performed. In the first ones, the arrested zone sinterings, long bars of sodium beta alumina with and without eutectic additive were fed into a hot furnace at different rates. The sintering was then arrested by quickly withdrawing the samples from the furnace. This procedure gives a sample along which length one can follow the microstructural evolution and local shrinkage as a function of position in the sintering schedule. In the second type of experiments, samples were fed through the hot zone of the furnace kept at 1640°C. The densities and 300°C D.C. ionic resistivities of the samples were compared as a function of post-annealing at 1375°C. The samples were in all cases packed in β-alumina powders of the same soda contents.

2.1 Arrested Zone Sinterings

The microstructural evolution and local shrinkage were examined for long, bar shaped samples partially fed into the hot zone of the furnace kept at 1700°C. Two different types of samples, prepared from preconverted Alcoa XB-2 "superground" sodium beta alumina, with 5 wt.% eutectic additive and without additive, were used. The feed rate was 0.63 cm.min^{-1}. At this feed rate thermal impedances were negligible and a direct comparison of local shrinkage, $\Delta L/L_O$, and microstructure was possible. The sintering history of each element of the specimen can then be found from the temperature (T)-time (t)

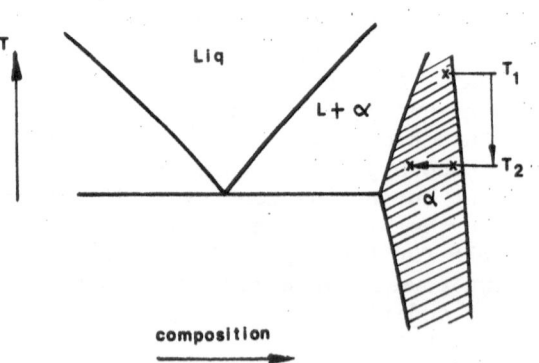

Figure 1. Principle of the eutectic additive method.

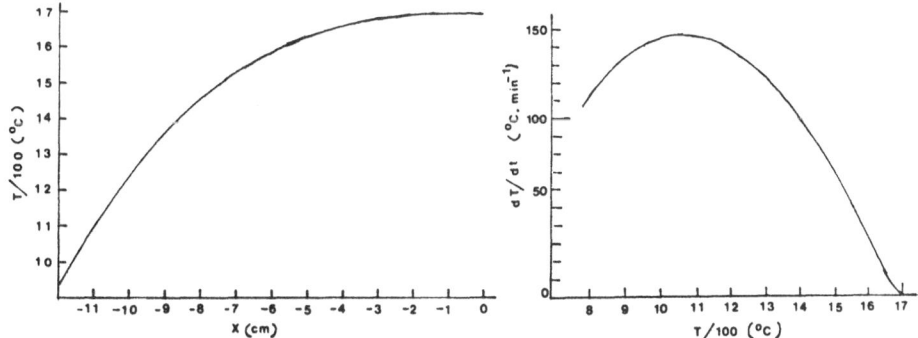

Figure 2.   (a) T-x diagram; (b) dT/dt-T diagram. From these data
            the sintering history of each sample element in the
            arrested zone sintering can be found.   x is the distance
            from the sample front.

diagram recorded at the sample front. The T-t history and the heating rates, dT/dt-T, are shown in Fig. 2a and 2b. Maximum heating
rates of about 150°C min^{-1} occur between 1000 and 1200°C. Near
1600°C the heating rates are about 30°C min^{-1}. The measured local
shrinkage as a function of place in the actual sintering schedule
is shown in Fig. 3 for the eutectic additive sample A, and for the
additive free sample B. The green densities of the samples were
60% of the theoretical density (3.26 g.cm^{-3}). A significant, rapid
increase in density is noted for the additive sample, A, between
1550°C and 1600°C. Such an effect is absent for the additive free
sample, B. We attribute this anomalously high sintering activity
in the vicinity of the NaAlO$_2$-β alumina eutectic temperature (1580°C)
to the presence of the transient liquid eutectic. The microstructures in the region of interest are compared in Fig. 4. The micrographs labeled Aa-Ac are fractographs of the eutectic additive sample A, where a-c correspond to the positions indicated in Fig. 3,
while those labeled Ba-Bc are from the additive free sample. The
difference in micromorphology between A and B is rather pronounced.
Especially the sudden increase in grain size from Ac to Ab is
striking. Note, however, that the rapid densification and grain
growth does not seem to entail the development of pronounced grain
facets. We do not see this as incompatible with the activity of
minor amounts of a transient liquid phase. If the grains are not
surrounded completely by a liquid in which they are partly soluble,
there is no particular reason for developing such facets. In sodium
beta alumina, faceting results from high temperature grain growth in
which size increase in the direction normal to the basal planes

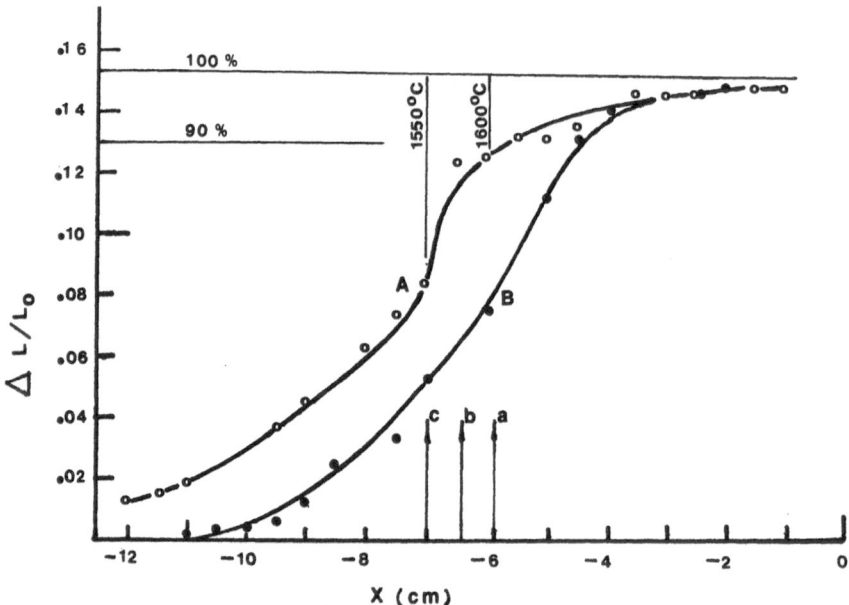

Figure 3.   Local shrinkage $\Delta L/L_O$ as a function of position.   A:
            eutectic additive sample; B: additive free sample.   The
            green densities are 60% of the theoretical density.

proceeds by propagation of single or multiple spinel block ledges.
Evidence for such a mechanism may be seen in transmission electro-
micrographs of samples prepared at 1750°C, Fig. 5.   Samples sintered at lower temperatures invariably show more wavy boundaries
and grain aspect ratios closer to 1.   The grain morphology thus is
mainly dependent on kinetic factors rather than grain boundary
energy anisotropy.   Classical behavior in which the nucleation
rate/growth rate ratio increases with decreasing temperatures may
thus simply account for the observed grain morphologies as a function of temperature.

     Fracture surfaces of samples of Ab and Bb (Fig. 4) are com-
pared at a higher magnification in Fig. 6.   Curiously, the additive
free sample, Bb, contains needle-like features.   Such needles are
not found in the additive samples.   It is possible that such needles might be produced as a result of gaseous Na transport down
the temperature gradient during rapid zone sintering.   If very steep
temperature gradients exist, it is not inconceivable that gas phase
transport of Na in porous samples might lead to transient phases

Figure 4. Aa-Ac: fractographs of sample corresponding to positions
           a-c marked in Figure 3. Note the rapid microstructural
           evolution from Ac to Ab. Ba-Bc: fractographs of addi-
           tive free sample.

with Na contents higher than the $NaAlO_2$. Such phases might be
partly responsible for densification below 1500°C at high feed
rates in zone sintering. These effects would tend to obscure the
action of transient $NaAlO_2$-β alumina eutectics at high feed rates
in zone sintering, as was observed for samples fed into the furnace
at 2.5 cm min^{-1}. Thermal impedances were very significant at these
feed rates and microstructure/T-t history could not be reliably correlated.

When sample densities exceed 90% theoretical, a state of closed
porosity is reached, as can be noted from the lack of penetration
of liquid into samples during Archimedes density measurements. The
arrested zone sinterings thus suggest that the additive samples
might be sintered at temperatures as low as 1600 to 1650°C. They
also show that the advantages of eutectic additive sintering are
negligible for sinterings above 1700°C.

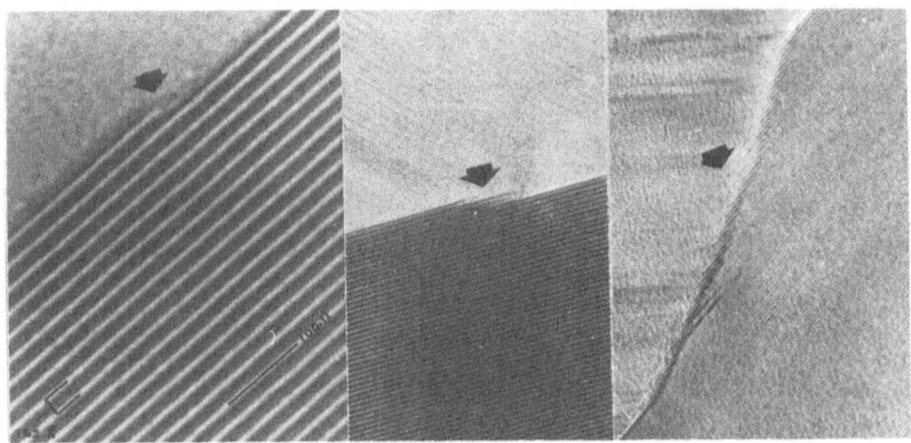

Figure 5.   Single and multiple spinel block ledges observed in
            electrolyte prepared at 1700°C.   The ledges are parallel
            to the basal plane (00.1).   The fringe spacings in all
            micrographs is 11.3Å (transmission electron microscopy).

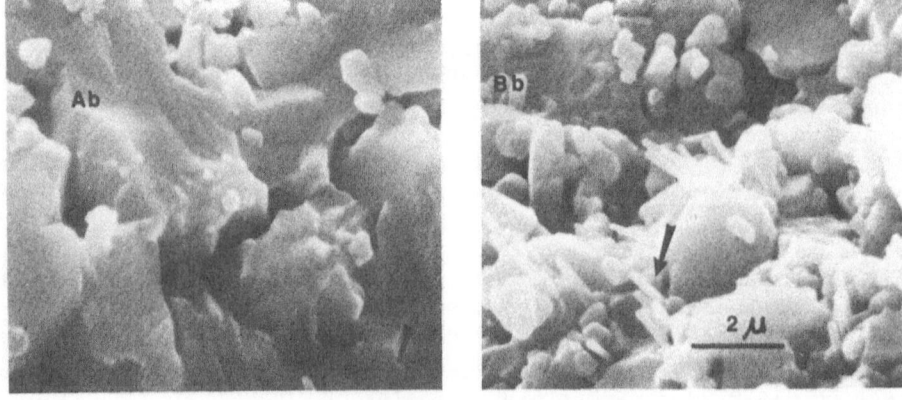

Figure 6.   Comparison of the microstructure of Ab and Bb (Figure
            9). Note the needle-like features in Bb marked with
            an arrow.

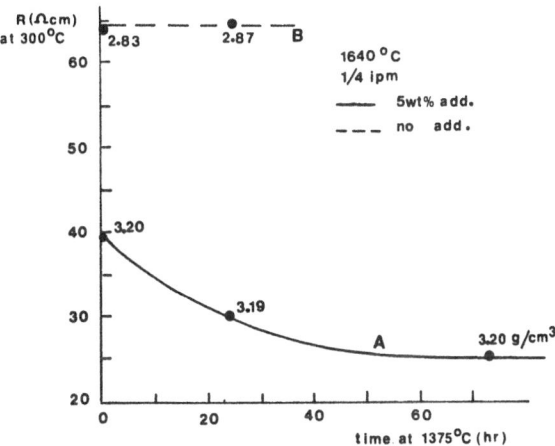

Figure 7.   Densities and 300°C D.C. ionic resistivities of samples
            sintered at 1640°C. A: eutectic additive sample; B:
            additive free sample.

2.2   Complete Zone Sintering at 1640°C

As suggested by the arrested zone sintering, complete zone
sinterings were carried out with the furnace at 1640°C and with a
feed rate of 0.63 cm min^{-1}, for samples with 5 wt.% eutectic additive and for additive free samples. The densities and 300°C D.C.
Na^+ ionic resistivities were determined as a function of post annealing at 1375°C. The results are shown in Fig. 7. Curve A is
for the additive sample, while B is for the additive free sample.
Clearly, when the starting density of the solid electrolyte is too
low, post annealing at 1375°C, while slightly improving the density, does not have marked effects on the ionic resistivity. The
additive sample, however, has an initial density of about 98%
theoretical and does improve the resistivity to an acceptable 300°C
value for pure β alumina. In this case the density was unaffected
by the post-annealing treatment. At the same time, little over-all
grain growth was observed. The main microstructural changes were
found in pockets of micron or submicron grains where significant
local grain growth seemed to have occurred. It is thus possible to
improve the sinterability of commercial, prereacted sodium beta
alumina powders between 1500–1600°C by the addition of a minor
amount of metastable $NaAlO_2$–β alumina eutectic.

III. SUMMARY

     The presence of an active transient eutectic during zone sin-
tering of sodium beta alumina to which 5 wt.% of the $NaAlO_2$-β
alumina eutectic had been added, has been demonstrated when heating
rates are high.

     Rapid zone sintering of commercial, preconverted β-alumina to
which extra soda was added in the form of the metastable $NaAlO_2$-β
alumina eutectic gives high density electrolyte. A post-annealing
treatment is, however, necessary to bring the 300°C ionic resistivity to an acceptable level.

ACKNOWLEDGEMENTS

     This work was supported by the Electric Power Research
Institute.

REFERENCES

1. J. T. Kummer and N. Weber, Trans. Soc. Automotive Engrs. 76,
     1003-1007 (1968).
2. I. Wynn Jones and L. J. Miles, Proc. Brit. Ceram. Soc. 19,
     161-178 (1971).
3. A. K. Ray and E. C. Subbarao, Mat. Res. Bull. 10, 583-590 (1975).
4. T. J. Whalen, G. J. Tennenhouse, and C. Meyer, J. Amer. Ceram.
     Soc. 57, 497-498 (1974).
5. "Sodium-Sulphur Battery Development for Bulk Power Storage",
     Interium Report, R. P. 128-2, Electric Power Research
     Institute, September 1975. Prepared by J. B. Bush, Jr.,
     General Electric Company Research and Development Center.
6. R. M. Cannon and U. Chowdry, this Volume.
7. "Research on Electrodes and Electrolyte for the Ford Sodium
     Sulphur Battery", RANN, NSF-C805 (AER-73-07199), January
     1975, Prepared by S. Weiner, Ford Motor Company.
8. L. C. De Jonghe and H. Chandan, Ceram. Bull. 55, 312-313, (1976).
9. L. C. De Jonghe and H. Chandan, U. S. Patent 3,959,022, May 25,
     1976.
10. Y. LeCars, J. Théry and R. Collongues, Rev. Int. Hautes Tempér.
     et. Réfract. 9, 153-160 (1972).

DISCUSSION

Jim Reed (Alfred University): Do you control the partial pressure
of alkalis in the sintering atmosphere? Is this of significance
in sintering β and β" alumina?

Author: Soda loss during sintering can be an important factor in
the properties of the electrolyte. Usually it is minimized by
sintering in dry oxygen,[5] by encapsulating in platinum,[7] or by
packing in coarse powders of the same composition. We use this
last method and find no loss of soda during sintering. If such
precautions are not taken, a high resistivity surface layer is developed that contains α-alumina.

W. J. Huppmann (Max Planck Institute for Metals Research): Can
you give any information on how long the liquid phase is present
during sintering and how fast homogenization occurs?

Author: The shrinkage data in Figure 3 suggests that the liquid
phase is active for approximately ½ to 1 minute at these heating
rates.

MICROSTRUCTURAL EVOLUTION DURING THE PROCESSING OF SODIUM

β-ALUMINA

Uma Chowdhry* and R.M. Cannon

Department of Materials Science and Engineering
Massachusetts Institute of Technology
Cambridge, Massachusetts 02139

ABSTRACT

Critical examination of the microstructural evolution at each stage in the processing of sodium β-alumina has revealed a set of controllable parameters and broad guidelines that are necessary to the understanding of the solid state sintering behavior of the material. The importance of the mode of powder preparation and the resultant powder characteristics, including the green compact microstructure, in controlling the sintering behavior are discussed. If processing variables are carefully controlled, β-aluminas can be sintered to high densities by a solid state mechanism. Encouraging results using additives for prevention of discontinuous grain growth are also described.

INTRODUCTION

The processing of sodium β-alumina, a fast ion conducting, solid state electrolyte to be used in the high energy density sodium-sulfur battery is an area of considerable interest. The potential problems associated with attempting to sinter multi-component materials having hexagonal symmetry, high vapor pressures and anisotropic properties are well known and, in fact, β-aluminas are generally regarded as difficult to densify without the aid of a transient liquid phase. To achieve the desirable properties of high conductivity, mechanical strength, and chemical durability, the optimal microstructural requirements

---

* Presently at CRD, E.I. duPont de Nemours and Co, Wilmington, Del.

would be: low porosity, preferably present as fine, uniformly
distributed pores within grains; uniform, 5-10μ grain sizes
without the presence of large grains which tend to act as current concentrators and lower mechanical strength; avoidance of
grain boundary phases which lower conductivity and are susceptible
to aqueous or other chemical corrosion; avoidance of discontinuous grain growth by incorporation of additives that go into solution rather than remain as second phases which decrease cell
lifetime; and absence of flaws introduced in the early stages of
processing which tend to grow during densification.

     With these goals in view, several methods of preparing homo-
geneous powders with and without additives have been investigated
and various techniques for improving their sinterability have
been examined. As the objective was to study the solid state sintering and grain growth, compositions   were selected to yield
the β phase at sintering temperatures. Although the β" phase
has a higher conductivity at high temperatures, it only exists
in equilibrium as a ternary compound and so is less amenable to
controlled study. Preparation of powders was followed by calcination to the β and/or β" phases prior to sintering.

                      EXPERIMENTAL PROCEDURE

     In the synthesis of multicomponent ceramics, mechanical mix-
ing of the component oxides followed by calcination usually results in a mixture of several non-equilibrium phases or else undesirably coarse powders. In this study, undoped and doped reactive β alumina powders were synthesized by a variety of chemical
techniques aimed at obtaining intimate mixing of the constituents.
Their sinterability was compared to that of the only commercially
available powder[†]. Most compositions were selected to be in the
β single phase field [1] with an Al:Na ratio of 8.5 or 9, chosen
to avoid the presence of Na rich liquid phases during sintering.
A few doped powders were made with higher soda contents, Al:Na
being 6. The techniques investigated were 1) decomposition of
alums using Al alum with Na oxalate or with Na and Ni acetates;
2) freeze-drying of sulfates or nitrates of Al and Na and with
addition of Ni or Co sulfates; 3) decomposition of gels [2] prepared by polymerization of Al and Na citrates in a solution of
ethylene glycol and water and with additions of Fe or Co citrates.
Transition metal oxide additives up to 1 wt % were incorporated
directly using the above techniques or were introduced by vapor
transport resulting in lower concentrations of approximately
0.1 wt %. The stages in the formation of β-alumina from these
powders were followed by x-ray diffraction subsequent to calcination at different temperatures.

[†]Alcoa XB-2 Superground, Al:Na ratio 7.5, single phase β-alumina.

Calcination to the β phase was followed by light grinding of the powders with a mortar and pestle using ethylene glycol or amyl alcohol to wet the powders. Compaction by die pressing was done at 20,000 psi. More extensive grinding was accomplished by vibro-milling using high purity Al_2O_3 balls* in polyethylene jars with the powder suspended in dichloromethane with trichloroacetic acid added as a deflocculant[3]. The suspensions were then centrifuged at successively higher speeds to separate various size fractions. Using speeds of 3600 rpm, particle sizes < 0.3 μm were obtained. The as-deposited powder compacts were dried and sintered directly without further compaction. The sintering studies were done in an Al_2O_3 tube furnace with the samples packed in coarse β-alumina powder in an Al_2O_3 crucible to minimize soda loss. Doping by vapor transport was accomplished by packing a pellet of doped powder near the subject undoped compact.

EFFECTS OF POWDER CHARACTERISTICS AND ORIGIN:

The commercial XB-2 β-alumina powder in the superground version possesses angular particles many of which are submicron sized crystallites agglomerated into 2-5 μm sized clusters. In addition, there is an appreciable fraction of particles in the 2-5 μm size range, Fig. 1a. This powder, if cold pressed without binder or additives, sinters to a low density compact (ρ < 90%) at relatively high temperatures, Fig. 1b.

(a)   (b)

Fig. 1.   XB-2 Superground Powder: (a) as-received, (b) fractograph of sample cold pressed and sintered at 1700°C/1hr.

* Coors AD-995.

The calcination of alums at 900°C-1000°C resulted in a very friable powder, the predominant phase being α-Al_2O_3 with weak traces of $NaAlO_2$ and β''. Complete conversion to β'' and some β occurred only at higher temperatures. In the doped powders, color variations indicated that cation segregation had developed. As the powders were well mixed before calcining and the alum melts before decomposing, it is presumed that the segregation develops during the decomposition of molten salt into several solid sulfate or oxide phases. SEM micrographs of both the doped and undoped powders revealed interconnected chains of needles and plates about 1 to 2 μm in major dimension, in clusters of 5 to 6 μm. The particle shape was not conducive to good densification and final densities obtained after sintering were low.

The freeze dried sulfates reacted to form primarily β'' with residual sulfates and traces of α-Al_2O_3 and aluminates at 900°C to 1000°C. They were fully converted to β'' and β mixtures at 1100°C to 1200°C. The powders subsequently converted to β during sintering at high temperatures. The calcined undoped and doped sulfates initially gave very fine, ~1000Å, particles which were agglomerated in a structure which was a relic of the freeze-dried morphology, Fig. 2a. During sintering, they coarsened into platey particles and densification ceased at densities below 90%, Fig. 2b. Nitrates were less satisfactory as the powder is deliquescent and freeze drying without melting the hydrated salts is difficult.

Using the gel technique, an amorphous phase was reproducibly obtained after calcination below 800°C. This is indicative of cation homogeneity on an atomic scale. Subsequent calcination of the pure and Co-doped gels resulted in a λ phase[4] at 900°C

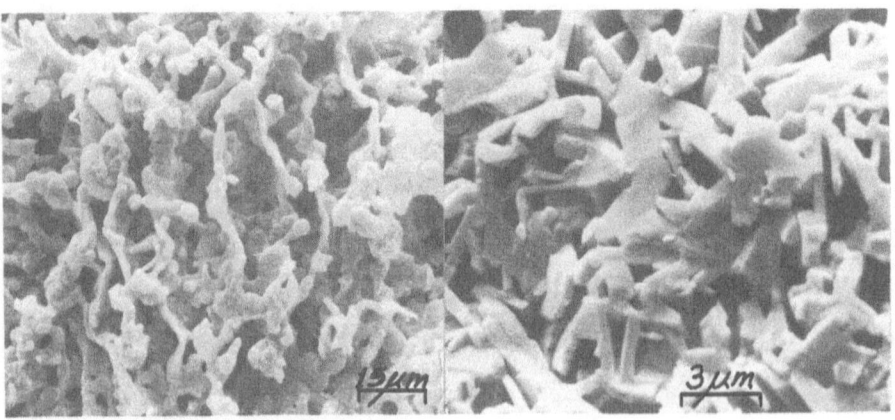

(a)   (b)

Fig. 2. Freeze-dried powder of Al, Na, and Ni sulfates, (Al:Na = 6.1, 1% NiO): (a) as freeze-dried from 20% solution, (b) fractograph of sample cold pressed and sintered at 1700°C/2 hr.

(a)   (b)

Fig. 3.   Gel derived powder (Al:Na = 9): (a) as calcined at
          1275°C/2 hr., β" phase, (b) fractograph of sample
          sintered at 1680°C/30 min.

(a)   (b)

Fig. 4.   Gel derived powder with 0.2% Fe_2O_3, (Al:Na = 5.6):
          (a) as-calcined at 1275°C/4 hr., β" phase, (b)
          fractograph of sample sintered at 1740°C/30 min. using
          powder ground and classified to < 0.5 μm.

which converted to β" above 1200°C. The Fe-doped gel gave β"
and traces of β at 800°C. The platey particle morphology after
conversion to β" is shown in Fig. 3a. The resultant densification
was poor with final densities of about 85%, even when using ground
and classified submicron powder, Fig. 3b. The resulting β" powder
from the Fe doped gel appeared to have a higher aspect ratio,
Fig. 4a, than that from the undoped gel, Fig. 3a. The final sintered densities obtained from this ground and classified powder
were also low for the plates grew to impingement and densification
ceased, Fig. 4b. Locally where the plates were nearly aligned
good densification resulted. At the stage where angular pores
remain between elongated grains, further firing will coarsen the
structure with little change in morphology, but will not produce
further densification.

In summary, most of the chemical techniques resulted in
platey particles which were agglomerated to some degree. The
chemical decomposition techniques all tended to result in the
formation of β" with small traces of β after calcination. After
sintering above 1500°C, the samples had converted to the β phase.
The poor particle shape, wide particle size distribution and the
presence of agglomerates are the root causes of the low sinterability observed in the various powders examined.

SOLUTIONS

One approach used to overcome the problems caused by
platey particles was to inhibit the growth of the high aspect
ratio plates seen in Fig. 3a. The precursor λ phase in the
calcined gels was composed of extremely fine, equiaxed particles
1000 Å in size, Fig. 5a. The λ powder was ground, centrifugally
classified and deposited, and in situ transformation to the β/β"
phase during sintering was allowed to occur. Relatively high
densities (~97%) were obtained before discontinuous grain growth
terminated densification, Fig. 5c. Examination of the partially
sintered compact, Fig. 5b, reveals that although the phase transformation results in platey particles, their aspect ratio is
much lower than that for the gel derived powder calcined to the
β" phase, Fig. 3a. The particle packing evidently restricts the
lateral growth of the plates and this allows sintering to higher
densities.

An example of a centrifugally cast submicron compact of β
alumina powder is shown in Fig. 6a. Although it has the desirable
features of a monodispersed, equiaxed powder, there is evidence
for some agglomerates which may preferentially densify and act as
seeds for discontinuous grain growth. This compact, only 35%
dense in the green state, sinters to 97% at 1530°C, Fig. 6b. At
higher temperatures, densities in excess of 99% can be obtained,
Fig. 6c, without discontinuous grain growth if the time-temperature schedule is carefully controlled. Perhaps the fact that

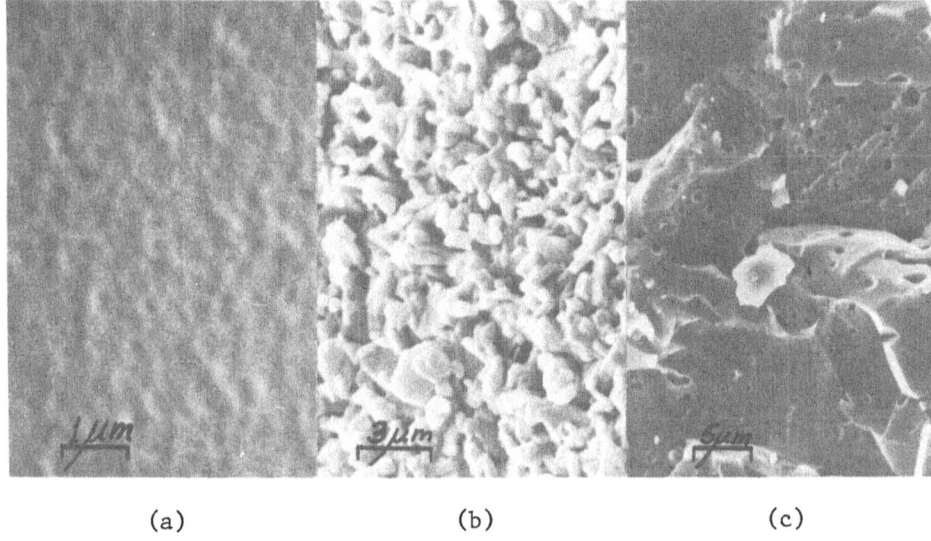

(a)   (b)   (c)

Fig. 5.   Samples from submicron, gel derived, λ phase powder:
          (Al:Na = 9): (a) λ powder obtained after calcining at
          900°C/2 hr.; (b) as-fired surface of sample sintered
          at 1660°C/10 min., and (c) fractograph of sample sin-
          tered at 1250°C/4 hr. plus 1680°C/2 hr.

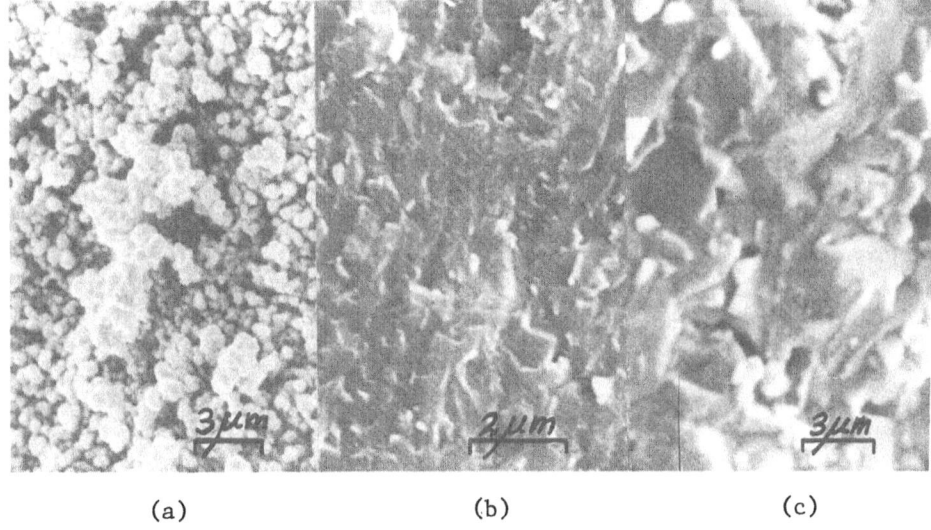

(a)   (b)   (c)

Fig. 6.   Samples from submicron (< 0.3 μm), centrifugally cast
          XB-2 powder (Al:Na = 7.5): (a) as-cast compact, fracto-
          graphs of samples sintered at (b) 1530°C/2 hr. and
          (c) 1680°C/30 min.

this powder resulted in higher densities than are achieved for the
gels may be attributed to its lower purity, higher soda content,
or more equiaxed morphology. The process is, however, undesirably
sensitive to the time-temperature schedule, and as a result, reproducibility of the final grain size was difficult.

As samples made from both the undoped gel and submicron Alcoa
powders were susceptible to discontinuous grain growth, Ni, Co or
Fe doping by vapor transport was investigated as a means of inhibiting grain growth. Figs. 7a, b show the effect of using 0.1 wt
% Ni as a grain growth inhibitor in a vapor doped, gel derived
compact of λ powder. In contrast to results with undoped samples,
densities in excess of 99% were obtained, but discontinuous grain
growth did occur and the optimum solute concentration remains to
be determined.

Transition metal oxide dopants were also added to submicron
Alcoa powder compacts, Figs. 8a, b. These additives all allowed
higher densities to be achieved before the onset of secondary
grain growth. In addition, Co lowered the densification rate
appreciably. Fe additions were more effective than Ni or Co in
preventing discontinuous grain growth. For both the doped and
undoped materials where discontinuous grain growth was avoided,
the final grain sizes were in the 1-3 μm range. Thus, high densities and fine grain sizes without pore entrapment are possible.
Provided these additives do not excessively raise the electronic
conductivity, they would be useful in allowing a relaxation of
the stringent time-temperature sintering schedules required for
the pure samples.

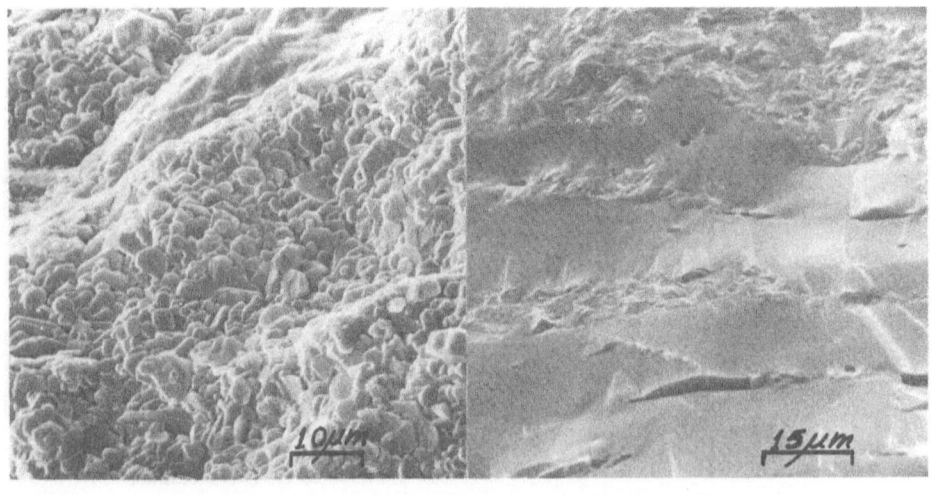

(a)   (b)

Fig. 7. Fractographs of classified, gel derived powders
         (Al:Na = 9) vapor doped with NiO during sintering
         at 1705°C/1 hr.

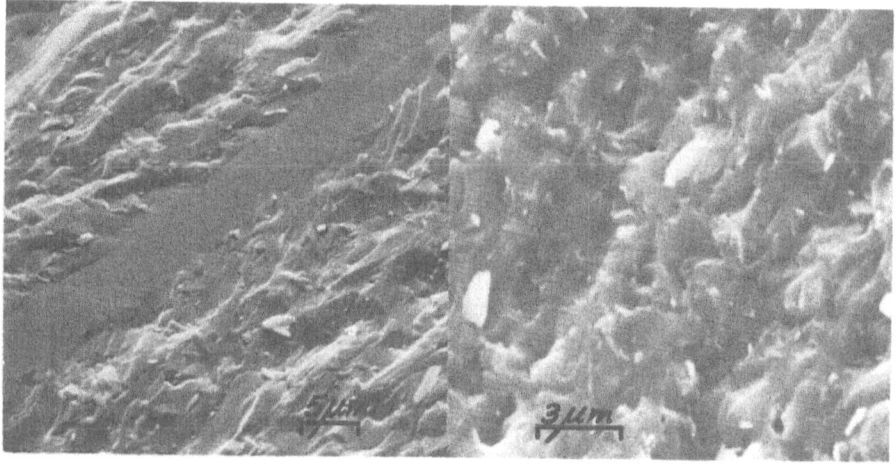

(a)   (b)

Fig. 8. Classified (< 0.3 μm) XB-2 powders vapor doped during
         sintering: (a) fractograph of NiO doped sample sintered
         at 1680°C/2 hr. (b) fractograph of Fe_2O_3 doped sample
         sintered at 1740°C/30 min.

DISCUSSION

The powders produced by chemical decomposition frequently
resulted in the formation of one or more precursor phases such
as λ, a mullite-like structure with a wide range of homogeneity[4],
or α-Al_2O_3 and traces of Na aluminates. Higher calcining temperatures always resulted in mixtures of β" and β; these could be
intergrowths of the two phases mixed on a very fine scale[5]. The
kinetics of decomposition for these various processes are not known
and the intermediate phases which result often depend on details
of the low temperature precursors and preparation methods. Decomposition of alums, for instance, has been reported to result
in an amorphous phase which converts directly to the β phase at
high temperatures[6], which is in disagreement with our studies.
Similarly, formation of the λ phase has been thought to depend
upon the degree of homogeneity of the cations in the precursor
salts[4] and perhaps the precursor itself. It has been reported
to result from decomposition of nitrates[4] and from coprecipitated
oxalates[2] with the observation that high soda concentrations and
ternary cation additions appear to suppress its formation[2]. Using
gels of aluminum triisopropylate and various sodium salts,Morgan
found that decomposition frequently gave δ or η forms of Al_2O_3
plus aluminates as intermediates rather than λ.[5] The results
were very sensitive to cation as well as anion additives, such

as F, and the decomposition time and temperature.

Most decomposition methods yield powders with poor sinterability which is apparently related to the powder morphology and
the agglomerate structure. Although the nonequiaxed morphologies
are related in detail to the decomposition path and agglomerate
structure of the precursor, their frequent appearance suggests
the growth kinetics are strongly anisotropic for the hexagonal
or rhombohedral β or β" structures. In this regard no obviously
beneficial effects of Al:Na ratio or of additives were observed
in this investigation. Nevertheless, better understanding of the
decomposition processes and kinetics may allow a choice of appropriate precursor phases, decomposition procedures, and additives to improve the sinterability of the powder.

Except for the obvious effect on particle morphology, no
other effects of the concurrent conversions from λ to β" and β"
to β on the sintering behavior were evident. A few constant
heating rate dilatometry experiments[7] indicated that the results
were similar for β", λ and XB-2 powders, but that the β" powders
sintered at a slower rate - this is thought to be a particle shape
effect. No unusual temperature effects indicative of liquid phases
or other unexpected transformation rate effects were observed in
the dilatometry data.

With poor powders densification ceased at densities below
90% although coarsening continued with little change in morophology.
The better powders densified until discontinuous grain growth occurred, Fig. 5c and 9a. Theory[8] indicates that the susceptibility
to grain growth with pore entrapment exists when the volume fraction
porosity shrinks below a critical level. This critical porosity
level is very sensitive to grain size distribution. It occurs at

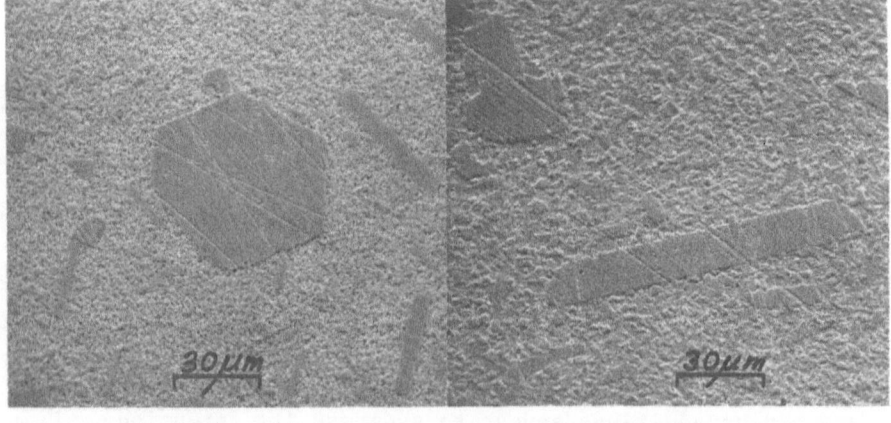

(a)   (b)

Fig. 9. Abnormal grain growth in samples of classified (<0.3µm)
         XB-2 powder sintered at 1680°C/2 hr.: (a) undoped,
         (b) NiO vapor doped. Etching with hot H_3PO_3 tends to
         exaggerate the apparent porosity.

1% for a narrow size distribution, but increases to 9% for a
wide size distribution if pore drag alone controls the boundary
mobility as is apparently the case for the pure gel, Fig. 5c.
Thus, for pure material sintering to full density cannot be
achieved without grain growth (usually abnormal) causing pore
entrapment within the grains. Theoretical[8], as well as experimental work on sintering has indicated that additives (or impurities) can have beneficial results if used judiciously. They may
enhance diffusivities, inhibit grain growth by solute drag or
by second phase pinning, inhibit particle coarsening and perhaps
enhance pore mobility which in turn lessens the tendency for pore
entrapment by discontinuous grain growth. The higher densities
achieved using Fe_2O_3 or NiO suggest these additives either reduce boundary mobility by solute drag or perhaps increase the
pore mobility.

The type and amount of additive chosen for use as a grain
growth inhibitor appears to be important. Beneficial effects of
MgO, ZrO_2, Y_2O_3, CaO and SrO as sintering aids for β alumina
have been observed [9,10,11] and although not understood, are probably due to liquid phase effects in some instances. All of these
can result in second phases. Conversely, MgO, Cr_2O_3, V_2O_5, CoO,
MnO_2, Fe_2O_3 and NiO have been reported to inhibit [12] or to not
affect sintering[10,12,13]. In contrast, our results suggest that
in small concentrations, Fe_2O_3 and to a lesser extent NiO are
effective grain growth inhibitors. These are only effective if
the powders are sufficiently good to allow sintering to densities
over 90%. These solutes cannot compensate for poor powder morp-
phology and may, in fact, exaggerate the tendency for formation
of such undesirable morphologies during chemical preparation.
These are apparently soluble in the β structure although at large
concentrations may cause stabilization of the β" system.

It is also reported that high soda concentrations improve
sinterability[9]. As with other additives, this may be due to compensation by defects in the spinel block which result in enhanced
diffusivities or may simply be due to liquid phase effects[14].
Owing to the uncertainties in the $NaAlO_2$-Al_2O_3 phase diagram [7],
it is extremely difficult to be certain that no liquid is present
during the sintering of β alumina even though low soda contents
were used to attempt to prevent liquid formation. The lack of
sharp discontinuities in the dilatometry data and the fact that
nearly complete densification could be achieved below the binary
eutectic (~1585°C) lends support for a solid state sintering mechanism. The existence of sharply faceted grains as in Fig. 9a is
often taken as evidence for the presence of a liquid phase. Recent work[15] has shown, however, that this faceting can also occur
if boundaries migrate at intrinsic velocities. It was observed
that in the 0.1 wt % Ni doped sample, Fig. 9b, the boundaries
around the periphery of the abnormal grains were irregular or cusp-
shaped rather than faceted. This indicates that a solute drag

mechanism may be slowing the boundaries of discontinuously growing grains, and in this case the faceting may not be a result of liquid phase effects.

SUMMARY

β-aluminas can be sintered to high densities at relatively low temperatures by a solid state mechanism if sufficient control is exercised over the particle shape, size and size distribution. Most of the chemical preparation techniques yield powders which are platey and/or agglomerated such that partially densified structures with elongated grains and angular pores evolve; these are resistant to further densification, although they do continue to coarsen. Problems associated with platey powders can be minimized if transformation from the precursor to the β or β" phase is allowed to occur in situ during sintering. The onset of discontinuous grain growth can be delayed by the use of grain growth inhibitors such as NiO or Fe_2O_3, the latter having been found to be the most effective.

ACKNOWLEDGEMENTS

The financial support from General Electric Co., RDC, Schenectady, N.Y. as well as helpful discussions with S. P. Mitoff and R. W. Powers are gratefully acknowledged.

REFERENCES

1. R.C. Devries and W.L. Roth, J. Am. Ceram. Soc. 52, 364 (1969).
2. S.A. Weiner et al, "Research on Electrodes and Electrolytes for the Ford Sodium-Sulfur Battery". Annual Report Contract No. NSF-C805 (AER-73-07199) (Jul. 1975).
3. J.H. Kennedy and A. Foissey, J. Electrochem. Soc. 122, 482 (1975).
4. A.G. Elliott and R.A. Huggins, J. Am. Ceram. Soc. 58, 497 (1975).
5. P.E.D. Morgan, Mat. Res. Bull. 11, 233 (1976).
6. S.P. Mitoff, "Synthesis, Fabrication and Characterization of Solid Electrolytes in Fast Ion Transport in Solids", ed. W. Van Gool, North Holland, (1973).
7. R.M. Cannon and U. Chowdhry, "Sintering of Prereacted β and β" Aluminas", Final Report to G.E. Co., July 1977.
8. M.F. Yan, R.M. Cannon, U. Chowdhry, H.K. Bowen, "Effects of Impurities and Pores on Grain Boundary Mobility", presented at the Annual Meeting of the American Ceramic Society, April 1977, to be published.

9. D. Chatterji, et al, "Development of Sodium-Sulfur Batteries
        for Utility Application", EPRI EM-266, Project 128-3
        Annual Report, (Dec. 1976).

10. J. Fally et al, J. Electrochem. Soc. 120, 1292, 1296 (1973).

11. I. Wynn Jones and L.J. Miles, Proc. Brit. Ceram. Soc., No. 19,
        161 (1971).

12. J.H. Kennedy and J.R. Akridge, J. Am. Ceram. Soc. 59, 279
        (1976).

13. J.H. Kennedy and A.F. Sammells, J. Electrochem. Soc. 119,
        1689 (1972).

14. L.C. DeJonghe and H. Chandan, Am. Ceram. Soc. Bull. 55,
        312 (1976).

15. M.F. Yan, R.M. Cannon and H.K. Bowen, "Grain Boundary Migra-
        tion in Ceramics", in Ceramic Microstructures '76 ed.
        R.M. Fulrath and J.A. Pask, Westview Press, Boulder,
        Colo. p. 276, (1977).

FABRICATION AND PERFORMANCE OF MHD ELECTRODES

W. Roger Cannon, R. L. Pober, H. Kent Bowen

Massachusetts Institute of Technology

Cambridge, Massachusetts

INTRODUCTION TO MHD ELECTRODE MODULES

A brief presentation of the requirements for electrode module materials will perhaps afford some insight as to the reasons for selecting the development program for materials preparation and fabrication techniques described herein. A simple module may consist of three parts: a water cooled, metallic base which serves as the current collector and connector; an electronic conductor electrode which transfers electricity to the plasma; an electrical insulator to separate modules. The selection of electrode material is most critical and difficult. This material must have good electronic conductivity and should be a good electron emitter. It must be chemically resistant to and/or compatible with potassium *seed*, coal slag, and the insulator material at temperatures above about 1900°K; as well as being compatible with the base material at temperatures up to perhaps 1300°K. In addition to having a melting point above 2300°K, the material should exhibit good thermal shock resistance. Although the *ideal* electrode material has yet to be proven as part of a *useful* electrode module, progress has been made toward this goal.

The methods available for fabricating electrode and insulator materials include *traditional* processes such as sintering and hot pressing, as well as *newer* processes such as CVD and arc plasma spraying. Because the ceramic electrode need only be very thin for high heat flux electrodes, arc plasma spraying was selected as the primary process for fabricating electrode and insulator materials for MHD electrode modules. This process allows two dozen electrodes and insulators to be fabricated (onto metallic bases) in just a few hours. Geometry of the metallic base is not critical, as the plasma spray coating replicates the surface configuration; this allows freedom in the design of the modules. Changing coating materials is quite simple, so it is easy to fabricate varying composition (multi-layered) electrodes. Porosity in the electrode material can be controlled through the arc plasma spray operating parameters. Coatings near theoretical density are achievable. The primary limitation for plasma sprayed electrodes is the maximum thickness attainable while still maintaining good bonding to the base material.

Electrode materials prepared, fabricated and tested are of several chemical compositions and represent several families of crystal structure. Specific examples are iron-aluminum spinel, lanthanum ferrite and lanthanum ferrite-strontium zirconate perovskites, and zirconia-ceria, zirconia-calcia and zirconia-yttria fluorites. These materials were selected because of their anticipated performance towards specific goals, such as coal slag compatibility, high temperature stability, and *ideal* (all purpose) performance.

INTRODUCTION TO FABRICATION VIA ARC PLASMA SPRAYING

The basic procedure for fabricating ceramic materials by arc plasma spraying is quite simple. The substrate is grit blasted to produce a clean, rough surface. The flowable, ceramic powder is introduced into a powder feeder which transmits the powder to the plasma gun nozzle. The powder mixes with the plasma jet and then impinges on the substrate, developing a bond and accumulating material. Composition of the plasma gas and carrier gas may be varied independently, and are typically *inert*.

Properties of the plasma sprayed materials are affected by the plasma gun operating parameters and the particle size distribution of the powder. Multi-layer coatings are produced by simply changing the powder being fed to the plasma gun nozzle. Continuously varying composition materials could be fabricated by utilizing a powder feeder with provision for mixing two powders in varying proportion. In most instances, the plasma environment appears to be slightly *reducing*, so one must be aware of the possibility for compositional change from powder to coating. The deposition rate used for most materials tested is about twenty to fifty micrometers per pass. With each subsequent pass, the substrate becomes hotter. As the substrate temperature varies, so may the density (porosity) and composition of the material. Thus, one may find that as the coating gets thicker, the material may not be uniform in composition and properties.

ARC PLASMA SPRAYING PARAMETERS - GENERAL DATA

The plasma spray facility at M.I.T. has dust and fume control provided by a water curtain spray booth of about 1.3 by 2 meters cross section, with an average exhaust velocity of 1 meter per second at the position occupied by the plasma gun. The spray device is a Bay State Abrasives PlasmaGun Model PG-100 with a forty kilowatt (dual voltage) power supply capable of supplying one thousand amperes. The powder feeder is a Model PF-500-2, utilizing interchangeable cannisters and screw feed of the powder to the powder/carrier gas mixing device. The arc jet is directed vertically downward and the gun is attached to an automatic traversing device which provides for a reproducible, uniform coating. Substrates are supported on an adjustable, hydraulic lift table so that the gun to substrate distance can be maintained for various size substrates. Typical spray parameters are as follows:

Arc Plasma Gas - Argon 5.85×10^{-4} meter^3/second
Powder carrier gas - Argon 0.633×10^{-4} meter^3/second
Powder Feed Rate - $0.3 - 0.8 \times 10^{-3}$ kilogram/second
Arc Voltage - 30 volts
Arc current - 700 amps
Gun to Work Distance - 9 centimeters
Traverse Velocity - 0.3 meters/ second
Offset per Pass - 2.5 millimeters

EXPERIMENTAL PROGRAM

The variable parameters for this series of experiments were: gun-to-work spray distance, 6 to 12 centimeters; arc current, 700 to 900 amperes; coating thickness, 0.5 to 2 millimeters; and powder particle size distribution, 40 to 100 micrometers. Effects of varying these parameters were determined by measuring the density of the coating produced for a given set of conditions.

Process variable changes are reported for three materials: iron-aluminum spinel, lanthanum ferrite, and zirconia-yttria. These powders were prepared by compounding oxides with other materials (e.g., zirconates, carbonates, metals) and blending. The materials were then calcined and/or fired. A few processing details should be noted: iron-aluminum spinel is fired in a closed muffle, so that oxygen partial pressure can be controlled; lanthanum oxide is calcined to eliminate water before weighing. After firing, the powder cake is crushed, milled, and sieved to produce the desired particle size.

Substrates were prepared from low carbon steel as 3.7 centimeter square plates, 1 centimeter thick.

The automatic traversing machine was set to cover an area larger than that occupied by the specimens being sprayed. This provided
a uniform coating since the specimen was within the constant velocity
area of the spray pattern. The plasma gun was ignited and allowed
to stabilize, and the powder feeder was started while the gun was
positioned away from the specimens. Spraying was continuous until
the desired thickness was obtained for a particular specimen, at
which time that specimen was removed from the spray area; the other
specimens remained to be sprayed for thicker coatings.

Density of the coatings was determined by an immersion (buoy-
ancy) technique, after removing the material from the substrate.
The surfaces of the specimens were sealed to prevent water absorb-
tion during the immersion procedure. Some specimens were measured
with micrometers to determine the volume; then density was calculated
for comparison with the data obtained by immersion.

RESULTS OF PLASMA SPRAYING EXPERIMENTS

Figure 1 reports density values as a function of plasma arc
current, spray distance, coating thickness and particle size. Results shown in Fig. 1, indicate a definite trend in density variations with each of these parameters. A possible wider variation in
parameters and the variation of additional parameters may have further increased the density, but very reasonable densities were achieved, and the experiments performed illustrate some of the fundamental
principles governing plasma spraying.

It is clear from Figs. 1(a) and (c) that increasing the plasma
enthalpy by increasing the arc current results in a denser coating.
According to the literature[1] a maximum enthalpy will be achieved
beyond which extensive vaporization of the particles will limit the
density, but results in Fig. 1 indicate that such a high enthalpy
has not been achieved.

The results showing increased density of the coating with decreased spraying distance in Figs. 1(a), (c) and (e) is undoubtedly
related to the cooling of the plasma with distance from the spray
nozzle by mixing of the plasma gas with the ambient gas. The ceramic particles, however having a high heat capacity and heat of
fusion and low heat transfer, etc., do not reach their maximum temperature for some distance of travel from the nozzle. It might be concluded from Fig. 1 then, that the particles have reached their maximum
temperature at less than the 6.4 cm and continue to cool with distance.
To test this hypothesis, a glass slide was introduced rapidly into and
out of the plasma to collect a number of splat-cooled particles for
microscopic examination. An example of a micrograph of a splat-cool-

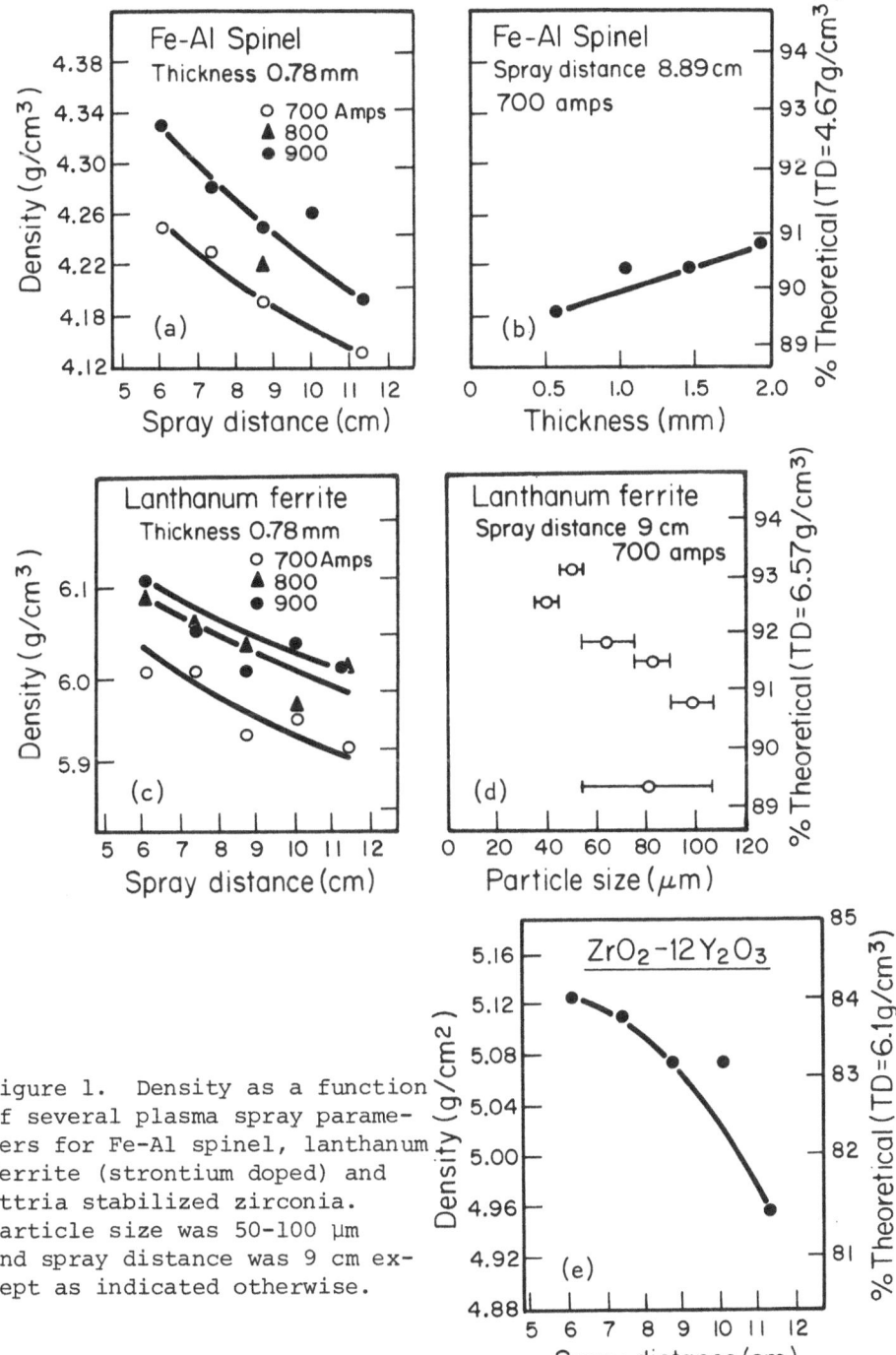

Figure 1. Density as a function of several plasma spray parameters for Fe-Al spinel, lanthanum ferrite (strontium doped) and yttria stabilized zirconia. Particle size was 50-100 μm and spray distance was 9 cm except as indicated otherwise.

ed particle which retained a low viscosity during cooling is shown in
Fig. 2. These particles must have reached a temperature well above
the melting point before striking the glass. The low viscosity is
particularly evidenced by the fact that the centers are transparent
to transmitted light. At greater distances than 6.4 cm most particles
appeared as in Fig. 2. At 6.4 cm, however, particles did not splat completely, and thus were apparently cooled below their melting point before completely spreading across the glass slide.

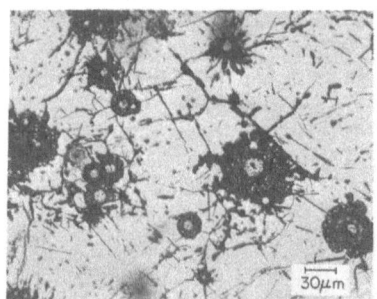

Figure 2. Particles of Fe-Al
spinel splat-cooled on a glass
slide.

    The high density coatings sprayed at 6.4 cm is thus unexpected
(Fig. 1). Two explanations may be offered. The substrate is heated
to a higher temperature by the plasma as the torch is closer to the
working surface and the velocity of the particles are greater at
distances closer to the torch, i.e., the splat time is smaller. Temperature measurements were, in fact, made with a brightness pyrometer
by attempting to match the filament to the spot under the nozzle as
the plasma torch moved by. The temperature difference between the
spot at 6.4 cm and 11.5 cm was ∿100°K (neglecting emittance correc-
tions).

    Results shown in Fig. 1(b) indicate that the density of the sub-
strate increases with plasma sprayed thickness. The explanation to
this result is similar to the above. That is, the substrate temperature increases as the coating thickness increases, both because of
the overall module heating and because of the increased insulation
thickness. No temperature measurements, however, were made to substantiate this claim.

    Finally, the results shown in Fig. 1(d) indicate that the coat-

ing density increased with decreasing-particle-size and also with a
decreasing width of the particle-size cut. The former is explained
by the fact that the short dwell times in the flame are needed for
melting fine particles, and that high particle velocities are achieved with fine particles due to the lower inertial force. Glass slide
experiments described earlier indicate complete melting of fine particles even at 6.4 cm. Lower densities with a wider particle size
cut are thought to have resulted from separation of various particle
sizes in the flame producing a non-homogeneous coating.[1,2]

MICROSTRUCTURAL FEATURES

Grain structure, porosities and microscopic defects were studied
by examining fracture surfaces of plasma sprayed coatings. Fig. 3
shows that the microstructure consists of layers of columnar grains,
each representing a rapidly quenched liquid particle. In general the
more highly dense specimens consisted of uniform parallel layers of
columnar grains perpendicular to the surface, whereas, in the less
dense materials the layers were more random.

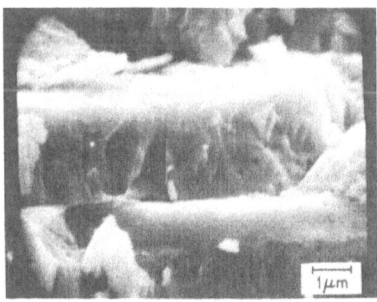

Figure 3. Fracture sur-
face of plasma sprayed
$La_{.9}Sr_{.1}FeO_3$, showing
layers of columnar grains.

Figure 4. Fracture surface of plasma
sprayed $La_{.9}Sr_{.1}FeO_3$ showing a large
defect between layers.

Two types of porosity could be observed: fine shrinkage pores
visible within the layers (Fig. 3) and large defects usually between
layers. An example of a large interlayer crevice is shown in Fig. 4.
The abundance of these large defects would indicate that they controlled the density. Perhaps the ultimate density achievable for
the plasma sprayed coatings on the other hand, is determined by the
fine porosity.

The abundance of large defects offers plenty of sites for fracture initiation as will be discussed in the next section. Two defects afford special mention: (1) an example of a particularly large defect filled with unsplat-cooled particles is shown in Fig. 5. The particles have evidently melted in the plasma and recooled before striking the substrate. (2) An example of a partially unmelted particle is shown in Fig. 6. These unmelted particles were not found in great abundance.

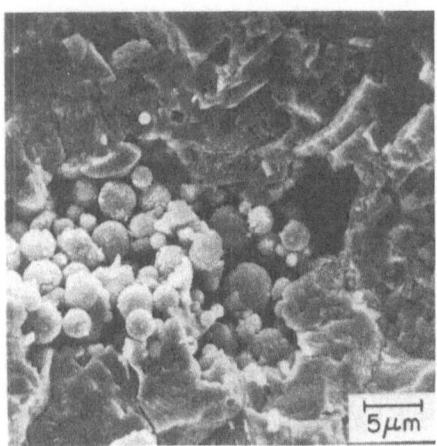

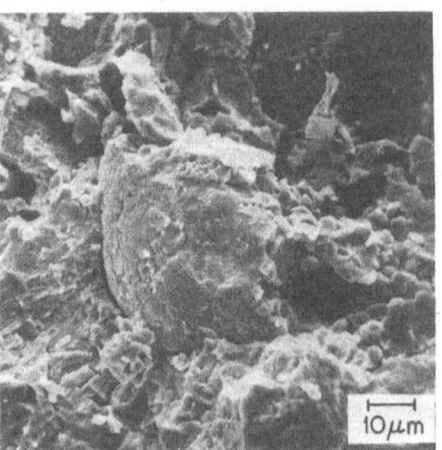

Figure 5. Fracture surface of plasms sprayed La$_9$Sr$_1$FeO$_3$ showing a large defect containing resolidified particles.

Figure 6. Fracture surface of a strontium zirconate-lanthanum ferrite (strontium doped) solid solution showing an unmelted particle.

An example of an MHD material which has a high vapor pressure near its melting point is 25m/o ZrO$_2$·75m/o CeO. This material does not plasma spray well. Densities of only 85% theoretical and low spray efficiencies are achieved. An examination of the microstructure (Fig. 7) reveals poor bonding between particles. This is due to rapid vaporization at the time of bonding.

THERMAL SHOCK RESISTANCE

One of the qualities MHD electrode materials must possess is good thermal shock resistance since the channel heat up and cool down may be reasonably rapid. Plasma sprayed coatings are thought to possess these characteristics. J. Shao and R. M. Cannon[3] performed some thermal shock measurements on both plasma sprayed and sintered iron-aluminum spinel materials by the technique of Hasselman.[4] Coatings were sprayed on blocks which had been prepared with only a light grit blast and the coatings were removed during cooling. Sintered

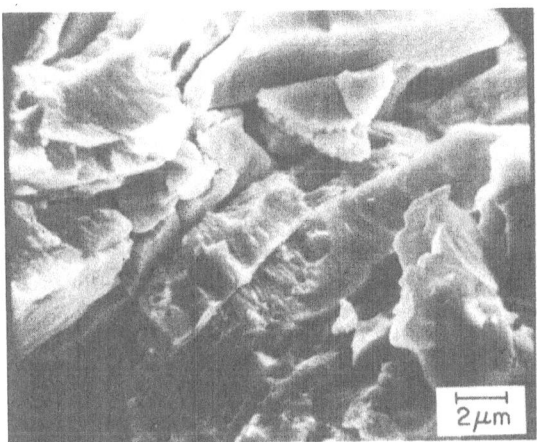

Figure 7. Fracture sur-
face of a 25m/o ZrO_2-
75m/o CeO plasma sprayed
coating.

material was prepared with a wide range of densities. Bars of 2.5
mm x 5 mm x 3 cm were cut from the plasma sprayed material and from
some isostatically pressed and sintered material. Five to ten samples were quenched from various temperatures in water and then rupture tested in four point bend tests. The fracture strength did not
decrease with quenching until a critical temperature quench was
reached at which point the strength dropped off suddenly. This temperature difference (ΔT_c) represents the value required to produce
cracks of the critical length. Results are listed in Table I. Although the fracture strength of the plama sprayed material was not
the highest of the samples tested, the critical ΔT_c was much higher
than the rest. These results are explained easily from the discussion of the micrographs noting that the number of potential sites
for crack nucleation is large in the plasma sprayed material and
thus the fracture strength is low. Also the critical quenching temperature is expected to be high if a large number of flaws are present.[4]

Table 1. Thermal Shock Characteristics of 64% $FeAl_2O_4$
36% Fe_3O_4 Sintered and Plasma Sprayed Bars.

Sample #	Fabrication Technique	Density (% T.D.)	MOR Unquenched (MPa)	ΔT_c (°K)	MOR Drop at ΔT_c (MPa)
1	Sintered	48	4.0	200	1.3
2	Plasma Sprayed	83	19.3	900	9.6
3	Sintered	95	94.4	140	44.8
4	Sintered	70	9.6	500	1.4

ELECTRODE MODULE TESTING AT MHD SIMULATION FACILITY

The eventual use of high temperature electrode modules requires materials developments consistent with the MHD environment. The compatibility requirements of the several materials needed in the complete electrode-insulator module is especially acute. The coupling of laboratory experiments to results in the simulation facility is a key point in the successful design and construction of modules for actual MHD systems. The M.I.T. Simulation Test Facility has been designed for the final module design and materials proof test. Separate steps to prove thermal/mechanical, chemical, and electrical performance are used to separate the critical problems. For example, the facility allows for continuous visual inspection of electrodes during a run.

When testing electrodes in a simulated environment, a large number of test variables have very significant effects on the performance of the electrodes. Some of the variables are directly related to the test configuration. The observed ultimate failure of an electrode is often preceded by a less obvious condition, such as a cool boundary layer from up-stream cold walls, thermal or mechanical inadequacy (though not outright failure) of an adjacent or opposing electrode, or internal manufacturing defects not manifest until the specimen is subjected to severe test conditions.

Electrode modules capable of operating at T >1700°C have been tested at conditions of: $\dot{Q}$ >150 w/cm^2, J $\sim$1.5 A/cm^{-2} (electrode), potassium seed concentration of 1.5m/o, plasma temperature >2225°C, plasma conductivity in excess of 1 mho/m, for up to 4 hours of test time. Electrodes have also been tested in plasma with 1w/o slag, for up to 3 hours, at somewhat lower temperatures and heat fluxes. Normally, the test configuration in this facility consists of two stacks of electrodes (one anode and one cathode) 4.2 cm apart and 7.5 cm in active wall width, installed on an insulated mounting (6.3 cm, measured along the stream). This mounting is made of water-cooled copper blocks with insulating partitions to reduce current leakage to the burner. For precise results it is desirable to use at least 5 electrodes per stack, where the upstream-most two and the downstream-most two electrodes act as guard electrodes for the others. The purpose of the guard electrodes is to protect the middle electrode(s) from end effects and, therefore, to realize a uniform electric field in the plasma stream between opposite electrodes.

Electrode modules were tested in three *environments:* thermal, plasma composed only of combustion products; thermal + chemical, plasma *seeded* with approximately one m/o potassium and/or one w/o slag; thermal + chemical + electrical, electrode modules connected as anodes and/or cathodes for a direct current passing through plasma via generator (i.e., no magnetic field). The several electrode materials were tested in a variety of electrode module configurations in an effort to define some critical problems in electrode materials performance. The results of this effort are summarized in Table 2, and indicate the relative performance of some materials tested under various plasma conditions. Although none of the materials and configurations yet tested in this program have performed well in all phases of testing, there has been progress in modifying module designs to improve performance.

Table 2. Characteristics and Performance of Some Plasma Sprayed Materials

	$64FeAl_2O_4$ $36Fe_3O_4$	$FeAl_2O_4$	$La_{.9}Sr_{.1}FeO_3$	$88ZrO_2$ $12Y_2O_3$	$25ZrO_2$ $75CeO_2$	$75SrZrO_3$ $25La_{.9}Sr_{.1}FeO_3$
Typical Coating Densities (G/cm^3)	4.18 (89%)	3.87 (89)	5.92 (90)	–	6.86 (85)	4.90 (86)
Phases from X-ray	2 phases	single phase	single phase	–	2 cubic phases	single phase
Thermal	good[1]	good[1]	good[2]	good	good	good
Thermal + Chemical	good	good	–[3]	good[4]	good[4]	good[4]
Thermal + Chemical + Electrical	good	good	–[3]	poor[5]	good	untested

(1) At $< \sim 1700°K$
(2) $< \sim 1900°K$
(3) Tested only as a sub-surface layer
(4) Tested in a non-slag environment
(5) High oxygen conductivity possibly reason for failure

CONCLUSIONS

Ultimately the life of MHD electrodes will be limited by their ability to withstand the chemical, thermal and electrical conditions of the MHD environment. As improved materials are being developed to withstand these environments, improved fabrication techniques are necessary to reduce the porosity and increase the strength, and thus to

reduce the rate of degradation by the environment. Plasma sprayed
coatings may be improved by adjusting the enthalpy of the flame and
the powder characteristics. An understanding of the effects of these
changes on the physical nature of the coating provides useful guides
for processing improvements.

REFERENCES

1. D. A. Gerdeman and N. L. Hecht, Arc Plasma Technology in Materials
    Science, Springer-Verlag, New York, Wien, 1972.
2. G. Perugini, Ceramurgica, 2, [4], 191-199 (1976).
3. J. Shao and R. M. Cannon, unpublished results.
4. D. P. H. Hasselman, J. Amer. Ceram. Soc. 52, [11], 600-604,
    (1969).

Acknowledgement: The writers wish to acknowledge J. Centorino and
W. Siegfried who performed much of the experimental work, and wish
to thank J. Shao and R. M. Cannon for the use of unpublished data.
This work was performed under ERDA Contract No. ER-76-C-01-2215.

FABRICATION AND PROPERTY CONTROL OF LaCrO$_3$ BASED OXIDES

Harlan U. Anderson

Department of Ceramic Engineering
University of Missouri
Rolla, Missouri 65401

INTRODUCTION

In recent years the search for materials which can be used as electrical conductors at elevated temperatures in corrosive environment has lead us to investigate many of the high melting point oxides. Unfortunately most of these oxides have quite wide band gaps and therefore possess little electrical conductivity. This is particularly true for those oxides whose bonding orbitals involve only s & p electrons (e.g., MgO, Al$_2$O$_3$). However oxides whose cations have d-electron states available do show rather high electrical conductivity (e.g., NiO, CoO). In these oxides, conductivity can be attributed to electrons or holes moving through the d-electron energy levels of the cation either by a localized hopping process or by a narrow conduction band mechanism.[1][2] Therefore most of the compounds under consideration contain transition metal cations.

Among the compounds being studied are the perovskites (ABO$_3$) and spinels (AB$_2$O$_4$) where the A site contains a rare earth or alkaline earth cation and the B site is a transition metal cation. The electrical conductivity of such compounds can be increased by substituting either acceptor or donor type cations on the A or B lattice sites. The type of dopant required depends upon whether the native conductivity is p or n type.

The rare earth and alkaline earth perovskites and spinels encompass a large number of compounds which find usage for a number of applications such as insulators (MgAl$_2$O$_4$), capacitors (BaTiO$_3$), MHD electrodes (LaCrO$_3$), fuel cell interconnects (LaCrO$_3$), fuel cell electrodes and catalysts (LaCoO$_3$, LaFeO$_3$).[3-6] The material to be used depends upon the particular temperature, electrical or catalytic properties that the application requires.

One critical application is that for electrodes in the hot wall, open cycly MHD power generator. Such as application requires the electrode to maintain both chemical and mechanical integrity at temperatures exceeding 1600°C for long periods of time (>1000 hours) in oxidizing atmospheres ($PO_2 \simeq 10^{-3}$ atm) in the presence of seed (K^+) and coal slag as well as possess electronic conductivity at room temperature of $>10^{-1}$ (ohm-cm)$^{-1}$. [There are additional requirements, but they are well-documented elsewhere.][4,7] Unfortunately, these property requirements are so severe, that there are few, if any materials that can fill them all.

Some of the most promising materials are the rare earth and alkaline earth chromites. Therefore, they have been subjects of a number of investigations.[4,8-10] These studies have shown that materials based on $LaCrO_3$ and $MgCr_2O_4$ meet many of the requirements but fall short in such properties as volatilization and chemical (and electro-chemical) corrosion.

A second important application is that of the interconnect material in the high temperature, solid oxide fuel cell. This application requires a material to have a good thermal expansion match with the zirconia electrolyte, high electronic conductivity [$>10^{-1}$ (ohm-cm)$^{-1}$], stability over the oxygen activity range 10^{-19} to 1 atm at 1000°C, and, in addition, must be capable of being fabricated into a nonporous layer.[6] Again the materials that appear to best fulfill these requirements are $LaCrO_3$ based oxides.

Thus, Cr-containing perovskite-like compounds such as $LaCrO_3$ are finding uses as electrodes and interconnectors in hot wall, open cycle MHD power generators and high temperature solid oxide fuel cells respectively. These applications require the materials to be dense and to meet the property requirements such as thermal expansion, electrical conductivity, and volatilization which are presented by these two processes. In order for compounds like $LaCrO_3$ to meet these requirements, the basic properties must be altered. It is the intent of this report to discuss (1) how Cr-containing oxides are sintered to obtain the densities required for these applications and (2) how one goes about altering the properties of perovskite-like compounds such as $LaCrO_3$ in both a predictable and a reproducible manner to meet a given set of requirements.

DENSIFICATION OF Cr CONTAINING OXIDES

In general oxides which contain substantial quantities of Cr are difficult to pressureless sinter to high density. In fact under ambient atmospheric conditions it has been demonstrated that compounds such as Cr_2O_3, $MgCr_2O_4$ and $LaCrO_3$ show very little increase

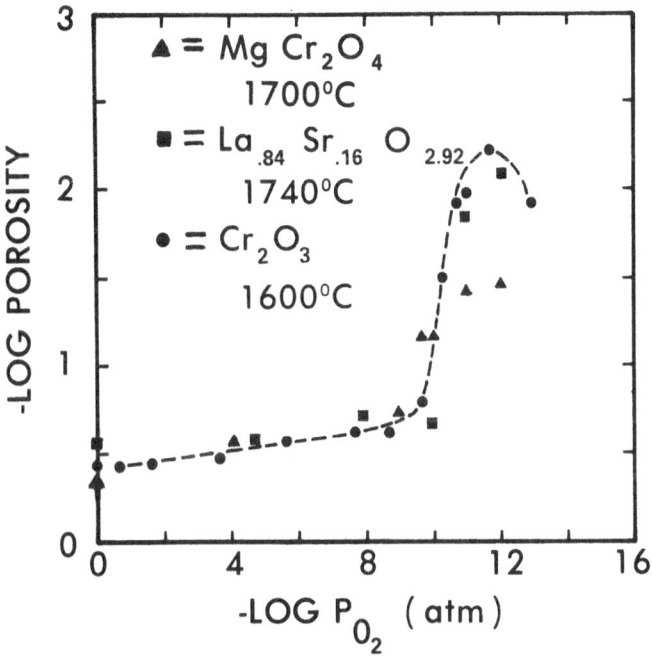

Figure 1: Dependence of porosity of sintered chromites on oxygen
activity; Sintering time = 1 hr; porosity = 1- sintered density/
theoretical density.

in density from that of the unsintered unit.[11-15] However, in
these studies, densities of greater than 90% of theoretical could
be achieved if the sintering atmosphere contained oxygen of activity
near that specified by the Cr/Cr_2O_3 phase boundary. Figure 1 illustrates the strong dependence that oxygen activity has on final
densification which they observed.

The influence of oxygen activity on the densification rate was
demonstrated by Halloran and Anderson [13] in an experiment where
they measured the shrinkage rate at constant shrinkage (3%) as
function of oxygen activity at 1125°C. They found that maximum
shrinkage rate occurred only near the oxygen activity specified by
the Cr/Cr_2O_3 phase boundary. Higher oxygen activity resulted in
essentially no densification.

The maximization of densification rate of these compounds may
be related to the stabilization of the Cr ion in its trivalent state.
At oxygen activities higher than that specified by the Cr/Cr_2O_3

equilibrium, the Cr^{3+} ion can be oxidized into states whose oxides are unstable and an increase in volatilization is observed. As a result of loss of Cr oxide compounds in oxidizing atmospheres densification may be impeded. It was shown by Anderson[14] that substantial grain growth occurs in oxidizing atmospheres. This suggests that the transport mechanism is controlled by either evaporation-condensation or surface diffusion in oxidizing conditions whereas it is controlled by a densification mechanism at the lower oxygen activities where maximum densification occurs.

The results of these studies leads one to the conclusion that in order to sinter Cr-containing oxides to high density it is necessary to control the oxygen activity in the furnace. This means that atmosphere control must be provided on the sintering furnace. One way of doing this is to flow CO/CO_2, H_2/CO_2, or H_2/H_2O gas mixtures through the furnace. Darken and Gurry showed that if a linear flow of about 1 cm/sec is established in a tube furnace, it is possible to calculate the oxygen activity from the knowledge of the flow rates of each gas and thermodynamic data of the gas reactions.[16]

If available, a solid electrolyte cell can be used as a direct measurement of oxygen activity in the furnace. Unfortunately, there are no electrolytes available which can be used in the temperature and oxygen activity ranges required to sinter Cr-containing oxides, T = 1600-1800°C and $PO_2 = 10^{-12}$ to 10^{-9} atm, respectively. However, cells can be used to establish the gas ratios in the gas stream by monitoring the gas before and after it flows through the furnace. This allows better control of the gas content than is possible from flowmeters alone.

PROPERTY CONTROL OF $LaCrO_3$

Electrical Conductivity

As mentioned previously the conduction in $LaCrO_3$ can be attributed to hole motion in the d-electron energy levels of the Cr.[1,2] The substitution of a divalent ion on either the La or Cr lattice sites can be electronically compensated by (1) the formation of Cr^{4+} [holes] or (2) the formation of oxygen vacancies. The former case increases the electrical conductivity whereas the latter forms a defect which does not contribute to conduction and is an insulator when compared to the former. Due to these two competing compensation mechanism, the conductivity of $LaCrO_3$ is dependent upon the atmosphere in which it is equilibrated. Since reducing atmosphere favors the formation of oxygen vacancies, densification of the material either by atmosphere sintering or by hot pressing causes it to be a rather good insulator. After subsequent equilibration in oxidizing atmospheres, the

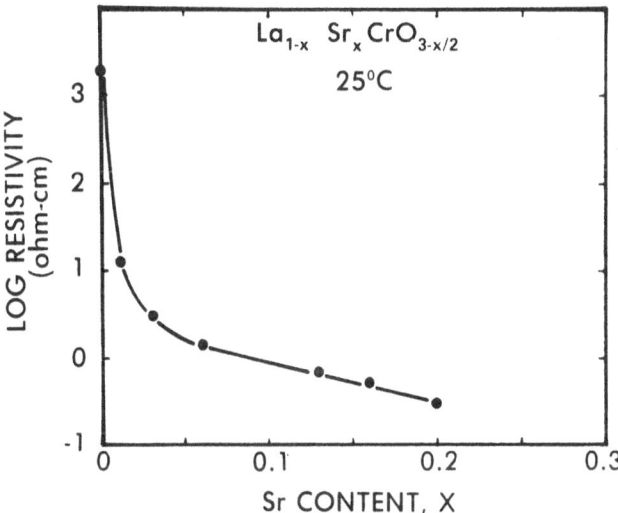

Figure 2.   Electrical Resistivity of $La_{1-x}Sr_xCrO_3$ as function of Sr
content (from Meadowcroft).[4]

conductivity at room temperature has been observed to increase by
as much as eight orders of magnitude.[17]

Figure 2 is an example of the influence of the substitution
of Sr^{+2} for La^{+3} on the electrical resistivity. These data were
taken from one of Meadowcroft"s reports for samples which were
sintered in oxidizing atmospheres; thus, the electrical resistivity
is quite low.[4] Equilibration of similar compositions in a reducing
atmosphere results in an increase in resistivity by six to eight
orders of magnitude. To control properly the electrical resistivity,
and to insure that the substitution ion is of the proper ionic radii
to fit into the selected lattice site, it is essential that care must
be taken to equilibrate the material in an oxidizing atmosphere to
minimize the concentration of oxygen vacancies. For example, in
order to obtain maximum and reproducible electrical conductivities,
sintered units of $LaCrO_3$ must be annealed in the temperature range
1400 to 1700°C for several hours in an oxidizing atmosphere. The
temperature and time required is a function of the size of the unit
being annealed.

Volatilization

Due to the loss of Cr oxide compounds from LaCrO , considerable
weight loss and electrode recession occurs at temperatures in the

1700 to 1800°C range. In a dry atmosphere a recession rate of
1 cm/10⁴h is to be expected whereas in a moist atmosphere the rate
might be as much as 10 times higher. The increased weight loss in
moist atmospheres is from the formation of $CrO_2(OH)$ which is more
volatile than the oxides of Cr.[18] Since the weight loss or recession
rate greatly limits the usefulness of $LaCrO_3$-based materials, it is
desirable to reduce the volatilization rate.

It has been found that the substitution of Al for Cr greatly
reduces the weight loss. As can be observed in Figure 3, the substitution of 10-to-20 m% of the Cr by Al reduces the weight by as
much as an order of magnitude.[8,10] At the present time, there is
no good explanation for this effect. At first the possibility was
considered that the Al formed a barrier on the surface which reduced
the volatility of Cr. However, if this were the case, a change in
electrical conductivity with time should be expected. This has not
been observed.

It is also possible that the Al tends to lower the activity of
Cr in the $LaCrO_3$ structure. The data of Sasamote and Sata show
that this is apparently what happens when Cr is placed in $LaCrO_3$
as compared to Cr_2O_3.[19] The Cr appears to be more stable in the

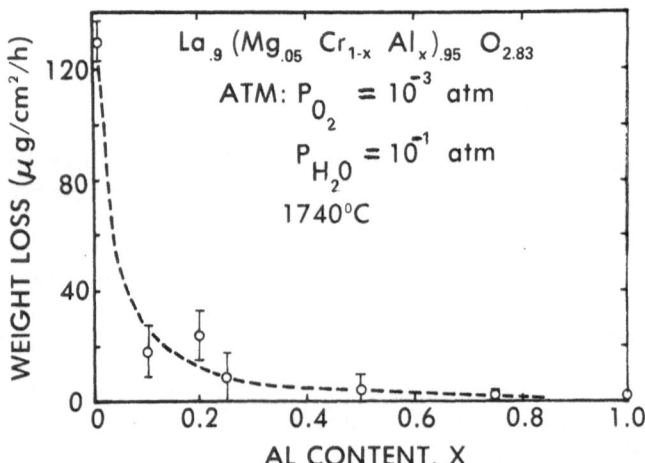

Figure 3. Weight loss ($Mg/cm^2/h$) of $La_{.9}(Mg_{.05}Cr_{1-x}Al)_{.95}O_{2.83}$ as
function of Al content: Temperature = 1740°C; Atm: Flowing
gas mixture (0.001 atm O_2+0.1 atm H_2O+N_2); Flow rate = 1
linear cm/sec.

perovskite structure of $LaCrO_3$ than in the corundum structure of Cr_2O_3.[18] Thus, it might not be unreasonable to consider that the additions of Al increase the stability of Cr in La CrO_3. Until more data are collected the selection of the proper mode of stabilization can not be made.

Since the electrical conduction occurs through the d-electron energy levels of the Cr, the replacement of Cr by Al has a detrimental effect of the electrical conductivity. Typically the room temperature electrical conductivity decreases from 1 to less than 10^{-3} $(ohm-cm)^{-1}$ as the Al content increases.[8] It has been found that up to 50% of the Cr can be replaced by Al before the conductivity becomes too low for MHD purposes.

Thermal Expansion

Many of the applications of $LaCrO_3$-based oxides require that they have mechanical integrity with adjoining materials. For example, when $LaCrO_3$ is used as an interconnect material in solid oxide fuel cells it is essential that its thermal expandion closely matches that of zirconia. Most of the zirconia compositions have thermal expansion coefficients in the 10.5 to 12 x $10^{-6} m/m/°C$, however, $LaCrO_3$ (rhombohedral phase) has a coefficient of about 9.2 x $10^{-6} m/m/°C$.

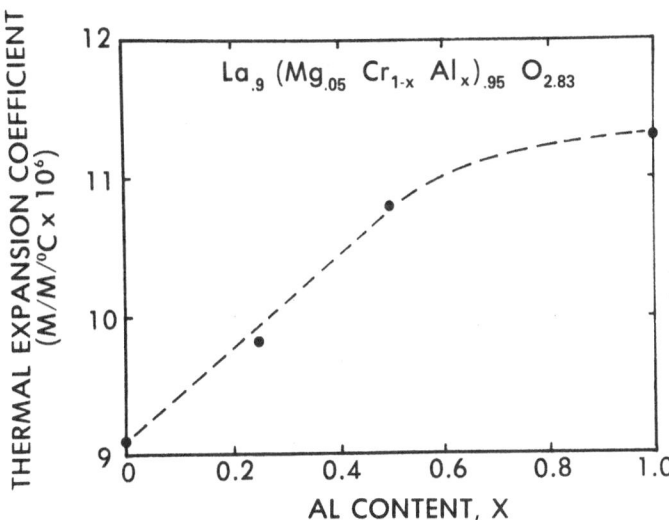

Figure 4. Thermal expansion coefficient of $La._9(Mg._05Cr_{1-x}Al_x)._95O_{2.83}$ as function of Al content: Temperature range 25 to 1000°C.

This difference in expansion coefficients is sufficient to cause cracking of the $LaCrO_3$. Thus a match of the expansion coefficient of $LaCrO_3$ to that of the zirconia is required.

Figure 4 shows the influence that the additions of Al have on the thermal expansion coefficient of $LaCrO_3$. As can be seen the expansion coefficient can be altered from 9.2×10^{-6} m/m/°C to 11.3×10^{-6} m/m/°C by the addition of Al. As much as 20 to 30% can be added without eliminating the usefulness as an electrical conductor.

Summary

In this report an effort was made (1) to show hoe Cr-containing oxides can be sintered to high density by proper control of the sintering atmosphere and (2) to show how the properties of ternary oxides can be altered in a systematic and reproducible manner. This was done by using Cr_2O_3, $MgCr_2O_4$ and $LaCrO_3$ as examples. However, the principles which were applied should be equally applicable to other perovskite and spinel oxides.

REFERENCES

1. I. G. Austin, N. E. Mott, Adv. in Phy. 18 41-102 (1969).
2. J. B. Goodenough, "Prog. in Sol. St. Chem." 5 145-400 Pergamon Oxford, (1971).
3. F. S. Galasso, "Structure, Properties and Preparation of Perovskite-type Compounds" Pergamon, Oxford (1971).
4. D. B. Meadowcroft, Brit. J. Appl. Phy. Ser. 2 2 1225 (1969).
5. D. B. Meadowcroft, Nature 226 847-848 (1970).
6. R. J. Ruka, "High Temperature Solid Oxide Fuel Cells" May 5-6 (1977) Brookhaven National Lab, Upton, N. Y. 11973.
7. H. K. Bowen and B. R. Rossing, "Materials Limiting Problems in Energy Production, Ed. C. Stein, Plenum Pub. N. Y. (1976).
8. H. U. Anderson, et al., "Conference of High Temperature Sciences Related to Open Cycle, Cool-fired MHD Systems" Argonne National Lab., April 4-6, 1977, Argonne, Ill.
9. H. U. Anderson, et al., "13th Rare Earth Research Conf." Oct. 16-20, 1977, Oglebay Park, Whelling W. V. (to be published, Plenum).
10. H. U. Anderson, et al., "High Temperature Solid Oxide Fuel Cells" May 5-6, 1977, Brookhaven Nat. Lab., Upton, N. Y.
11. P. D. Ownby pp. 431-37 in Materials Science Res. Vol. 6, Ed. by G. C. Kuczynski, Plenum Press, N. Y. (1973).
12. P. D. Ownby and Gw E. Jungquist, J. Amer. Cer. Soc. 55 [9] 433-36 (1972).

13. J. W. Halloran and H. U. Anderson, J. Amer. Cer. Soc. <u>57</u> [3]
        150 (1974).
14. H. U. Anderson, J. Amer. Cer. Soc. <u>57</u> [1] 34-37 (1974).
15. L. Groupp and H. U. Anderson, J. Amer. Cer. Soc. <u>59</u> [9-10]
        449-450 (1976).
16. L. S. Darken and R. W. Gurry, J. Amer. Chem. Soc. <u>67</u> [8]
        1398-1412 (1945).
17. H. U. Anderson - Unpublished (1977).
18. M. Yoshimura and H. K. Bowen, Private Communication (1977).
19. T. Sasamoto and T. Sata, Yogyo Kyoka Shi <u>74</u> [11] 408-18 (1971).

STRONTIUM CONTAINING PEROVSKITES AND RELATED

CONDUCTIVE ELECTRONIC CERAMICS

T. J. Gray

Olin Corporation - Metals Research Laboratories

91 Shelton Avenue, New Haven, Connecticut 06511

INTRODUCTION

Considerable interest attaches to conductive electronic ceramics, principally perovskites or perovskite-related materials, for use as electrodes in MHD, molten electrolyte and ambient temperature electrochemical applications. Predominant among these perovskites are those containing strontium. As early as 1968/9 Meadowcroft[1,2] advocated the use of lanthanum strontium chromite as a MHD electrode material, earlier investigations by Jonker[3,4] having identified the electrical and magnetic properties of rare earth perovskites and the advantages deriving from the addition of strontium. Galasso[5] has reviewed the preparation and properties of a very wide range of perovskites. More recently Takahashi[6] and others[7] have investigated oxygen ion conduction in nonstoichiometric perovskites.

Applicability of conductive perovskites, particularly lanthanum strontium cobaltite as an oxygen electrode for fuel cell applications has been patented by Meadowcroft while extensive research has related to other electrochemical applications. These materials possess the additional advantage of being catalytically active which leads to the possibility of their use in heterogeneous catalysis as well as many electrocatalytic processes.

Crystal Structure

The basic structure of the perovskites ABO_3 derives from $CaTiO_3$, the mineral perovskite, originally believed to be a cubic structure but now recognized after Megaw[8] as being orthorhombic.

$SrTiO_3$ however is strictly cubic. In this structure the A cation is coordinated with twelve oxygen ions and the B cation with six, a stable BO_3 skeletal structure being a prerequisite generally met when $r_B > 0.51 \text{Å}$ for oxides. Additional stabilization derives from a relatively large A ion. Goldschmidt[9] defined the size limitations in terms of a tolerance factor t, where

$$t = (r_A + r_0) / \sqrt{2}(r_B + r_0)$$

and where r_A, r_B and r_0 are the radii of the respective ions and $1 > t > 0.75$. The stability condition also requires that $r_A > 0.90 \text{Å}$ and $r_B > 0.51$.

Subject to these criteria a very wide range of compositions can be accommodated by the perovskite structure providing the net cationic condition of valence obtains. Some of these compositions can be grouped under the following general descriptions. Oxides of the type $A^+B^{5+}O_3$ typical of which are the alkali metal niobates and tantalates e.g. $KNbO_3$; $KTaO_3$. Related to these are the non-stoichiometric oxides $A_x^+B^{6+}O_3$ where x can vary over the range 0.33 to 0.95 of which the best known are the "tungsten bronzes", Na_xWO_3. Other non-stoichiometric compositions also exhibit perovskite structures such as oxygen deficient $SrTiO_{(3-x)}$, $SrVO_{(3-x)}$ and $SrFeO_{(3-x)}$.

The best known perovskites are those of the $A^{2+}B^{4+}O_3$ type particularly the alkaline earth titanates with their important ferroelectric properties. Their crystal structure has been characterized (after Roth[10]) as illustrated in Fig. 1a. Similarly, the very large group of $A^{3+}B^{3+}O$ compounds can be characterized (after Schneider[11]) as in Fig. 1b. These relationships are based purely on size and many exceptions are observed. The diagram for the $A^{2+}B^{4+}O_3$ has only limited application on account of ferroelectric distortions and structures with variable layer sequences.

In addition, there are a wide variety of complex compositions possible with several cations in the A or B sites (or both simultaneously) providing the net charge and size criteria are met. These are particularly important in respect of electronic ceramics based on strontium-containing perovskites since many relate to variations on $La_{(1-x)}Sr_xM\,O_3$ where M is one or more of the transition metals.

Preparation Of Strontium-Containing Perovskites

A variety of techniques can be utilized for the preparation of strontium-containing perovskites and the selection of the best technique for a particular compound will then depend on considerations concerning the chemical composition and the requirements of

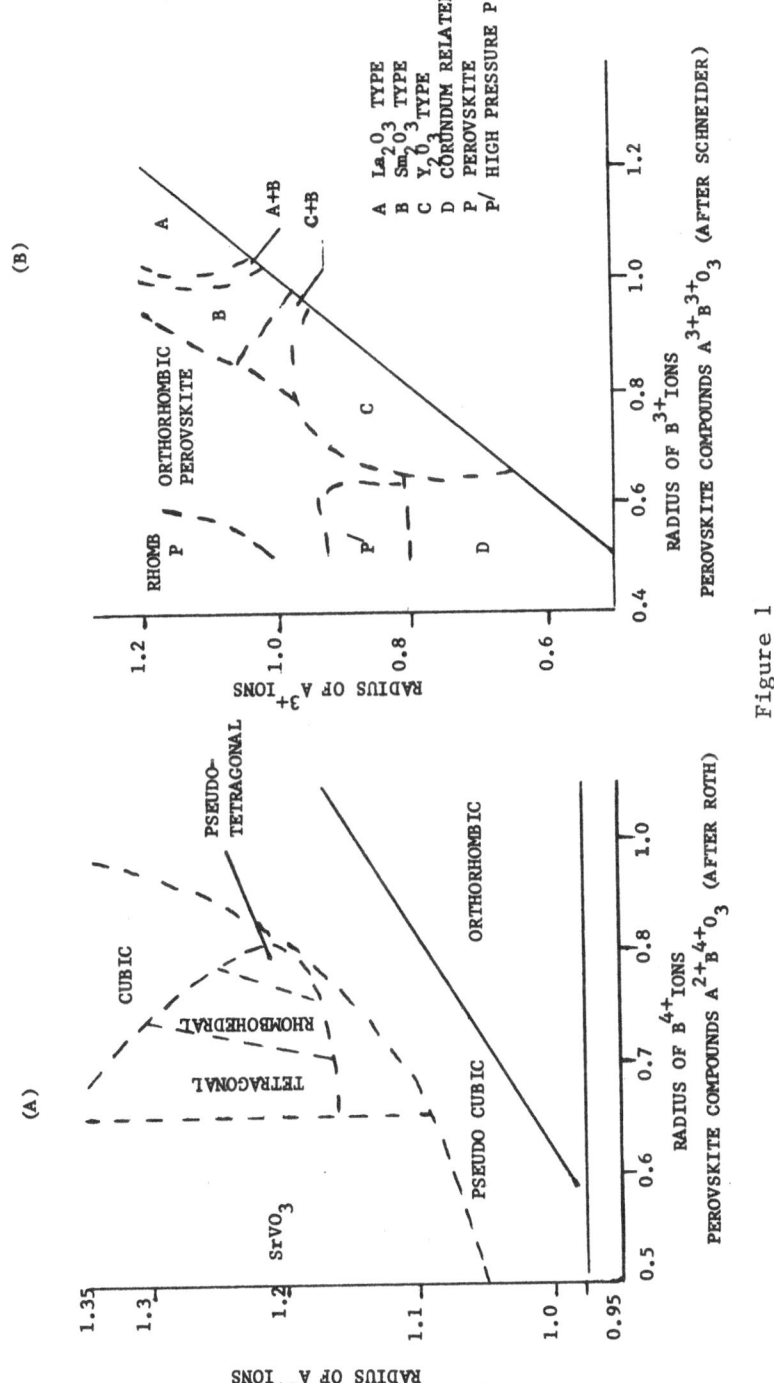

Figure 1

chemical purity, stoichiometry, homogeneity, degree of structure development, particle size and production costs.

They can be roughly divided into two groups according to the means of homogenization. The first group covers methods employing mechanical procedures which include the classical ceramic techniques of dry mixing and dry and wet ball-milling, while the second group is characterized by chemical processes which yield perfectly homogeneous final products or their calcinable precursors from solutions of suitable salts of the components. Among the latter methods the best known are precipitation, co-precipitation, spray-drying and freeze-drying, all widely used by the electronic ceramics industry. Various combinations of both types of methods (mechanical and chemical) are often found useful for the preparation of more complex perovskites, for example, the mechanical mixing of precipitated strontium titanyl oxalate with co-precipitated lanthanum and chromium hydroxides to product a mixture yielding $Sr_{1-x}La_xTi_{1-x}Cr_xO_3$ on calcination.

Mechanical Homogenization Techniques

The main advantages of mechanical homogenization are its simplicity, the relative ease of obtaining the desired composition and overall stoichiometry and the low cost. The disadvantages are the necessity to repeat the grinding and/or ball-milling at several intermediate states during the preparation to achieve reasonable homogeneity, the use of relatively high calcination temperatures which in turn cause normally undesirable grain-growth and the danger of contamination from the grinding media. One technical problem associated with ball-milling is the selection of a suitable liquid in relation to the solubility and chemical reactivity of the starting components. While water may be acceptable in some cases, for example mixtures of $SrCO_3$ and TiO_2 or $La_2(CO_3)_3:8H_2O$ and Cr_2O_3, an allowance for solubility is advisable for precise work. Organic solvents should be used when the water solubility is significant or where there is a danger of hydrolysis or other undesirable chemical reactions, thus ethanol or acetone should be used for ball-milling a $La_2O_3-Cr_2O_3$ mixture because of the strongly exothermic reaction of lanthanum oxide with water. Ball-milling of intermediate products also requires careful consideration. For example, while water may be suitable for ball-milling an unreacted $SrCO_3-TiO_2$ mixture, it can cause serious losses of strontium by hydrolysis if used with partially calcined mixtures which often contain significant quantities of Sr_2TiO_4 and $Sr_3Ti_2O_7$.

Oxides and carbonates are the most common materials used for preparation of strontium-containing perovskites by ball-milling and calcination. However, their reactivity is low which impedes the achievement of homogeneous products at temperatures low enough

to preserve fine particles. Substitution of selected precursors
such as oxalates can improve the reactivity considerably, yielding a more homogeneous product at lower temperatures.

Precipitation And Co-Precipitation Techniques

The distinction between precipitation and co-precipitation
should be made according to the character of the precipitate. If
a single phase product is desired such as barium titanyl oxalate,
normal precipitation may be justified whereas if two or more phases
are present in the precipitate, then co-precipitation is preferred.
At the Atlantic Industrial Research Institute, strontium titanate
has been prepared by an adaptation of the method originally devised
for barium titanate. A solution of titanium tetrachloride is prepared by the slow addition of oxalic acid after which the mixture
is slowly heated with continuous stirring. At the first appearance of opalescence, a solution of strontium chloride is added and
heating and stirring continued until all the precipitate redis-
solves. In contrast to barium titanate preparation, the strontium
analogue is soluble in dilute hydrochloric acid. The solution is
then neutralized with ammonium hydroxide and the fresh precipitate
separated by filtration, followed by repeated washing to eliminate
adsorbed chlorides. The precipitated strontium titanyl oxalate
containing minor amounts of strontium oxalate and hydrated titanium
dioxide yields strontium titanate on calcination with no foreign
phases detectable by x-ray analysis.

The preferred technique for the production of moderate to
large quantities of co-precipitated material initiated by Gray at
Bristol University in 1946 is illustrated in Figure 2. This
provides special precautions to prevent complex formation which
could occasion variation in composition and also prevents the subsequent adsorption of extraneous ions on the high surface area
hydrous oxides or other compounds, by operating at elevated temperatures with immediate washing of the precipitate as it is formed.
The mixed solution of components is fed at a controlled rate and
elevated temperature to a fine jet. As it merges it meets a cor-
responsing fine jet of hot precipitant adjusted to provide stoichiometric precipitation. This is oriented to occur in the
immediate vicinity of the surface of a rotating glass impeller
directed to project the precipitate upwards where it is met by a
fine spray of hot distilled water. The washed precipitate is carried down the walls of the collector vessel by hot distilled water
onto a rotating vacuum filter where further washing takes place.
Utilizing this technique permits the preparation of materials to a
compositional accuracy of better than 0.1%. The hydrous oxides or
other co-precipitates, by this technique, can be readily converted
to the oxides at relatively low temperatures providing a product
of very fine grain size and corresponding high surface area

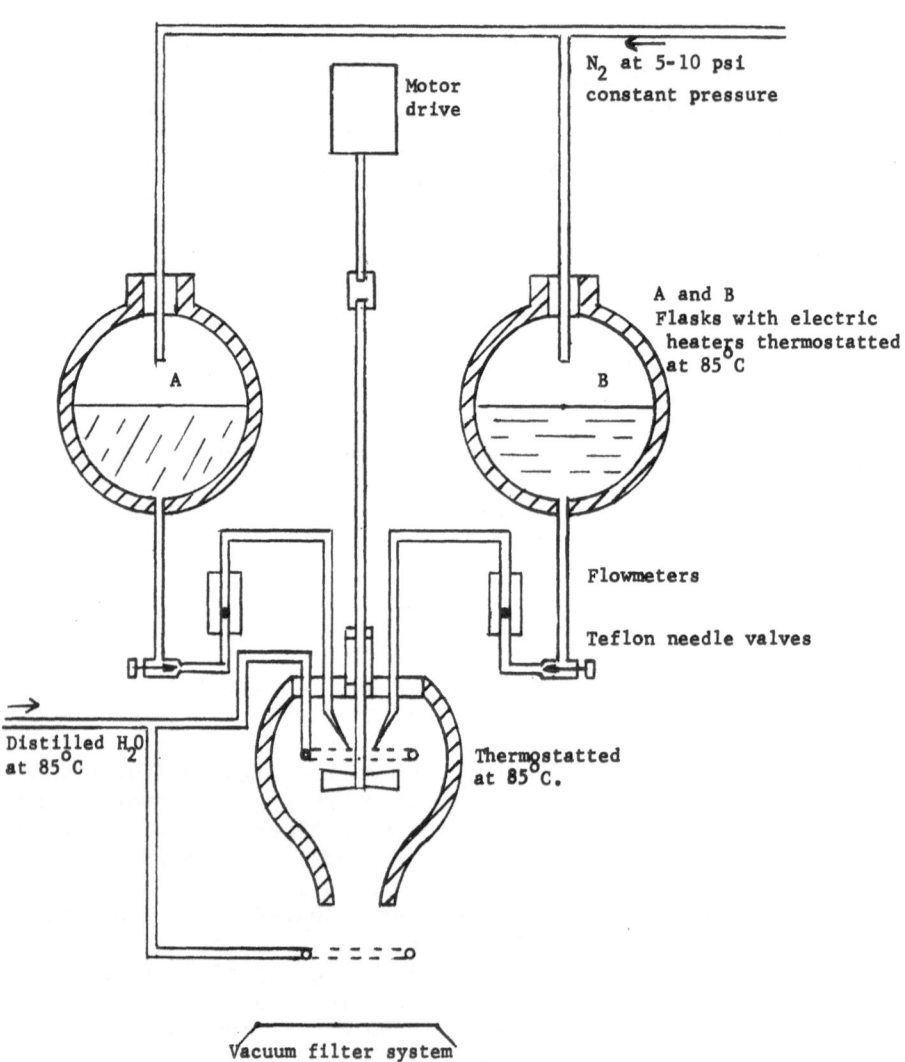

Figure 2

particularly appropriate for hot pressing and for use as catalysts.

The spray-drying and freeze-drying techniques used for the preparation of other electronic ceramics can also be applied to the preparation of strontium-containing perovskites without any significant difficulty. Nitrates are generally the best starting materials. These techniques are particularly appropriate for the preparation of ultrafine particles.

Calcination and Firing

Low calcination temperatures, preferably not exceeding 600°C are important for the preservation of submicrom particles and are particularly desirable for applications in catalysis. This excludes the use of mechanically prepared mixtures components with low reactivity such as oxides or carbonates, while mixtures of oxalates are somewhat more suitable. The best results are achieved with mixtures or precipitates prepared by chemical methods.

When full development of the perovskite structure is more important than the particle size higher calcination temperatures have to be used, either in the intermediate range 800-1000°C for chemically prepared samples or 1000°C and higher for mechanical mixtures. Repeated grinding and recalcination is usually necessary to bring the reaction in mechanically prepared mixtures to the desired degree of completion.

The formation of the perovskite phase is monitored periodically by x-ray crystallography after programmed firing on a time/temperature basis. Particular emphasis is accorded the establishment of a single phase if maximum conductivity is required or an optimum of a secondary "active" phase if catalytic activity is the primary consideration. The "active" phase has tentatively been identified with the K_2NiF_4 structure described by Rudorff[12] which is closely related to the perovskite structure. Isostructural compounds Sr_2TiO_4, $SrLaAlO_4$ and $CaMnO_4$ have been described by Ruddlesden & Popper.[13]

The forming techniques applied to strontium-containing perovskites include end-to-end pressing, isostatic pressing, hot pressing, extrusion and slip casting. The selection of the forming technique and subsequent firing schedule depends mainly on the desired density of the final product.

EXPERIMENTAL INVESTIGATION

The compounds $La_{(1-x)}Sr_xCrO_3$ and $La_{(1-x)}Sr_xCoO_3$ have received considerable attention, the former for MHD applications and the latter as a fuel cell (oxygen) electrode. They have also

engendered considerable interest as catalyst materials. In the
present investigations these materials have been studied from a
similar applicational interest in an endeavor to resolve and improve on their limitations.

It has been reported that while $La_{0.8}Sr_{0.2}CrO_3$ compositions
exhibit the electrical characteristics necessary for MHD application the material loses chromium probably by volatilization of
CrO_3 at high temperature. In a related study using this material
as an electrode in NaCl solution, coloration was initially observed
which did not recur on electrolysing in a fresh solution. This led
to the technique of prefiring $La_{0.8}Sr_{0.2}CrO_3$ isostatically pressed
electrodes at 1100-1200°C and machining to a size appropriate to
shrinkage at a higher firing temperature. Acid electrolytic
leaching was then utilized to remove the excess Cr_2O_3 presumably
localized at grain boundaries. After refiring at 1500°C the
material showed no further weight loss. This is in accord with
the subsequent observations of Khattak and Cox [14] who observed
diffraction lines of Cr_2O_3 if the La/Cr ratio was lower than 0.985.
These authors also proposed that the material in the fully oxidized
condition incorporated both trivalent and tetravalent chromium.

An alternate approach to establish greater stability at high
temperatures with acceptable although considerably poorer electrical conductivity has been adopted in an extension of these investigations. Jacobs et al.[15] have prepared mixed ceramics
structures containing $La_{0.84}Sr_{0.16}CrO_3$ and $SrZrO_3$. The latter is a
good insulator at room and moderate temperatures but exhibits
oxygen ion conduction at high temperatures. Preliminary data
indicates that a MHD temperatures 10-20% $La_{0.84}Sr_{0.16}CrO_3$ added
to $SrZrO_3$ brings the electrical conductivity close to the acceptable limit for MHD applications with high stability. However,
the conductivity is more than an order of magnitude lower than
the chromite alone.

Considerable interest attaches to the use of $La_{0.8}Sr_{0.2}CoO_3$
as a fuel cell electrode in which capacity it is equal to the
noble metals as an oxygen electrode. The unique contribution of
strontium was established by comparison of the pure compound
$LaCoO_3$ and those with alkaline earth metal substitution. The
data is illustrated in Table I.

The effect of progressive sintering of $La_{0.85}Sr_{0.15}CoO_3$ is
illustrated in Figure 3 showing the progressive grain growth while
Figure 4 shows the effect of electrochemical leaching of a specimen
fired at 1375°C.

TABLE I

Composition	c/a (hexagonal indexing)	Relative Resistivity
$LaCoO_3$	$\sqrt{5.805}$	12.0
$La_{0.8}Ca_{0.2}CoO_3$	$\sqrt{5.906}$	6.20
$La_{0.8}Ca_{0.1}Sr_{0.1}CoO_3$	-	3.88
$La_{0.8}Sr_{0.2}CoO_3$	$\sqrt{5.868}$	1.00
$La_{0.8}Sr_{0.1}Ba_{0.1}CoO_3$	-	1.44
$La_{0.8}Ba_{0.2}CoO_3$	$\sqrt{5.925}$	1.92

The numerical value of the hexagonal c/a ratio is a measure of the rhombohedral distortion which disappears when $c/a = \sqrt{6}$. It is evident that the rhombohedral angle is least affected by strontium substitution.

CONCLUSIONS

Several hundred perovskite compositions have been prepared, fabricated and characterized within this program, many exhibiting interesting characteristics of electrical conductivity, catalytic activity and piezoelectric properties. The possible variations are innumerable since the A and B ions can, subject to the general restrictions previously discussed, be substituted individually or simultaneously with two or more ions. Furthermore, the stoichiometry can be varied over fairly wide limits.

Much more research is needed to evaluate the many possible applications for these materials. Their investigation must embrace chemical, physical and structural and mechanical characterization before their potential can be fully realized.

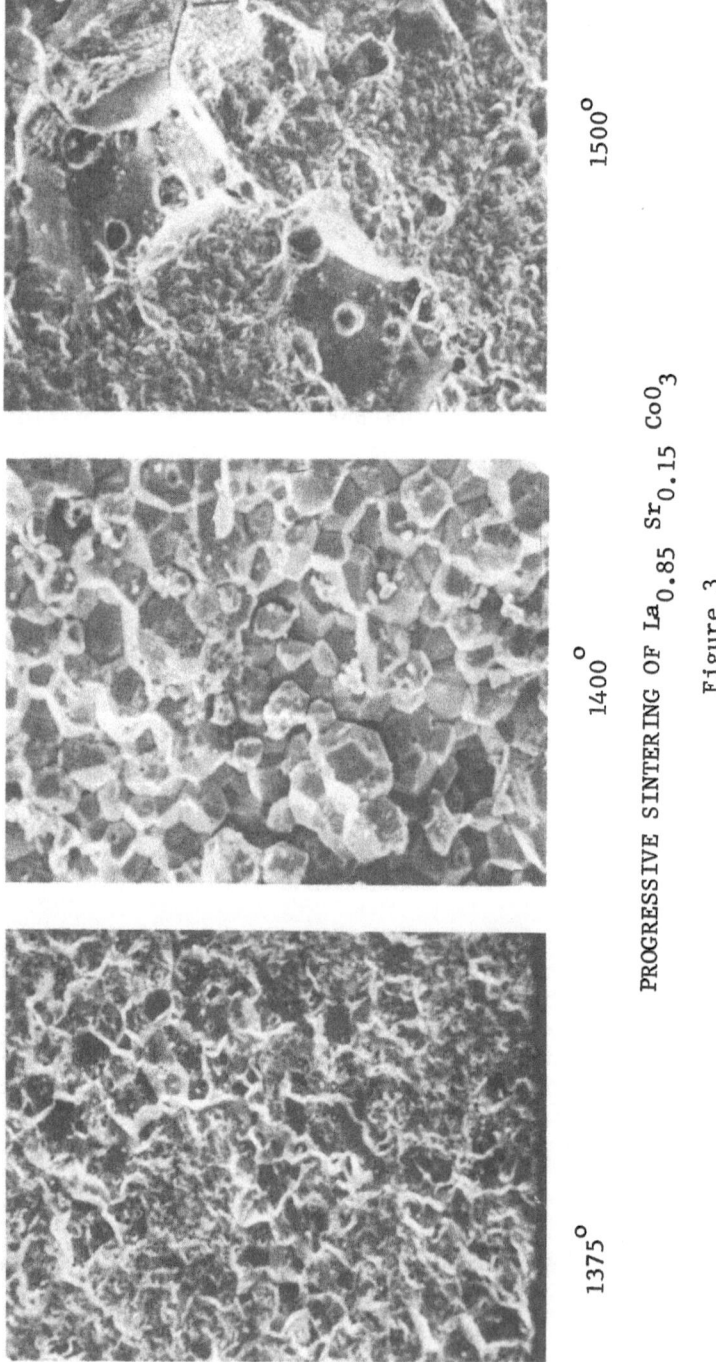

1500°

1400°

1375°

PROGRESSIVE SINTERING OF $La_{0.85}$ $Sr_{0.15}$ CoO_3

Figure 3

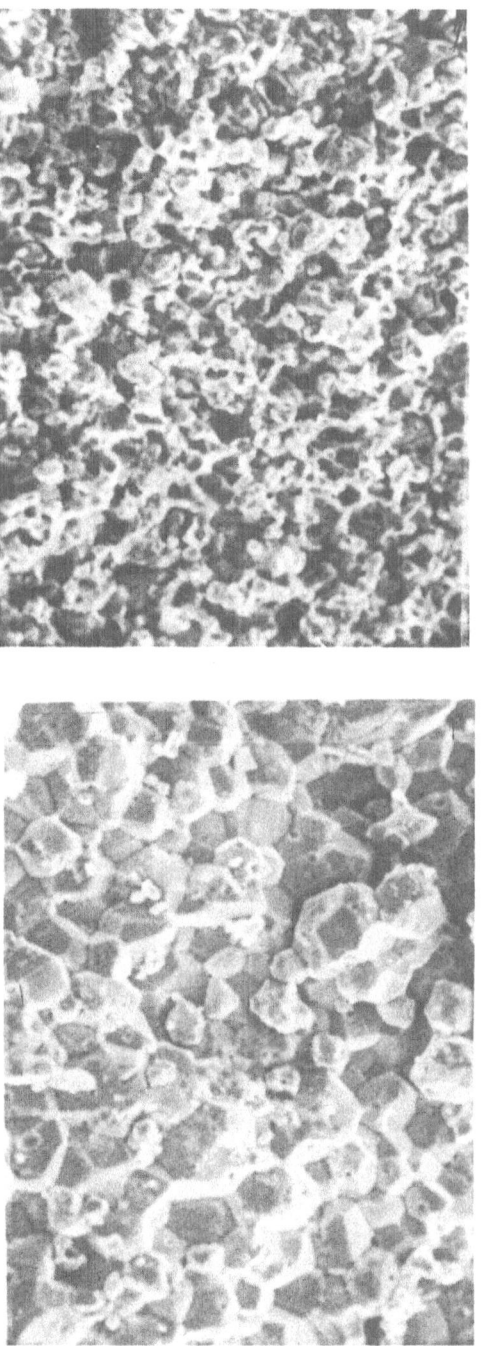

EFFECT OF ELECTROCHEMICAL LEACHING OF $La_{0.85}$ $Sr_{0.15}$ CoO_3

Figure 4

ACKNOWLEDGEMENT

These investigations were conducted as part of the "Special Project on Strontium Compounds" sponsored by the National Research Council, Ottawa, Canada. The assistance of R. Routil, K. M. Castelliz, M. Rockwell and W. Manuel of the Atlantic Industrial Research Institute, Halifax, Nova Scotia in preparing and characterizing the more than three hundred perovskites examined and of A. F. Beck, Olin Corporation, MRL, New Haven, Ct. in preparing the SEM micrographs is most gratefully acknowledged.

REFERENCES

1. Meadowcroft, D. B., Energy Conversion 8, 185 (1968).
2. Meadowcrift, D. B., Brit. J. App. Phys 2 1225 (1969).
3. Jonker G. H. and Van Santen, Physica 16 337; 599 (1950), ibid 19 120 (1963).
4. Jonker G. H., Physica 20 1118 (1954); 22 707 (1956).
5. Galasso F. S., Structure, Properties and Preparation of Perovskite Type Compounds, Pergamon (1969).
6. Takahashi, T. and H. Iwahora, Denki Kagaku 37 857 (1969); ibid 39 400 (1971), Energy Conversion 11 105 (1971).
7. Browall, K. W., O. Muller and R. H. Doremus, Mat. Res. Bull. 11 1475 (1976).
8. Megaw H. D., Proc. Phys. Soc. (London) 58 133, 326 (1946) Trans Faraday Soc A42 224 (1946).
9. Goldschmidt V. M., Geochemische Verteilungsgesetze der Elemente VII and VIII (1927-28).
10. Roth R. S., J. Res. Nat. Bur. Stds. RP 2736, 58 (1957)
11. Schneider S. J., R. S. Roth and J. L. Waring, J Res Nat Bur Stds 65A 345 (1961).
12. Rudorff W., J. Kandler and D. Babel, J Anorg Allgem Chemie 317 261 (1962).
13. Ruddlesden S. N. and P. Popper, Acta Cryst 10 538 (1957).
14. Khattak C. P. and D. E. Cox, Mat Res Bull 12 463 (1977).
15. Jacobs J, K. M. Castelliz, W. Manuel and H. King (In press).

DISCUSSION

David W. Johnson, Jr. (Bell Laboratories): Our experience at Bell Laboratories is that the La-Sr manganates are *more* active catalysts than the corresponding cobaltites for the oxidation of carbon monoxide.

Author: The catalytic activity of a compound or system must be defined relative to a particular reaction for a specific surface area and for a well-defined catalyst. The oxidation of carbon monoxide while commonly quoted on the basis of simplicity is not an adequate criterion for a good oxidation catalyst.

The activity of cobalt-containing catalysts including those with perovskite, K_2NiF_4, scheelite, and related structures depends critically on small deviations from stoichiometry. In the case of lanthanum strontium cobaltite, the stoichiometric compound is far less active than compositions deviating from stoichiometry and exhibiting the presence of K_2NiF_4 structure. Furthermore, the activity of a catalyst can change radically depending on pre-treatment and subsequent aging. Experience indicates that, in general, manganese-containing catalysts deteriorate rapidly and the initial apparent high activity is due to reaction with the catalyst material.

HOT PRESSED COMPOSITE CERAMIC MHD ELECTRODE DEVELOPMENT

J.J. Stiglich and L.A. Addington

Eagle-Picher Industries, Inc.

P.O.Box 1090 200-9th Ave.N.E., Miami, OK 74354

Hot pressing development work on various candidate MHD electrode materials has been going forward at Miami Research Laboratory
of Eagle-Picher for the last 12 months. Initial experiments were
supported by Battelle Northwest (Dr. J.L. Bates) and the current
work is sponsored by the Westinghouse R/D Center (Dr. B. Rossing).
Other experiments have been supported on Miami Research Lab. internal programs. The purpose of this paper is to present a brief
review of all fabrication efforts to date.

During the course of MHD electrode experiments it became apparent that a protective "cap" would be desirable on the hot wall type
electrode, as shown in Fig. 1. All work at MRL has been directed
toward fabricating electrode materials ($La(Mg)CrO_3$ from General
Refractories and from University of Missouri-Rolla, Fe doped spinel
MAFF-31 from Trans Tech) with various cap materials attached by
means of simultaneous hot pressing of multilayered composite samples.
Cap materials fabricated include ZrO_2 (doped with CeO_2 and Y_2O_3) from
A-T Research, HfO_2 (doped with CeO_2 and Y_2O_3) mixed at MRL using
HfO_2 from Wah Chang, CeO_2 from Alfa-Ventron, Y_2O_3 from Molybdenum
Corp. of America, and ZrO_2 (doped with 12 wt% Y_2O_3) from Zircar
Products. See Table I for electrode cap combinations and sources
of raw materials. Some of the parts were formed with a metal back
incorporated in the gradient structrue. Molybdenum from AMAX
Specialty Metals and Ti from Alfa Products, Ventron Corporation
were used in these cases.

Hot pressing was carried out in an Eagle-Picher designed semi-
continuous hot press which is resistance heated. The hot zone and
tooling are protected by a flowing inert gas. It is capable of
temperatures of $2300^{\circ}C$ over a hot zone of approximately 10.2 cm

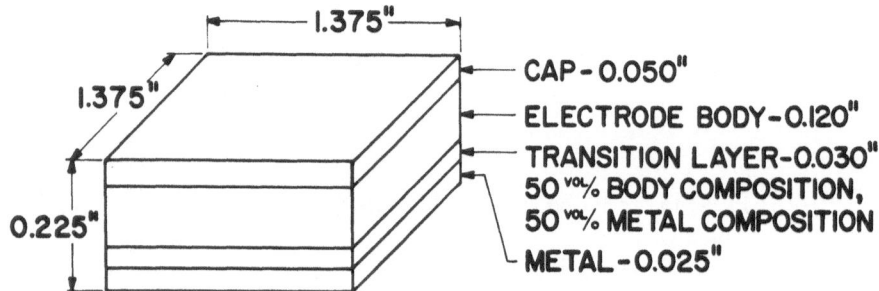

NOTE: PRESSED TO SHAPE & FINISH GROUND

Figure 1a.   BNWL project electrode segment with metal bottom layer.

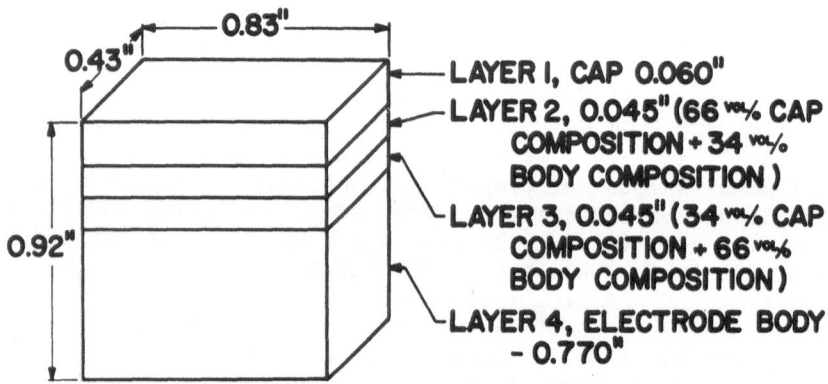

NOTE: PRESSED AS 1.5" DIA. DISCS, THEN
      SLICED & FINISH GROUND

Figure 1b.   Westinghouse project electrode segment.

diameter x 15.2 cm long. Graphite tooling was used in the BNWL
program and for the beginning of the Westinghouse program, but it
was found advantageous to switch to Al_2O_3 bore liners and punch
protectors when hot pressing the La-chromite materials. The reason for this was the tendency of the chromite to decompose at the
interfaces between the cap and the cap transition layers as evidenced by the bright phase presumed to be metallic Cr in Figure
2. A similar (ie. metallic precipitation) effect was noticed with
the MAFF-31, although it was hoped that changes in the hot pressing cycle might prevent or minimize it. These hopes were not realized on the BNWL project. The hydraulic pressure system can impose
34.5 MPa (5000 psi) on a 3.8 cm (1 1/2 inch) diameter disc. Fabrication parameters and results obtained in the various experiments
are as follows.

BATTELLE, NORTHWEST LABORATORY PROJECT.

Rectangular slabs 1.016 cm (0.400 in) x 3.492 cm (1.375) x
0.572 cm (0.225 in) were hot pressed in multibore graphite tooling
at temperatures ranging from 1550°C to 1650°C and at pressures from
29.3 MPa (4250 psi) to 37.9 MPa (5500 psi). The materials system
used in this study is designated in Table I as raw materials combination number 1 (RMCN 1), the Fe doped spinel electrode body with
a doped HfO_2 cap. Powder particle sizes were < 10 μm.

In addition to rectangular slabs, discs were pressed 4.128 cm
(1.625 in) diameter by 0.572 cm (0.225 in) thick for metal backed
parts and 0.432 cm (0.170 in) thick for parts without metal backs.
The discs were then sawed into rectangles of the shape shown in
Fig 1, A.   The materials used in disc experiments are those designated in Table I as RMCN 2 through 5, all containing $La(Mg)CrO_3$
in the electrode body.   Powder particle sizes were < 10 μm.

Results of the hot pressing experiments on both rectangular
slabs and discs are given below. (See Table II for summary). The
rectangular pressed parts, RMCN 1, exhibited delaminations within
the transition layer and between that layer and the pure metal and
pure spinel layers on either side. Parts fabricated at higher temperature (1650°C) showed a metallic precipitate in the delaminated
areas indicating reduction of Fe_3O_4 as previously discussed. The
volume change and thermal expansion mismatch resulting from the
reduction to the metallic phase appeared to aggravate the delamination. The CTE of metallic Fe is 1.5 to 2 times greater than
that of the spinel or the Mo/Ti mixture. The optimum pressing conditions were found to be 1600°C and 31.0 MPa (4500 psi). The overall bulk density obtainable under these conditions was 93% of theoretical. These conditions minimized but did not eliminate the
delaminations.

Table I - Electrode and Cap Combinations Investigated

RMCN*	CAP	ELECTRODE BODY	METAL BACK	**
1	HfO_2 80 Mole %(A)*** CeO_2 8 Mole % (B) Y_2O_3 12 Mole % (C)	$3MgAl_2O_4 \cdot Fe_3O_4$ MAFF-31 (D)	50 Wt % Mo (E) 50 Wt % Ti (B)	Fig. 1, A "
2	"	$La_{.95}Mg_{.05}CrO_3$ (F)	"	"
3	ZrO_2 88 Wt % Y_2O_3 12 Wt % (G)	"	"	"
4	Same as 2	"	None	"
5	Same as 3	"	"	"
6	ZrO_2 85 Mole % CeO_2 12 Mole % (H) Y_2O_3 3 Mole %	$La_{.95}Mg_{.05}CrO_3$, 80 Vol % (F) $La_{.95}Mg_{.05}Cr_{.5}Al_{.5}O_3$ 20 Vol % (H)	"	Fig. 1, B

Raw Material Particle Sizes Were All < 10 μm.

*	Raw Material Combination Number
**	Electrode Configuration, Fig. 1, A & B
***	Letter in Parentheses Indicates Vendor

(A)	Teledyne Wah Chang	(E)	AMAX Specialty Metals
(B)	Alfa Products, Ventron Corp.	(F)	General Refractories Co.
(C)	Molybdenum Corp. of America	(G)	Zircar Products, Inc.
(D)	Trans Tech, Inc.	(H)	A-T Research/UMR

Table II - Results of BNWL Project

RMCN	OPTIMUM CONDITIONS	% T D OBTAINED	REMARKS
1	1600°C 4500 PSI	93	Delaminations at either side of body/metal transition layer, minimized but not eliminated.
2	1650°C 4000 PSI	94.5	Negligible rejects
3	"	"	Progressive crazing of the cap observed to occur at room temperature. No spalling or part rejection.
4	"	"	Negligible rejects
5	"	"	Progressive crazing combined with rejectable delamination.

The parts pressed as discs (RMCN 2 through 5, La(Mg)CrO$_3$ electrode bodies) were more stable at higher fabrication temperatures than the Fe-doped spinel--based electrodes. Delaminations were rarely observed in the as pressed part with RMCN 2 through 4. There was, however, a problem with RMCN 5. Subsequent examination at BNWL indicated some reduction of La(Mg)CrO$_3$ at the interface between electrode body and cap in all these parts. The optimum pressing conditions were found to be 1650°C and 27.6 MPa (4000 psi). The overall bulk density obtained was 94.5% of theoretical.

WESTINGHOUSE PROJECT

Discs 3.9 cm (1.530 in) diameter by 2.3 cm (0.920 in) thick were hot pressed in graphite tooling at temperatures ranging from 1155°C to 1615°C and pressures from 29.6 MPa (4300 psi) to 33.8 MPa (4900 psi). One material system (RMCN 6) was developed in this study, and several other materials which were not ultimately utilized in that system were investigated. During the course of this development we concluded that the reducing effects of unprotected graphite tooling upon La(Mg)CrO$_3$ were beyond eliminating by manipulation of the hot pressing cycle. Alternative die liners were investigated, including BN aerosol, nickel, platinum, tantalum, Fiberfrax paper, MgO powder, Cr$_2$O$_3$ powder, and sintered Al$_2$O$_3$.

Various forms of Al$_2$O$_3$ proved to be the most effective barrier to reduction provided temperatures were kept low enough to preclude reaction between La(Mg)CrO$_3$ and Al$_2$O$_3$. High density parts (up to 97% of theoretical) were fabricated without free Cr metal at a temperature of 1475°C and a pressure of 4900 psi. This is demonstrated by comparing Figure 2, A which is a photomicrograph of an electrode segment hot pressed in an unlined graphite die with Figure 2, B a photomicrograph of an electrode segment pressed near the end of our development program.

The crazing or microcracking of the doped ZrO$_2$ cap is evident in Figure 2, B. This effect is similar to that observed in the ZrO$_2$ capped electrodes fabricated in the BNWL work and was not completely overcome. The raw materials from which the Westinghouse electrodes were fabricated were in many cases made in pilot production lots. Variances in pressability and bulk density were observed from lot to lot. The particle sizes were in general extremely fine, less than 5 microns ultimate crystalite size.

No rectangular slabs were attempted because it was felt that the principle of practical production rates had been demonstrated in the BNWL project. The writers are convinced that electrode shapes can be made cost effective by currently available production hot pressing techniques. It was therefore decided to slice the rectangular sections needed for the Westinghouse channel out

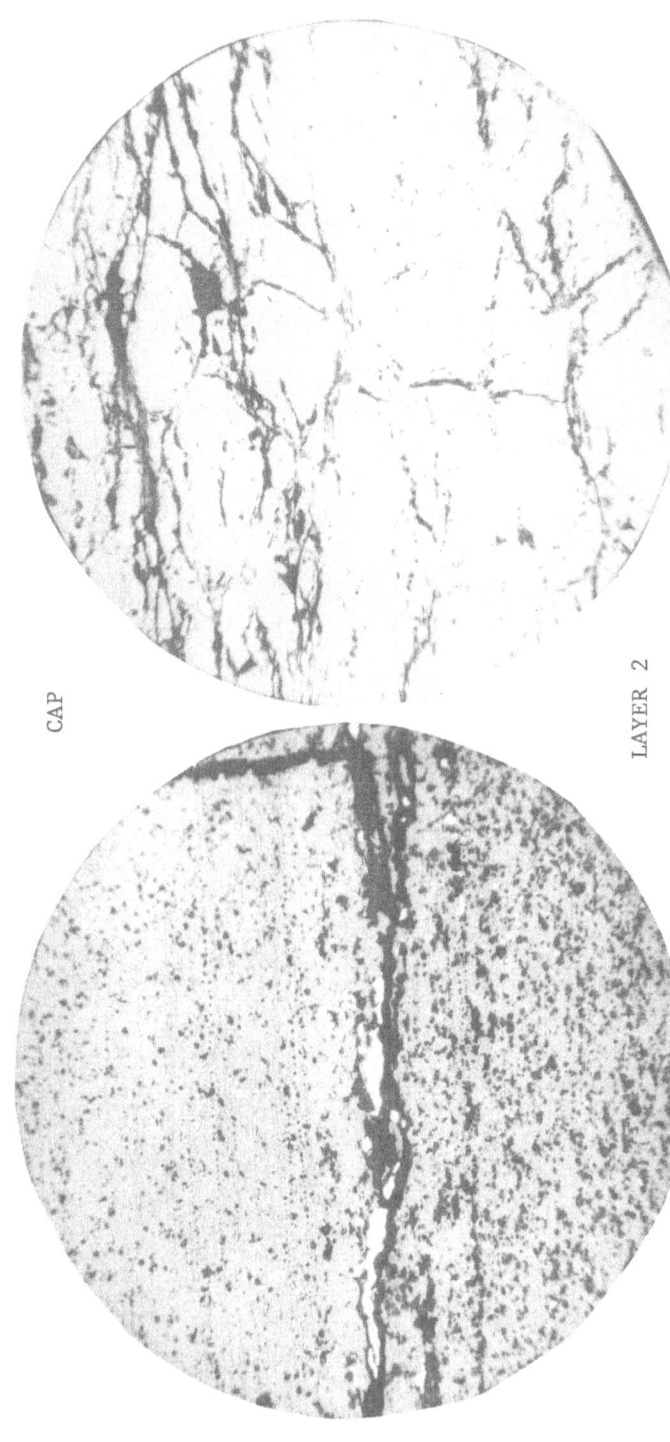

CAP

LAYER 2

FIG. 2, A – Cap to La(Mg)CrO$_3$ Bearing Layer Interface, Pressed in Unlined Graphite Die. (Note presence of bright metallic phase at interface.)

MAG. = 125 x

FIG. 2, B – Cap to La(Mg)CrO$_3$ Bearing Layer Interface, Pressed in Al$_2$O$_3$ Liner Near End of Program. (Note absence of bright metallic phase at interface.)

MAG. = 125 x

Table III. Results of Westinghouse Project.

RMCN	PRESSING CONDITIONS	% T D OBTAINED	REMARKS
6	1155°C to 1615°C 4300 to 4900 PSI	97	1. Reduction of La(Mg)CrO$_3$ to point of appearance of metallic Cr was eliminated. 2. ZrO$_2$ based cap exhibited microcracking.

of the developmental discs. To accomplish this, a Buehler slow
speed slitting saw was used.

The complex nature of the extreme conditions existing in a
coal fired MHD channel precludes simple tests or calculations of
the success of a candidate electrode. It must be tested under
conditions which reproduce as completely as possible the temperatures, material flows, chemistry, electrical, and other conditions which exist in the real channel. Unfortunately, the materials fabricated for BNWL were not tested under simulated MHD
channel conditions. However, the electrodes fabricated for
Westinghouse will be tested in such a channel at Westinghouse;
the tests are scheduled for approximately the same date as this
conference.

The Westinghouse project electrodes described in the body of
the paper were tested in the Westinghouse Materials Test Facility
shown schematically in Fig. 3. The test was conducted at 1750°C
for a duration of 21.5 hours at that temperature. Potassium seed
was injected to simulate the conditions to be encountered in a

NOTE ADDED IN PROOF:

The Westinghouse project electrodes described in the body of
the paper were tested in the Westinghouse Materials Test Facility
shown schematically in Fig. 3. The test was conducted at 1750°C
for a duration of 21.5 hours at that temperature. Potassium seed
was injected to simulate the conditions to be encountered in a

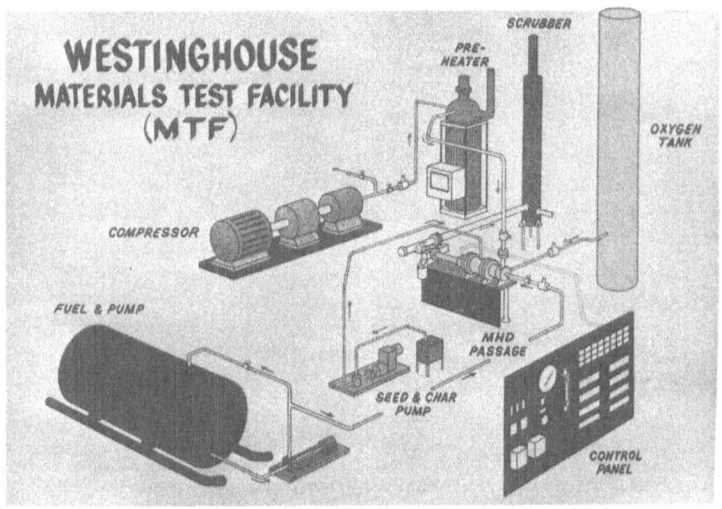

Figure 3. Westinghouse Materials Test Facility, Schematic.

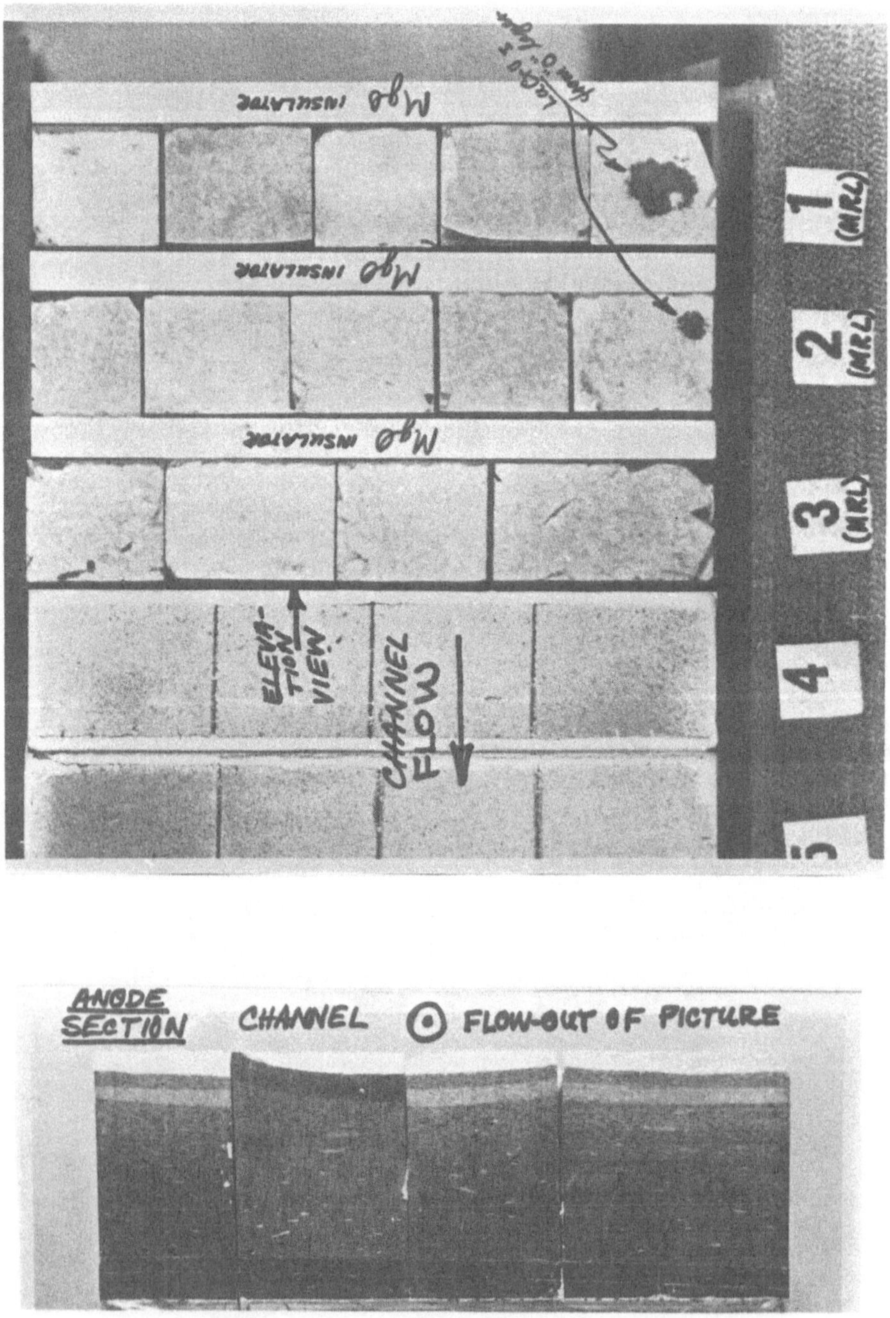

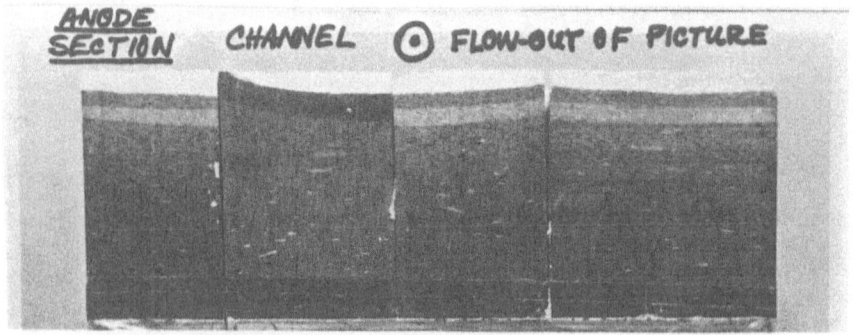

FIG.4- Anode Wall Section Prior to Test (Subject Electrodes No. 1,2,3).

working MHD channel. Current densities from 0.6 amp/cm^2 to
1.0 amp/cm^2 were imposed on the electrodes.

The visual inspection following testing was very encouraging.
The first cathode showed slight melting which could have been due
to a localized temperature excursion. The other cathode and anode
surfaces appeared undamaged. Measurements showed less than 0.1 mm
material loss. The caps did not exhibit any spalling tendencies.

Figure 4, shows a section of the assembled anode wall prior
to testing. The caps show some tendancy to chip in grinding, however,
this did not lead to spalling in the test. Post test photographs
were not available at the time of this writing.

ACKNOWLEDGEMENT

The authors would like to acknowledge helpful discussions and
technical support by Dr. B. Rossing and Mr. E. Kochka of the
Westinghouse R and D Center.

PROCESSING VARIABLES AFFECTING THE THERMOMECHANICAL DEGRADATION

OF MONOLITHIC REFRACTORY CONCRETES

Curtis E. Zimmer and Edward M. Anderson

Babcock & Wilcox Company

Lynchburg, Virginia

ABSTRACT

The mechanical reliability of monolithic refractory concrete linings for use in coal gasification process vessels must be improved if long life (>5 yrs) is to be achieved. High alumina (50-90+% Al_2O_3) and low silica and iron calcium aluminate bonded dense and insulating castables have a strong tendency to crack, shrink and spall during initial dry-out, heat-up and subsequent cool-down. Various parameters used in the installation of these linings and the critical nature of the processing steps during mixing, casting, curing and heat-up have a significant effect on the extent of phases developed, cracking and even explosive spalling of the material. This paper reviews how the thermomechanical degradation of dual component linings can be related to properties resulting from these processing variables.

INTRODUCTION

Many of the coal gasification pilot and demonstration plants under evaluation in the United States and overseas use single and dual component monolithic refractory concrete linings to protect the vessel shell. These linings are similar in design to those used in petrochemical industry process vessels. These refractories are expected to retain their physical and chemical integrity and to remain volume stable and relatively crack free for long periods of time during exposure to the gasifier environment. This environment includes corrosive and erosive conditions and temperatures of 815-1100°C (1500-2000°C).

These monolithic refractory linings have inherent disadvantages over brick type linings in errosive and corrosive environments but their lower material and installation costs and ease of installation, especially in complicated structures, are considered to be key advantages. Cracking, spalling and other forms of thermomechanical degradation are considered to be key problem areas with monolithic refractory concretes. The cracking and spalling of the lining leads to hot spots on the shell which can cause sudden shell failures and process shutdowns. Metal anchor and vessel shell interactions with the refractory lining compound this cracking and spalling tendency.

Wygant and Crowley[1] have attempted to study the factors which affect the cracking and thermomechanical degradation of single component monolithic refractory lined vessels. Considerably more work is needed in this area and on how the processing variables used to prepare these refractory concretes for service can be controlled.

The Department of Energy has several programs underway to study the integrity, stability and life of these materials in coal gasification environments. One program[2] is to improve the thermomechanical reliability of monolithic refractory linings for nonslagging coal gasifiers. The scope of this work[3] is to perform a systematic study on refractory lined vessels by monitoring the temperature profile, strains, stresses and acoustic emissions of linings, the vessel shell and anchors during the initial cure, dry-out, heat-up and cool-down of the whole system. This is being done to develop an understanding of degradation mechanics, improved lining designs, improved installation procedures and heat-up and cool-down guidelines. This program has involved designing and building a 1.5 m (5 ft.) diameter test vessel, developing a mathematical model along with test techniques, determining properties of these materials, testing 0.3 m (1 ft.) thick linings up to 1100°C (2000°F) and monitoring and analyzing lining and vessel results. Examples used later in this paper were taken from this study.

PROCESSING VARIABLES

Various parameters used in the installation of these refractory concrete linings and the sensitivity of these materials to the processing steps during mixing, casting, curing and heat-up have a significant effect on the extent of cement bond phases developed, cracking and even explosive spalling of the material. The refractory concretes of concern consist of tabular alumina or calcined kaolin aggregates bonded together with a calcium aluminate cement phase. The processing variables which are important to the installation and use of calcium aluminate bonded refractory concretes are water content, mixing time, casting temperature, cure temperature, surface smoothness of cast material, humidity and heating rate and schedule.

Depending upon whether the concrete is a dense or insulating type and whether it is cast, gunned or rammed, water additions can vary from 5 to 25 percent of the total batch weight. The water levels added will directly affect strength, shrinkage, permeability, porosity, thermal expansion and other properties. Generally, the lower the water content the greater the strength and the lower the shrinkage. Huggett[4] has noted that the drying and firing shrinkage can be excessive for a dense castable when the optimum water level is exceeded by only 2 to 5 percent. The shrinkage occurring during curing, drying and heat-up has a very significant effect on cracking tendency. Thus, the water content should be kept as low as possible to keep the strength high and reduce the amount of excessive water release occurring during the curing, dry-out and heat-up of mono-lithic linings. If these materials are mixed too long prior to installation, they set more quickly and have poor workability. This condition leads to a porous or weak structure. When compared with portland cement concrete, the working times of these refractory materials are quite short - on the order of 20 to 30 minutes versus 2 to 3 hours. This means one has to work fast and be totally pre-pared when beginning to install or work with these materials.

The casting and curing characteristics of these materials are affected by the temperature of the materials and equipment during installation. The higher these temperatures are, the faster the material will set up and the shorter will be its working time. The cracking and explosive spalling tendency of these linings also are affected by these conditions and the temperature experienced by the cement bonding phase during the cure. A requirement for satisfactory during is the retention of moisture within the concrete and the use of an acceptable curing temperature. Therefore, humidity plays an important role during the curing of these concretes to develop the proper phases and degree of crystallinity and hydration. Properties change when the curing temperature gets above 21-27°C (70-80°C). The use of castable curing temperatures below 21-27°C can produce a relatively impervious microstructure[5][6] which enhances explosive spalling. This has been found to be caused by a change in phase structure and in degree of crystallinity of the hydrated cement. A number of researchers[6,7,8] have found that the mineralogy of the hydrated cement phases vary depending upon the curing temperature used. Table 1 lists some of the major phases formed at various curing temperatures for a high alumina cement composed primarily of calcium aluminate. A hexagonal hydrated calcium aluminate phase will form when curing temperatures are below 24°C (75°F) as compared with the more desirabel cubic hydrated CA phase which forms between 27-46°C (80-115°F). The phases formed below 24°C are alumina gel and hexagonal hydrates CAH_{10} and C_2AH_8 which produce a low permeability structure. These phases transform to AH_3 and a cubic hydrate phase C_3AH_6 as the curing and/or drying temperature is progressively increased above 29°C (85°F). Curing at greater than

Table 1. Cement Phases Formed Vs. Curing Temperature

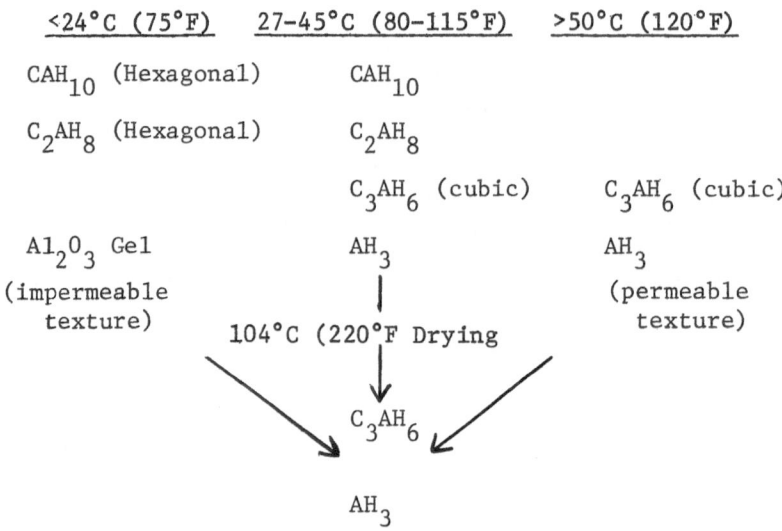

<24°C (75°F)	27–45°C (80–115°F)	>50°C (120°F)
CAH_{10} (Hexagonal)	CAH_{10}	
C_2AH_8 (Hexagonal)	C_2AH_8	
	C_3AH_6 (cubic)	C_3AH_6 (cubic)
Al_2O_3 Gel (impermeable texture)	AH_3	AH_3 (permeable texture)

104°C (220°F Drying

C_3AH_6

AH_3

50°C (120°F) causes C_3AH_6 and AH_3 to form initially, producing a more open porosity than the lower cured temperature systems. Upon heating, the cubic phases will form. The volume, and therefore, the resultant permeability and strengths of the hydrated phases, are different. In general, as curing temperature increases, the strength, permeability and explosion or cracking resistance increase.

    The surface condition of the installed concrete is important in that the surface pores must be left open to allow moisture to escape during heating. The two major methods of installing linings, casting and guniting, can give different surface textures and microstrucutes. Some observers[4] have even noted different types of cracks and crack patterns in various field units with linings installed by different techniques. This becomes more important where the process vessel has a vapor tight shell and uses insulating castable behind a dense castable service lining. Here, all the water must come out through the dense castable. Frequently, a workman slicks the surface of the lining to an artistically smooth finish which seals off the effective surface pores for moisture release. When this happens the lining becomes increasingly more sensitive to cracking and explosive spalling. To prevent this from happening, the hot face surface must have a coarse open texture to maximize the safe diffusion of the steam through the lining. Generally slow heating rates of ~28°C/hr (50°F/hr) are used to reduce or eliminate the cracking and explosive spalling caused by this effect.

The shrinkage occurring during curing, drying and heat-up has a very significant effect on cracking tendency. Drying shrinkage, which also depends on curing temperature, should be considered in the design of linings. However, curing shrinkage is generally considered to be very small and is normally ignored. During the early hours of setting, the refractory can shrink from 0.1 to 0.7 percent depending upon water content. If excessive mixing water is used and the lining is restrained such as by rigid forms or anchors or vessel geometry, cracking can occur before the lining is even heated. This point is not widely known and may account for cracking which occurs even though very conservative drying and subsequent heating schedules are followed for some linings. Strain analyses and post installation observations[2,3] have shown that significant stresses have been found in two component linings during curing due to this shrinkage as well as at temperatures of only (200°C) 400°F. During initial heat-up, normal thermal expansion is interrupted by considerable shrinkage between 200 and 370°C (400 and and 700°F). This shrinkage of around 0.1 to 0.4 percent, depending upon the material and water level used, is shown in Figure 1. Since the refractory concrete becomes brittle at this stage and has a low resistance to tensile stresses, the shrinkage occurring is also considered to be a major cause of cracking frequently observed in massive linings on initial heat-up. Figure 1 also shows that the water levels used during processing will alter the amount of shrinkage in this temperature range. There is also shrinkage on heat-up to higher temperatures. Not much shrinkage data is available in the literature, but what is found[1,9] indicates the shrinkage of 50 and 90+% Al_2O_3 castables will be around 0.2 percent at temperatures up to 1100°C (2000°F). This has been confirmed by our measurements with a 1.5 m (5 ft) diameter refractory lined vessel after firing to 650°C (1200°F). Huggett[4] has indicated that cracking can and does occur on cooling or due to the excessive shrinkage of castables when too high water levels are used. He suggests shrinkages should be kept below 0.1 percent to prevent cracking, but this is difficult with the many variations occurring during installation.

Cracking is also caused by stress interactions between refractory lining components, the lining and shell or the lining and metal inclusions. To examine the stress interactions in a 1.5 m (5 ft.) diameter refractory lined vessel, strain gages are embedded in the two lining components to confirm predictions by a math (computer) model. Figure 2 shows that if we look at predicted and experimental radial strains at midpoints of the dense and insulator component during heat-up to 650°C (1200°F), the experimental results indicate the dense component expands and pushes on the insulator. This causes the insulator to go into compression until the dense begins to cool and the insulator becomes hotter. The predicted and experimental radial strains agree well in some regions and poorly in others. The differences are thought to be due to cracking, to errors introduced

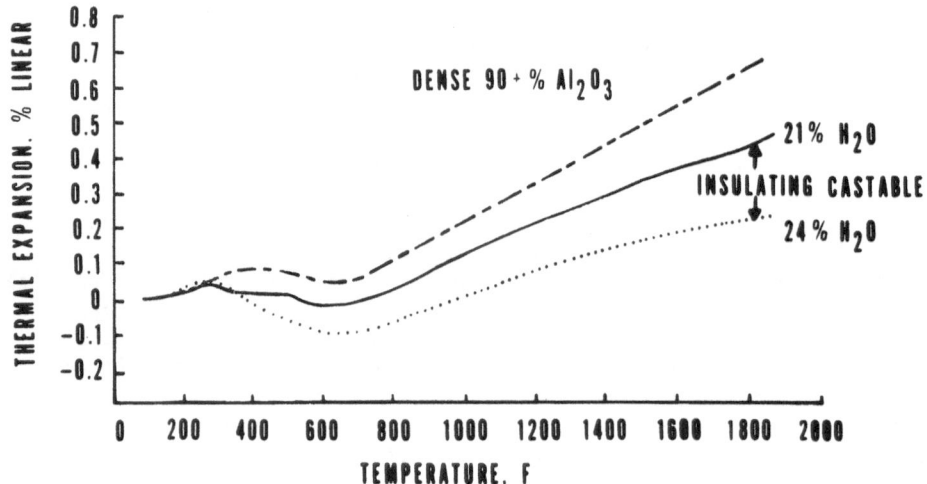

Figure 1.   Shrinkage and thermal expansion of dense and insulating
            castables on initial heating. Note increasing the level
            of mixing water increases the shrinkage.

by the data reduction equations used for the strain data, or else
to the properties that were used in the math model used to predict
the strains. Wygant and Crowley[1] developed a one dimensional
mathematical model for a single component refractory lining and
used it to theorize that stress relaxation through creep reduced
the tendency of the lining to crack. Their model did not include
the effects of the metal shell or of anchors except in the case of
thermal conductivity, nor did it consider multi-component lining
designs that are presently being used in coal gasification pilot
plants.

PRESENT GUIDELINES

    The sensitivity of the refractory to cracking and, in extreme
cases, to explosive spalling can be related to the curing temperature, the permeable porosity and the heat-up schedule used. It is
customary to recommend[5,6] that refractory castables be heated up
for the first time at a reasonably slow rate so that any steam
formed from the free or uncombined water can escape without causing

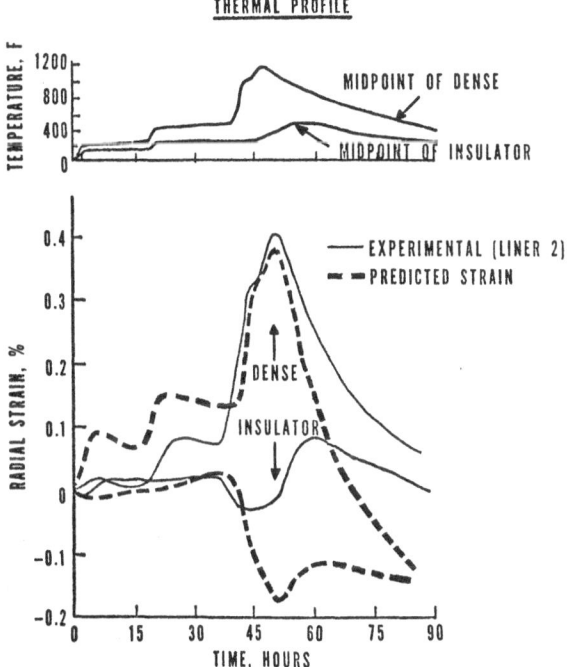

THERMAL PROFILE

STRAINS AT MIDPOINTS OF DENSE AND INSULATING COMPONENTS

Figure 2.   Predicted and experimental strains at mid-points of dense
and insulating components during heat-up to 650°C
(1200°F) of 1.5 m (5 ft.) dia. refractory lined vessel.

cracking or explosive spalling. Heating rates of 28-112°C/hr
(50 to 200°F/hr) are recommended with intermittent soaks for
safety. Rapid heating causes surface drying which does not readily
permit the passage of moisture from the internal regions of the
castable and results in the buildup of steam pressure. This pressure
can sometimes exceed the ultimate strength of the bond, and thus
explosive disintegration or spalling occurs. Figure 3 shows a dual-
component panel segment which explosively spalled the front 0.06m
2 1/2 in) of a 90[+] percent dense Al_2O_3 concrete with enough force
to blow this 136 kg (300 lb.) panel out of the furnace door. The
explosion was initiated by an unplanned heating rate of 120°C/hr
(250°F/hr), in the critical temperature range 200-550°C (400-1000°F).
However, there were other factors which contributed to this condition.
Among them was the fact that this panel was cured over a long week-
end in the winter when the room temperature was cut back to 10°C
(65°F). This gives credence to the findings of other investigators[5,6,7,10,11] who have pointed out that curing at cold temperatures less than 21-24°C (70-75°F) can lead to explosive spalling.
Another contributing factor was troweling the hot face smooth which

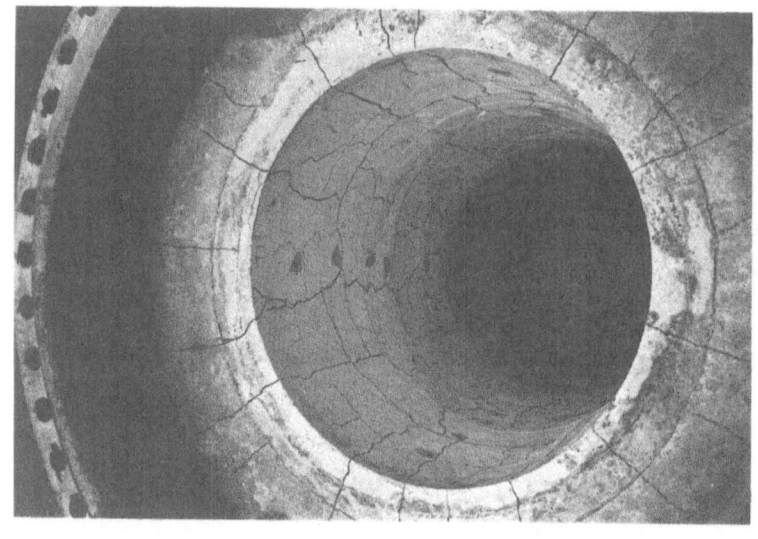

Figure 4. Top view of vessel/test
furnace showing crack
pattern in lining observed
after thermal cycles to
650 and 1100°C (1200 and
2000F). Note semi—regular
crack spacing.

Figure 3. Explosively spalled dual
component panel segment
after being blown out of
furnace door.

only served to close off surface pores which would allow the
moisture or steam to escape. Similar explosive spalling was seen
in another dual-component panel after an unprogrammed heating rate
of 238°C/hr (460°F/hr) in the range of 425-550°C (800-1000°F).
This occurred even though the panel was prepared by paying careful
attention to processing variables so as to reduce or avoid any
cracking or spalling. The forces generated during explosive
spalling were great enough to strip threaded nuts of bent anchor
extensions in the dense component. Thus, it took a heating rate
of 238°C/hr (460°F/hr (460°F/hr (460°F/hr) versus 120°C/hr
(250°F/hr) to explosively spall a panel made with good versus bad
processing variables.

Many anchor designs and spacings have been used to hold single
and multi-component monolithic lings in place.[10,12,13] The actual
configuration and spacing have varied from 0.3-1.0 m or more depending
on the lining thickness and shell curvature. Closer spacings of
less than 0.3 m (12 inc) have been found to cause severe cracking
and spalling. These findings indicate that anchor-refractory
interactions exist which are not well understood and must be reduced
or eliminated. As an example of the cracking tendency affected by
anchor interactions, Figure 4 shows crack patterns observed in a
lining tested to 1100°C (2000°F). V-type anchors were spaced at
0.3 m intervals. Although other factors contributed to cracking,
examination of the crack pattern indicates a relationship to anchor
spacing and location. The anchor spacings of 0.3 m (12 in.) in
this cylindrical geometry were too close.

CONCLUSIONS

The material effects of mixing water addition, low temperature
curing, low temperature shrinkage, permeability and surface condition,
as well as the design effects of anchor design and heating rate are
significant processing variables affecting the thermomechanical
degradation of monolithic refractory linings for coal gasification
process vessels. These must be carefully controlled if cracking
and even explosive spalling are to be reduced or eliminated. Designs
for linings of coal gasification vessels are still in the developmental stages, and a better understanding of cracking problems
resulting from the installation and initial heat-up need to be
developed if long life is to be achieved.

ACKNOWLEDGEMENT

This work was done under the auspices of the United States
Department of Energy (Contract No. EX-76-C-01-2218).

REFERENCES

1. J. F. Wygant and M. S. Crowley, Am. Ceram. Soc. Bull., 43, (3)
    173 (1964).
2. E. M. Anderson and C. E. Zimmer, "Improvement of Monolithic
    Refractory Linings for Coal Gasifiers", paper presented at
    the Refractories Division Fall Meeting of the American
    Ceramic Society, Bedford, Pennsylvania, October 6, 1977.
3. E. M. Anderson, et al., Quarterly Progress Reports for the
    Period July 1976-September 1977, prepared for U. S. Depart-
    ment of Energy, Contract No. EX-76-C-01-2218.
4. L. G. Huggett, "Lining of Secondary Reformers" in Materials
    Technology in Steam Reforming Processes, Permagon Press,
    New York, 1966.
5. M. S. Crowley and R. C. Johnson, Am. Ceram. Soc. Bull., 51 (3)
    226 (1972).
6. W. H. Gitzen and L. D. Hart, Am. Ceram. Soc. Bull., 40 (8)
    50 3 (1961).
7. G. V. Given, et al., Am. Ceram. Soc. Bull., 54 (8) 710 (1975).
8. M. S. Crowley and J. F. Wygant, Am. Ceram. Soc. Bull., 43 (1)
    1 (1964).
9. R. P. Heilich, et al., Am. Ceram. Soc. Bull., 50 (6) 548 (1971).
10. C. R. Venable, Am. Ceram. Soc. Bull., 48 (12) 11 14 (1969).
11. W. T. Bakker, "Materials for Coal Conversion Process: An
    Engineering Approach", Paper presented at the ERDA-FE
    Information Meeting, Gaithersburg, Md. 1975.
12. S. H. Vaughan, Iron & Steel Engineers, 5 64 (1972).
13. M. S. Crowley, Am. Ceram. Soc. Bull., 45 (7) 650 (1966).

UO_2-Gd_2O_3 SINTERING BEHAVIOR

Hubert H. Davis & Ralph A. Potter

Babcock & Wilcox
Lynchburg, Virginia
Lynchburg Research Center

ABSTRACT

The sintering characteristics of UO_2-Gd_2O_3 mixtures were investigated over the range 0 to 20 wt% Gd_2O_3. It was observed that as little as 0.5 wt% Gd_2O_3 will reduce the sinterability of powder mixtures. This effect is correlated to stoichiometry changes induced by the ionic substitution of Gd^{+3} for U^{+4} in the structure.

INTRODUCTION

The rare earth element gadolinium (Gd) is of considerable interest as a burnable poison in light water nuclear reactors. Two of the isotopes of the element are neutron absorbers: ^{155}Gd with a neutron absorption cross-section of 60,000 barns and ^{157}Gd with a cross-section of 240,000 barns. Each isotope occurs with an abundance of 15% in natural Gd, and neither undergoes a chain reaction to form other high cross-section isotopes. These factors result in a relatively fast burnout of the poison material.[1]

Gadolinium has been used as a burnable poison in nuclear reactors since about 1964, mostly in boiling water reactors,[2] and more recently in pressurized water reactors.[3] In all cases, the oxide Gd_2O_3 is integrally mixed in the starting UO_2 powder to become uniformly distributed in the UO_2 fuel pellet. Commonly, the Gd_2O_3 content is 1 to 3 wt%.

This study was performed to evaluate the effects of the Gd_2O_3 additions on the sinterability of the UO_2.

EXPERIMENTAL PROCEDURE

Pellets were pressed from mixtures of untreated UO_2 and Gd_2O_3 powders. The Gd_2O_3 powder (Type C, 99.99% pure, obtained from Molycorp*) was passed through a 400 mesh sieve to remove particle agglomerates. Two different types of UO_2 powders were used to look at the effect of Gd_2O_3. One UO_2 powder was manufactured via the ammonium diuranate (ADU) process and was obtained from Kerr-McGee Chemical Corporation. Previous experience with this batch of powder has shown it to be highly active (BET surface area = 3.94 m^2/g) with an O/U ratio of 2.16. Sinterability tests revealed that this powder easily densifies to 95-97% theoretical density (T.D.) when sintered in the temperature range 1550° to 1750^oC.

The second UO_2 powder was obtained from the ammonium uranyl carbonate (AUC) process. This was found to be of relatively low activity, sintering to only 91-92% T.D.

A master blend of 20 wt% Gd_2O_3 - 80 wt% UO_2 was prepared for each UO_2 powder by blending for one hour in a twin shell vee-cone blender. Portions of the master blend were then blended with various amounts of UO_2 to produce mixtures containing 1/2, 1, 2, 3, 4, 5, 10, and 20 wt% Gd_2O_3. A similar procedure was followed to prepare mixtures containing the AUC UO_2. Final compositions were 1, 2, 4, 10, and 20 wt% Gd_2O_3.

Pellets were uniaxially compacted using a hand press. Neither binders nor lubricants were added to the powder, although stearic acid was applied as a lubricant to the die walls. Green densities of the as-pressed pellets were about 5.5 to 5.7 gm/cc.

All blending, weighing, and pressing operations were carried out in a controlled humidity (45-50%) glove box. This was done as a precautionary measure to prevent segregation problems due to the hygroscopic nature of the Gd_2O_3 powder. A smear test on each blend gave no indications of segregation between the white Gd_2O_3 and the brown UO_2.

Sintering was carried out at 1650^oC in a forming gas mixture of 94% N_2/6% H_2. The heating cycle required 11 hours to reach 1650^oC, after which the temperature was held constant for 4 hours prior to cooling.

After sintering, all pellets were measured goemetrically for density. One pellet of each type was sectioned and prepared ceramographically to observe the microstructural characteristics.

One group of pellets with ADU-derived UO_2 was resintered using the same cycle as the sintering tests.

---

*Molybdenum Corporation of America, 6 Corporate Park Drive, White Plains, NY.

RESULTS AND DISCUSSION

A reduced sinterability was observed in the UO$_2$ powder containing Gd$_2$O$_3$ additions. This effect is present even for the
smallest addition (1/2 wt%) of Gd$_2$O$_3$, as can be seen in Figure 1.
The reduced sinterability occurred for both the ADU-derived and
the AUC-derived powders, even though these UO$_2$ powders had substantially different characteristics. In Figure 1, the difference
in the sintered densities (95% vs 90.5% T.D.) for the pellets
containing no gadolinia is due entirely to the different UO$_2$ powder
activities, since there was no difference in the processing of
the pellets.

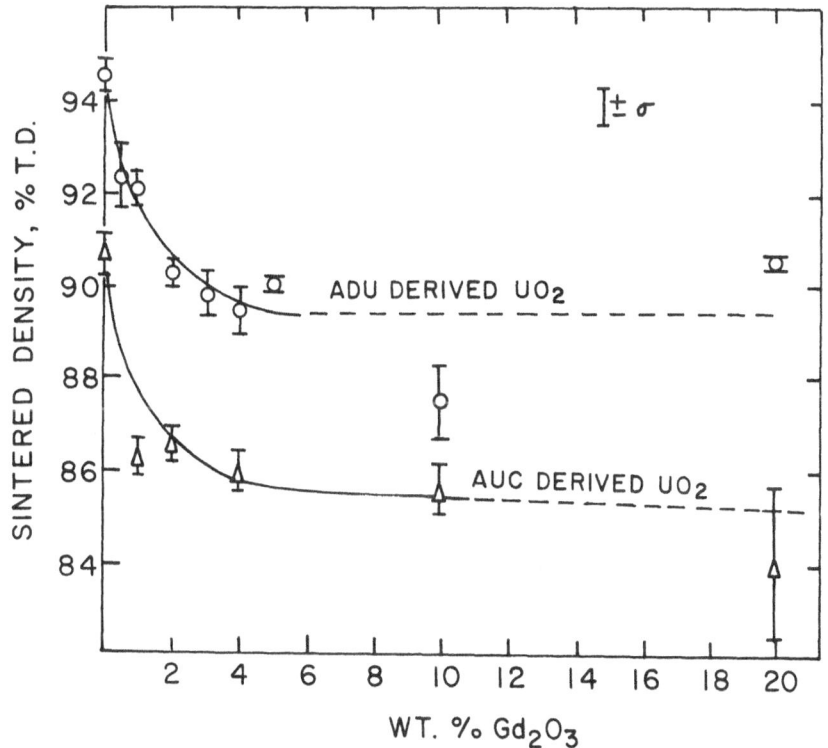

Figure 1.   Effect of Gd$_2$O$_3$ Additions on the Sintered Density of UO$_2$.

As the gadolinia content reaches about 2 to 4%, there is a
saturation of its effectiveness in reducing UO$_2$ sinterability.
Values for the densities of each mixture are shown in Table I.
Since the presence of Gd changes the theoretical density of the
material, the amount of Gd$_2$O$_3$ added must be compensated for in
calculating percent T.D. If all of the Gd$_2$O$_3$ remains as a second
phase distributed in the UO$_2$, then the theoretical density of the
pellet can be found from Vegard's Law, using x-ray densities of

UO$_2$ (10.97 gm/cc) and Gd$_2$O$_3$ (7.66 gm/cc) and the known proportions
of each phase. Percent theoretical densities based on this calculation are shown in Table I. However, some or all of the Gd$_2$O$_3$
dissolves into solid solution in the UO$_2$. When this occurs, the
UO$_2$ theoretical density is slightly reduced since Gd is a lighter
element than U and because of the charge compensation (probably a
loss of oxygen from the lattice) which must take place. Thus,
the actual theoretical density for the UO$_2$-Gd$_2$O$_3$ pellets is not
readily apparent, although it can be determined empirically by
crystallographic techniques.

Table I. Sintering Results for UO$_2$-Gd$_2$O$_3$ Mixtures

wt% Gd$_2$O$_3$	ADU-UO$_2$ Density		AUC-UO$_2$ Density	
	gm/cc	% T.D.	gm/cc	% T.D.
0	10.37	94.53	9.95	90.70
1/2	10.12	92.37	--	--
1	10.08	92.14	9.45	86.36
2	9.85	90.34	9.46	86.80
3	9.76	89.81	--	--
4	9.69	89.44	9.32	85.95
5	9.72	90.00	--	--
10	9.31	87.51	9.10	85.56
20	8.33	90.54	8.66	83.97

For small additions (<5%) of Gd$_2$O$_3$, the Vegard Law analysis
gives sufficient accuracy, particularly when compared to the reduced sinterability in various blends. Microstructures for UO$_2$
pellets manufactured using AUC-derived powder commonly [4] contain
intergranular porosity, as seen in Figure 2(A). With small
additions of Gd$_2$O$_3$ to the powder, the percent porosity increases,
but with no change in the porosity morphology (see Fig. 2(B)).

Figures 3 and 4 show the typical microstructures developed in
the pellets made from ADU-derived UO$_2$. Low magnification (30X)
photographs are shown for various Gd$_2$O$_3$ contents in Figure 3 while
higher magnification (100X) photographs are shown in Figure 4.
Normal UO$_2$ fuel fabricated from ADU-derived powder contains a fine,
evenly distributed porosity (intragranular) through the pellet, [4]
as illustrated in Figure 4(A). However, there is a distinct change
as Gd$_2$O$_3$ is added to the powder. Figures 3(B) and 4(B) show that
even 1/2 wt% Gd$_2$O$_3$ is sufficient to change the sintering characteristics of the powder. The result is a sudden increase in the
total amount of porosity (i.e., reduced pellet densities), and
the location of this porosity along grain boundaries, resulting in
a microstructure which shows dense "islands" of UO$_2$ surrounded by
the porous regions. Although not shown in these micrographs, each
"island" consists of numerous grains of UO$_2$. No quantitative

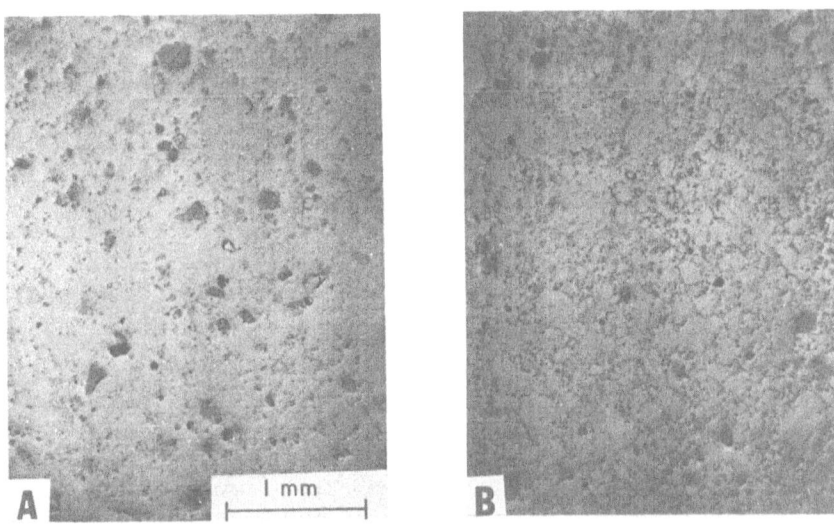

Figure 2. Transverse Section Through Pellets Prepared from AUC-
derived UO₂ Powder: (A) 0% Gd₂O₃; (B) 1.5% Gd₂O₃.

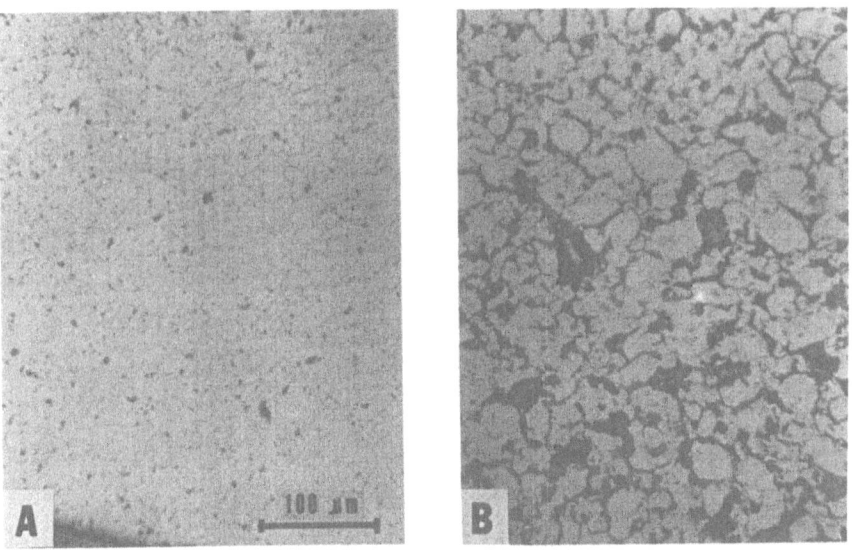

Figure 3. Transverse Sections Through Pellets Prepared from ADU-
derived UO₂ Powder; (A) 0% Gd₂O₃; (B) 1.5% Gd₂O₃.

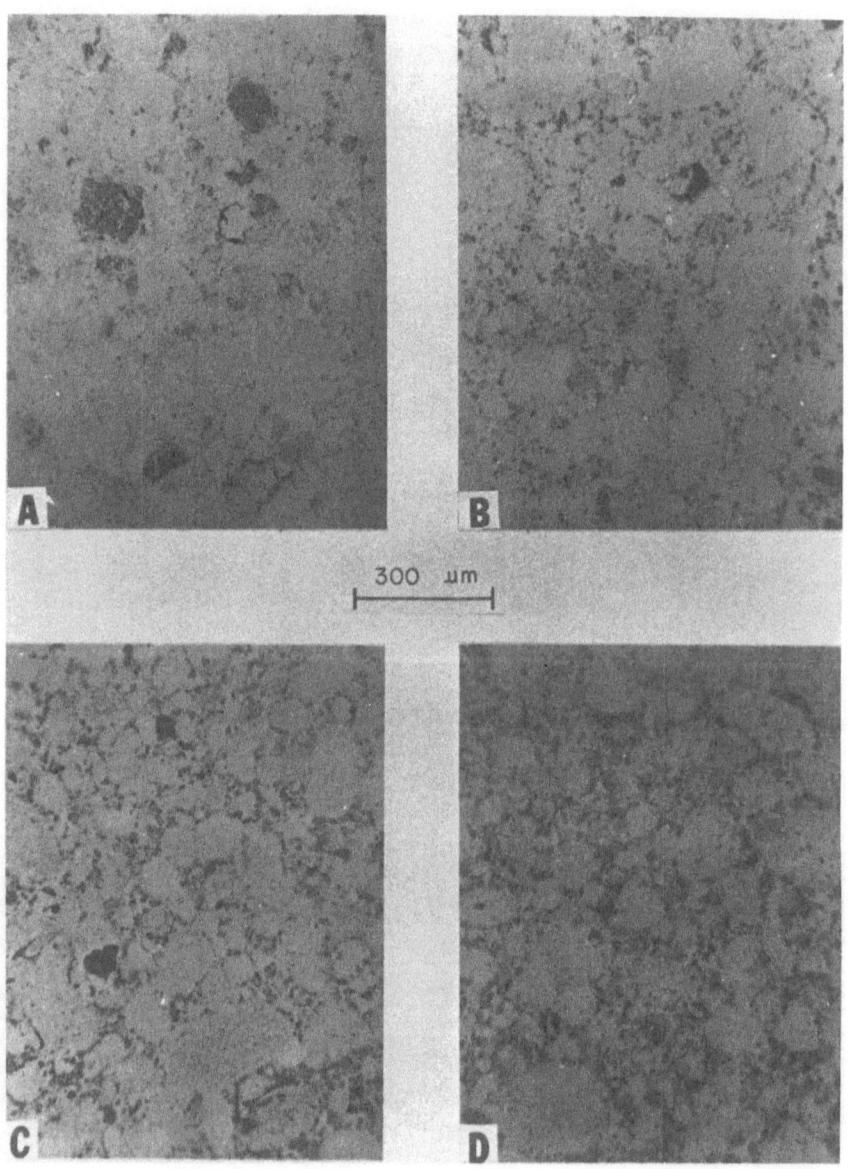

Figure 4. Porosity Morphologies in Pellets Prepared from ADU-derived
           UO_2 Powder: (A) 0% Gd_2O_3; (B) 0.5% GD_2O_3; (C) 1% Gd_2O_3;
           (D) 4% Gd_2O_3.

measurements were made, but the typical grain size appears to be in the range 5-15 μm while the size of each island is about 100-200 μm.

If the sintering process for pure UO_2 were to be interrupted so as to produce pellets with densities the same as those found in the UO_2-Gd_2O_3 pellets, the microstructures would appear entirely different. Thus, the presence of Gd in the fuel has not only reduced the powder sinterability, but in so doing has induced a different type of porosity morphology.

In all cases, the microstructures of the pellets revealed only $(U,Gd)O_2$ solid solution and porosity to be present. No Gd_2O_3 second phase was observed in any of the samples. The results are substantiated by x-ray diffraction analyses of the sintered pellets. For pure UO_2, a lattice parameter of 5.472 ± 0.002 Å was measured. This agrees well with the reported [5] value of 5.469 ± 0.002 Å. In comparison, a lattice parameter of 5.456 ± 0.005 Å was measured for the UO_2 - 20% Gd_2O_3 pellets, while a value of about 5.45 Å is reported [6] for the $(U,Gd)O_2$ solid solution prepared with 15 mole % Gd_2O_3 (∼20 wt% Gd_2O_3). In addition, a few of the sintered pellets were analyzed for percent Gd by x-ray emission analysis and atomic absorption analysis. The results of these analyses are shown in Tables II and III, and show reasonably good agreement between the amount of gadolinia added to the mixture and the amount determined by these analytical techniques.

Table II. Results of X-Ray Emission Analysis

Specimen No.	Amount Gd_2O_3 Added, wt%	Amount Gd_2O_3 by Analysis, wt%		
		Surface	Mid-Radius	Center
G-2	0.5	0.44	0.41	0.48
G-3	1.5	1.48	---	1.39
G-4	2.5	2.38	2.29	2.35

Table III. Results of Atomic Absorption Analysis

Specimen No.	Amount Gd_2O_3 Added, wt%	Amount Gd_2O_3 by Analysis, wt%
1-1	0.5	0.50
1-2	0.5	0.49
2-1	1.5	1.45
2-2	1.5	1.42
3-1	2.5	2.43
3-2	2.5	2.44

The explanation for the reduced sinterability is apparently related to the stoichiometry changes which accompany the incorporation of Gd into UO_2. Most simple metal oxide structures can be built up on the basis of a close-packed arrangement (or nearly so) of oxygen ions with cations placed in available interstitial sites. However, in UO_2, the large size of the uranium ion dictates that the oxygen ions have a simple cubic arrangement, with the U^{+4} ions located in the eight-fold coordination site at the center of the cube. In order to maintain electrical neutrality in this arrangement, only half as many U^{+4} ions are required as O^{-2} ions. Thus, only half of the available eight-fold coordination sites (for U^{+4}) are occupied. This is the structure of fluorite, CaF_2. A notable feature of this structure is the large number of vacant sites on the uranium sublattice.

In pure UO_2, large deviations in stoichiometry in the direction UO_{2+x} can occur. Hyperstoichiometry in uranium dioxide occurs by the introduction of excess oxygen into the structure rather than the formation of metal ion vacancies.[7] This conclusion is supported by magnetic susceptibility measurements and density measurements. Most likely, the excess oxygen enters the structure as oxygen ion interstitials on the vacant lattice sites, with an accompanying oxidation in some U^{+4} to U^{+6} ions to maintain charge neutrality.

In stoichiometric UO_2, oxygen ion diffusion is $\sim 10^6$ times faster than uranium ion diffusion.[8,9] Therefore, diffusion of the uranium ion (since it is the slower moving species) will control solid-state reactions such as sintering in UO_2. However, the rate of chemical equilibration of the O/U ratio with the atmosphere will be rapid relative to sintering.

The activation energy for the diffusion of uranium in UO_{2+x} is considerably reduced[7,9] compared to that for stoichiometric UO_2. This means that uranium ion diffusion is faster in UO_{2+x} than in $UO_{2.0}$. However, it remains slower than that of the oxygen ion. Thus, excess oxygen in the structure permits the uranium ion to diffuse much faster, primarily due to the increased distortion and strain in the lattice.

The addition of Gd_2O_3 to UO_2 results[6] in the formation of a homogeneous solid solution up to about 50 wt% Gd_2O_3. This occurs by the ionic substitution of Gd^{+3} for U^{+4} in the cubic fluorite lattice. Since the replacement of U^{+4} by Gd^{+3} results in a charge misbalance, compensating changes must also occur. Sintering of uranium fuels is commonly performed in reducing atmospheres, thus it is most likely that the loss of excess oxygen ions will be the mechanism by which charge neutrality is maintained.

Most uranium dioxide powders vary significantly from stoichiometry. Typical deviations are x = 0.05 to 0.2 in UO_{2+x}. After sintering, the material is nearly stoichiometric, with x = 0.002 to 0.005 being obtained in most cases. It is assumed that, in Gd-bearing fuel, the ionic substitution of Gd into the structure will contribute toward stabilizing a lower O/U ratio than required by the atmosphere alone. This leads to a reduced mobility of the U^{+4} ions and thus a slower sintering rate for the material.

The UO_2-Gd_2O_3 pellets were resintered using a firing schedule similar to the initial sintering run. Densities of the pellets increased about 1% T.D. during resinter and independent of the Gd level. Normally, fuel pellets with densities as low as 91% would be expected[10] to increase in density by >3% during thermal resinter.

The reason for the reduced resinter behavior in the Gd-bearing pellets can only be indirectly attributed to the gadolinium. These pellets have a comparatively stable porosity morphology as a result of the effect of Gd during initial sintering. Thus, it is difficult to remove much of the void space in these pellets, and the degree of resinter is not as great as in normal UO_2. The porosity morphology of the Gd-containing pellet was not changed by the resinter exposure.

SUMMARY

Fuel pellets of UO_2-Gd_2O_3 were prepared in a humidity-controlled environment and sintered at $1650^\circ C$. Two types of UO_2 powder were used, and Gd_2O_3 contents ranged as high as 20 wt%. The results show that the Gd_2O_3 reduces the powder sinterability and induces a change in the porosity morphology of the pellet. These changes are attributed to the ionic substitution of Gd^{+3} for U^{+4} in the structure.

REFERENCES

1. W. K. Anderson and J. S. Theilacker, Editors, Neutron Absorber Marerials for Reactor Control, U. S. Government Printing Office, 1962.
2. Docket 50-254-15, Quad-Cities Station Units 1 and 2, Amendment 9, Contained Burnable Neutron Absorber as Supplementary Control.
3. Docket 50-261-333, H. B. Robinson Unit 2, License OPR-23, Description and Evaluation of Test Assemblies Containing Gadolinium-Bearing Fuel Rods - H. B. Robinson Unit No. 2 Cycle 3.
4. H. H. Davis and L. J. Ferrell pp. 389-403 in Ceramic Microstructures '76. Edited by R. M. Fulrath and J. A. Pask, 1977.
5. R. P. Elliott, Editor, Constitution of Binary Alloys, First Supplement, McGraw-Hill, Inc., 1965, p. 701.

6. R. J. Beals and J. H. Handwerk, J. Am. Ceram. Soc., <u>48</u> (5)
    271-274, (1965).

7. I. Amato, R. L. Colombo, and A. M. Protti, J. Nucl. Mat. <u>11</u> (2)
    229-235, (1964).

8. K. W. Lay and R. E. Carter, J. Nucl. Mat. <u>30</u>, 74-87, (1969).

9. J. F. Martin and P. Contamin, J. Nucl. Mat. <u>30</u>, 16-25 (1969).

10. H. H. Davis, L. J. Ferrell, and J. L. Harrison, Am. Ceram. Soc.
    Bull., <u>52</u>, (9) 721 (1973).

PROCESSING REQUIREMENTS FOR PROPERTY OPTIMIZATION OF Eu_2O_3-W CERMETS FOR FAST REACTOR NEUTRON ABSORBER APPLICATIONS*

A. E. Pasto and V. J. Tennery

Oak Ridge National Laboratory

Oak Ridge, Tennessee 37830

Europium sesquioxide is a candidate fast reactor neutron absorber material. It possesses several desirable characteristics for this application, but has a low thermal conductivity. This gives rise to pellet cracking during reactor operation. To increase the thermal conductivity without great sacrifice in nuclear worth, addition of tungsten to Eu_2O_3 has been evaluated. Synthesis and fabrication techniques described allow preparation of high-density compacts of Eu_2O_3—15 vol % W possessing favorable thermal conductivity and thermal expansion.

INTRODUCTION

Europium sesquioxide is an attractive neutron absorber material for fast reactors. For this application it possesses the following favorable properties: high beginning-of-life nuclear worth (large neutron capture cross section of europium),[1] compatibility with potential cladding material,[2] high melting point,[3] and ease of fabricability.[4] Further, unlike boron-containing materials, its neutron capture processes do not lead to generation of He or ^{3}H, and its chain-type absorption mechanism results in a slower decrease in nuclear worth with burnup.[3] However, its relatively low thermal conductivity[5] and high gamma heating rate lead to large thermal gradients during reactor operation. High tensile stresses are built up at the surface, which result in cracked pellets, observed both in irradiation[5] and out-of-reactor tests.[5] To prevent crack formation

---

*Research sponsored by the Department of Energy under contract with the Union Carbide Corporation.

and possible resultant pellet-cladding mechanical interaction, the pellets must be fabricated with small diameters. This requires production of large numbers of pellets and lengths of cladding, with attendant high assembly costs.

Consequently, we have evaluated means of incorporating the europium into higher thermal conductivity materials. A preliminary assessment[6] indicated that europia-tungsten cermets offered potential. A summary of that work and follow-on studies form the basis of this presentation.

SYNTHESIS AND FABRICATION

Initial studies concentrated on mixtures of pure tungsten and monoclinic europia. The monoclinic form was used because it is denser than the cubic low-temperature form, and its phase stability range extends to above 2000°C, allowing preparation at high temperatures, if necessary. These studies were pursued by C. S. Morgan of our group.[6] He showed that cermets could be synthesized by several methods, including chemical-vapor-deposition (CVD) of tungsten from WF_6 or $W(CO)_6$ onto europia grains or *in situ* coating of grains with tungsten from $W(CO)_6$-Eu_2O_3 or WO_3-Eu_2O_3. Synthesis was followed by densification into pellets through sintering or hot-pressing. Densities above 95% of theoretical were obtained for selected compositions. Near room temperature cermets containing about 10 vol % W had about 2.4 times the thermal conductivity of Eu_2O_3.

A comprehensive evaluation of the effect of preparation technique and volume-fraction tungsten on thermal conductivity was carried out by S. J. Mitchell and R. K. Williams.[7] To determine the best fabrication process, cermets prepared by different tungsten coating techniques were fabricated and compared. The coating processes used were tungsten oxide (WO_3) decomposition, tungsten carbonyl [$W(CO)_6$] decomposition, tungsten powder blending, and chemical vapor deposition (CVD). Cermets containing nominally 5, 10, 15, 20, and 25 vol % W were prepared by each of these techniques. The europia starting material used in each coating process was monoclinic europia prepared by crushing and sieving hot-pressed pellets. The feed powder for europia preparation was Molycorp product code 5000, a 99.99% pure material. $W(CO)_6$, WO_3, WF_6, and W powder were commercially obtained products exceeding 99% purity. Procedures used for synthesis were as follows:

WO_3 Decomposition:

1. blend 44–250 μm grains of monoclinic Eu_2O_3 with WO_3 powder in acetone until dry,

2. fire at 1200°C for 15 min in hydrogen,

3. repeat with successive additions of WO$_3$ as necessary to obtain a well-coated sample with the desired tungsten content.

W(CO)$_6$ Decomposition:

1. dry blend 44—250 μm monoclinic Eu$_2$O$_3$ with W(CO)$_6$ powder,

2. fire at 1200°C for 10 min in argon in a preheated furnace to prevent excessive volatilization of the carbonyl,

3. repeat with additions of W(CO)$_6$ as required to obtain a well coated sample with the desired tungsten content.

Tungsten Powder Blending:

1. weigh the Eu$_2$O$_3$ (44—250 μm) and tungsten ($\sim$9 μm) powders,

2. blend for 1 hr in an oblique blender.

Chemical vapor deposition:

1. use Eu$_2$O$_3$ powder with a particle size range of 105 to 177 μm (—80 +140 mesh),

2. deposit tungsten on the Eu$_2$O$_3$ in a fluidized-bed coater using tungsten hexafluoride (WF$_6$) and hydrogen.

These coated powders were hot-pressed in 25-mm-diam (1-in.) graphite dies at 1425°C in vacuum for 30 min with 41 MPa (6000 psi) pressure on the sample.

Microstructures of the four types of cermet are shown in Fig. 1. Each contains about 10 vol % W. In the CVD product, the tungsten network is well defined and dense, as shown in Fig. 2, but some tungsten was apparently deposited in the open porosity in the Eu$_2$O$_3$. Also, some large voids remained between the coated particles, and this problem was more severe at higher tungsten contents. The WO$_3$ decomposition and powder blending processes produced samples with similar microstructures and densities. In both types of samples the tungsten formed a somewhat diffuse, lower density network. In addition there was some evidence of reacted zones within the Eu$_2$O$_3$ grains in the WO$_3$ product. The densities of the W(CO)$_6$ samples were very high, but the tungsten network was not as well defined as in the CVD product. Also, much of the tungsten was contained within the Eu$_2$O$_3$ grains, where it would not be expected to significantly enhance the thermal conductivity. These latter two processes were not pursued further because of the difficulty involved and the relatively poor microstructural quality of the product.

PHYSICAL PROPERTY MEASUREMENTS

Physical property data on selected cermets are presented in Table 1. The electrical resistivity ratio (RRR) values give an

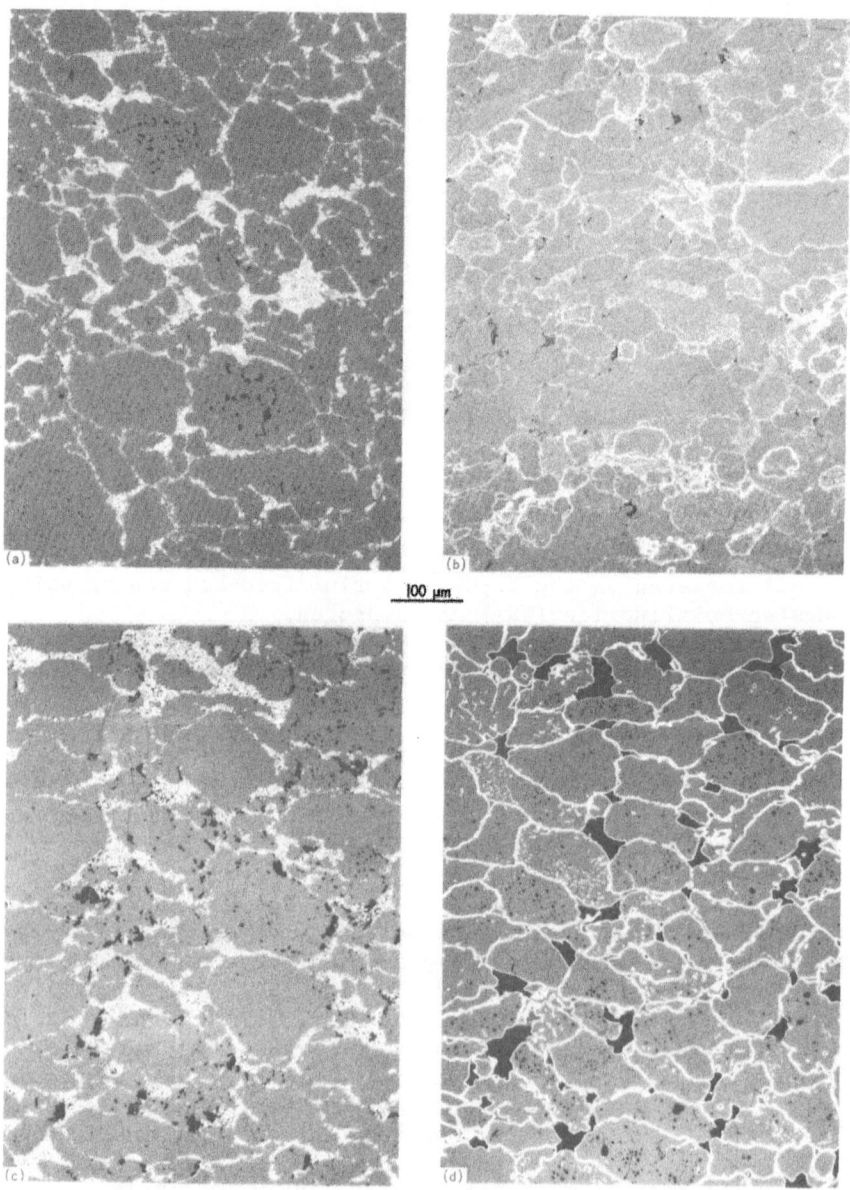

Fig. 1.   Typical areas of tungsten-Eu_2O_3 cermets produced by hot-
          pressing four different types of compacts at 1425°C. As
          polished.   (a) WO_3 decomposition, 10% W.   (b) $W(CO)_6$
          decomposition, 8.3% W.   (c) Tungsten powder blend, 10% W.
          (d) Chemical vapor deposition, 9.5% W.

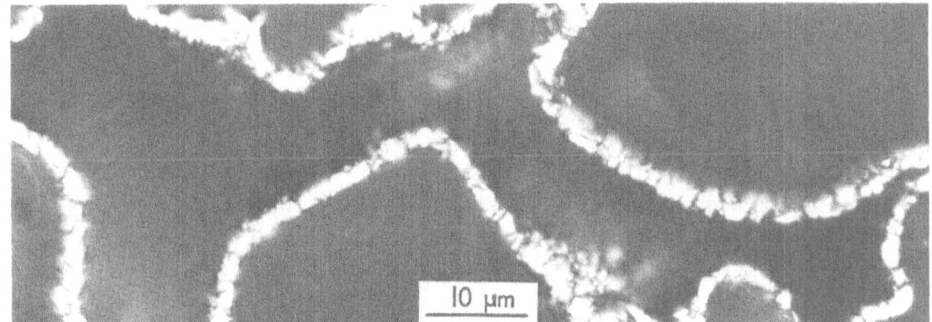

Fig. 2.   Cermet of Eu_2O_3-9.5 vol % W made by hot-pressing CVD-
          coated particles at 1425°C. Specimen was etched to reveal
          the microstructure of the tungsten. Dark areas are Eu_2O_3
          and voids.

Table 1. Physical property values for hot-pressed
europia-tungsten cermet samples prepared
by four different processes

Tungsten (vol %)	Density (% of Theoretical)	Thermal Conductivity[a] ($W\ m^{-1}\ K^{-1}$) at:			Electrical[b] Resistivity at Room Temperature ($\mu\Omega$-m)	Resistivity Ratio: Room Temperature / Liq. Helium Temp
		30°C	60°C	90°C		
Blended Tungsten and Europia Powders						
5	98.3	4.65	4.47	4.30	318	8.45
15	92.6	7.82	7.58	7.34	2.90	14.5
20	90.8	10.8	10.5	10.2	1.47	16.1
Decomposition of WO_3 Powder						
5	96.8	3.98	3.80	3.66	295	3.91
15	90.5	4.75	4.60	4.45		
20	85.7	4.46	4.30	4.23		
Decomposition of Tungsten Carbonyl						
6	99.2	5.68	5.51	5.34	11.6	4.92
8.3	97.4	6.83	6.58	6.45	5.95	5.92
11.6	98.2	8.44	8.22	8.00	2.87	5.48
Chemical Vapor Deposition						
6.5	96.6				2.06	20.23
9.5	92.8	9.95	9.67	9.44	1.21	21.78
17.3	88.2	15.2	15.0	14.6	0.64	

[a]Obtained by a comparative longitudinal flow method. Values are believed to be
uncertain by less than ±5%.

[b]Average of two values obtained by the four-probe dc method.

indication of the purity of the tungsten, and all the values are
high enough to indicate that the tungsten was not seriously contaminated. Tungsten quality decreases in the order: CVD, powder
blend, $W(CO)_6$, and WO_3 decomposition, but the increase noted in RRR
with increasing tungsten content is probably due to boundary
scattering and not purity differences.

The thermal conductivity values are compared with data previously obtained for pure monoclinic Eu_2O_3 in Fig. 3. The CVD process
yields the highest relative thermal conductivity, undoubtedly due
to the uniformity and density of the tungsten matrix.

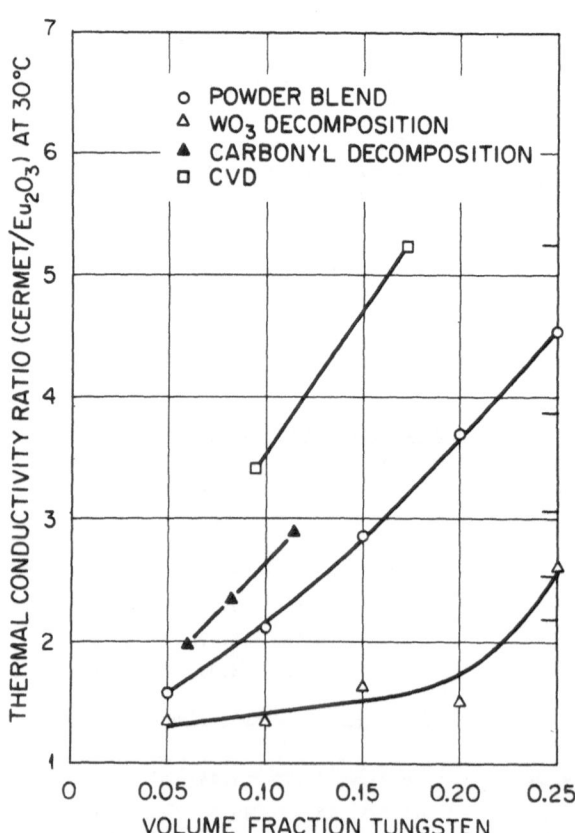

Fig. 3. Thermal conductivities relative to Eu_2O_3 of Eu_2O_3-W
         cermets produced by four processes.

From values of thermal conductivity (λ) for tungsten and Eu$_2$O$_3$
as functions of temperature, and knowledge of the low-temperature
cermet λ and electrical resistivity (ρ), a set of thermal conductivity values for a CVD-prepared 95%-dense Eu$_2$O$_3$-15 vol % W cermet was
calculated.[8] Results are presented in Fig. 4. Our goal of increased
λ relative to Eu$_2$O$_3$ has apparently been attained. The roughly five
fold increase in λ means that much larger pellets of absorber
material could be used in reactor control assemblies, thus lowering
fabrication costs. More recently, we have attempted to increase
the density of cermets prepared by the powder-blending (PB) technique.
Four means of doing so were evaluated: (1) hot-pressing at higher
temperatures, (2) increasing initial density of Eu$_2$O$_3$ grains,
(3) using a finer tungsten powder, and (4) using sintering aids
(e.g. palladium) in the tungsten.

Cermets prepared combining techniques (1) and (2) were
successful. The finer tungsten particle approach was of some small
significance in gaining increased thermal conductivity, but not
density. Table 2 illustrates these effects. It should be noted
that even at densities higher than for CVD cermets the powder blend
(PB) cermet thermal conductivities are not as high. Electrical
resistivity data suggest that the reason for this is the inclusion
of Eu$_2$O$_3$ fines in the tungsten matrix. Small particles are in fact
often found microscopically in the tungsten matrix. These could
perhaps be eliminated through use of carefully screened Eu$_2$O$_3$,
removing all particles finer than about 44 μm.

Attempts to sinter PB cermets to high density at 1400°C in H$_2$,
with and without additions of palladium to the tungsten, were largely
unsuccessful. Densities about 78% of theoretical were obtained in
both cases for a 15 vol % W cermet, while at this temperature
sintering of pure tungsten with and without palladium addition
yielded pellet densities of >98% and 66%, respectively.

One other parameter of importance to control rod design is
thermal expansion. This was measured on a 76-mm-long (3-in.) rod
specimen of Eu$_2$O$_3$—15 vol % W cermet prepared by the PB technique
and hot-pressed to 90% of theoretical density. A high-precision
differential dilatometer[9] recorded length of the sample at various
preprogrammed temperatures. Data are presented in Table 3 along
with our previous data for Eu$_2$O$_3$ and literature values for tungsten.

Using the average coefficients of linear thermal expansion for
tungsten and Eu$_2$O$_3$ for the range 300—1000 K, applying the mixture
rule,

$$\overline{\alpha}_{cermet} = \overline{\alpha}_E V_E + \overline{\alpha}_W V_W \; ,$$

where

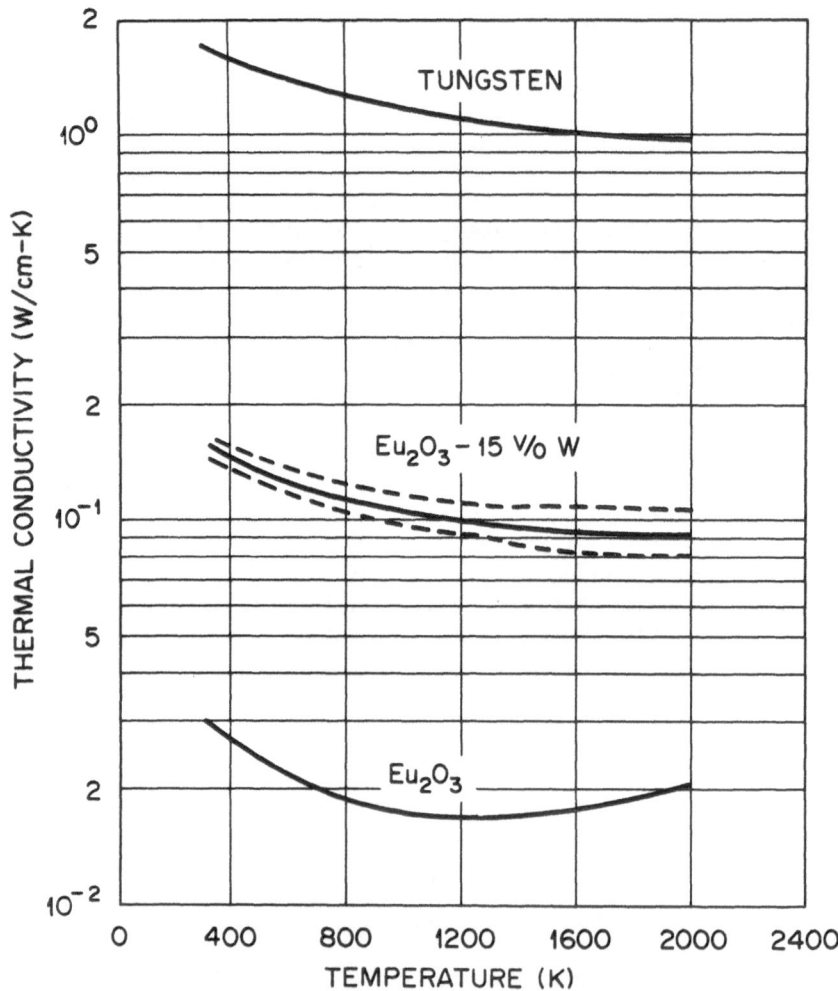

Fig. 4. Calculated thermal conductivity of Eu_2O_3—15 vol % W Cermet.

$\overline{\alpha}$ = average coefficient of linear thermal expansion,

V_E = volume fraction Eu_2O_3,

V_W = volume fraction W,

and solving for V_W, we calculate a volume fraction of tungsten of approximately 0.41. Since the cermet actually contains only 0.15 volume fraction tungsten, the tungsten appears to play a stronger

Table 2.   Transport properties of some Eu$_2$O$_3$-tungsten cermet samples

Hot-Pressing Temperature (°C)	Density (% of Theoretical)	Tungsten Content (vol %)	Thermal Conductivity at 300 K (W/m K)	Electrical Resistivity at 300 K (μΩ m)
Prepared by Powder Blending Process				
1425[a]	92.6	15	7.85	2.95
1500[a]	92.9	15	9.13	2.23
1600[a]	97.1	15	9.99	1.93
1600[b]	97.0	15	10.3	1.95
Prepared by Chemical Vapor Deposition				
1425	88.2	17.3	15.2	0.65
1500	89.6	17.9	18.2	0.58
1600	94.5	12.8	14.4	0.87

[a]Prepared from 9 μm average particle size tungsten powder.

[b]Prepared from 3 μm tungsten powder.

Table 3.   Thermal expansion of tungsten, Eu$_2$O$_3$,
and Eu$_2$O$_3$–15 vol % W cermet

Temperature		Linear Expansion at Temperature, %		
°C	K	W[a]	Monoclinic Eu$_2$O$_3$[b]	Monoclinic Eu$_2$O$_3$–15 vol % W
27	300	0	0	0
127	400	0.043	0.082	0.069
227	500	0.083	0.170	0.137
327	600	0.125	0.259	0.207
427	700	0.169	0.351	0.279
527	800	0.215	0.445	0.353
627	900	0.263	0.543	0.429
727	1000	0.313	0.644	0.507
$\bar{\alpha} \, \frac{300}{1000} =$		4.47×10^{-6}	9.20×10^{-6}	7.24×10^{-6}

[a]Data for powder-metallurgically prepared tungsten sheet, taken from Proceedings of 5th Conference on Thermal Conductivity, Vol. 2, October 1965.

[b]Data from ref. 5.

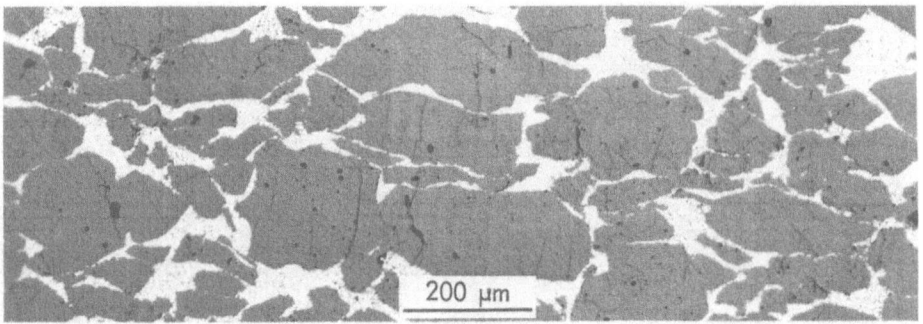

Fig. 5. Microstructure of Eu_2O_3—15 vol % W cermet prepared by
the powder-blending process. Cracks in Eu_2O_3 grains
due to thermal stresses are visible.

role in determining the cermet thermal expansion than predicted by
the rule of mixtures. This is to be expected since tungsten is
the continuous matrix phase and expands less than the europia grains
that it surrounds. The europia grains, however, are microcracked,
as shown in Fig. 5 for one of the test specimens. This occurs on
cooling the cermet from its 1600°C fabrication temperature, during
which a relatively strong mechanical Eu_2O_3-tungsten bond is formed.
During cooling, the europia tries to contract faster than the
matrix, resulting in tensile stresses in the grains, leading to
cracking. On subsequent heating to temperatures below 1600°C, the
expansion curve for the cermet is dominated by the tungsten because
part of the Eu_2O_3 thermal dilation is accommodated by expansion into
the previously generated cracks.

SUMMARY AND CONCLUSIONS

Europium oxide, a candidate fast reactor control material, can
be fabricated into a material of higher thermal conductivity through
inclusion of a tungsten matrix. This is best accomplished by
blending sized monoclinic Eu_2O_3 grains with fine tungsten powder or
by chemically vapor depositing tungsten on europia, followed by hot-
pressing in vacuum. An Eu_2O_3—15 vol % W cermet so produced possesses
a thermal conductivity of four to five times that of pure Eu_2O_3 and
a linear thermal expansion coefficient 20% lower than that of the
oxide alone. These two factors should yield an absorber material
exhibiting better in-reactor performance, with lower initial assembly
fabrication costs.

ACKNOWLEDGMENT

The authors wish to thank C. S. Morgan, S. M. Tiegs, and R. K. Williams of the Metals and Ceramics Division, ORNL, for their contributions to this study.

REFERENCES

1. H. Takahashi, *Evaluation of the Neutron and Gamma-Ray Production Cross-Sections of* ^{151}Eu *and* ^{153}Eu, BNL-19455 (ENDF-213) (November 1974).

2. J. R. DiStefano, *Compatibility of* Eu_2O_3 *with Type 316 Stainless Steel and Sodium*, ORNL/TM-4780 (January 1975).

3. A. E. Pasto, *Europium Oxide as a Potential LMFBR Control Material*, ORNL-TM-4226 (September 1973).

4. A. E. Pasto, M. M. Martin, and R. G. Donnelly, pp. 14-22 in *Proc. 11th Rare Earth Research Conference*, Oct. 7-10, 1974, CONF-741002-P1 (1974).

5. A. E. Pasto and M. M. Martin, Eu_2O_3: *Properties and Irradiation Behavior*, ORNL-5291 (August 1977).

6. A. E. Pasto and C. S. Morgan, *Status and Potential of Europium-Based Fast Neutron Absorber Materials*, ORNL/TM-5244 (February 1976).

7. S. J. Mitchell and R. K. Williams, *LMFBR Materials Development Program Quart. Prog. Rep. March 31, 1976*, ORNL-5157, pp. 129-39.

8. R. K. Williams, *LMFBR Material Development Program Quart. Prog. Rep. Sept. 30, 1976*, pp. 175-90.

9. T. G. Kollie, D. L. McElroy, J. T. Hutton, and W. M. Ewing, p. 129 in *Thermal Expansion - 1973, AIP Conf. Proc. 17*, American Institute of Physics, New York (1974).

CERAMIC PROCESSING OF BORON NITRIDE INSULATORS*

C. S. Morgan and R. W. McCulloch

Oak Ridge National Laboratory

Oak Ridge, Tennessee 37830

INTRODUCTION

Fuel pin simulators (FPS), an example of which is shown in Fig. 1, are the prime elements of several test facilities at the Oak Ridge National Laboratory (ORNL). These experimental facilities are used to conduct out-of-reactor thermal-hydraulic and mechanical interaction safety tests for both light-water and breeder reactor programs. The FPS units simulate the geometry, heat flux profiles, and operational capabilities of a reactor core element under steady-state and transient conditions. They are subjected to temperatures as high as 1600°C (2900°F) and power levels as high as 57.5 kW/m (17.5 kW/ft) as well as severe thermal stresses during transient tests.

The insulating material in the narrow annulus between the heating coil and the FPS sheath is subjected to very rigorous conditions. Accuracy of the reactor safety test information and validity of the test data depend on the heat flux uniformity under all test conditions and on the reliable operation of all fuel pin simulators and their internal thermocouples. Boron nitride (BN), because of its high degree of chemical inertness combined with its relatively unique properties of high thermal conductivity and low electrical conductivity, is the most suitable insulating material for FPS. The important BN properties, thermal conductivity and electrical resistance, are strongly influenced by crystallite orientation and by impurities. This article describes new BN powder processing techniques, which optimize these properties. First prior methods of preparing BN-insulated FPS are described.

---

*Research sponsored by the Department of Energy under contract with Union Carbide Corporation.

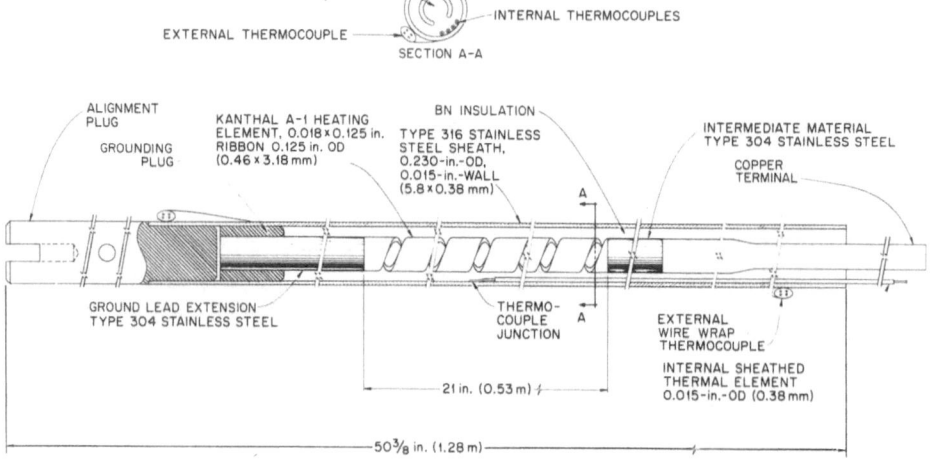

Fig. 1. Fuel Pin Simulator, Showing Boron Nitride Insulation.

PRIOR METHOD OF FABRICATING BN INSULATED FPS

The common technique of FPS manufacture involves introducing powder into the annular area, tamping to a predetermined density, and swaging to final density. The final density and annular thermal conductivity are fixed by the initial filling density, the amount of swaging, and the orientation of BN platelets. Nonuniformity in the density of the BN and heating element eccentricity are easily introduced by powder tamping and filling variations, tool tolerances, and possible operator errors. In addition, tool wear introduces metal particles into the BN, often causing failure of a unit by short circuiting when voltage is applied. Further, crystallographic orientation of BN, the importance of which is described below, was found to be very unfavorable in powder filled and swaged FPS.

BORON NITRIDE

The crystal structure of BN, which is shown in Fig. 2, resembles that of graphite except that the hexagonal arrays are directly aligned.[1] In the crystallographic a direction BN has the highest thermal conductivity of any known electrical insulator, while the thermal conductivity in the c direction is only 0.025 of this amount. It is evident, therefore, that crystallite orientation is of great importance. The degree of preferred orientation; that is, a direction parallel to the radial direction in the FPS, can be determined by measurement of the x-ray diffraction intensity from the hexagonal planes and from the (100) planes.

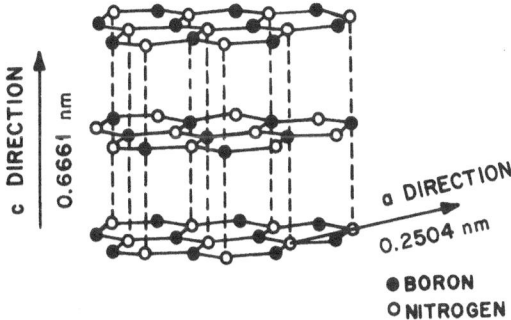

Fig. 2. Crystal Structure of Boron Nitride with Dimensions of the Hexagonal Cell.

Samples of as-received BN powders had a measured intensity ratio close to the ASTM standard value for a randomly oriented BN powder of $I_{0002}/I_{100} = 100/17 = 5.88$. Intensity ratios much greater than 5.88 are desired in order to have more favorable heat transfer properties in FPS. Ratios as unfavorable as 100/57 = 1.75 were found in FPS prepared by filling with BN powders. Greatly improved orientations, e.g., as high as 100/0.4 = 250 can be obtained in pressed BN forms.

A number of BN powders are commercially available, but this investigation was centered on two types of BN powders produced by Union Carbide Corporation: HCM and TS-1325. The first is a high-purity, hot-pressed and ground powder with a tap density of approximately 1 Mg/m^3 (45% of theoretical density) and a particle size distribution ranging from below 10 μm to agglomerates of 350 μm. The other is a high-purity powder with a tap density of about 0.65 Mg/m^3 (30% of theoretical density) but with a wider particle size distribution (≈1 mm to 7 μm) and a higher percentage of small particles. Both powders, as received, had oxygen contents from 0.1 to 0.3 wt % and carbon contents from 0.04 to 0.1 wt %.

Even though the received powders were purer than 99.5%, both contained small amounts of metal and other electrically conducting impurities dispersed as small particles. These impurity particles are potentially detrimental to FPS operation since they can provide shorting paths in the insulated gap between the heating element and sheath wall. This distance is small, being about 0.5 mm over lengths as long as 2.15 m. The metallic particle content of the powders was reduced by about 90% by magnetic separation performed by Magnetic Engineering Associates, Inc., Cambridge, Massachusetts. The TS-1325 powder was further cleaned before use by seiving out larger particle aggregates and separating visibly contaminated particles.

ORIENTATION IMPROVEMENT

The basic improvement in crystallite orientation is obtained in the preform pressing step, as platelets tend to become oriented with the hexagonal plane perpendicular to the pressing direction. Further improved preferred orientation during pressing was obtained by reducing the particle size of the BN powder. The as-received powder, after being cleaned to remove metallic particles, was ball-milled overnight in hexane in a ceramic mill using alumina balls. The ball milling served primarily to break up aggregates. For instance, in the case of TS-1325 powder, which had a smaller initial particle size than the HCM powder, the average particle size after milling overnight was reduced only about 1 to 7 μm, as determined with a Coulter Counter.* However, the fraction of this powder passing a 100-mesh screen (149-μm openings) increased from 35% to essentially 100%. Diffraction data obtained on preforms pressed from this ball-milled powder showed intensity ratios of 100/0.4 = 250 compared with 100/1.1 = 91 on preforms pressed from as-received TS-1325 powder and 100/4 = 25 from as-received HCM powder.

Thermal conductivity measurements on the cold-pressed preforms have not yet been obtained, but enhanced radial thermal conductivity due to the preferred orientation is evident by qualitative room-temperature measurements. Since BN purity is of such paramount concern, great care was taken to keep fibers and lint out of the powder during handling. Use of ball-milling suspension agents such as ethyl alcohol was avoided because traces of water in these liquids resulted in unacceptable oxidation of the BN powders.

APPLICATION OF LUBRICANTS AND PREPRESSING

Although BN preforms can be pressed without lubricating additives, die life, initial preform density, and density uniformity are much improved by use of a lubricant because of the reduced friction. In addition, automatic production, needed to produce the required thousands of pellets for FPS production, is not possible without lubricants. Lubricants can be added either into the powder or on the die surface. Both methods have been used successfully, with the latter method preferred because it attains higher density in the finished preform. However, only the former technique has been fully developed, so the discussion will be limited to the use of lubricants in the powder.

Lubricants, dissolved in hexane, are added to a thick hexane slurry of the BN powder. The mixture is allowed to evaporate, with occasional stirring to ensure that the lubricant is distributed

---

*Registered trademark of Coulter Electronics, Inc., Hialeah, Florida.

homogeneously on the particle surfaces. Many organic compounds
were tested, and nearly all gave some improvement in powder pressability. However, two additional factors must be considered. The
final strength after lubricants have been removed must be adequate
for preform handling during loading of FPS, and the preform density,
which is usually reduced as the lubricant is removed, must be as
high as possible for good thermal conductivity. The most suitable
lubricant was determined to be a composite consisting of 0.5 wt %
(based on the weight of BN) dodecane plus 0.5 wt % dodecanol. Amyl
alcohol at a concentration of 4 wt % was also an effective lubricant but was more difficult to work with because of its higher
volatility. Camphor, although not an optimum lubricant for die
operation, produced preforms with the highest final density and
strength.

BN powders were prepressed and sieved to obtain a particle
size range that would flow into the die and to obtain higher powder
density in the die. Prepressing was carried out in 100-g batches
at a pressure of 34.5 MPa (5000 psi). After granulation the
powders were sieved to obtain a —30 +100 mesh fraction. The
presence of lubricants in the powder facilitated preparation of
sized powder with desirable flow characteristics by strengthening
prepressed aggregates.

PREFORM PRESSING AND PURIFICATION

The prepared BN powder was pressed into tubular shapes that
fit the annular gap of fuel pin simulators. Solid cylindrical
pellets were also pressed to fill the volume inside the coiled
FPS heating elements (Fig. 1). Typical length to wall thickness
ratio of tubular preforms was about 12.5 and the length to diameter
ratio of solid and tubular preforms was about 4.0. Density, before
purification, was about 1.91 Mg/m^3 for pressing pressures of 310 MPa
(45 ksi). Axial density variation was less than 2% with suitably
prepared powders.

After pressing, preforms were put in a vacuum oven for several
days with the temperature slowly raised to 200°C to remove the bulk
of lubricating additives. Purification was completed by treating
the preforms in anhydrous ammonia at 930°C for a minimum of 8 hr.
For example the anhydrous ammonia treatment reduced the oxygen
content from 810 to 225 ppm and carbon from 471 to 100 ppm. The
final oxygen and carbon levels were equivalent to or lower than
the initial values of magnetically cleaned BN powder. High-
temperature electrical conductivity tests of FPS have yielded
typical values of 10^4 Ω-m at 1200°C.

Figure 3 shows a group of as-pressed preforms. Purified preforms were weaker than as-pressed preforms but were sufficiently
strong for fabrication of FPS. Removal of lubricating additives

Fig. 3.   Boron Nitride Preforms Prepared for Installation in an FPS.

tended to reduce the density of preforms, apparently by allowing or causing the particle structure to relax. Typically the density decreased 8 to 10% for preforms in which the additives were 0.5 wt % dodecane plus 0.5 wt % dodecanol or amyl alcohol. The density loss was much greater with use of some additives, such as stearic acid, and was less for camphor. Outside diameter increases of about 3% accompanied the 8 to 10% density decrease. Crystallographic orientation of preforms was not significantly affected by lubricants or by their removal.

PREFORMED FPS FABRICATION

The BN preforms are grooved, if necessary to accommodate thermocouples, and are individually inserted and crushed to assure complete filling of the FPS annular volume. Crushing moderately reduces the degree of preferred orientation. For example, in one test the I_{0002}/I_{100} value changed from 100/1 = 100 to 100/3 = 33. Swaging of the assembled FPS, which is needed to improve thermal contact of the BN with the FPS sheath and heating element, results in further reduction of preferred orientation. With respect to thermal conductivity, this is offset by the density increase due to swaging. The overall effect of density increase accompanied by reduction in preferred orientation on thermal conductivity has not yet been determined. However, since initial orientation is higher with preforms than with powder and the amount of swaging required is reduced significantly because of higher initial BN density of preforms, better orientation and increased thermal conductivity should result.

The use of preforms affords several advantages in production of FPS units in addition to preferred crystallographic orientation. Typical preform lengths are about 10 mm, so that preform filling is much simpler and quicker than powder filling. In addition, metal spalling from tamping tools is practically eliminated, the chance of introducing other fabrication impurities is reduced, and heating element eccentricity (now determined by preform dimensions) is typically too small to measure by standard gaging. The reduced swaging required results in less heating element and thermocouple distortion and improved dimensional tolerances.

Infrared scans were used to evaluate the effect of annular BN density uniformity and thermal conductivity on the uniformity of the heat flux profile of FPS. Figure 4 shows infrared scans after transient power application to the cladding for two FPS, where one is powder filled and one is preform filled. Since the heat generation in the sheath is uniform, infrared readings vary axially only as a result of thermal conductivity and density variations across the insulation (including conductivity of interfaces). The nearly flat temperature profile of the scan of the preform-filled FPS attests to the uniformity of the insulation. By comparison the insulation uniformity afforded by powder packing is distinctly inferior.

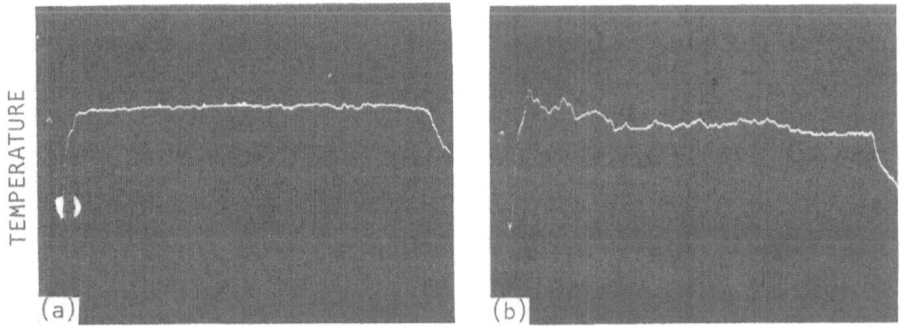

DISTANCE ALONG FPS

Fig. 4. Infrared Scans of Two FPS After a Transient Cladding Current. (a) Preform filled. (b) Powder filled.

CONCLUSION

The use of boron nitride preforms with preferred crystallite orientation, while relatively new, promises better control of FPS fabrication and critical properties affecting FPS operation. The increased accuracy and reliability needed of FPS units operating under severe test conditions can more realistically be met by this boron nitride fabrication method.

ACKNOWLEDGMENT

X-ray diffraction determination of BN orientation was performed by O. B. Cavin. Sigfred Peterson edited the manuscript and Julia Bishop prepared it for publication.

REFERENCE

1. Pease, R. S., *Acta Cryst*. 5, 356 (1972).

PART VI

ENERGY RELATED CERAMICS II:

Non-Oxide Ceramics

GRAIN BOUNDARY ENGINEERING IN NON-OXIDE CERAMICS

R. Nathan Katz, and George E. Gazza

Army Materials and Mechanics Research Center

Watertown, Massachusetts 02172

ABSTRACT

The deliberate selection and control of composition, structure and processes occurring in the grain boundary, both during and after fabrication, with the aim of specific property modification may properly be defined as Grain Boundary Engineering. We will review various strategies for doing Grain Boundary Engineering and give examples of the use of these strategies in the development of hot pressed silicon nitride. Research needs for further development of Grain Boundary Engineering (GBE) in non-oxide systems are discussed.

INTRODUCTION

The deliberate selection and control of composition, structure and processes occurring in the grain boundary, (or the grain boundary affected region), during or after fabrication, aimed at specific property modification, may properly be termed Grain Boundary Engineering (GBE). It is frequently the case that the nature of the grain boundary controls critical optical and/or mechanical properties. In these instances GBE as a strategy is often more fruitful than the more conventional microstructure development or compositional control approaches. Indeed in past papers the authors have discussed examples of GBE as applied to improving the optical properties of ionic ceramics.[1,2]   A summary of GBE techniques, in the broadest sense, is presented in Table I.

Table 1. Summary of GBE Technique

Techniques	Examples	Mechanisms	References
•Liquid Phase Sintering			
•Fugative	MgO + LiF ——	Sol'n-Pcpt? ——	[3] [4]
•Non-Fugative	Si_3N_4 + MgO ——	Liq Phase Sint ——	[5] [6]
	Mg Al_2O_4 + Li_2O & SiO_2 ——	Liq Phase Sint ——	[7]
•Compound Formation in GB			
•Via Liq Phase (with poss. further react.)	Si_3N_4 + Y_2O_3 ——	Cryst. of GB Liq ——	[8] [9] [10]
•Without Liq Phase	Lucalox (Al_2O_3 + MgO) ——	Spinel @ GB	[11]
•Promotion of Volume Diffusion	SiC ——	C to remove SiO_2 B changes γ_{SV}	[12]
•Press Forge Single Crystals	CaF_2, KCℓ ——	Disloc Generation & Organization	[13] [14]
•Impurity Control	Si_3N_4 + MgO Si_3N_4 (RBSN)	Reduce Ca ——	[15] [16]
• Second Phase Particle Additions			
•Mechanically Added	BeO + ZrO_2		
•Pcpt. at GB			

As discussed below and in several following chapters, the production of fully dense silicon nitride (either hot pressed or sintered) requires the use of densification aids which form a liquid phase during processing, resulting in a grain boundary phase which in turn controls the high temperature mechanical properties of the material. During the past six years an extensive amount of research and development effort has been focused on manipulating this grain boundary phase primarily to improve the high temperature properties of hot pressed silicon nitride (HPSN). Thus, the development of fully dense, high temperature HPSN provides an excellent ongoing case history in GBE. This case history will be reviewed, highlighting several of the GBE techniques (strategies) listed in Table I. This review also emphasizes that the GBE strategies used to develop improved HPSN relied on inferential evidence as to the nature of the grain boundary phase(s) present. While much was accomplished using such inferential evidence, as described below, direct, quantitative information relative to the grain boundary would have considerably accelerated these developments. Therefore, the second portion of this paper deals with recent progress in several key areas of materials science relevant to GBE in non-oxide ceramics. Needs for additional capabilities or information critical to the future development of GBE will also be discussed.

GBE AND THE OPTIMIZATION OF HOT PRESSED Si_3N_4

Since the late 1950's, considerable interest has grown in the development of Si_3N_4 and SiC materials which have high potential as component materials in various propulsion and power generation systems [1] State-of-the-art high strength, hot pressed Si_3N_4 has resulted from a series of developments involving: empirical selection of densification aids, control of high temperature strength-limiting reaction products in the grain boundary via a variety of strategies, and selection of new families of densification aids specifically chosen to yield grain boundary phases with improved properties. In contrast to materials improvements in many other ceramic systems, essentially all of the improvements in hot pressed Si_3N_4 as an engineering ceramic have resulted from grain boundary phase manipulations. These developments will now be reviewed.

MgO: Its Role and Importance

Since Si_3N_4 dissociates before significant sintering can occur, an additive was required which would permit densification at temperatures below those at which significant dissociation occurs. Early attempts to obtain a densification aid which would produce fully dense, high strength, hot pressed Si_3N_4 were successfully carried out by Deeley, et al.[5] who demonstrated that MgO additions gave results superior to the many other additives considered. Although X-ray diffraction analysis of the hot pressed samples indicated only Si_3N_4 present, primarily beta phase, it was assumed that the MgO additive reacted with the surface silica on the Si_3N_4 particles to form a vitreous glass which produced a continuous bonding boundary phase. Further optimization of the system was carried out by Lumby and Coe,[6] who showed the significance of using high alpha phase Si_3N_4 powder as starting material and the strength dependence on hot pressing time. This work implied the importance of the $\alpha \rightarrow \beta$ transformation in developing high strength material, as later rationalized by Lange.[17] Further attempts to define the role of MgO during densification were carried out by Evans and Sharp[18] who confirmed the existence of a glass phase at triple points by transmission electromicroscopy. However, a similar uniformly distributed grain boundary phase was only inferred. Indirect evidence for a non-crystalline grain boundary phase of $MgSiO_3$ composition was obtained by Wild, et al.[19] from annealing experiments on hot pressed Si_3N_4 containing 10% MgO. Visual evidence for glassy phases existing in high MgO content hot pressed Si_3N_4 by a transmission electron microscope study were subsequently obtained

by Drew and Lewis.[20]   This study also confirmed the role of the
glassy phase in the $\alpha \rightarrow \beta$ transformation, thought necessary for the
attainment of high strength Si_3N_4.

In attempting to define the composition of the observed glassy
phase, studies with Auger spectroscopy and electron probe microanalysis were performed [21,22]   and results suggested the existence
of various grain boundary glass compositions, i.e., $xSiO_2 \cdot yCaO \cdot zMgO$.
These analyses also indicated that a proportion of the smaller
cations, (i.e., Mg, Al) have diffused into the Si_3N_4 structure
during hot pressing.

While hot pressed Si_3N_4 produced in this way had properties
which where encouraging for use in high temperature applications,
e.g., gas turbines, indepth studies of creep and high temperature
modulus of rupture (MOR) at ~1200 C indicated that, in contrast to
reaction bonded Si_3N_4, hot pressed material exhibited a significant
fall off in properties. Much inferential evidence from the results
of Auger spectroscopy, TEM, creep behavior, etc. pointed to the
glassy grain boundary phase being responsible for this behavior.

Studies[21,15]   initiated to determine the influence of residual
impurities in the starting materials on the high temperature properties
of the boundary phase have shown that the apparent viscosity
of the boundary phase is purity dependent. In particular, the Ca
cation was found to be detrimental to high temperature properties
of Si_3N_4 which exhibited increased sub-critical crack growth[23]
and decreased creep resistance.[21]   Evidence to date suggests that
this is due to the presence of the boundary phase which becomes
viscous at high temperature allowing grain boundary sliding and
associated slow crack growth.

To summarize the above, the importance of MgO as an additive
to hot pressed Si_3N_4 was that it permitted the attainment of full
density and possibly provided the medium through which an $\alpha \rightarrow \beta$
solution-reprecipitation transformation could occur, with con-
comitant high RT strength. The principal drawback of the conventional use of MgO as an additive was the limitation it imposed
on high temperature strength and creep behavior.

At this point the central problem faced by researchers was to
find a way to create a more refractory grain boundary phase (GBP)
or to eliminate it. In spite of the experimental difficulties in
direct characterization of this grain boundary phase (namely that
the small volume percent of this "amorphous" phase in a crystalline
matrix is not amenable to X-ray diffraction methods and its small
dimension rules out conventional microprobe and similar direct

techniques) and having to rely on inferential results, significant improvements in hot pressed Si_3N_4 have been made by applying the GBE strategies shown in Figure 1.

1. Reduce Ca, Na, etc., impurities to make the grain boundary $MgSiO_3$ glass more refractory.

2. Develop a densification aid to yield a more refractory glass than $MgSiO_3$.

3. Develop a nonglass grain boundary.

4. Eliminate the grain boundary phase by promoting sintering via volume diffusion.

Figure 1. Strategies to increase the high-temperature behavior of hot-pressed Si_3N_4.

Compositional Control of the Magnesium-Silicate Grain Boundary Glass Phase

The first approach to compositional control of the magnesium-silicate grain boundary glass phase was to reduce the amounts of impurities which would lower the softening temperature of this phase. This approach was initially pursued by Kossowsky[21] and Lange[23] at Westinghouse and Richerson[15] at the Norton Company. These investigators elucidated the detrimental role of Ca, as well as other alkali and alkaline earth elements, on the high temperature strength and creep resistance of the magnesium silicate grain boundary phase.

Efforts were then made to produce higher purity Si_3N_4 starting powder in order to reduce detrimental impurity levels. Lange also demonstrated that control of the MgO/SiO_2 ratio* could beneficially influence the high temperature properties of HPSN[24] Strength values at 25C and 1400C were measured for molar ratios

*Some SiO_2 is always present on the surface of Si_3N_4 powder. Hence it is essential to account for the precise amount of SiO_2 in any given powder when doing GBE, in spite of the difficulties in doing so.

from 1-10. It was shown that maximum values were obtained at 1400C for MgO/SiO_2 ratios of 3-4. This was attributed to either a change in refractoriness or volume content of the grain boundary phase. The change in refractoriness may be the result of increasing nitrogen solubility in the phase or shifting away from glass forming regions in the $MgO-SiO_2$-impurity system.

Efforts to produce higher purity Si_3N_4 powders were pursued in attempting to minimize the residual detrimental impurity content. This produced significant improvement in high temperature properties but the properties of the specific $MgO-Si_3N_4$ reaction products formed, i.e., $MgSiO_3$ or Mg-Si-O-N, then became limiting on strength and creep resistance. It thus became apparent that significant increases in the high temperature properties of hot pressed Si_3N_4 would require other approaches.

Densification Aids Other Than MgO

Another approach for increasing high temperature properties would be to induce the formation of more refractory reaction products between the additive, and the SiO_2 layer on the Si_3N_4, to form the boundary phase. Gazza,[8,9] explored this approach and found that yttria was an effective additive for this purpose (Figure 2). It was reasoned that the probable formation of $xY_2O_3 \cdot ySiO_2$ glasses and compounds would be more refractory than $xMgO \cdot ySiO_2$ glasses. Further, that the large ionic radius of the Y cation would tend to keep it in the boundary vicinity rather than diffusing away into the Si_3N_4 grains. Also, it might be possible to crystallize these "glassy" phases. At low Y_2O_3 additive levels, <5 w/o, these glasses appear to form and some improvements in high temperature properties were observed. However, it was found that maximum strengths were obtained with 10-15 w/o Y_2O_3 required to completely react with the surface silica on Si_3N_4 powder, and a more complex role of Y_2O_3 additions to Si_3N_4 was explained by work at Newcastle [25] These studies have shown that the refractory nature of the $Si_3N_4-Y_2O_3$ reaction products are due to the formation of $xSi_3N_4 \cdot yY_2O_3 \cdot zSiO_2$ compounds which may further react with Si_3N_4 to produce $Si_3N_4 \cdot Y_2O_3$ composition. In addition to producing high melting point refractory phases, residual impurities such as Ca, Al, and Mg are accommodated into solid solution by these compounds thus limiting their detrimental effect. Similar reaction products were reported by Tsuge et al.[10] in concurrent work. Although enhanced properties have been demonstrated for the $Si_3N_4-Y_2O_3$ additive system, a problem with anomalous strength degradation at approximately 1000C has been

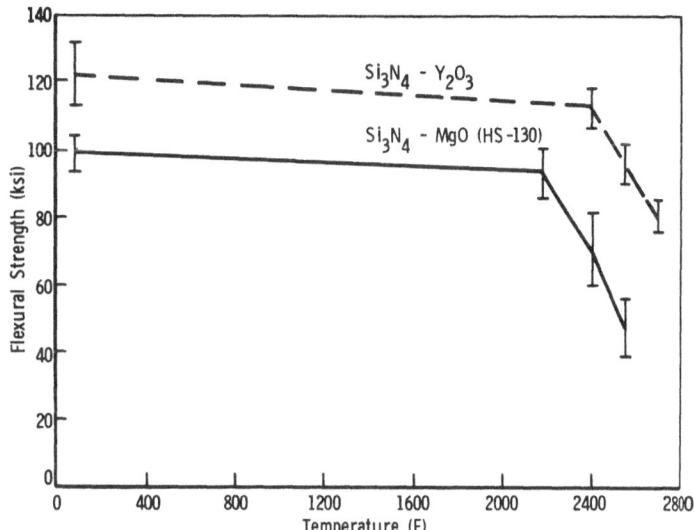

Figure 2. Strength versus temperature for Si_3N_4 $+Y_2O_3$ versus Si_3N_4 + MgO.

reported and further studied by several investigators [26-28]
Lange[26]   found that this behavior is related to the lack of
oxidation resistance of certain Si_3N_4-Y_2O_3 reaction phases at
1000C and severe cracking of specimens containing these phases
occurs. More recent work of Lange's also indicates that if one
remains within the $Si_2ON_2 \cdot Y_2Si_2O_7 \cdot Si_3N_4$ compatibility triangle in
the $SiO_2 \cdot Y_2O_3 \cdot Si_3N_4$ phase diagram one obviates the intermediate
temperature stability problem.[27]   High yttria additions (>10 w/o)
to Si_3N_4 were reported to form phases which had poor oxidation
resistance and exhibited linear type kinetics rather than parabolic.
Si_3N_4-Y_2O_3 compositions containing these phases cracked severely
during oxidation at 1000C but not at higher temperatures. It was
found that by altering the composition so that phases formed were
within the $Si_3N_4 \cdot Si_2N_2O \cdot Y_2Si_2O_7$ compatibility triangle, exellent
oxidation resistance is observed. It was suggested that this was
due to the formation of $Y_2Si_2O_7$ which exists in equilibrium with
SiO_2 and preserves a protective surface coating. Brennan[28]   is
investigating Si_3N_4-15 w/o Y_2O_3 materials using high and low purity
Si_3N_4 starting powders, and thusfar, finds the anomalous 1000C
behavior to be purity dependent. Tsuge, Nishida and Komatsu[29]
found that for their particular Si_3N_4-Y_2O_3 additive (~5%) system,
strength problems at intermediate temperatures could be overcome by
crystallizing the glass phase formed by the additive reaction with
Si_3N_4.

Other additives, such as CeO_2 [30,31] other rare earth
oxides [32] and ZrO_2 [33,34] have also been studied to determine
their effect on pressure sintering of Si_3N_4 and potential for
enhancing high temperature properties of Si_3N_4.

Crystallization of Grain Boundary Phase

The Mg-silicate liquid acts as a densification aid in hot
pressing Si_3N_4 with MgO additions. The resultant grain boundary
phase is vitreous and tends to creep at high temperature thus
limiting its usefulness above this temperature. Glassy phase has
also been inferred in yttria doped (~5 w/o) Si_3N_4 resulting in some
strength reduction at high temperature. High temperature strength
behavior was improved by crystallizing the glass phase. Further
development of high temperature strength by grain boundary
crystallization (GBC) was accomplished by using Y_2O_3 and Al_2O_3 as
additives to promote densification. Thermal shock resistance
measurements by quenching into water indicated ΔT_c values > 1000C
could be obtained. Strength retention of 140 KSI at 1000C was
also achieved. [29,35]

Promotion of Volume Diffusion

Thusfar, additions used for enhancing the sinterability of
Si_3N_4 have resulted in the formation of separate boundary phases
which have controlled properties. A further approach to sintering,
successfully used by Prochazka [12] with SiC, would be to use
additions which promote greater degrees of volume diffusion, remove
inhibiting species from the system by volatilization, form solid
solutions and produce "clean" grain boundaries. While various teams
are pursuing research on Si_3N_4 using this concept, no one has yet
achieved the goal. The achievement, however, may result in differences in microstructural morphology and fracture mode of Si_3N_4
which may benefit some properties and be detrimental to others.

GBE RESEARCH TOOLS--PROGRESS AND NEEDS

At present GBE in non-oxide ceramics is more of an art than a
science. The further development of several areas of research is
critical to future progress on GBE. These areas include:

- Grain Boundary Characterization Techniques
- Phase Equilibria Data
- Diffusion Data
- Improved Processing Technology

While much progress has been made in improving silicon nitride and silicon carbide materials based on inferential models of the grain boundary, the need for direct observational techniques is clearly evident. In the past two years lattice imaging transmission electron microscopy has emerged as a powerful tool for obtaining such information.[36] Figure 3 shows the grain boundary phase in HPSN with a MgO additive. By coupling lattice imaging TEM with microanalysis and microdiffraction techniques[37] direct evidence has been obtained to support liquid phase sintering via solution-reprecipitation mechanism models for HPSN with Y_2O_3 additions. This work has also confirmed that Y does not migrate into the grains, but remains in the grain boundary as initially hypothesized by Gazza.

Phase equilibria studies, particularly in the $MgO-Si_3N_4-SiO_2$ and $Y_2O_3-Si_3N_4-SiO_2$ systems are contributing to improved strategies for compositional optimization in silicon nitride systems. However, detailed diagrams for other systems i.e., $CeO_2-Si_3N_4-SiO_2$ or $ZrO_2-Si_3N_4-SiO_2$ are required. Also, these systems are in reality not

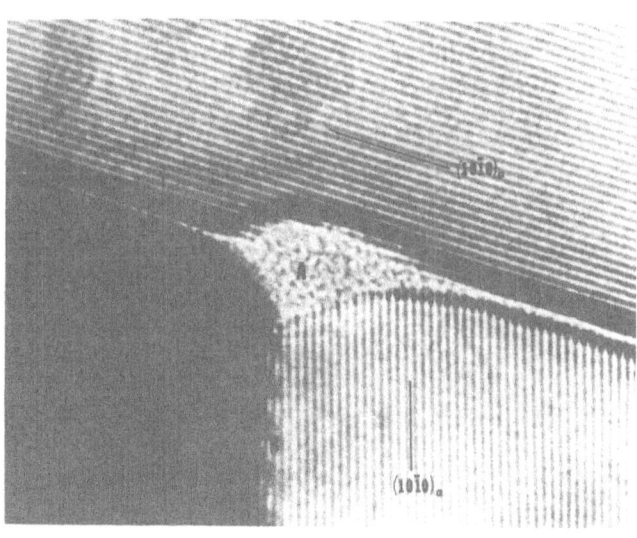

Figure 3. Grain boundary phase (a) in HPSN with MgO addition.[36]

ternaries but are at least quarternaries (and most probably much more complex). Therefore, there is a great need for multicomponent phase equilibria data, particularly in the Si_3N_4-rich portions of the phase diagrams.

Diffusion data is critical if sintering by volume diffusion is to be attained in Si_3N_4, and similar non-oxide ceramics. Such data for Si_3N_4 is limited to self diffusion studies carried out by Wuensch and Vasilos [38]   This is clearly an area requiring more emphasis.

The area of improved processing technology is very broad, but can be focused on those issues which would tend to make grain boundary compositional control easier to attain. For example better techniques to assure uniform distribution of additives are needed. Similarly, ball milling or other mixing procedures which must be designed so that "tramp" impurities are not introduced into the system. For example, when milling MgO-Si_3N_4 mixtures in alumina mills one may have to deal with the system Al_2O_3:MgO:Si_3N_4: SiO_2 rather than MgO:Si_3N_4:SiO_2. Similarly if more than one additive is used, it may be preferable to pre-react or pre-mix the additives to promote homogeneity. Improved processing control will help limit the "unknown" variables.

SUMMARY

The above review has demonstrated that GBE provides a useful perspective for the development of useful engineering materials, where the grain boundary properties are performance limiting. It is also clear while much progress in grain boundary characterization techniques has been made in the past several years, much remains to be done. It is the authors' view that future developments and materials optimization studies in the development of sintered silicon nitrides, "sialons", and silicon carbides, will be significantly aided by the combination of the GBE perspective, with the "new" characterization techniques.

REFERENCES

1. Katz, R. Nathan, "Recent Developments in High Performance Ceramics", Materials Technology -- 1976, AIP Conf. Proc. No. 32, Amer. Inst. Phys., New York (1976).

2. Katz, R. Nathan and Gazza, G. E., "Grain Boundary Engineering and Control in Nitrogen Ceramics," Proc. of the Conf. on Nitrogen Ceramics, Canterbury, England, 1976, In press.

3. Rice, R. W., Bull. Amer. Ceram. Soc., 41, (1962), p. 271.

4. Atlas, L. M., Jour. Amer. Ceram. Soc., 41, (1957), p. 196.

5. Deeley, G. G., Herbert, J. M. and Moore, N. C., Powder Met 8, (1961), p. 145.

6. Lumby, R. J. and Coe, R. F., Proc. Brit. Ceram. Soc., 15, (1970), p. 91.

7. Gatti, A., Mehan, R. L. and Noone, M. J. Naval Air Systems Comm. Rpt., Contract No. N00019-71-C-0126, Dec. 1971.

8. Gazza, G. E., Jour. Amer. Ceram. Soc., 56, (12), (1973), p. 662.

9. Gazza, G. E., Amer. Ceram. Soc., Bull., 54, (9), (1975), p. 778.

10. Tsuge, A., Kudo, H. and Komeya, K., Jour. Amer. Ceram. Soc., 57, (6), (1974), p. 269.

11. Coble, R. L., Jour. App. Phys., 32, [5], (1961), pp. 787-99.

12. Prochazka, S., "Sintering of Silicon Carbide," Ceramics for High Performance Applications, Chapter 12, ed. J. J. Burke, A. E. Gorum, and R. N. Katz, Brook Hill Publishing Co., Chestnut Hill, MA (1974).

13. Rice, R. W., in Ultrafine Grain Ceramics, Syracuse Univ. Press, NY, (1970), p. 203.

14. Harrison, W. B., AFML-TR-109, July 1975, p. 210.

15. Richerson, D. W., Amer. Ceram. Soc. Bull., 52, (7), (1973), p. 560.

16. Mangals, Jr., "Development of a Creep-Resistant Reaction-Sintered Si_3N_4," Ceramics for High Performance Applications, Chapter 9, ed. J. J. Burke, A. E. Gorum, and R. N. Katz, Brook Hill Publishing Co., Chestnut Hill, MA (1974).

17. Lange, F. F., Jour. Amer. Ceram. Soc., 56, (10), (1973), p. 518.

18. Evans, A. G. and Sharp, J. V., Jour. Mat. Sci., 6 (1971), p. 1292.

19. Wild, S., Grieveson, P., Jack, K. H., and Latimer, M., in Special Ceramics 5, ed. P. Popper, Brit. Ceram. Res. Assoc. (1972), p. 377.

20. Drew, P. and Lewis, M. H., Jour. Mat. Sci., 9, (1974), p. 261.

21. Kossowsky, R., Jour. Mat. Sci. 8, (1973), p. 1603.

22. Powell, B. D. and Drew, P., Jour. Mat. Sci., 9, (1974) p. 1867.

23. Lange, F. F. and Iskoe, J. L., High Temperature Strength, "Behavior of Hot-Pressed Si_3N_4 and SiC: Effect of Impurities," Ceramics for High Performance Applications, Chapter 11, ed. J. J. Burke, A. E. Gorum, and R. N. Katz, Brook Hill Publishing Co., Chestnut Hill, MA (1974).

24. Andersson, C. A., Lange, F. F. and Iskoe, J., "Effect of the MgO/SiO$_2$ Ratio on the Strength of Hot-Pressed Si$_3$N$_4$," Contract No. N00014-74-C-0284, Tech. Rpt. No. 3, Oct. 15, 1975, Westinghouse Elec. Corp., R & D Center.

25. Rae, A. W. J. M., Thompson, D. P., Pipkin, N. J. and Jack, K. H., in Special Ceramics 6, ed. P. Popper, Brit. Ceram. Res. Assoc., 1976.

26. Lange, F. F., Singhal, S. C., and Kuznicki, R. C., Westinghouse Electric Corp. Tech. Rpt. 6, Contract No. N00014-74-C-0284, Apr. 1976.

27. Bratton, R. J., Andersson, C. A., and Lange, F. F., "Hot Pressed Si$_3$N$_4$ Developments," Ceramics for High Performance Applications II, Ed. J. J. Burke, R. N. Katz, and E. Lenoe, Brook Hill Publishing Co., Chestnut Hill, MA (1977). In press.

28. Brennan, J. J., United Tech. Research Center Tech. Tpt. R75-912081-2, Contract N62269-75-C-0137, Sept. 1975.

29. Tsuge, A., Nishida, K., and Komatsu, M., Jour. Amer. Ceram. Soc., 58, (7-8), (1975), P. 323.

30. Huseby, I. C., and Petzow, G., Powder Met. Int., 6, (1), (1974), p. 17.

31. Mazdiyasni, K. S. and Cooke, C. M., J. Amer. Ceram. Soc., 57, (12), Dec. 1974), p. 536.

32. Andersson, C. A. and Bratton R. J., "Ceramic Materials for High Temperature Turbines," Final Report under ERDA Contract E (40-1) 5210, Aug. 1977.

33. Rice, R. W. and McDonough, W. J., Amer. Ceram. Soc. Bull. 54 (8), (1975), p. 753.

34. Vasilos, T., AVCO Corp., Lowell, MA. personal communications.

35. Komeya, K., Tsuge, A. Hashimoto, H. Kubo, T., and Ochiai, T., "Silicon Nitride Ceramics for Gas Turbine Engines," Gas Turbine Society of Japan, Paper No. 65, Tokyo Joint Gas Turbine Congress, Tokyo, Japan, May 1977.

36. Clarke, D. R. and Thomas, G., "Grain Boundary Phases in a Hot-Pressed MgO Silicon Nitride," ERDA Contract W-7405-ENG-48, Rpt. No. LBL-6004, Jan. 1977. In press with Jour. Amer. Cer. Soc.

37. Clarke, D. R., and Thomas, G. "Microstructure of Y$_2$O$_3$ Fluxed Hot-Pressed Silicon Nitride," Lawrence Berkeley Laboratory Rpt. No. 6272. Submitted for publication J. Amer. Ceram. Soc.

38. Wuensch, B. J. and Vasilos, T. "Self Diffusion in Silicon Nitride, Final Rpt. (ARPA) Contract No. N00014-73-C-0212, March 1975.

DISCUSSION

J. Thomas Smith (GTE Laboratories, Inc.): Dr. Katz's review of
Si_3N_4 hot pressed with a range of Y_2O_3 levels included discussion
of specimens oxidized at $1000^{\circ}C$ which showed anomalous distortion
and gross microcracking after several hours of exposure. This behavior was encountered with compositions outside the compatibility
triangle, Si_3N_4-Si_2N_2O-$Y_2Si_2O_7$. Research at GTE Laboratories with
pressureless sintered Si_3N_4 plus Y_2O_3 for additions of 4 and 6
weight % Y_2O_3 lying in the Si_3N_4-Si_2N_2O-$Y_2Si_2O_7$ compatibility triangle and for additions of 8 to 12 weight % Y_2O_3 in the adjoining
compatibility triangle, Si_3N_4-$Y_2Si_2O_7$-$Y_{10}Si_7O_{23}N_4$, is in marked
contrast to the data reviewed in the subject presentation.

Our results[1] for sintered samples showed that all Si_3N_4-Y_2O_3
compositions in both compatibility triangles exhibited parabolic
oxidation kinetics during air oxidation over the temperature range
1000° to $1350^{\circ}C$ for times of 300 to 800 hours. The data analysis
was consistent with oxygen diffusion through the surface oxide as
the rate controlling mechanism. After oxidation, the sintered
specimens exhibited tightly adherent, crack-free oxidation coatings and were not distorted nor did they contain the gross cracking as observed with the materials reported in the present review
paper. Room temperature strength measurements were reported for
sintered Si_3N_4-Y_2O_3 compositions from both compatibility triangles
utilizing specimens oxidized at 1000° and $1200^{\circ}C$ for periods to
700 hours. There was a slight strength decrease (about 15%) for
these oxidized specimens relative to the as-sintered values, but
no catastrophic failures were found.

Authors: The authors appreciate the comments from Dr. Smith re-
iterating that GTE Si_3N_4-Y_2O_3 sintered material with compositions
within the Si_3N_4-$Y_2Si_2O_7$-$Y_{10}Si_7O_{23}N_4$ compatibility triangle have
not catastrophically oxidized at $1000^{\circ}C$. Sintering studies reported by Rowcliffe and Jorgensen,* using GTE SN402 Si_3N_4 powder
with Y_2O_3 additions up to 17 w/o, indicate that catastrophic oxidation is not observed at $1000^{\circ}C$ with their sintered material.
The work of Brennan,** cited in our review, also suggests that
there is further need for more detailed investigations of stability and strength retention in the Si_3N_4-Y_2O_3-SiO_2 system, particularly where the influence of various sources of well characterized

---

[1] J. T. Smith, "Properties of Fully-Dense, Sintered Si_3N_4 Compositions," Paper 61-BN-77F, Basic Science and Nuclear Division
Meeting, Am. Cer. Soc., Hyannis, Mass., September 25-28, 1977.

starting Si_3N_4 powders and processing techniques may be studied and compared.

*Rowcliffe, D. J. and Jorgensen, P. J., "Sintering of Silicon Nitride," Proceedings of the Workshop on Ceramics for Advanced Heat Engines, Energy Res. and Dev. Adm., CONRT, Orlando, Florida, January 1977.

**Brennan, J. J., United Tech. Research Center Tech. Rpt. R75-912081-2, Contract N62269-75-C-0137, September 1975.

THE FABRICATION OF DENSE NITROGEN CERAMICS

K.H. JACK

Crystallography Laboratory
The University of Newcastle upon Tyne

THE DENSIFICATION OF SILICON NITRIDE AND SIALONS

In fabricating nitrogen ceramics the objective is to produce a shaped component with high strength, oxidation resistance, negligible creep and good thermal shock properties all at temperatures above 1400°C. Densification, whether by hot-pressing or by pressureless sintering, requires an additive the function of which is to provide conditions for liquid-phase sintering. Thus, magnesium oxide reacts with the silica that is always present as a surface layer on the nitride powder to give what was at first thought [1] to be a liquid near the $MgSiO_3$-SiO_2 eutectic composition and which on cooling gives a low softening-temperature glass. Impurities lower the glass viscosity and so the strength and creep resistance of the high-density, hot-pressed product decrease rapidly above 1000°C. It is now well-established that the liquid phase is an oxynitride. Indeed, bulk samples of nitrogen-containing glasses in the Mg-Si-O-N and Mg-Si-Al-O-N systems [2] and in the corresponding yttrium systems have been prepared.

Liquid-phase densification occurs by the three stages described by Kingery, [3] i.e. (i) particle rearrangement, (ii) solution-precipitation and (iii) coalescence, and the relative contributions of the first two will vary with the additive. Accompanying densification there is always some phase transformation from α- to β-Si_3N_4 but although Brook et al. [4] state that the rates of transformation and densification follow the same first-order kinetics and are both rate-limited by diffusion through a grain-boundary phase, they do not consider transformation to be a necessary factor for densification. The requirements for complete

561

densification are agreed as (i) an appreciable amount of liquid, (ii) an appreciable solubility of the solid, and (iii) complete wetting.   The rate of material transport by solution of α-Si_3N_4 and precipitation of β will depend upon the difference in solubility of the two forms and hence upon the free energy change $\Delta G(\alpha-\beta)$. The onset of densification is expected to depend on the temperature of liquid formation, and with different additives the liquid viscosity and the degree of wetting will vary.   Furthermore, the liquid characteristics and composition will change with time and temperature, with the effectiveness of mixing and the initial distribution of the additive, and with the particle size and reactivity of the mixed powders.   It is not surprising that the role of additives is not always clear.

For liquid-phase densification an externally applied pressure is merely additive to the effective pressure derived from capillary forces.[5]   Thus, the same considerations should apply equally to pressureless sintering except, of course, that the rates will be correspondingly lower than for hot-pressing.   In the densification of sialons compared with silicon nitride, the general difference is in the more extensive formation of liquid phases and in the increased free energy change, $\Delta G(\alpha-\beta')$.

Experimental

To compare the densifying characteristics of different additives, samples of the appropriate powder mix were cold-compacted in a steel die to approximately 60% theoretical density and the pellet so-produced was trimmed to a cylinder 22mm diameter x 25mm high for hot-pressing or 10mm x 10mm for pressureless sintering.   For the initial appraisal of hot-pressing behaviour the pellet was placed between graphite plungers inside a graphite die with all surfaces protected by a deposited boron nitride layer and with the pellet embedded in boron nitride powder.   A pressure of 31.7 MNm^{-2} (4600 psi) was applied from cold and maintained constant during each run.   Heating to 1800°C required about 30 minutes with an almost linear rise in temperature.

In subsequent measurements of hot-pressing kinetics, the pressure was applied cold to compress the pellet and its surrounding boron nitride and was then reduced to a small value during heating. On reaching the required hot-pressing temperature the full pressure was applied instantaneously and the shrinkage during isothermal densification was followed continuously.

For pressureless sintering, the pellet was embedded in boron nitride powder within an alumina crucible and heated in a slow stream of nitrogen at one atmosphere pressure.

Hot-pressing Silicon Nitride

Figure 1 compares the relative shrinkages for 5w/o of each of the additives MgO, Y_2O_3, ZrO_2 and CeO_2 during the 30 minutes required to reach 1800°C. With all four additives full density is reached after a further 30 minutes at not less than 1800°C but the rate of densification and the density on first reaching the maximum temperature are roughly related to the appearance of liquid in the system. With yttria, densification is slower than with magnesia because of the smaller amount of liquid and because of its higher viscosity. It is impossible to account for the behaviour of ZrO_2 and CeO_2 in terms of large volume liquid-phase formation at the low temperatures at which slow densification is observed. The rates seem characteristic of a particle rearrangement process involving only small amounts of liquid.

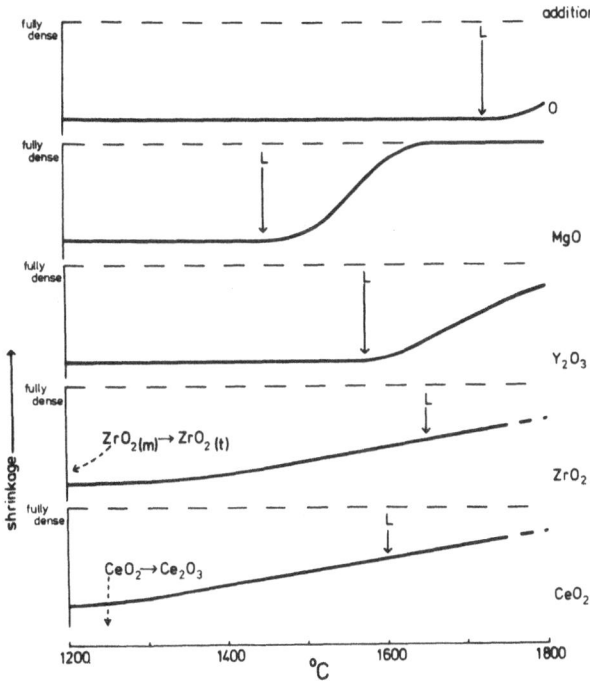

Figure 1.   Shrinkage v. temperature for 5w/o additives in hot-pressed Si_3N_4

Pressureless Sintering of Silicon Nitride

Terwilliger & Lange [6] showed that silicon nitride powder
($90\alpha:10\beta$) could be pressureless sintered with 5w/o MgO by heating
in the range 1500-1750°C either for long periods at the lower
temperatures or for short periods at the higher ones. In the
present work, almost full densification (i.e. **less than** 5% closed
porosity) is achieved with 5w/o MgO at temperatures as low as
1450°C. Above 1700°C decomposition with weight losses prevents
complete sintering. Table 1 shows that at 1350°C, below the
$\alpha-\beta$ thermodynamic transformation temperature,[7] no transformation
is in fact observed and there is negligible densification. Also,
under conditions which give almost complete densification and
appreciable transformation (40% transformation with 5w/o MgO at
1450°C) with high α-powder, there is no densification with high
β-powder and, of course, little transformation. With yttria
instead of magnesia there is much more transformation but much
less densification.

Table 1.   Pressureless sintering of silicon nitride for 3h at
           1350° and 1450°C ($-\Delta V/V_o$ for zero porosity $\sim$40%)

starting powder $\alpha-\beta$	additive w/o	% change at			
		1350°C		1450°C	
		$-\Delta V/V_o$	β	$-\Delta V/V_o$	β
90:10	none	0	0	1	+5
	5MgO	3	0	37	+40
	$7Y_2O_3$	2	0	13	+80
30:70	none	0	0	1	0
	5MgO	3	0	2	+15

Hot-pressing β'-Sialon Compositions

In the Si-Al-O-N system (see Figure 2) the small silica-rich
liquid field at 1650°C extends as the temperature rises and
includes X-phase at about 1720°C. There then exists a considerable
two-phase field, β' + liquid. Mixtures of Si_3N_4, Al_2O_3 and AlN
or other appropriate starting powders can be hot-pressed to give a
pure, theoretical-density β'-sialon provided that compensation is
made for the surface oxides that are always present on the nitrides.
Chemical reaction starts at about 1450°C and the densification that
also starts at this temperature is probably due to particle
rearrangement accompanying the chemical reaction. The densification
is complete at 1750°C but completion of the $\alpha-\beta'$ transformation
requires about one hour at this temperature and, if the composition
is perfectly balanced, all liquid phase should disappear just when

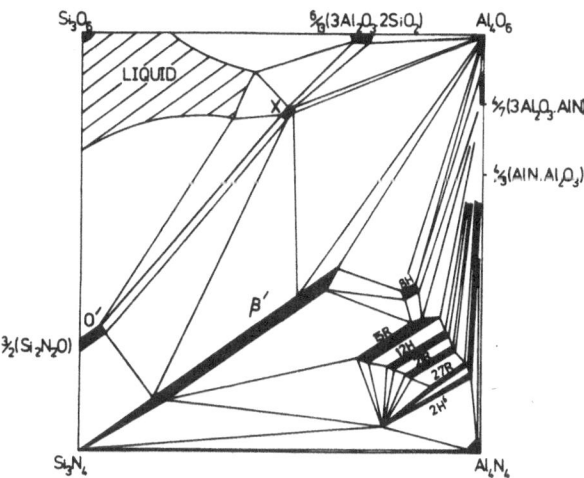

Figure 2.   The Si-Al-O-N behaviour diagram at 1700°C.

the reaction is finished.

Densification and transformation are facilitated by using an additive because liquid metal-alumino-silicates containing nitrogen are formed and so allow rearrangement as well as the solution-precipitation transport of material.   Also, the driving force for the transformation to β' is much greater than that for α-β.   For example, if the chemical activity of "β-silicon nitride" is reduced to 0.1 of its value by its reaction to give a β' solid solution, the transformation temperature is reduced by 200°C.

Pressureless Sintering of β'-Sialons

The additives MgO, Y_2O_3, ZrO_2, CaO and B_2O_3 promote β' formation during the pressureless sintering of a z = 3 β'-sialon composition and all, except B_2O_3, give considerable densification at 1600°C and 1650°C.   5w/o MgO is the most effective at 1650°C for 30 minutes but 4w/o Y_2O_3 and 3w/o ZrO_2 also give theoretical density if maintained long enough at 1650°C or if the temperature is raised to 1700°C.   Above 1700°C loss of silica occurs by volatilisation or oxidation, or both, and the remaining composition moves towards the AlN corner of the system to produce polytype phases.

The Kinetics and Mechanism of Densification

These general observations suggested that in both hot-pressing and pressureless sintering liquid-phase solution-precipitation is

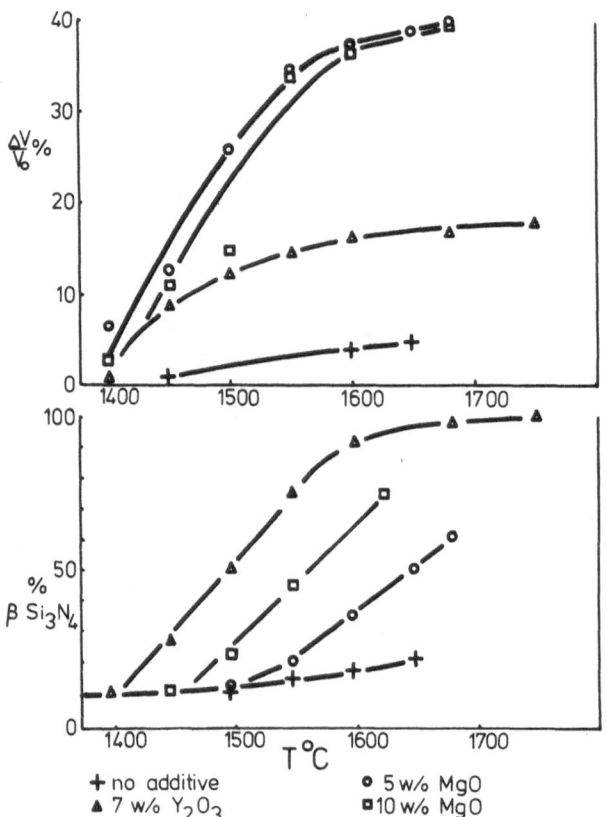

Figure 3.   Shrinkage and α-β transformation for Si_3N_4 sintered
            for 30 minutes at different temperatures

an important but not necessarily the main contributing stage of
the process.   Following Terwilliger and Lange[8] and Mitomo[9] it
seemed that the same mechanisms might be applied to sintering with
and without pressure by using Kingery's model.[3]

        Figure 3 shows the shrinkage of silicon nitride ($\Delta V/V_0$ for
zero porosity ∼40%) pressureless sintered for 30 minutes at
different temperatures.   It supplements Table 1 by illustrating
that extensive α-β transformation occurs with little densification
for yttria and vice versa for magnesia.

        Figures 4 and 5 are log shrinkage versus log time plots for
sintering at 1450° to 1600°C with 5w/o MgO and 7w/o Y_2O_3.   In
both cases the rapid densification of the rearrangement process
is obvious, but for Y_2O_3 it is responsible for only 7-9% shrinkage

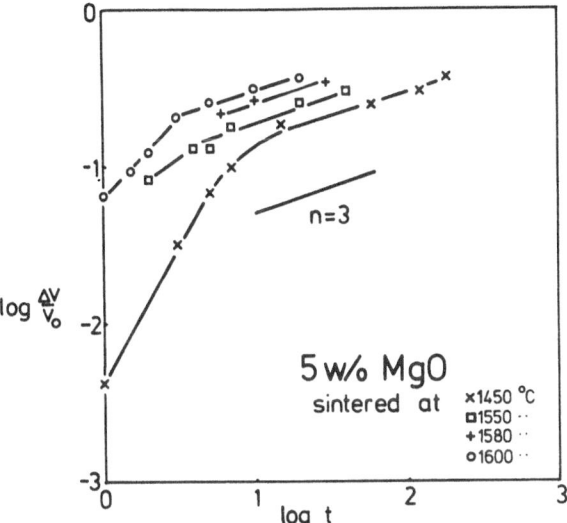

Figure 4.   Log shrinkage ($\Delta V/V_o$) v. log time (t, minutes) for Si_3N_4 sintered with 5w/o MgO at 1450-1600°C.

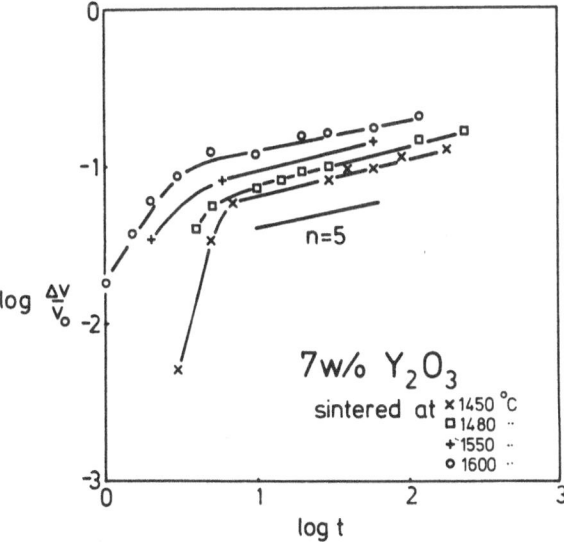

Figure 5.   Log shrinkage ($\Delta V/V_o$) v. log time (t, minutes) Si_3N_4 sintered with 7w/o Y_2O_3 at 1450-1600°C.

(i.e. less than 25% of the total densification) whereas with magnesia the first stage is responsible for about one-half of the full densification. As might be expected, this is due to the larger liquid volume obtained with magnesia.

During the second-stage solution precipitation, $\Delta V/V_0 \propto t^{1/n}$ where n = 3 or 5 depending on whether the rate-controlling step is respectively (a) solution into or precipitation from the liquid and (b) diffusion through the liquid. For MgO, n = 3 and for Y_2O_3, n = 5. With yttria, where diffusion is slow, the solution of α and precipitation of β will be relatively fast and so the transformation α-β can occur with only little transport and hence limited densification. With magnesia there is negligible α-β transformation during the initial rearrangement that accounts for about half of the densification, and then during the second stage the relatively rapid transport of material by diffusion and the slow precipitation of β ensures that any α-β transformation is accompanied by further densification.

The main features of hot-pressing as well as pressureless sintering are explained. Full densification never occurs without some transformation because the second stage of the process involves solution of α and precipitation of β. On the other hand, solution of α and precipitation of β can occur without major transport of material and so some transformation takes place without densific- ation. The amount of densification without transformation during the first stage depends on the quantity of liquid formed and hence on the amount of additive, on the kind of additive, and on the amount of surface silica present on the nitride. The amount of transformation without densification during the second stage depends on whether solution or diffusion is rate-controlling and hence upon the characteristics of the liquid, i.e. again upon the specific additive used. No densification during this stage can occur without some α-β transformation but, as pointed out by Brook et al.[4] the amount required for the removal of (say) 25% porosity need only be 12%.

Figure 6 compares hot-pressing with pressureless sintering for Si_3N_4:7w/o Y_2O_3. The density reached during the first-stage rearrangement is naturally greater with pressure, but in both cases the second-stage solution-precipitation is diffusion controlled and the mechanism seems the same. With 5w/o Y_2O_3:5w/o Al_2O_3, phase investigations show larger amounts of liquid formation at lower temperatures and hence lower viscosities under comparable conditions; the rate controlling step during the second stage is solution into or precipitation from the liquid in the same way as for MgO.

It should be noted that the densification behaviours of

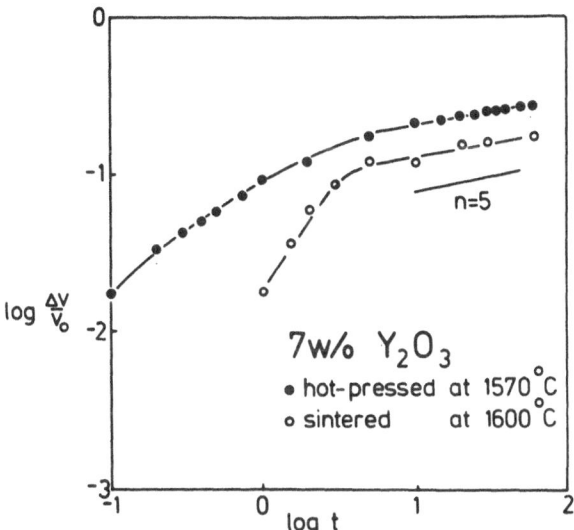

Figure 6.   Comparison of hot-pressing Si_3N_4 with 7w/o Y_2O_3 at
1570°C with sintering at 1600°C

silicon nitrides from different sources are different in detail.
Lucas silicon nitride containing about 4w/o surface silica
densifies more readily under given conditions than Starck Berlin
silicon nitride containing less than 3w/o silica.   This is due
to the different volumes and rates of liquid formation, and it is
significant that the densifying characteristics of the two powders
become almost identical by adding about 1w/o of finely divided
silica to the Starck Berlin silicon nitride.

The effects on properties of using oxide additives for
densification are now considered.

MAGNESIA

Figure 7 shows the Mg-Si-O-N behaviour diagram; the liquid
region at 1700°C is in dashed outline and homogeneous glass
compositions found on cooling are within the shaded area "G".
Lange [10] states that maximum strength and oxidation resistance at
1400°C occur with $MgO:SiO_2$ ratios that approach zero or infinity,
that is, along the joins Si_3N_4-Si_2N_2O and Si_3N_4-MgO.   It is
clear from the composition of the glass region that silicon
nitride with its usual surface silica (~4w/o) will contain glass
after hot-pressing or sintering with added magnesia.   Only when
the MgO content is low enough to bring the overall composition

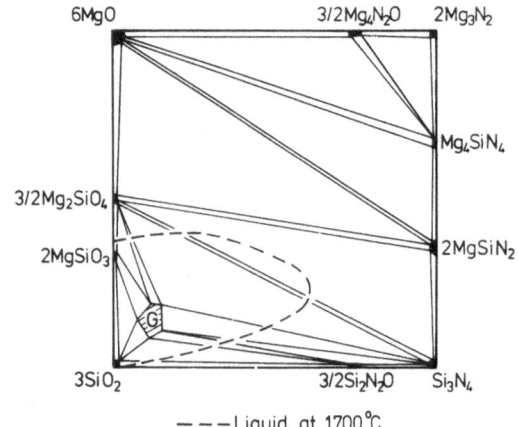

Figure 7.   The Mg-Si-O-N behaviour diagram

within the Si_3N_4-Si_2N_2O two-phase region or is high enough to bring
it into the three-phase triangle Si_3N_4-$MgSiN_2$-Mg_2SiO_4 will glass
be avoided.

Liquid and glass-forming regions extend into the Mg-Si-Al-O-N
system and here the 3M:4X plane is important (Figure 8) because
the β'-sialon phase extends from the Si_3N_4-Al_3O_3N join towards
forsterite, $2MgO.SiO_2$.   By addition of appropriate amounts of
Al_3O_3N (Al_2O_3+AlN) with just sufficient MgO to react with the
surface silica, a homogeneous, single-phase β'-magnesium sialon
can be obtained.   The first-formed Mg-Si-Al-O-N liquid that
assists densification is subsequently incorporated into the β'
solid-solution by heat-treatment.

MgO-BeO Additions

Because Be_2SiO_4, with the same atomic arrangement as β-Si_3N_4,
has a higher solubility in β and β' than Mg_2SiO_4, mixtures of
BeO+MgO extend the homogeneity range of β' and allow more
compositional flexibility for the incorporation of all additives
and impurities into a single-phase material.   Preliminary
observations suggest that some β'-Be-sialons can be melted and
re-solidified without decomposition, and Oda [11] reports near-
theoretical densities for silicon nitride pressureless-sintered
with BeO:MgO and BeO:MgO:CeO_2 mixtures.

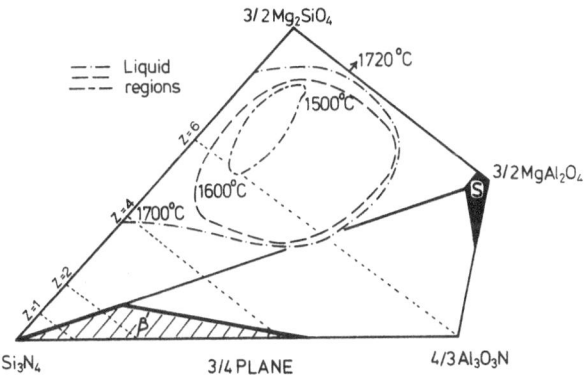

Figure 8.   The 3M:4X plane of the Mg-Si-Al-O-N system with liquid
regions at 1500-1700°C

YTTRIA

The behaviour diagram of the Y-Si-O-N system (Figure 9) shows
the following four quaternary phases:

(i)   N-melilite, $Y_2O_3.Si_3N_4 = Y_2Si_2O_3N_4$

(ii)   N-apatite, with a homogeneity range $(Y, \square)_{10}(SiO_4)_6(O,N)_2$

(iii) N-YAM, $2Y_2O_3.Si_2N_2O = Y_4Si_2O_7N_2$

(iv)   N-α-wollastonite, $Y_2O_3.Si_2N_2O = 2YSiO_2N$

all isostructural with the corresponding silicates or aluminates.[12]
With silicon nitride, yttria reacts initially with the surface
silica and some nitride to give liquid which promotes densification.
As reaction proceeds, the liquid combines with more Si_3N_4 to give
one or more of the quaternary refractory bonding phases in which
are accommodated in solid solution the impurities that would
otherwise degrade properties by cooling to give a glass.

Products containing 15w/o Y_2O_3 oxidise only very slowly at
1400°C (see Figure 10) to give cristobalite and γ-$Y_2Si_2O_7$ in a
protective glaze but at 900-1100°C the quaternary phases oxidise
rapidly to cristobalite and y-$Y_2Si_2O_7$ with a large change in
specific volume.   There is extensive cracking and this exposes
fresh surfaces for further attack;   see Figure 11.   Lange et al.
have pointed out [13] that materials free from the oxynitrides,
i.e. with compositions within the compatibility triangle
Si_3N_4-Si_2N_2O-$Y_2Si_2O_7$, show good oxidation resistance at both 1000°
and 1400°C.   Figure 12 is the Si_3N_4-corner of the Y_2O_3-Si_3N_4-SiO_2
system on which lines representing 2, 5 and 10w/o surface silica
are shown.   Clearly the amount of added yttria to bring the

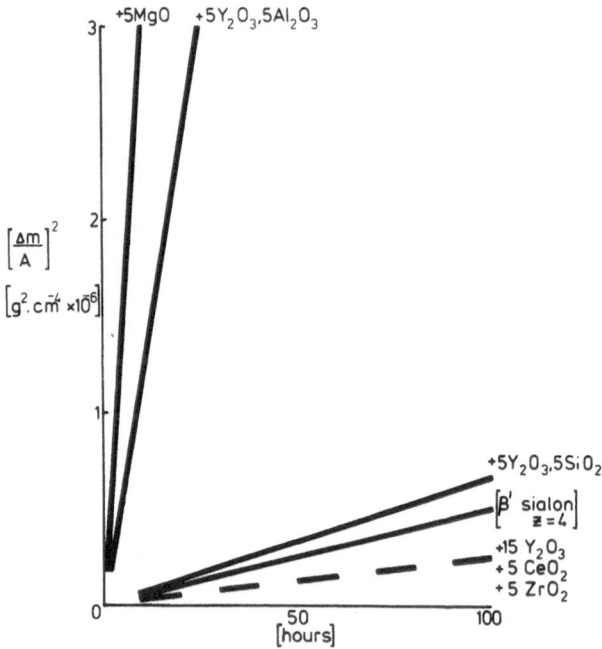

Figure 10.   Oxidation of hot-pressed silicon nitride at 1400°C

Figure 11. Si_3N_4 bars oxidised for 120h at 1000°C

upper: hot-pressed with 5w/o SiO_2 : 5w/o Y_2O_3

lower: hot-pressed with 15w/o Y_2O_3

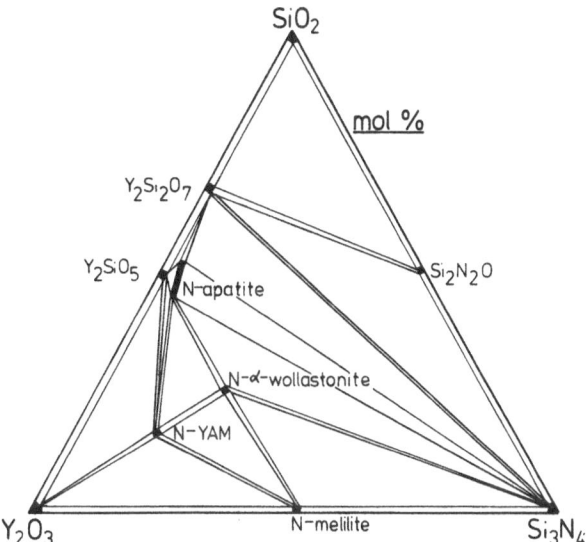

Figure 9. The Y_2O_3-SiO_2-Si_3N_4 behaviour diagram at 1700°C

composition within the favourable triangle depends very critically upon the amount of surface silica. In preventing the formation of oxynitrides to improve oxidation resistance, impurities such as calcium are no longer incorporated in a refractory phase but remain in a grain-boundary glass; this impairs the high-temperature creep resistance.

The Y-Si-Al-O-N System

In silicon nitride hot-pressed and pressureless-sintered with Y_2O_3+Al_2O_3 mixtures no new N-containing crystalline phases occur but appreciable volumes of low-viscosity liquid wet the solid completely and migrate rapidly through the whole microstructure.[14] Within the Si_3N_4-Y_2O_3-Al_2O_3 triangle (Figure 13) there is a wide range of compositions, expanding with increasing temperature, where the cooled products are β'-sialon and a Y-sialon glass of composition near $Y_{10}Si_{12}Al_{17}O_{57}N_5$. These materials show none of the catastrophic oxidation at 1000°C of some of the Y_2O_3:Si_3N_4 compositions but at 1400°C the surface glaze, containing some cristobalite, is not completely protective and the parabolic rate constant of about $10^{-1}mg^2cm^{-4}h^{-1}$ is nearly 10^3 times greater than that of the best Si_3N_4+Y_2O_3(+SiO_2); see Figure 10.

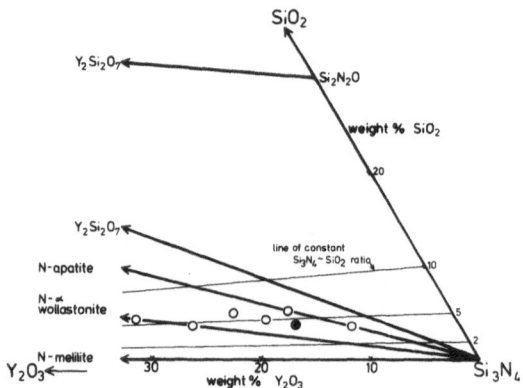

Figure 12.   The Si_3N_4 corner of the Y_2O_3-SiO_2-Si_3N_4 system
             showing 2, 5 and 10w/o surface silica

The lowest liquidus in the Y_2O_3-SiO_2-Al_2O_3 system is at
1350°C and at temperatures above this the oxidation of yttrium
sialons will involve liquid formation and so will accelerate.

CERIA

Fully dense products are obtained by hot-pressing Si_3N_4 with
ceria and with up to 10w/o CeO_2 the only crystalline phase
observed is β.   As with yttria, the oxide combines with surface
silica and some nitride to give a liquid which cools to an
oxynitride glass. CeO_2 decomposes to Ce_2O_3 at about 1230°C, but
when silicon nitride is present the latter is oxidised with loss
of nitrogen and with the formation of silicon oxynitride or
silica according to the equations:

$$2Si_3N_4 + 6CeO_2 \rightarrow 3Si_2N_2O + 3Ce_2O_3 + N_2 \tag{1}$$

$$Si_3N_4 + 12CeO_2 \rightarrow 3SiO_2 + 6Ce_2O_3 + 2N_2 \tag{2}$$

Contrary to a previous report,[15]   the further reactions of
cerous oxide give four quaternary oxynitrides that are isostructural or nearly so, with the corresponding yttrium compounds
(melilite, apatite, YAM and wollastonite) plus a fifth phase,
$Ce_2O_3.2Si_3N_4 = Ce_2Si_6O_3N_8$, that has no yttrium analogue;   see
Figure 14.   This correspondence between the Ce-Si-O-N and
Y-Si-O-N systems suggests that the behaviour and properties of
silicon nitride and sialons densified by ceria will be similar to
those using yttria.   The advantage of ceria is that it should
have a built-in safeguard to assist in maintaining the

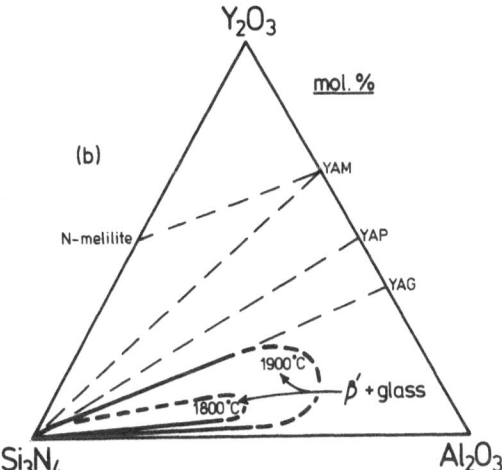

Figure 13.   The Y_2O_3-Al_2O_3-Si_3N_4 behaviour diagram showing glass
              formation after heating at 1800-1900°C

composition within the desirable Si_3N_4-Si_2N_2O-silicate triangle.
For the same initial surface silica content on the silicon nitride,
more than twice the weight of ceria can be added compared with
Y_2O_3 before any N-apatite appears in the product.   As expected,
the catastrophic oxidation at 1000°C for compositions exceeding
a critical Y_2O_3 concentration is not found at corresponding CeO_2
compositions.   However, oxidation resistance with CeO_2 and
CeO_2+Al_2O_3 is generally worse than with similar yttrium-containing
materials.   With 5w/o CeO_2:5w/o Al_2O_3 silicon nitride at 1000°C
gives a net-work of surface cracks and sub-micron holes due to
gas evolution, and with 10w/o CeO_2:10w/o Al_2O_3 layers of scale
flake off to expose new surfaces for further oxidation.   At 1400°C
Si_3N_4 with 2w/o CeO_2 shows little surface degradation but higher
ceria specimens are seriously corroded and show that the surface
has been liquid at temperature.   Alumina-containing samples show
cristobalite growing in a surface glaze containing large gas
bubbles and cracks.

     The explanation for this behaviour is that cerium in its
silicates, aluminates, aluminosilicates and oxynitrides is
trivalent but under conditions of high oxygen potential it becomes
tetravalent and all its compounds decompose to give ceric oxide,
CeO_2, when heated in air to about 1400°C.   Oxynitrides decompose
with nitrogen evolution and this breaks up the oxidised surface.

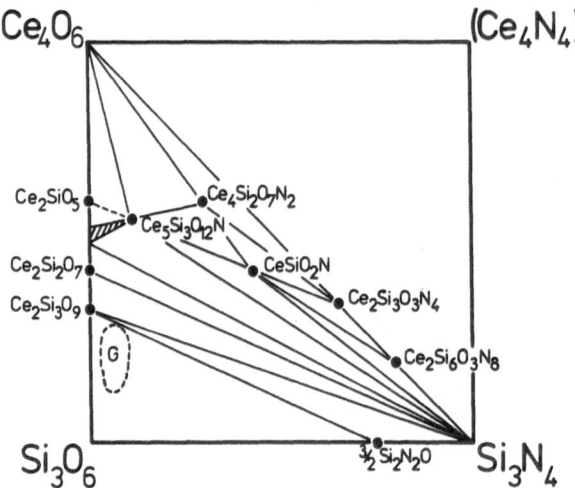

Figure 14.   The Ce-Si-O-N behaviour diagram at 1700°C

ZIRCONIA

From the Zr-Si-O-N behaviour diagram (Figure 14), a small amount of liquid which dissolves nitride will be formed locally at silica-rich grain boundaries when Si_3N_4 is hot-pressed with 5w/o ZrO_2.   The final product within the triangle Si_3N_4-Si_2N_2O-ZrO_2 will contain little glass and so should show good oxidation resistance.   In fact, after 120h at 1000° and 1400°C, bar specimens show no detectable weight increase and are unchanged except for a thin coherent surface glaze at the higher temperature.

With 5w/o ZrO_2:5w/o Al_2O_3 densification is more effective and oxidation resistance is no different at 1000° and 1400°C from that with ZrO_2 alone.   Only one new phase is found in the Zr-Si-Al-O-N system and although its oxidation is accompanied by complete disintegration, its composition (near $ZrAl_4O_5N_2$) makes its occurrence unlikely in nitrogen ceramics.

It is reported [16] that reaction of ZrO_2 with Si_3N_4 can lead to the formation of zirconium nitride.   Given an initial composition within the Si_3N_4-Si_2N_2O-ZrO_2 triangle of Figure 15, loss of volatile SiO+N_2 moves the composition into the three-phase field Si_3N_4-ZrN-ZrO_2, i.e. the reaction

$$2Si_3N_4 + 2SiO_2 + 2ZrO_2 \longrightarrow 2ZrN + 8SiO + 3N_2 \qquad (3)$$

is thermodynamically possible if SiO and N_2 are removed by volatilisation.

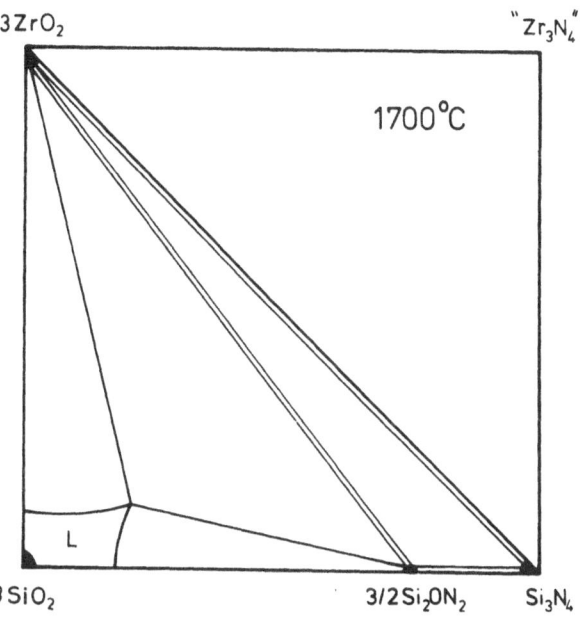

Figure 15. The Zr-Si-O-N behaviour diagram at 1700°C

CONCLUSIONS

Oxide additives used in hot-pressing and pressureless sintering of nitrogen ceramics promote liquid-phase densification interpreted by a Kingery model in which the contributions by the successive stages (i) rearrangement and (ii) solution-precipitation vary with the particular additive. The $\alpha-\beta$ phase transformation in Si_3N_4 occurs only in stage (ii). Liquid phases, on cooling, give additional crystalline or vitreous phases that affect chemical and mechanical properties. Only from a detailed knowledge of phase relationships in the appropriate system is it possible to select and control the additives to produce a fully dense, high-strength, oxidation-resistant material. So-far, this combination of properties in an engineering ceramic suitable for prolonged use in an oxidising environment at above 1400°C has not been achieved.

REFERENCES

1. S. Wild, P. Grieveson, K. H. Jack and M. J. Latimer, Special Ceramics, 5, 377 (1972).
2. K. H. Jack, Proc. NATO Advanced Study Institute, "Nitrogen Ceramics", Canterbury 1976, to be published.
3. W. D. Kingery, J. Appl. Phys. 30, 301, (1959).

4. R. J. Brook, T. G. Carruthers, L. J. Bowen and R. J. Weston,
    Proc. NATO Advanced Study Institute, "Nitrogen Ceramics",
    Canterbury 1976, to be published.
5. W. D. Kingery, J. M. Woulbroun and F. R. Charvat, J. Amer. Ceram.
    Soc. 46, 391, (1963).
6. G. R. Terwilliger and F. F. Lange, J. Mat. Sci. 10, 1169, (1975).
7. A. Hendry, Proc. NATO Advanced Study Institute, "Nitrogen
    Ceramics:, Canterbury 1976, to be published.
8. G. R. Terwilliger and F. F. Lange, J. Amer. Ceram. Soc. 57,
    25, (1974).
9. M. Mitomo, J. Mat. Sci. 11, 1103, (1976).
10. F. F. Lange, paper presented at the Basic Science Fall Meeting
    of the Amer. Ceram. Soc., San Francisco (1976).
11. I. Oda, Proc. NATO Advanced Study Institute "Nitrogen Ceramics",
    Canterbury 1976, to be published.
12. A. W. J. M. Rae, Ph.D. Thesis, University of Newcastle upon
    Tyne (1976).
13. F. F. Lange, S. C. Singhal and R. C. Kuznicki, J. Amer. Ceram.
    Soc. 60, 249, (1977).
14. J. D. Venables, D. K. McNamara and R. G. Lye, Proc. NATO
    Advanced Study Institute "Nitrogen Ceramics", Canterbury
    1976, to be published.
15. R. R. Wills and J. A. Cunningham, J. Mat. Sci. 12, 208, (1977).
16. P. F. Becher and S. A. Halen, Proc. 5th Army Materials
    Technology Conference "Ceramics for High-Performance
    Applications II", Newport 1977, to be published.

DISCUSSION

P. E. D. Morgan (University of Pittsburgh): We recently looked
(using a diffracted beam monochromator diffractometer technique)
at the XRD of as hot pressed and oxidized silicon nitride containing small additions of ZrO_2 and MgO. The samples were made and
oxidized by the Ceramics Branch Group at the Naval Research Laboratory in Washington. We noted that, after oxidation at $1430^{\circ}C$ for
100 hours, some Si_2N_2O had been produced throughout the body
(which was ~ 2 mm thick). It appears that the grain boundary oxide
phase used for hot pressing is providing an easy path for oxygen
into the bulk of the sample. Thus the notion of using nitrides
because of their extremely low diffusion values is vitiated by
providing for the ready access of oxygen via an oxygenatious grain
boundary phase. As determined by XRD, samples containing initially 4% ZrO_2 or 3% MgO developed Si_2N_2O in the center to about the
same degree. The 4% ZrO_2 sample initially contained a few percent
of silicon which was still partially present in the center after
the oxidation. Very minor amounts of unknown phases were also
present.

CERAMICS IN THE Si-Al-O-N SYSTEM FABRICATED BY CONVENTIONAL POWDER PROCESSING AND SINTERING TECHNIQUES

Arno Gatti and Michael J. Noone[*]

General Electric Company, Space Sciences Laboratory

P. O. Box 8555, Philadelphia, Pa. 19101

I. INTRODUCTION

This paper summarizes the results of a program designed to investigate methods of fabricating ceramic materials with structurally useful thermal and mechanical properties from solid solutions of silicon, nitrogen, and metal oxides. The investigations included studies of the relationship of precursor powders and processing methods to the resulting composition and microstructure of fabricated specimens and, in turn, the relationship of these compositions and microstructures to the mechanical properties of the fabricated specimens. Emphasis was placed throughout the program on the use of conventional but controlled ceramic powder processing, fabrication of green shapes, and firing procedures, to achieve high density in the components using pressureless sintering techniques. Compositions within the silicon-aluminum-oxygen-nitrogen system, generally referred to as "sialons", were studied.

The stimulus for this work was the data reported almost simultaneously by Jack [1,2,3] and Oyama [4,5,6] that the extensive solid solutions which exist in the $Si_3N_4-Al_2O_3$ system result in materials which may have unique properties and may be readily fabricated to high density by pressureless sintering techniques. Studies of sialons at Lucas in England [7,8] showed that materials with attractive mechanical and thermochemical properties could be produced, at least by the hot-pressing route, and that single-phase materials existed over only a narrow range of compositions represented by the formula: $Si(6-z)AlzN(8-z)Oz$.

*Present address: CertainTeed Corp., Valley Forge, Pa. 19482

The potential for development and exploitation of sialon materials particularly by the pressureless sintering approach led to the initiation of this work and several similar programs. [9-12] It has been demonstrated in this work and elsewhere, particularly at Lucas,[13] that dense materials can be obtained by pressureless sintering from a wide range of compositions and with properties amenable to exploitation in hazardous thermal, chemical, and friction and wear environments.

No specific phase equilibria studies were performed in our work and the compositions and fabrication procedures were designed as far as was practical to be consistent with the formula established by Lumby and the narrow phase field for achievement of single-phase structures observed by him and developed in more detail by Gauckler.[14]

The observed rapid formation of the β' solid solution in sialon systems and the ease with which impure "sialon" compositions could be densified by sintering were manifestations of the operation of a liquid phase sintering mechanism. A primary objective of our work was to achieve densification with a minimum of impurities or densification aids and to eliminate, or at least minimize, the presence of a liquid phase at the conclusion of the densification process. The present authors had previously exploited this technology to develop dense sintered spinels in which additives were used to encourage liquid phase sintering in the initial densification stages and were then removed or included into the solid solution by careful control of temperature and atmosphere in the final stages of densification.[15,16]

The present work explored fabrication variables in the sialon system to obtain an understanding of the means to achieve dense single-phase materials. Materials in the sialon system were readily fabricated by conventional ceramic processing techniques, and were fired under controlled conditions to produce structural ceramics which, except for lower as-fabricated strengths, have properties comparable to Si_3N_4-based materials produced by more complex and expensive methods. Properties of these materials were not optimized by concentrating on the production of any single composition. However, sufficient data were attained to indicate that materials in the sialon family can be produced using low-cost, large-scale, fabrication processes insensitive to specific starting materials and that the properties may be optimized to suit the requirements of several potential applications.

Simple mixing techniques initially led to glass-bonded materials with inferior high temperature properties and glassy oxidation products. It was, nevertheless, found that simple processing techniques could be employed to produce sizeable components of essentially single crystalline phase (β'-Si_3N_4 solid solution with about 50% Si_3N_4). Later studies employing sialon milling media demonstrated that compositions rich in Si_3N_4 could be readily fabricated with a minimum of glassy grain boundary phase and hence with properties appropriate to several potential structural applications. These materials also possessed essentially only one crystalline phase (β'-Si_3N_4 solid solution) and could be processed reproducibly using controlled atmosphere sintering. Compositions from this phase of the program are being studied further in ongoing programs both at General Electric Company and at Air Force Materials Laboratory.

II. EXPERIMENTAL PROCEDURES

Three basic methods of powder processing were used: (i) mixing of component powders in a Waring Blender[R], (ii) milling of powders separately in a high alumina mill and combining them into appropriate compositions, and (iii) milling of powder mixtures using a polyure-thane-lined mill with sialon grinding media.

The Waring Blender technique provided a rapid means for making low contaminant powder mixtures when using an aluminum blade. Materials with sintered densities of 95-98% were common for MgO-doped sialon bodies processed by this method. However, the Si_3N_4 powder (-325 mesh 85% α-phase)* and AlN* was not ideal for incorpor-ation into the sialon mix since other constituents such as the Al_2O_3*, and MgO* were in the submicron range (0.3-3 μm). Thus the mixing and sintering behavior of these materials were not optimum for production of dense ceramics by simple mixing. It was initially considered that further comminution of the AME powder would lead to contamination and degradation in terms of inhomo-geneous microstructure and mechanical properties. It was evident, however, that maximum densification would not be achieved unless starting powders were of about equal particle size and sufficiently fine to promote active sintering, despite the liquid phase mechanism. Therefore, sialon mixtures were prepared by milling in high-alumina mills to reduce the particle size of the AME Si_3N_4 by up to an order of magnitude. The more favorable particle size distribution reduced sintering times, sintering temperatures, and compositional inhomo-geneities. Mill wear, however, was excessive and, as expected, led to large variations from the "starting" compositions.

The most acceptable materials were obtained by milling in a polyurethane-lined mill with sialon balls and using a mineraliza-tion technique to control the amount and composition of the liquid phase. Rather than rely on mill wear to add glass-forming contamin-ants such as silica, magnesia and calcium, these materials were added to the alumina component of the sialon composition, calcined for "mineralization" (and homogenization) and reduced to a fine

* Obtained from AME (England), Cerac Inc., Linde Air Products Div., (Linde A), and Baker Chemical Company, respectively.

particle size during sialon milling. This technique led to sialon
compositions with a broad firing range and with "liquid" phases less
susceptible to low temperature AlN reactions which would otherwise
reduce the total liquid (SiO_2 and silicates) volume available to
aid densification. As the study progressed, the amount of liquid
formers was reduced to a minimum in order to maximize high temperature properties. This final phase of the work utilized only the
wear of "sialon" balls in the mill to furnish the "liquid"
necessary for sintering to high density, i.e., no sintering aids
were "added" separately.

During processing there was a reduction of available AlN due
to oxidation and water reaction to form NH_3 (gas) and $Al(OH)_3$. It
was found, and has been shown by others (17), that the reduction
of available AlN in the system results in greater liquid content
since the SiO_2 is not reduced by AlN reaction and remains available
to form liquid silicates which enhance sintering. This inhibition
of AlN activity occurs both by direct conversion of AlN to $Al(OH)_3$
and, indirectly, by coating each particle with an $Al(OH)_3(Al_2O_3)$
reaction layer.

Powder forming techniques used during the study included
hydrostatic pressing at pressures up to 400 MPa (50,000 psi) slip
casting, jiggering, and injection molding.

Firing of sialon compositions was performed in a molybdenum
box packed loosely with "sialon" powder and covered with a molybdenum lid to control weight loss during firing. The molybdenum
box was placed into a molybdenum retort the mouth of which was
covered with molybdenum foil and through which flowed nitrogen at
a controlled rate. This assembly was fired in a hydrogen ambient
to temperatures up to 1770°C for various times. The sialon packing
material adheres to the surfaces of fired sialon but can be readily
removed. Intricate shapes such as turbine blades are less readily
cleaned-up if adhering scales persist after firing - particularly
if the components have thin walls. A parting compound was
developed for these components to permit them to be extracted
cleanly after firing. This parting compound consisted of a thin
layer (~ .050") alumina-alcohol slurry sprayed or painted onto the
unfired sialon component followed by another thicker layer of a
blended slurry consisting of 1/3 Si_3N_4-1/3 Al_2O_3-1/3 AlN (by weight).
The part was then packed in loose previously fired, coarsely ground,
sialon mix and fired in the molybdenum retort.

III. RESULTS AND DISCUSSION

A. Simple Mixing

Table I is a partial list of MgO-containing compositions
which were evaluated by the Waring Blender technique. Fifty gram

batches of each composition were mixed using powders of Linde A
alumina, AME Si_3N_4, Bakers MgO in an aluminum bladed Waring Blender
with isopropyl alcohol as the mixing vehicle. The Si_3N_4 content
was maintained constant at 60 w/o and the alumina-magnesia ratio
was varied. Compositions with 10.8 and 13.3 w/o MgO were chosen to
provide a stoichiometric magnesium aluminate spinel with the
available Al_2O_3 in the case of the 10.8 material and a magnesia-rich
spinel for the 13.3 material. Additions greater than 5% MgO do not
show appreciably greater shrinkage after firing.

Examination of polished sections of fired specimens showed
multi-phase structures, particularly in specimens induced to densify
by the higher MgO additions. Areas of varying reflectivity, of much
greater size than the initial particle size, were observed and all
specimens contained remnant porosity which had a frothy appearance
in specimens with excessive glass phase. AlN was added to these
compositions in amounts varying from 0 to 20% (in 5% steps) and
reduced densification was observed. Premature SiO_2-AlN reactions
and the inability of this technique to produce homogeneous mixes
or reduce the particle size of the coarser constituents led to the
study of milling techniques for material preparation.

2. Milling

A sialon composition 60/40 Si_3N_4/Al_2O_3 + 8%/MgO was pre-
reacted at 1780°C for one hour, crushed and milled for 72 hours
in propanol. The mill and balls were Borundum[R] (Norton Company):
approximate composition 85% Al_2O_3, 11% SiO_2, 2% MgO, 1.2% CaO,
balance TiO_2, Fe_2O_3, Na_2O and K_2O (.89%). After milling the batch
had more than doubled in weight, (a typical 100 gm charge weighed
215 grams). This mill wear was fairly constant throughout the study
when Si_3N_4, or β' sialon, or AlN was milled. These initial milling
experiments were thus more a study of mill contamination effects
than of comminution but nevertheless the results provided clues
to the nature of the achievement of high density in sintered sialons.

Table I. Compositions of Sialons Containing MgO in Weight Percent
with Shrinkage and Weight Loss Data Firing at 1700°C

(wt. percent)			% Linear Shrinkage at 1700°C			Wt. Loss %
Si_3N_4	Al_2O_3	MgO	1 h	2 h	3 h	1 h/1700°C
60	40	1/4	7	10	11	13
60	39	1	9	10	10	—
60	37	3	10	11	12	12
60	40	5	13	15	16	10
60	32	8	14	16	16	8
60	29.2	10.8	14	16	16	5
60	26.6	13.3	14	15	15	2

Discs of this material (20 mm diam. X 6 mm thick) were pressed at 267 Pa (20 TSI) and shrinkage data, micro-structure and x-ray diffraction data were obtained. This material (with a probable composition of now 20%Si_3N_4, 60%Al_2O_3, 5%MgO, 6%SiO_2, and 6%CaO based on mill compositon and pick-up) sintered to full density at 1550°C with a linear shrinkage of 19%. Material sintered above 1600°C experienced bloating which was no doubt a result of reaction between Si_3N_4 and the glassy phase with N_2 being liberated at higher temperatures. The additional alumina, silica, calcia, and magnesia picked up during milling greatly enhanced the glass supply. It was clear that such glasses could not be tolerated by Si_3N_4 systems much above 1600°C.

MgO was a powerful sintering aid for these Si_3N_4-based materials, presumably by combining with surface SiO_2 on the Si_3N_4 particles and alumina to form low viscosity, low melting MAS glass. The MAS glass formed during the initial sintering process enhances densification but can remain at higher temperatures with the Si_3N_4 to cause bloating. This may be avoided by addition of AlN [8,10,17] which reacts with the available silica to form additional "sialon" and reduces the quantity of glass considerably if not totally. Indirectly adding SiO_2 and MgO during milling produced excessive MAS glass in the body and bloating resulted. The milled material sintered to full density at relatively low temperatures before the bloating reaction was observed. When AlN was added to this readily sinterable milled material, inhibition of initial sintering was not observed. However, the bloating during final high-temperature firing was stopped. These results provided the first clue that "mineralizing" of the sintering aid would retard the formation of "x-phase" (*) in the presence of AlN additions. Otherwise, AlN would simply limit the amount of liquid phase and effectively retard sintering as was observed for the simply-mixed components.

It was clear that milling of highly abrasive Si_3N_4 in Borundum mills would lead to excessive contamination of the batch. It was imperative that the composition of sialon be brought under greater control so that mixtures could be more precisely formulated to minimize additional phases, excess glass content, etc. and thus produce strong, dense material of controlled composition. Accordingly, large quantities of sialon material were milled for 72 hours to produce a starting powder of approximately 40% Si_3N_4, 50% Borundum and 10% AlN by weight. Large quantities of various diameter cylinders were isostatically pressed at 67 MPa (10,000 psi) from these powders so that "sialon" grinding balls could be made. The balls, actually round-ended cylinders, were fired at 1700°C

---

* The silicon-oxynitride phase first reported by Jack[1] and Oyama.[4]

for 1 hour in N_2/H_2 and were dense. Sufficient balls were made to give a 55% ball charge in either a low-ash rubber-lined mill or a polyurethane lined mill (Norton Company). Particle size reduction after 24 hours running in these mills was comparable to Borundum milling.

Of course, the sialon material of which these balls were made was fabricated using mill-contaminated material so that some carry-over was to be expected. However, the level of contamination was reduced considerably with no free SiO_2 pick-up (since the additions of AlN to the original mix converted SiO_2 to 'X'-phase and/or β' sialon). The MgO pick-up was at low level since it was only about 1% in the balls themselves.

3. Milling with "Sialon" Media

Studies to utilize the mineralizing approach were begun initially by "mineralizing" the total alumina charge used. (The Al_2O_3, MgO, SiO_2 were pre-reacted). A series of compositions (see Table 2) were milled with sialon balls, pressed, and fired. Changes in SiO_2, MgO additions did not affect the sinterability or fired crystalline compositions of these materials which were all apparently single-phased β' sialon (by x-ray diffraction). It was hypothesized that at least the minimum amount of liquid necessary to enhance the conversion of the body to sialon was present in these materials for every composition studied. This tended to confirm the hypothesis that the volume and distribution of glass phases present were most important and that their composition was not necessarily critical. This led to further speculation that additives other than MgO would also be effective densification aids using the "mineralized" approach. Thus a mechanism for employing liquid-phase sintering was available which could be tailored to ensure that the final product would contain only refractory crystalline phases and thus have attractive high temperature properties.

Table 2.   Compositions of Sialon Bodies Processed with Varying
           Amounts of "Glass-Former"

Sample #	Composition: Weight Percent					
	Si_3N_4	Al_2O_3	SiO_2	MgO	AlN	Balance*
#102	40	51	6.6	1.2	15	1.2
#108A	40	48	9.0	1.8	15	1.2
#108B	40	53.4	4.2	1.8	15	0.6
#108C	40	51	6.6	1.8	15	0.6
#110	40	49.8	6.6	3.0	15	0.6

*Balance is made up of CaO, Fe_2O_3, TiO_2, etc.

Table 3. Compositions Firing Schedule and Final Sintered Density
          of Sialons of High Si_3N_4 Content

Spec #	(wt. percent)			Milling Time (Hours)	Mill Pick-up (Grams)	Firing Schedule	Final Density (gm/cc)
	Si_3N_4	Al_2O_3	AlN				
128	66	24	10	8	1.6	1740°C	2.10
128	66	24	10	72	20	1 hour	3.10
129	88.7	8	3.3	72	21	↓	3.10
130	94.5	4	1.5	72	20		2.96

Composition corrected for mill "pick-up" ater 72 hrs.

Spec #	Si_3N_4	Al_2O_3	AlN	SiO_2	MgO	CaO
128	64.0	26.0	8.3	0.9	0.6	0.1
129	83.0	13.0	2.7	0.9	0.6	0.1
130	87.0	10.0	1.9	0.9	0.6	0.1

The success of the "mineralizing" approach permitted further studies of reduced additive concentration. Accordingly, compositions shown in Table 3 were studied using wear of sialon balls as an additive mechanism (the mill was polyurethane-lined with propanol vehicle. Material milled for 72 hours readily achieved high density when fired). The odor of ammonia indicated that AlN and Si_3N_4 were reacting with the propanol through hydrolization of water impurities thus affecting the milled composition. Introduction of sintering aids during milling has also been used by others [18]   and is probably unavoidable if a long milling time is required. These results showed that excessive amounts of liquid formers had been used in the earlier mineralizing studies.

These compositions were reproduced by adding 5% by weight synthetic mineralized "Borundum" to a mill charge and milling for 24 hours with sialon balls. Little ball wear resulted but sinterable sialons containing up to 90% weight percent Si_3N_4 were produced which contain in effect only about 0.1% MgO as the active sintering aid.

IV. CONCLUSIONS

Initial experiments identified a "mineralization" mechanism which is effective in the sintering of sialon compositions. Mineralization was accomplished in several ways including pre-reaction of sialon composition and regrinding to pick up "mineralized" material through contamination: by mineralizing the alumina addition and grinding with sialon balls; and by contamination of the mixture through hydrolization of AlN and Si_3N_4 constituents while simultaneously adding mineralized contaminants from wear of grinding media. It was **difficult** to separate these processes and gain complete control of their effect on the sialon processing. However, an appreciation that such processing variables exist and a knowledge of their magnitude was employed to manipulate their influences on the ultimate batch composition and was utilized to achieve high densities in sintered sialon bodies.

It was also shown that if contamination were not completely controlled, then the role of "sintering aids" could not be correctly interpreted: very small amounts of mineralized MgO can be extremely effective even for sialons of high Si_3N_4 content when the mixes are highly milled (and presumably hydrolized). It is further apparent that strict adherence to "phase diagram" composition is virtually impossible without awareness of these influences, and the direction in which they all, separately and collectively, can move the assumed composition toward the eventual fired composition. It was shown that pressureless consolidation of Si_3N_4 systems was difficult and perhaps not possible without a liquid phase so that reactions which restrict liquid formation must be retarded (at least in the initial stages of sintering) in order to permit the material to densify.

The very mechanisms which allow the attainment of high density in Si_3N_4 systems also affects the high temperature behavior of the products and most of these materials suffer from excessive creep at temperatures above 1200°C because of the viscous behavior of residual glassy second phases. The properties of the sialon compositions fabricated during the program were comparable to those in the literature in terms of thermal expansion, oxidation resistance, strength and modulus. Exceptionally high strength was not consistently achieved. However, it is to be noted that the primary goal of this program was to develop processing techniques for sialon materials which could lead to their use in practical large-scale operations.

Ceramic processing techniques were developed which were based on conventional ceramic technology and were of sufficient scale to fabricate examples of radomes, IR-domes, and turbine blade shapes without regard to starting powders. Further, no size limitations were encountered other than those imposed by available hydropressing facilities and furnace chamber capacity. Specimens suitable for further evaluations of rain erosion, radar response, reflectivity, hot corrosion, etc. were readily fabricated and are the subject of on-going test programs.

ACKNOWLEDGEMENTS

This paper is based on work performed under Air Force sponsorship on contract AF33615-74-C-4073 "Methods of Fabricating Ceramic Materials" with Drs. James M. Wimmer and Peter L. Land as program monitors (Final Report AFML-TR-77-135). The authors gratefully acknowledge the support of William Laskow in the fabrication studies, T. A. Harris, and M. I. Birenbaum for assistance in materials fabrication and P. J. Grosso for assistance in mechanical testing.

REFERENCES

1. K. H. Jack and W. I. Wilson, Nature Phys. Sci. 238, 28-9 (1972).
2, A Correspondent, Nature 238, No. 5360), 128-9 (1972).
3. K. H. Jack, Trans. and J.Brit.Ceram.Soc. 72, 376-84 (1973).
4. Y. Oyama and O. Kamigaito, Japan J.Appl.Phys. 10, 1637 (1971).
5. Y. Oyama, Japan J.Appl.Phys. 11, 760-1 (1972).
6. Y. Oyama, Yogyo-Kyokai-Shi 82, No. 7, 352-7 (1974).
7. W. J. Arrol, 729-38 "Ceramics for High Performance Applications"
    729-38, Eds. J. J. Burke, A. E. Gorum, and R. N. Katz, Brook
    Hill Publishing Co. (1974).
8. R. J. Lumby, B. North, and A. J. Taylor, Special Ceramics 6,
    283-98 by P. Popper, The British Ceramic Research Assn. (1974).
9. F. F. Lange, Final Report on Contract N00019-73-C-0208, Feb.'74.
10. G. K. Layden, Final Report on Contract N00019-75-C-0232, Feb. '76.
11. G.K. Layden, Quarterly Prog. Rpt., Contract NA53-19712,
    June 1975 to date.
12. P. L. Land, J. M. Wimmer, R. W. Burns, and N. S. Choudhury,
    Interim Report covering Oct. 1973 to Oct. 1975 on Contract
    F33615-71-C-1463, April 1976. (To be published in J. Am.
    Ceram. Soc., 1977).
13. R. J. Lumby, B. North, and A. J. Taylor, "Ceramics for High
    Performance Applications-II", Proceedings of the Fifth Army
    Materials Technology Conference, Newport, R. I., March 1977
    (to be published).
14. L. J. Gauckler, H. L. Lukas, and G. Petzow, J. Am. Ceram.
    Soc., 58, Nos. 7-8, 346-7 (1975).
15. A. Gatti, Final Report on Contract N00019-69-C-0133 (1969).
16. A. Gatti, R. L. Mehan, and M. J. Noone, Final Report on
    Contract N00019-71-C-0126 (1971).
17. L. J. Gauckler, S. Boskovic, I. K. Naik, and T. Y. Tien,
    Proceedings of the Workshop on Ceramics for Advanced Heat
    Engines, Orlando, (Jan. 1977). ERDA Report: Conf-770110, (1977).
18. R. R. Wills, R. W. Stewart and J. M. Wimmer, J. Am. Ceram.
    Soc., 60 (1) 64 (1977).

POLYTYPISM IN MAGNESIUM SIALON

David R. Clarke

Rockwell International Science Center
Thousand Oaks, California 91360

T. M. Shaw

University of California
Berkeley, California 94720

ABSTRACT

The crystallography and morphology of the 12H and 21R polytypes in a magnesium sialon ($Mg_{1.86}Si_{1.67}Al_{2.47}O_{3.19}N_{3.81}$) have been investigated by lattice fringe imaging and electron diffraction. Intergrowths of the two polytypes occur, but since they have different crystal sizes, mismatch strains are introduced which may have a detrimental effect on mechanical properties of the material. It is concluded that careful control of the starting powders and long hot-pressing times will be necessary to avoid the formation of these strained intergrowths.

INTRODUCTION

Recent investigations by Jack[1,2] have explored the properties of a complex series of nitrogen ceramic alloys, based on cation and anion substitutions in silicon nitride, with a view to their use in high temperature applications. These materials have since come to be known as the Sialons.

X-ray analysis of phases present in a number of Sialons has revealed the presence of a class of compositionally dependent polytypes based on the wurtzite structure within the $Be-Si\ O-N$,[3] $Si-Al-O-N$,[2] $Mg-Si-Al-O-N$ and $Li-Si-Al-O-N$[4] systems. Following Jack,[2] the polytypes in the $Si-Al-O-N$ system can be described as a regular insertion of a cubic stacked layer into the basic hexagonally stacked wurtzite structure. In the cubic stacked layer, since the upward and downward pointing tetrahedral sites do not share a common base, they can both be occupied simultaneously by non-metal atoms. This gives a layer composed of two non-metal atoms per metal atom (MX_2). In the remaining wurtzite structure, upward and downward pointing tetrahedral sites share a common base and consequently remain mutually exclusive sites for non-metal atoms. The compositions here thus cannot exceed MX. For example, a 12H (Ramsdell notation) polytype can be described as a regular insertion of a MX_2 cubic stacked layer every six close packed MX layers to produce an overall structure having hexagonal symmetry consisting of two blocks each of six close packed layers and an overall composition M_6X_7. Furthermore, each polytype extends along a line of constant metal to non-metal atom ratio and has a narrow range of homogeniety in the phase diagram with respect to changes in the M:X ratio. On substitution of MgO into these polytypes, Jack[2] found that each phase is also extended into the corresponding planes of metal to non-metal atom ratio in the MgSi-Al-O-N phase diagram when represented by Janeck's triangular prism.

In the present investigation microstructural and crystallographic features of a hot pressed 12H Mg-Si-Al-O-N polytype in the M_6X_7 plane have been investigated using high resolution electron microscopy. In particular, the technique of lattice fringe imaging has been used to examine directly the stacking periodicities of the close packed planes in the polytype, just as the stacking periodicities in the Be-Si-N system have been studied.[5]

EXPERIMENTAL

The material, a 12H magnesium-sialon having a composition of $Mg_{1.86}Si_{1.67}Al_{2.47}O_{3.19}N_{3.81}$ (M_6X_7), was kindly supplied by Professor K. H. Jack. It was prepared by hot pressing a mixture of MgO, Al_2O_3, AlN and Si_3N_4 powders at 1800°C for two hours, and x-ray analysis showed it to contain 99% of the 12H polytype structure.

Electron transparent samples were prepared by the standard methods of grinding and subsequent ion beam thinning. The microscopy was performed with a Philips EM301 microscope operating at 100keV. Details of the technique of lattice fringe imaging have been described fully elsewhere.[6]

MICROSTRUCTURE

The microstructure consists of long interlocking laths or plates several microns in length and about one micron in width, as illustrated in Fig. 1(a). Also present at three and four grain junctions were pockets of a glassy phase (see accompanying paper by Clarke and

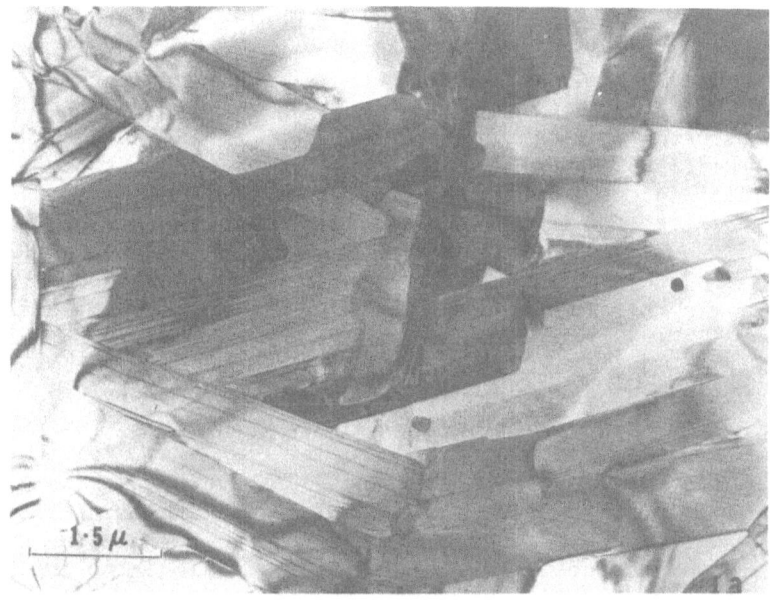

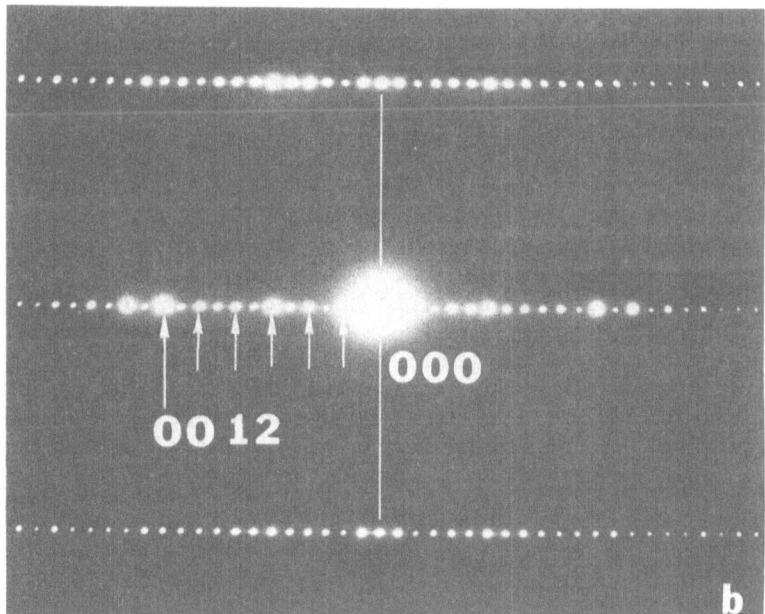

Figure 1. (a) Low magnification bright field transmission electron
          micrograph of the microstructure of a magnesium sialon
          illustrating the long interlocking nature of the grains.
          (b) Electron diffraction pattern indicating that the
          crystal structure is 12H.

Thomas.[7] Selected area electron diffraction patterns taken from the
individual grains oriented so that the close packed planes lie
parallel to the electron beam showed that the long axes of the plate
were perpendicular to the C-axis of the crystal structure.

The electron diffraction patterns also provided a convenient
means of identifying the polytype structure, since with the beam
perpendicular to the C-axis, the spacing of reflections along the
001 row indicates the stacking periodicity, and the symmetry of the
non-central rows of reflections with respect to the central (001)
row indicates the symmetry of the structure, i.e., rhombohedral or
hexagonal. In this manner it was found that the majority of the
grains had the 12H structure, in agreement with the x-ray analysis.
For example, the diffraction pattern of Fig. 1(b) is from one of
the 12H grains; it has hexagonal symmetry, and there are twelve
reflections from the transmitted beam to the reflection corresponding
to the close packed spacing (0012). On tilting away from this exact
symmetric orientation so that only the (001) systematic row of
reflections is excited, the 001, where 1 = odd integer, reflections
completely disappear; demonstrating that they occur only by double
diffraction. Thus the largest real stacking periodicity in the
close packed direction is equivalent to six close packed layers.

The close packed layers of the 12H structure can be imaged in
this orientation by lattice fringe imaging, as is shown in Fig. 2(a).
The spacing of the close packed layers is 2.7Å. In the image, every
sixth fringe is noticeably darker than its neighbors, which indicates,
under the conditions in which the micrograph was recorded, that there
is a change is local potential such as caused by a change in composition or stacking. These are believed to be the sites of the
cubic stacked layers having the composition MX_2. Also noticeable in
the fringe image is an additional change in contrast that occurs
midway between the darker fringes. (This can best be seen by viewing
the image obliquely and parallel to the fringes). This slightly
darker fringe is seen most easily in the microdensitometer trace of
Fig. 2(b) and occurs in every block. Its origin is unconfirmed, but
seems to result from a change in local atomic coordination due to
the accomodation of the additional magnesium atoms. Confirmation
from calculated lattice fringe images of postulated structures is
underway, as are x-ray structure determinations.[4]

INTERGROWTHS

Although the majority of grains had the 12H structure, several
grains having a 21R polytype structure were also observed. These
grains correspond to the adjacent M_7X_8 plane of the magnesium-sialon
phase diagram.

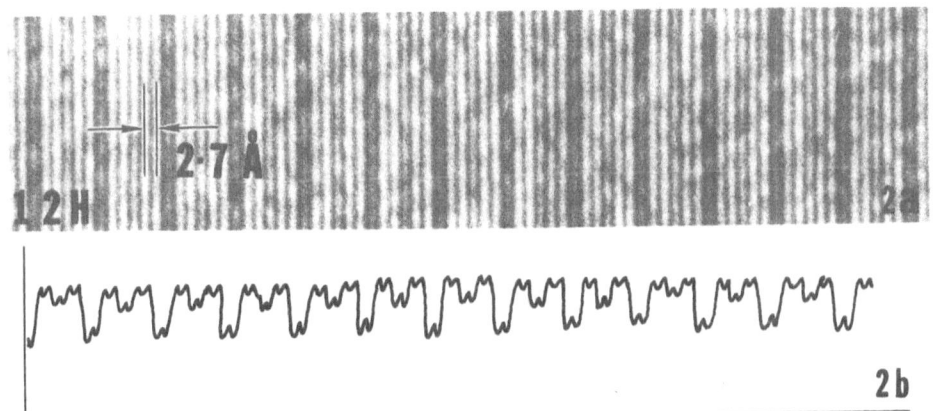

Figure 2.   (a) Lattice fringe electron micrograph of the 12H structure
            in which the individual close-packed planes (2.7Å spacing)
            are revealed. Change in contrast every sixth fringe
            corresponds to the regular insertion of a MX_2 layers every
            six planes. (b) Microdensitometer trace taken of the
            image (a) suggests that another change in atomic structure
            occurs midway between the MX_2 layers.

In addition, several grains were seen that were partially transformed from one polytype structure to another, as in Fig. 3(a). In
the region of the left, a regular periodicity of 18.9Å has been
imaged, which the elctron diffraction pattern of Fig. 3(b) indicates
to be a well ordered 21R polytype region. The electron diffraction
pattern from the region to the right (Fig. 3(c)) identifies it as
the 12H polytype. The 32Å periodicity imaged results from the
asymmetry of the block interfaces. The boundary between the two
structures lies parallel to the close-packed planes.

In the electron diffraction pattern taken from both regions
simultaneously, reflections from both polytypes can be seen
(Fig. 3(d)). Close examination of the reflections from the close-
packed planes in each structure (arrows) shows that they are not
coincident, which indicates that there is a difference in the close
packed layer spacing of the two polytypes. The non-central rows of
reflections from each polytype are also not coincident, thereby
indicating that there is also a mismatch between the two structures
perpendicular to the close packed planes. In the image, strain
contrast can be seen at several points along the boundary implying
that there are misfit dislocations present to accomodate this mismatch.

An additional feature of the boundary are the unit cell high
ledges arrowed in Fig. 3(a). Their presence suggests that a ledge
migration mechanism is responsible for the transformation from one
structure to the other.

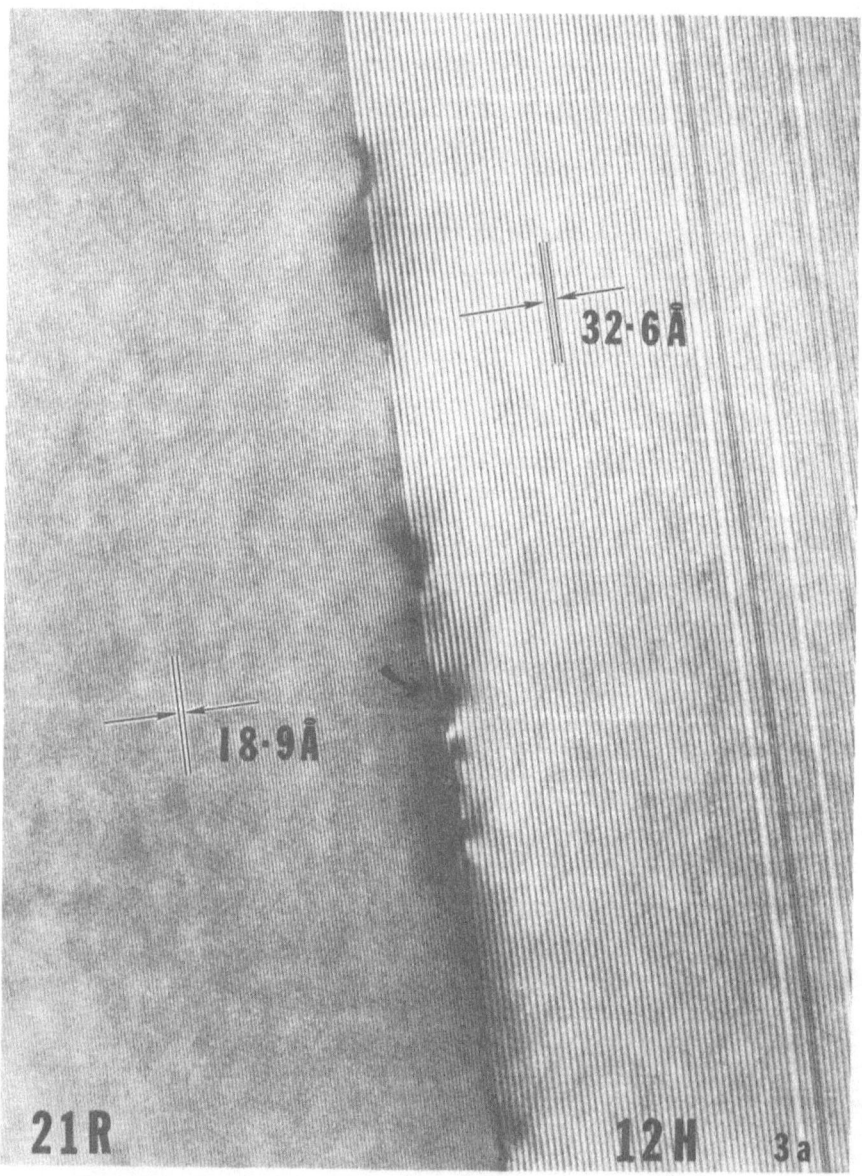

Figure 3. (a) An intergrowth between the 21R polytype (left) and the
12H polytype (right) in which the polytype spacing has been
directly imaged.

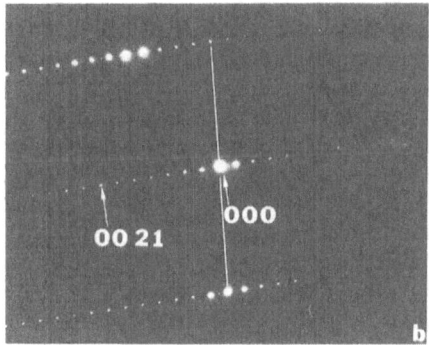
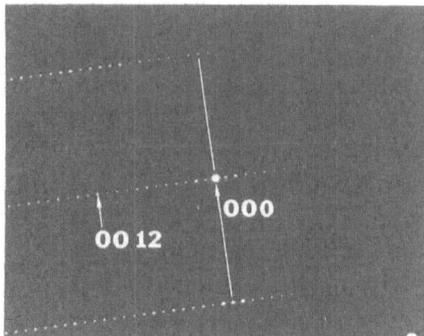
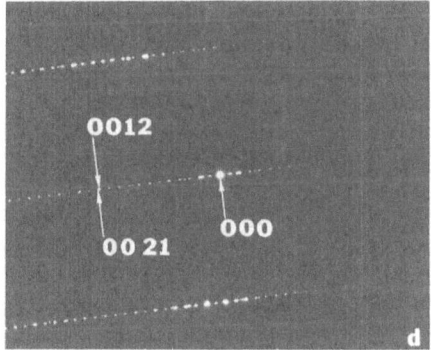

Figure 3 (continued) Selected area diffraction patterns confirm the 21R structure (b) and the 12H structure (c). The diffraction patterns of 3(d) was taken from across the interface.

DISCUSSION

These intergrowths are probably undesirable, since the strained interface could act as a possible source for crack initiation. However, they will be difficult to avoid in hot pressed materials since the polytypes exist over such a narrow range of homogeniety with respect to changes in the metal-to-nonmetal ratio. Such changes in composition may arise from two causes. First, if the exact M_6X_7 composition is not produced by the starting mixture of powders, small amounts of other polytype structures will arise. Second, local variations in composition in the sample produced by local inhomogenieties in the initial powder mix will result in the formations of other polytypes structures, such as 21R, as intermediate structures which may subsequently transform by diffusional processes to form a single homogeneous phase material. Thus, to avoid such intergrowths, careful control of the starting powders and long hot pressing times will be required.

Probably more serious will be the effect of long term oxidation or nitridation atmospheres on the structural stability of the material. Either process under active conditions would promote a change in the local nonmetal/metal atom ratio. If a similar change in crystal structure size as that on going from the 12H to 21R occurs, considerable strains would be produced to the detriment of the strength of the material.

CONCLUSIONS

The microstructure and crystallography of a 12H magnesium-sialon polytype material have been investigated by high resolution electron microscopy. In addition to the grains of 12H structure, there are a number having the 21R structure and also some that display coherent intergrowths of both. Since these intergrowths are seen to be accompanied by mismatch strains, and because each polytype can exist over only a narrow homogeniety range, it is unlikely that a single phase, strain free material can be produced.

ACKNOWLEDGEMENTS

Support for this work from the National Science Foundation (DMR-75-11352) for one of the authors (T.M.S.) is gratefully acknowledged. D.R.C. wishes to thank the Rockwell International Independent Research and Development Program for support of this work.

It is a pleasure to thank Professor K.H. Jack of Newcastle University for kindly providing the magnesium-sialon material and in communicating the results of his work on the Sialon polytypes prior to publication. We also wish to acknowledge the encouragement given by Professor G. Thomas.

REFERENCES

1. K. H. Jack and W. I. Wilson, Nature Phys. Sci. 238, 28 (1972).
2. K. H. Jack, J. Mater. Sci. 11, 1135 (1976).
3. D. P. Thompson, J. Mater. Sci. 11, 1377 (1976).
4. K. H. Jack, Private communication.
5. D. R. Clarke, T. M. Shaw and D. P. Thompson, J. Mater. Sci., 1977, in press.
6. D. R. Clarke, J. Am. Ceram. Soc., Proc. Symp. Electron Microscopy Applied to Ceramics Practice, in press.
7. D. R. Clarke and G. Thomas, these proceedings.

DENSE SILICON NITRIDE CERAMICS: FABRICATION AND INTERRELATIONS

WITH PROPERTIES

F. F. Lange

Rockwell International Science Center

Thousand Oaks, California 91360

INTRODUCTION: GENERAL TECHNOLOGY

Powder routes are used to fabricate dense Si_3N_4. Efforts to sinter pure Si_3N_4 powder have not been successful,[1][4] presumably due to insufficient volume diffusivity, decomposition at temperatures >1850°C[2] and volatilization caused by active oxidation[3] in the low oxygen partial pressures required to prevent the formation of SiO_2. Deeley et al.[4] discovered that Si_3N_4 could be fabricated by hot-pressing powders containing a densification aid. Today, many metal oxides and some nitrides are known densification aids. Although hot-pressing currently results in a superior product, the feasibility of pressureless sintering with a densification aid has been demonstrated.[5-8]

Densification of Si_3N_4 powder with the aid of a metal oxide is generally attributed to the presence of a liquid formed at high temperatures according to the general reaction:

$$Si_3N_4 + SiO_2 + \text{metal oxide} = \text{impurities} \rightarrow Si_3N_4 = \text{Liquid} \quad (1)$$

where SiO_2 is present either as a surface layer on each particle or as Si_2N_2O. Neglecting possible mass losses caused by volatilization, the composition of the liquid and equilibrium fraction of solid Si_3N_4 will depend on the composition of the starting powder, the phase equilibria of the composite system, and the densification temperature. The apparent role of the liquid is to promote mass transport by the solution-reprecipitation of Si_3N_4, which results in the disappearance of the voids and thus, densification.

Upon cooling, the liquid solidified: Si_3N_4 + Liquid → Si_3N_4 + secondary phases. The number, chemistry and content of the secondary phases depend on the composition of the starting powder, and the phase relations in the composite system. Non-equilibrium phases, e.g., glassy silicates, are observed.[9-12] In addition, the crystal structure of Si_3N_4 can be expanded or contracted by the concurrent substitution of certain metal cations for silicon and oxygen for nitrogen.[13] As expected, the secondary phases and the solid-solution alloying of Si_3N_4 can significantly influence all properties.[14-16]

The present day manufacture of dense Si_3N_4 can be divided into three steps: (1) manufacture of Si_3N_4 powder by reacting silicon with nitrogen, (2) preparation of powder for densification by adding the required densification aid and reducing the particle size by milling, and (3) densification of the composite powders by either hot-pressing or sintering. The exact mechanics of each step have been developed through individual experience; commercial practice is proprietary. The principal objective here will be to show the known and/or hypothetical interrelation between these three steps through the development of microstructure and properties.

POWDERS

Silicon: The Raw Material

Silicon, the second most abundant terrestial element, is commercially produced in carbon-electrode furnaces by a thermochemical reaction between crushed quartzite rock and high purity coke. Excess SiO_2 prevents the formation of SiC. The molten Si is tapped, cast, cooled and pulverized to produce a crude powder that is ~98% pure. Acid washing can increase the purity to ~99.5%. Further purification requires the synthesis of a volatile silicon compound. Needless to say, as indicated by chemical analysis, most current producers of Si_3N_4 start with acid washed, 'crude' silicon.

The major cation impurities in 'crude' silicon are Fe, Al, Ca and Mg which reflect the major impurities in the raw materials and the iron implements used for pulverizing. Oxygen is the major impurity in silicon powder. Of the major impurity cations listed above, only Al does not form a silicide. It can be presumed that the major impurities can be present as silicides, oxides or complex silicates.

Si_3N_4: The Starting Powder

Although Si_3N_4 powder can be produced by vapor phase reaction using volatile silicon compounds, only the current, commercial process of reacting silicon powder with nitrogen will be discussed.

It is obvious that the oxygen partial pressure in the nitriding environment should be less than that required to form SiO_2. A less obvious corollary is that the proper nitriding environment defines the conditions of active oxidation,[13] i.e., the formation of volatile SiO. As seen below, SiO appears to play an important role in the nitriding process.

Experience has shown that the reaction of nitrogen with pure, semiconductor grade silicon does not go to completion, viz, very little Si_3N_4 is formed over reasonable reaction periods. The work of Atkinson et al.[18] has shown that Si_3N_4 nuclei first form on the pure silicon surface. As summarized in Fig. 1, these nuclei grow across the surface by a vapor phase reaction which is evident by the concurrent growth of surface pits between the nuclei. Reaction is prematurely terminated when the growing nuclei impinging on one another closing the Si surface from further reaction. Atkinson et al.,[18] suggest that Si is the volatile species, but Lange[19] pointed out that SiO vapor is more probable. The proposed cyclic reaction, which involves the active oxidation of Si, the reaction of SiO with N_2 to form Si_3N_4 with the concurrent release of oxygen is shown in Fig. 1.

Reaction kinetics are greatly enhanced when one of a number of different metal nitriding aids are added to semiconductor grade silicon.[20-22] High reaction kinetics are also observed for acid-washed 'crude' silicon which contains the desired contaminant. The role of the nitriding aid is still uncertain, but several rules are apparent. First, the nitride of the metal additive must have a higher free energy of formation than Si_3N_4. For example, the formation of Mg_3N_2 and ALN is more favorable than Si_3N_4 and neither Mg nor Al are nitriding aids; conversely, both Fe and Mn are good

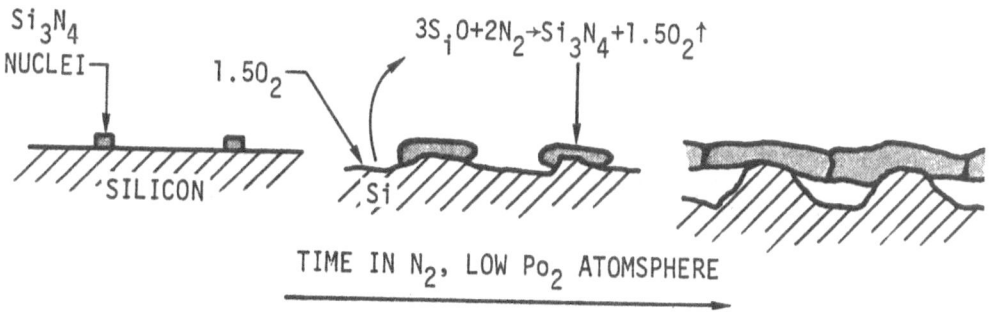

FIGURE 1

nitriding aids. Second, the nitriding aid is only effective above
the metal-Si eutectic temperature. These rules suggest that particles
of the metal nitriding aid react with the surrounding silicon
particles to form a silicon-rich liquid in equilibrium with the
remaining silicon. The active oxidation of the metal-silicon liquid
produces volatile SiO which reacts with N_2 to form Si_3N_4. As active
oxidation depletes the metal-Si liquid of silicon, the liquid
further dissolves the surrounding silicon to maintain an equilibrium
composition. In this manner, the metal-Si liquid 'eats' its way
through the surrounding silicon unhampered by surface closure,
producing volatile SiO to form Si_3N_4 by a vapor phase reaction.

Most Si_3N_4 powders produced by nitriding contain the two hexagonal crystal structures, α- and β-Si_3N_4. Since either structure
can be produced from the other by a 180° rotation of two Si containing stacking planes (viz. β (abab...)→α (abcd...= abqe ...)),
the transformation requires reconstruction (e.g., through solution
and reprecipitation). Although both structures are well known,[23-26]
their thermodynamic interrelation is still in question. This is a
critical question since high α-phase powders are required to produce
the tough, strong material.[27] Experience suggests that nitriding
temperatures <1325°C result in high α/β ratios. Since the reaction
is exothermic, the volume of the material is important for temperature control.[28] Other factors also appear important. Most investigators agree that α-Si_3N_4 forms through vapor phase reactions.

The oxygen content (the major impurity in most powders, with
the possible exception of unreacted silicon) is controlled by the
nitriding environment (furnaces with porous refractories usually
result in high oxygen contents). Oxygen contents have been
observed to range between 0.4 to 4.0 wt%.[29] If it is assumed that
the oxygen is in the form of SiO_2, this range converts to 2-14 mole %
SiO_2. That is, SiO_2 is a powder constituent that cannot be neglected.

Powder Preparation

The cake of Si_3N_4 formed by nitriding must be reduced to powder
by pulverizing and milling. The metal oxide (nitride) densification
aid(s) are usually mixed with the Si_3N_4 powder by liquid milling
with tungsten carbide milling media. Polymer mill jars help to
eliminate contaminants other than carbon. Silicon nitride can be
hydrolyzed to some extent by the liquid milling agent, e.g., Si_3N_4
milled in water produces strong ammonia odors. Dry alcohols reduce
this tendency.

The dried, milled powder can be air-classified to minimize the
number of large, un-milled agglomerates and much of the dense tungsten
carbide contamination. Large contaminant particles or - Si_3N_4

agglomerates left in the composite powders can be carried through to the dense material and act as flaws to increase the observed scatter in strength values.[30]

PHASE RELATIONS

Densification Procedures

The composite powders are usually hot-pressed in graphite dies at temperatures between 1650°C and 1750°C for several hours under 28 MPa (4000 psi) pressure. Boron nitride slurries are used to coat the graphite parts to ease material removal. Liners made of graphite paper are more desirable for this purpose since the BN becomes embedded in the surface of the hot-pressed billet. Cold-pressing the composite powder in steel dies prior to placing them in the graphite dies eliminates the possibility of contamination with loose graphite particles.

Pre-pressed powder shapes to be pressureless-sintered are embedded within either Si_3N_4 powder or loose powder of the same composition.[3,6] This technique minimizes weight losses due to volatilization to values $\leqslant 2$ wt%. At temperatures <1800°C, volatilization appears to be caused by active oxidation.[3] Thus, techniques which minimize the availability of oxygen to the powder are required to minimize compositional changes and to maximize the sintering phenomena.

Densification Kinetics and the $\alpha \to \beta$ Conversion

Densification kinetics of Si_3N_4 powder hot-pressed with the aid of ~15 mole% MgO (~5 wt%) has been obtained by Terwilliger and Lange,[31] and Brook et al.[32] Terwilliger and Lange obtained apparent activation energies in the range of 700 kJ mole^{-1} (1500°C to 1700°C) and illustrated that the densification kinetics of both high α and high β powders are similar for comparable powder processing. Brook et al. reports activation energies of 700 kJ mole^{-1} at temperatures >1550°C. They suggested that liquid phase sintering occurs in the higher temperature regime. Both groups show that the densification kinetics increase linearly with applied pressure (7 to 35 MPa) and with the MgO content (3 to 28 mole% MgO). Neither group investigated the effect of SiO_2 (e.g., in terms of the MgO/SiO_2 molar ratio) on the densification kinetics. Adequate correlations have not been made between densification kinetics and the content of the liquid phase responsible for densification.

The $\alpha \to \beta$ conversion is concurrent with densification. Brook et al.[32] obtained the same activation energies for the conversion as they obtained for densification, strongly suggesting that the mechanisms responsible for mass transport are the same for both phenomena. Both Iskoe and Lange[33] and Brook et al. have shown that the $\alpha \to \beta$ conversion occurs at a slower rate relative to densification, i.e., full densification can be achieved prior to full $\alpha \to \beta$ conversion. Full densification only requires the mass transport of a portion of the initial powder, therefore much of the $\alpha \to \beta$ conversion should take place after densification.

Lange[27] was the first to recognize that an equiaxed grain morphology was obtained with high β starting powders and a fibrous grain morphology resulted from a high α starting powder. The fibrous microstructure results in a higher fracture toughness and strength relative to the equiaxed microstructure.[27] Thus, high $\alpha-Si_3N_4$ starting powders are required to obtain optimum mechanical properties. Iskoe and Lange[33] observed that the growth of a fibrous microstructure is concurrent with the $\alpha \to \beta$ conversion. Based on these observations and the assumption that grain growth perpendicular to the fiber axis (c-axis) could be neglected, it was shown[32] that the aspect ratio (R) of the fibers would depend on the initial α/β ratio:

$$R = 1 + \alpha/\beta. \qquad (2)$$

This relation was obtained by assuming that the growth of the grains occurred by the preferential solution of the α particles and repre-cipitation of Si_3N_4 on pre-existing β particles. In addition, the distribution of fiber diameter should be the same as the initial particle size distribution. This model suggests that both the fiber diameter and aspect ratio and thus the microstructure of hot-pressed Si_3N_4 can be controlled by the size distribution and the α/β ratio of the starting powder.

Sub-Solidus Tie Lines and Known Eutectics

Initial phase studies neglected the SiO_2 content of the Si_3N_4 powder and assumed a simple binary, Si_3N_4/metal oxide, system. Many of these initial results are erroneous. Gauckler and Petzow[34] pointed out that phase relations should be considered in terms of the reciprocal reaction;

$$Si_3N_4 + \frac{6}{x} M_2O_x \to 3SiO_2 + \frac{4}{x} M_3N_x \qquad (3)$$

and thus, for the general case (neglecting impurities and valence changes),

$$zSi_3N_4 + ySiO_2 + \frac{6}{x}M_2O_x \to (y+3)SiO_2 + \frac{4}{x}M_3N_4 + (z-1)Si_3N_4 \to \text{reaction products} \quad (4)$$

Therefore, the reaction products should be represented in the Si_3N_4-SiO_2-M_2O_x-M_3N_x pseudo-quaternary system (x = metal's valence state).

Compositional changes due to volatilization and the solidification of nitrogen-silicates as glasses are two problems encountered in determining the phase equilibria of Si_3N_4 systems. Because of these and other difficulties, many investigators refer to their results as behavioral diagrams, i.e., indicative of what would be observed using methods of fabricating dense Si_3N_4.

Although different investigators are not in exact agreement, apparent sub-solidus phase relations (i.e., behavioral diagrams) are known for limited, but important compositional areas in the following systems: Si_3N_4-SiO_2-Al_2O_3-ALN,[34,35] Si_3N_4-SiO_2-BeO-Be_3N_2,[36,37] Si_3N_4-SiO_2-Al_2O_3-ALN-BeO_2-Be_3N_2,[13] Si_3N_4-SiO_2-Y_2O_3-YN,[14,35,39] Si_3N_4-SiO_2-Ce_2O_3-CeN,[40] and Si_3N_4-SiO_2-MgO-Mg_3N_2.[15] The latter three, represented by the conventional, partial mole fraction plot, are illustrated in Figs. 2 and 3a. The Si_3N_4-M_3N_x-M_2O_x portion of these three systems are presently unknown.

$$Si_3N_4\text{-}SiO_2\text{-}T_2O_3 \text{ System}$$

The Si_3N_4-SiO_2-T_2O_3 system has been investigated by three groups.[14,38,39] The composition of the apatite phase is still in question. Figure 2a shows the apatite phase with the vacancy composition $Y_9 \square (SiO_{3.5}N_{.5})_6 O_2$ which is consistent with the generalized vacancy, site-occupancy/charge balance formula observed for rare-earth, oxy-apatites: $Ln_{8+2x+.67y} (SiO_{4-x}N_x)_6 O_y$, where $0<y<2$ and $0<2x+.67y<1.33$. Placement of the nitrogen in the tetrahedral sites is consistant with Morgan's bonding studies for nitrogen-apatites.[41]

$$Si_3N_4\text{-}SiO_2\text{-}Ce_2O_3 \text{ System}$$

Compositions in the Si_3N_4-SiO_2-Ce_2O_3 system can be obtained by the generalized reduction/oxidation reaction:[40]

$$12CeO_2 + ySiO_2 + zSi_3N_4 \rightarrow 6Ce_2O_3 + (3+y)SiO_2 + (z-1)Si_3N_4 + 2N_2 \rightarrow$$
$$\text{reaction products} \qquad (5)$$

In this case, the CeO_2 is reduced to Ce_2O_3 and the oxygen released during this reduction oxidizes some Si_3N_4 to SiO_2. Thus, the use of CeO_2 limits the composition of the products to Ce_2O_3/SiO_2 molar ratios <2. Due to this compositional limit, the two expected phases (based on an analogy with the Si_3N_4-SiO_2-Y_2O_3 system, $Ce_2Si_3N_4O_3$ and $Ce_4Si_2O_7N_2$ have not been fabricated. The compound $CeSiO_2N$, has the same crystal structure observed for $YSiO_2N$.[42] The exact composition of the nitrogen apatite, represented in Fig. 2b as $Ce_{10}(SiO_{3.67}N_{.33})_6O_2$ (=$Ce_5(SiO_4)_3N$) is still in question.

a)

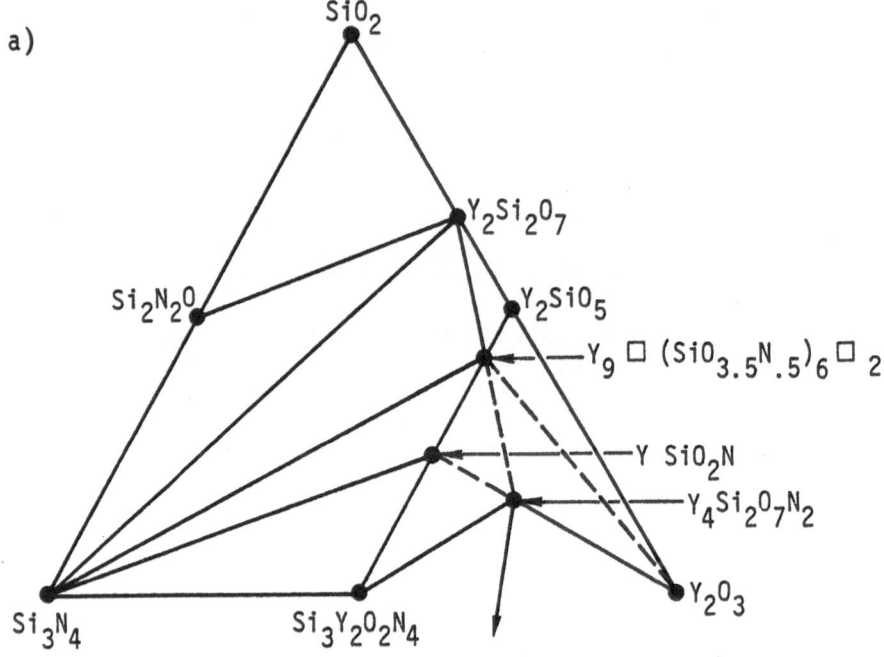

b)

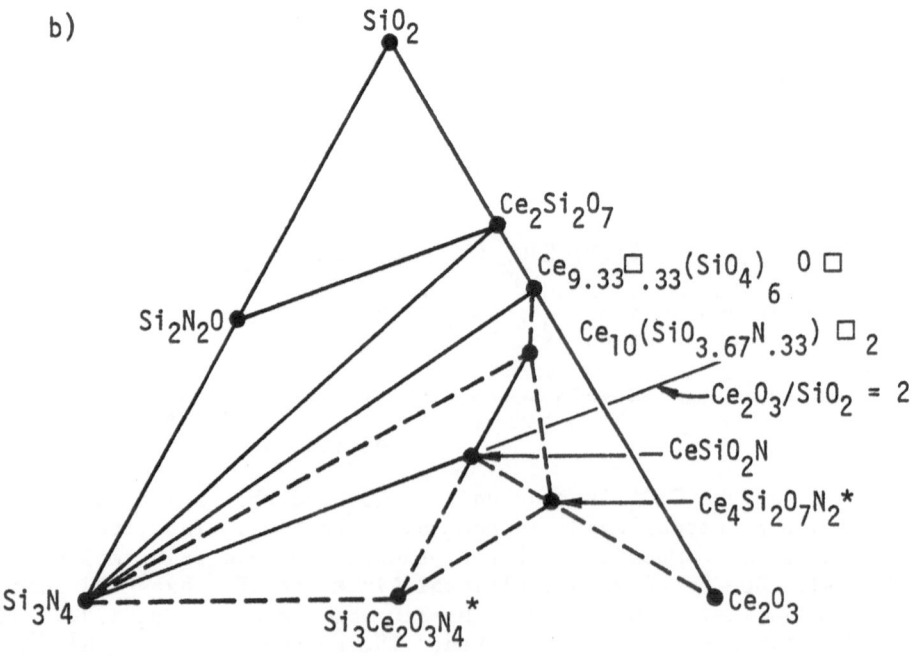

*(expected, but unobserved)

FIGURE 2

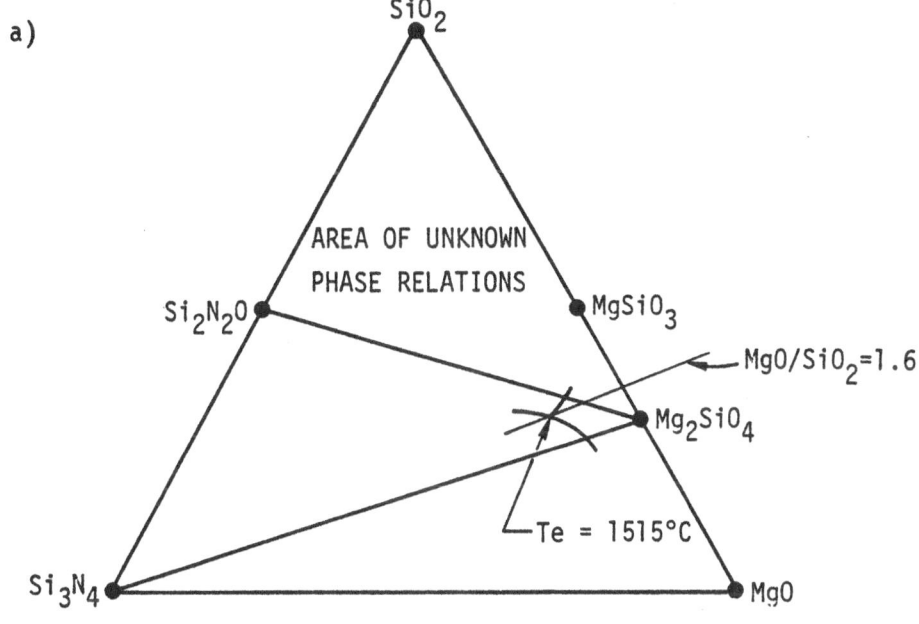

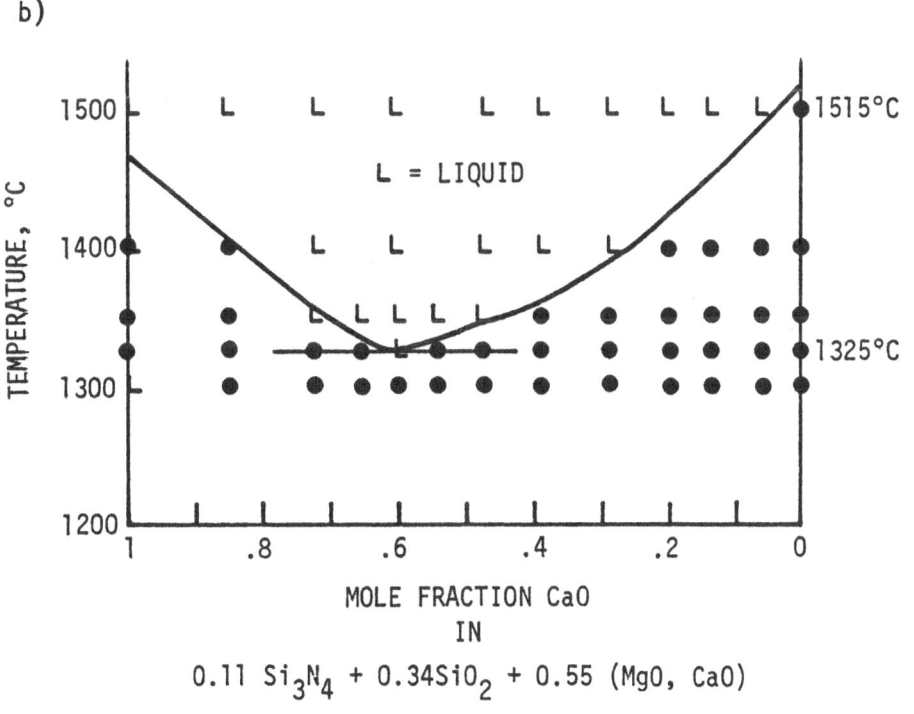

FIGURE 3

Si_3N_4-SiO_2- MgO System

The sub-solidus phase relations for the Si_3N_4 SiO_2-MgO system is shown in Fig. 3a. Recent melting experiments[43] have established three important eutectics in this system: (1) the Si_3N_4-Mg_2SiO_4 binary eutectic composition, 0.2 Si_3N_4 + 0.8 Mg_2SiO_4 at 1560°C; (2) the Si_2N_2O-Mg_2SiO_4 binary eutectic composition, 0.4 Si_2N_2) + 0.6 Mg_2SiO_4 at 1525°C; and (3) the Si_3N_4-Si_2N_2O-Mg_2SiO_4 ternary composition, 0.1 Si_3N_4 = 0.3 Si_2N_2O = 0.6 Mg_2SiO_2 at 1515°C.

In addition, the effect of CaO on lowering the ternary eutectic melting temperature has been investigated.[43] This work, summarized in Fig. 3b, was performed by mixing the ternary eutectic with a similar Si_3N_4-Si_2N_2O-Ca_2SiO_4 composition and determining melting temperatures for the composite powders. As illustrated in Fig. 3b, at temperatures >1325°C, a liquid should be present in dense compositions within the Si_3N_4-Si_2N_2O-Mg_2SiO_4 compatibility triangle when CaO is present as an impurity.

PROPERTIES: RELATIONS TO COMPOSITION

Mechanical Properties

The need for α-Si_3N_4 powders to develop the fibrous microstructure important for high toughness and strength has already been discussed above. The development of the fibrous grain structure under an applied axial pressure during hot-pressing leads to texturing,[27] viz, a greater proportion of the fibers are aligned perpendicular to the hot-pressing direction. All bulk properties exhibit anisotropy caused by texturing, e.g., flexural strengths are ~20% greater for bar specimens cut perpendicular to the hot-pressing direction relative to parallel specimens.[27]

Mechanical property degradation occurs at high temperatures.[44] The temperature where degradation begins depends on composition. Three compositional effects are known. Historically, the effect of impurities was observed first.[44,45] As clearly demonstrated by Iskoe et al.,[46] of the major cation impurities found in Si_3N_4, Ca produces the most significant degradation for material densified with the aid of MgO. The second compositional effect is that due to changes in the major powder constituents as expressed by the MgO/SiO_2 molar ratio[15] for compositions in the Si_3N_4-SiO_2-MgO system. As illustrated in Fig. 4, the flexural strength at 1400°C for a series of materials containing a fixed molar content of Si_3N_4 exhibits a minimum at MgO/SiO_2 ~2, i.e., for compositions close to the Si_3N_4-Mg_2SiO_4 tie line (see Fig. 3a). These data (and similar results obtained for two other series) clearly show the need to know the SiO_2 content, particularly when small amounts of MgO are

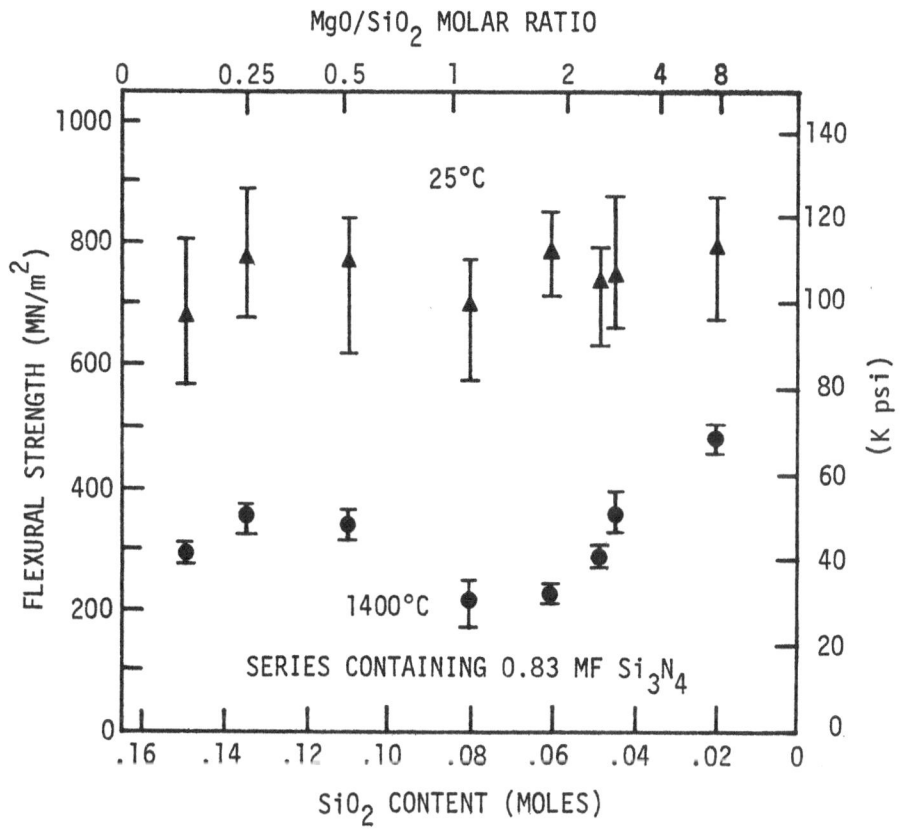

FIGURE 4

used for densification. The third compositional effect is caused by major changes in the densification aid, e.g., higher strengths can be obtained at 1400°C for certain compositions within the Si_3N_4-SiO_2-Y_2O_3 and Si_3N_4-SiO_2-Ce_2O_3 systems relative to the compositions that have been fabricated in the Si_3N_4-SiO_2-MgO system.

Many investigators have proposed that the degradation of mechanical properties at high temperature is caused by a viscous liquid located between the elastic Si_3N_4 grains. Models proposed to explain creep and subcritical crack growth indicate that both the viscosity and volume content of the liquid are important parameters.[47] Since a glassy phase is frequently observed at triple points and occasionally between the Si_3N_4 grains,[12] it has been assumed that degradation will begin at a temperature where the glass is soft enough to act as a viscous fluid, e.g., viscosities <10^7 poise. This assumption appears consistant with all facts. For example,

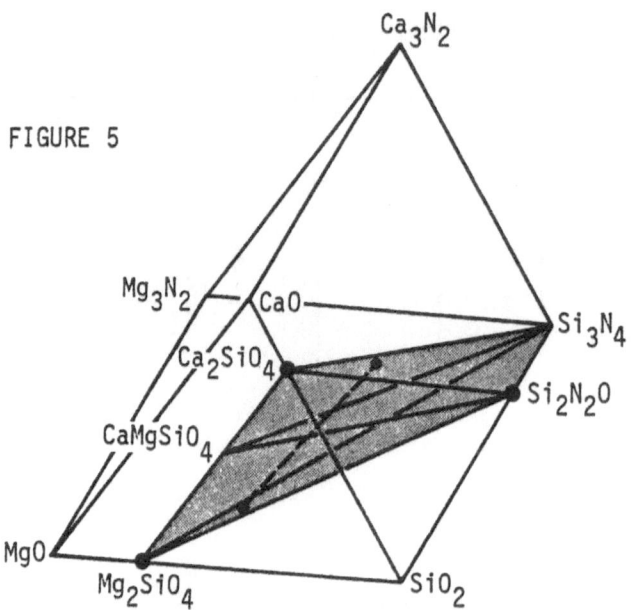

FIGURE 5

recent observations have shown that subcritical crack growth (a
phenomenon associated with strength degradation in polyphase Si_3N_4)
occurs by cavitation and that a tacky phase is observed between
the grains associated with extensive cavitation.[48] Since the
softening point of a silicate glass is usually ~200°C below its
melting temperature, the question of greatest concern is the
composition and melting temperature of the glass. An answer to
this question might be obtained by studying the phase equilibrium
of the relevent system.

Since the eutectic composition solidified last, it might be
expected that the glass composition is close to the eutectic
composition. Thus, both the composition and melting temperature
of the glass can be approximated by that of the eutectic. If it
is assumed that the glass composition is the same as the eutectic,
then its volume content can be calculated using the lever rule.
With this train of thought, we can now examine what might be
expected for composition in the Si_3N_4-SiO_2-MgO system.

Let us choose a composition in the Si_3N_4-Si_2N_2O-Mg_2SiO_4 equilibrium is reached at the fabrication temperature, as evident by
complete $\alpha \rightarrow \beta$ conversion. During cooling, Si_3N_4 and one of the other
two phases will precipitate from the liquid until the eutectic
temperature of 1515°C is reached. At this temperature the eutectic
liquid is assumed to solidify as a glass. The volume content of

this glass will depend on the total composition relative to the eutectic composition. Using the lever rule, it can be shown that the volume of glass is maximum for compositions with a MgO/SiO_2 molar ratio of 1.6. If CaO were added as an impurity, the eutectic of interest would lie in the Si_3N_4-Si_2N_2O-Mg_2SiO_4-$CaMgSiO_4$ compatibility element shown in Fig. 5 and the last drop of liquid would not solidify until ~1325°C as shown by the data in Fig. 3b*.

This reasoning can be used to interpret the effects of both impurities and the MgO/SiO_2 ratio on the high temperature strength of compositions within the Si_3N_4-SiO_2-MgO system. Impurities such as CaO will both lower the eutectic temperature and change its composition to promote larger volume contents of the viscous phase. Compositions that exhibit the lowest strength at 1400°C (MgO/SiO_2 ~2, see Fig. 4) are approximately the same as those that contain the largest amount of viscous phase (MgO/SiO_2=1.6), indicating a close correlation between observed and predicted behavior.

It should be noted that in the absence of a glassy phase a liquid would form at the eutectic temperature. Its volume content will be governed by the lever rule and the temperature above the eutectic. Degradation of mechanical properties as a function of temperature will be more rapid relative to the slow degradation observed for materials containing a glass (the viscosity of the glass will decrease as the eutectic temperature is approached, whereas in the absence of a glass, a liquid will not appear until the eutectic temperature). Thus, even in the absence of a glass phase, degradation should be expected above the eutectic temperature.

Oxidation

The high temperature passive oxidation kinetics of polyphase Si_3N_4 materials can exhibit extreme variations. These variations can be attributed to compositional effects. Three general compositional effects have been observed involving (1) impurities, (2) unstable secondary phases and (3) reactions between the SiO_2 formed during oxidation and the secondary phases.

The effect of impurities can be divided into two categories: homogeneously distributed impurities and hetrogeneously distributed impurities. Oxidation experiments[49] with Si_3N_4 densified with MgO show that the homogeneously distributed impurities such as Ca and Al diffuse from the bulk to the silicate surface scale in an attempt to minimize chemical potential differences. The contamination level within the bulk material can significantly influence the oxidation kinetics as indicated by the different oxidation rates of relatively impure HS110 Si_3N_4 and the purer form of this commercial material manufactured by the Norton Co., NC132 Si_3N_4.[50] Evidence indicates that a portion of the oxide scale was liquid at the oxidation temperature, suggesting that the impurities promote the formation of low temperature eutectics.[49] Heterogeneously distributed impurities in the form of aggregated inclusions of (W,Fe)-silicides rapidly react with the silicate surface scale on Si_3N_4 materials hot-pressed with MgO to form surface pits.[51] These pits can drastically reduce the material's flexural strength.[52] Since a similar reaction is not observed for Si_3N_4 fabricated with Y_2O_3,[51] it can be concluded that the reaction of the Fe with the Si_3N_4-MgO polyphase material produces a low temperature eutectic which locally accelerates the oxidation kinetics to produce a surface pit.

All of the quaternary phases in the Si_3N_4-SiO_2-Y_2O_3 system exhibit non-passive oxidation kinetics, e.g., $Y_2Si_3N_4O_3$ exhibit linear oxidation kinetics.[14] These compounds are unstable in high temperature oxidizing atmospheres relative to Si_3N_4. In addition, all Ce-compounds in the Si_3N_4-SiO_2-Ce_2O_3 system, including the Ce-silicates, oxidize to CeO_2+SiO_2 at relatively low temperatures.[40] The presence of the unstable yttrium phases in polyphase Si_3N_4 can cause considerable material degradation in an elevated temperature regime where Si_3N_4 exhibits very little oxidation, i.e., between 800-1200°C. In this regime, the molar volume change of the oxidized, unstable phases produces surface cracks which expose more material to oxidation and quickly result in a general disintegration of the bulk. The Si_3N_4 remains unoxidized. In this case, the unstable yttrium phases exhibit ~30% increase in molar volume during oxidation. In the Si_3N_4-SiO_2-Ce_2O_3 system, polyphase materials containing small amounts of the unstable Ce-silicate phases do not degrade upon oxidation despite the observed surface oxidation of the unstable phases. For this case, the molar volume increase of the Ce-silicates is ~8%. These observations suggest further studies into the relations between volume changes produced by oxidation and the stress distributions that arise in these polyphase materials.

Of all the polyphase Si_3N_4 materials examined, materials fabricated in the Si_3N_4-Si_2N_2O-$Y_2Si_2O_7$ compatibility triangle (see Fig. 2a) exhibit the greatest resistance to oxidation.[14] The apparent reason for this is the compatibility of SiO_2, the oxidation product of Si_3N_4, with the secondary phases, $Y_2Si_2O_7$ and Si_2N_2O, and the relatively high eutectic temperatures within this compositional area. When this compatibility does not exist, as in the Si_3N_4-SiO_2-MgO system (see Fig. 3a), the secondary phases, e.g., Mg_2SiO_4, will take part in the oxidation reaction to form silicates richer in SiO_2 as evident by the $MgSiO_3$ phase observed in the oxide surface scale of Si_3N_4 fabricated with MgO.[49] The diffusivity during the SiO_2-second phase reaction should increase the oxidation kinetics of Si_3N_4 as evident by the decrease in oxidation resistance of materials fabricated with increasing MgO/SiO_2 molar ratios.[15]

CONCLUDING REMARKS

It is obvious that phase equilibria are very powerful tools in understanding the relations between fabrication, microstructure development and properties. In developing these interrelations, all chemical constituents must be considered. For example, it has been shown that small amounts of impurities can produce large effects. Knowing how the material is made is essential in anticipating improvements.

ACKNOWLEDGMENTS

The fabrication section of this work was supported by a Rockwell International Independent Research and Development program. The property section was supported by the Air Force Office of Scientific Research, Contract No. F49620-77-C-0072.

REFERENCES

1. C. Greskovich and H. H. Rosolowski, J. Amer. Ceram. Soc. 59, 337 (1976).
2. S. C. Singhal, Ceramurgia Inter. 2, 123 (1976).
3. F. F. Lange, unpublished work.
4. G. G. Deeley, J. M. Herbert and N. C. Moore, Powder Met. 8, 145 (1961).
5. G. R. Terwilliger and F. F. Lange, J. Mat. Sci. 10, 1169 (1975).
6. L. G. Gauckler, S. Boskovic, I. K. Naik and T. Y. Tien, Proc. Workshop on Ceramics for Adv. Heat Engines, Orlando, Fla., 1977, p. 321, NTIS Conf. 770110, UC95a.
7. H. F. Priest, G. L. Priest and G. E. Gazza, J. Amer. Ceram. Soc. 60, 81 (1977).
8. M. Mitomo, J. Mat. Sci. 11, 1103 (1976).
9. J. V. Sharp and A. G. Evans, J. Mat. Sci. 6, 1292 (1971).
10. R. Kossowsky, J. Mat. Sci. 8, 1603 (1973).
11. P. Drew and M. H. Lewis, J. Mat. Sci. 9, 261 (1974).
12. D. R. Clarke and G. Thomas, J. Amer. Ceram. Soc. (in press).
13. L. J. Gauckler, H. L. Lukas, and T. Y. Tien, Mat. REs. Bull. 11, 503 (1976).
14. F. F. Lange, S. C. Singhal and R. C. Kuznicki, J. Amer. Ceram. Soc. 60, 249, (1977).
15. F. F. Lange, J. Amer. Ceram. Soc., (in press).
16. F. F. Lange, H. J. Siebenneck and D. P. H. Hasselman, J. Amer. Ceram. Soc. 59, 454 (1976).
17. C. Wagner, J. Appl. Phys. 29, 1295 (1958).
18. A. Atkinson, A. J. Moulson and E. W. Roberts, J. Mat. Sci. 10, 1242 (1975).
19. F. F. Lange, Discussion; NATO ASI on Nitrogen Ceramics, Canterbury, England, Aug. 1976, in press.

20. D. R. Messier and P. Wong, J. Amer. Ceram. Soc. 56, 480 (1973).
21. Sin-Shong Lin, J. Amer. Ceram. Soc. 60, 768 (1977).
22. F. F. Lange and J. L. Iskoe, unpublished work.
23. D. Hardie and K. H. Jack, Nature 180, 332 (1957).
24. R. Marchand, Y. Laurent, J. Lang and M. T. LeBihan, Acta. Crys. 25, 2157 (1969).
25. I. Kohatsu and J. W. McCauley, Mat. Res. Bull. 9, 917 (1974).
26. K. Kato, Z. Inque, K. Kiljima, I. Kawada, H. Tanaka and T. Yanane, J. Amer. Ceram. Soc. 58, 90 (1975).
27. F. F. Lange, J. Amer. Ceram. Soc. 56, 518 (1973).
28. A. Atkinson and A. D. Evans, Trans. and J. Brit. Ceram. Soc. 73, 43 (1974).
29. F. F. Lange, "Task I: Fabrication, Microstructure and Properties of Selected SiAlON Compounds", Final Report, NAVAIR Systems. N00019-73-C-0208, Feb. 1974.
30. H. R. Baumgartner and D. W. Richerson, Fracture Mechanics of Ceramics, Vol. 1, p. 367, Ed. by R. C. Bradt, D. P. H. Hasselman and F. F. Lange, Plenum, 1974.
31. G. R. Terwilliger and F. F. Lange, J. Amer. Ceram. Soc. 57, 25 (1974).
32. R. J. Brook, T. G. Carruthers, L. J. Bowen and R. J. Weston, Proc. NATO ASI on Nitrogen Ceramics, in press.
33. J. L. Iskoe and F. F. Lange, pp. 669-678 in Ceramic Microstructures '76, R. M. Fulrath and J. A. Pask, ed., Westview Press, Boulder, Colo., 1977.
34. L. J. Gauckler, H. L. Lukas and G. Petzow, J. Amer. Ceram. Soc. 53, 346 (1975).
35. K. H. Jack, J. Mat. Sci. 11, 1135 (1976).
36. I. G. Huseby, H. L. Lukas and G. Petzow, J. Amer. Ceram. Soc. 58, 377 (1975).
37. D. P. Thompson and L. J. Gauckler, J. Amer. Ceram. Soc. 60, 470 (1977).
38. D. P. Thompson, J. Mat. Sci. 11, 1377 (1976).
39. R. R. Wills, S. Holmquist, J. M. Wimner and J. A. Cunningham, J. Mat. Sci. 11, 1305 (1976).
40. F. F. Lange, "Fabrication of Si_3N_4-Ce_2O_3-SiO_2, Materials: Phase Relations, Sinterability, Strength and Phase Stability", to be published.
41. P. E. D. Morgan, Bull. Amer. Ceram. Soc. 56, 300 (177).
42. P. E. D. Morgan, P. J. Carroll and F. F. Lange, Mat. Res. Bull. 12, 251 (1977).
43. F. F. Lange, unpublished work.
44. F. F. Lange, J. Amer. Ceram. Soc. 57, 84 (1974).
45. D. W. Richerson, Bull. Amer. Ceram. Soc. 52, 560 (1973).
46. J. L. Iskoe, F. F. Lange and E. S. Diaz, J. Mat. Sci. 11, 908 (1976).
47. F. F. Lange, Deformation of Ceramic Materials, p. 361, Ed. by R. C. Bradt and R. C. Tressler, Plenum Press, 1975.

48. F. F. Lange, "Evidence for Cavitation Crack Growth in Si_3N_4", to be submitted for publication.
49. S. C. Singhal, J. Mat. Sci. <u>11</u>, 500 (1976).
50. A. F. McLean, E. A. Fisher and R. J. Bratton, Brittle Materials Design, High Temperature Gas Turbine, Interim Report AMMRC-CTR-72-19, p. 158, Sept. 1972.
51. F. F. Lange, "Reaction of Iron with Si_3N_4 Materials to Produce Surface Pitting", to be published.
52. S. C. Singhal, "Effects of Oxidation on Strength Reduction in Si_3N_4 and SiC", to be published.

DISCUSSION

W. J. Huppmann (Max Planck Institute for Metals Research): You mentioned solution-reprecipitation as one of the densification mechanisms. Do you have a physical picture of this sintering mechanism? Also, I would like to draw attention to a model which we recently proposed for sintering of W in the presence of liquid Ni (W. J. Huppmann, H. Riegger: The Int. J. of Powder Met. and Powder Techn., Vol. 13 (1977) in press) which may be relevant for the Si_3N_4 systems discussed here. In this model dissolution of small particles enables rearrangement and dense packing of other particles.

Author: Liquid-phase sintering is the working model I use to explain densification. A liquid first forms at the lowest eutectic temperature by a reaction between all end-member constituents. The equilibrium amount of liquid present during all but the initial period (where non-equilibrium liquids may come and go) will depend on the initial composition relative to the eutectic composition (including impurities), the explicit function describing the liquidus curve, and the temperature above the eutectic. Rearrangement will occur once a liquid is formed, and it may be the major contributor to densification at high temperatures ($\geqslant 1750^{\circ}C$) as evident by attainment of full densification with very little $\alpha \rightarrow \beta$ conversion. Solution-reprecipitation certainly takes place as evident by the $\alpha \rightarrow \beta$ conversion both during *and after* densification. It may contribute more to densification at lower temperatures ($< 1750^{\circ}C$) where less liquid is present. The active oxidation of both Si_3N_4 *and* the liquid nitrogen-silicates to produce volatile SiO complicates this naive physical model.

P. E. D. Morgan (University of Pittsburgh): I like the idea that, in the examples you gave, the elongated β grains have grown from preexisting β grains, because in our work* on hot pressing of amorphous $\rightarrow \alpha \rightarrow \beta$-$Si_3N_4$ we always saw **equiaxed** structures because there was no preexisting β present.

Author: The analysis leading to $R = 1 + \alpha/\beta$ neglects nucleation and growth and therefore breaks down as $\alpha/\beta \rightarrow \infty$. As $\alpha/\beta \rightarrow \infty$ nucleation and growth of β grains should predominate. Thus if no β grains (or particles) pre-exist (i.e., $\alpha/\beta = \infty$) one would expect the nucleation of many β grains throughout the body and a relatively low aspect ratio after the $\alpha \rightarrow \beta$ conversion.

---

*P. E. D. Morgan, Office of Naval Research Report, December 1973, AD-778-373 NTIS.

HIGH-PRESSURE HOT-PRESSING OF SILICON NITRIDE POWDERS

Svante Prochazka and William A. Rocco

General Electric Company
Corporate Research and Development Center
Box 8, Schenectady, New York   12301

I.   INTRODUCTION

Silicon nitride has been a prime candidate material for active components of future high temperature heat engines, particularly gas turbines. Its high temperature properties such as oxidation resistance, strength, creep and stress-rupture are critically dependent upon additives which are necessary for consolidation of Si_3N_4 powders by hot-pressing into pore free ceramics. Typical additions, MgO, Y_2O_3, ZrO_2 at a few percent level, degrade high temperature properties severely.

Attempts to consolidate silicon nitride powders without sintering aids by hot-pressing have shown that essentially no densification takes place (Deeley 1961, Terwilliger 1974). Under typical conditions - 1750°C and 4 MPa - usually poorly bonded, friable compacts result which show fractional density nearly equivalent to that obtained by cold compaction. Only one exception has been reported by Morgan (1973) who observed densification of a very pure amorphous Si_3N_4 powder derived from silicon diimide at 1800°C and 4 MPa to about 90% of theoretical density. The absence of densification may be attributed to the low diffusivity of nitrogen in Si_3N_4. Kijima and Shirasaki(1976)determined D_N in β-Si_3N_4 and extrapolating their data to 2000°K gives 3×10^{-16} cm^2/sec. which is about an order of magnitude higher than D_c in SiC according to data by Ghoshtagore and Coble (1966). Considering the uncertainties involved in the measurements and extrapolations it is reasonable to expect similar behavior of these two solids in phenomena controlled by internal mass transport, such as sintering and creep. Indeed, it has been shown that pure SiC also does not densify under these conditions

(Prochazka 1974, Alliegro 1956). However contrary to Si_3N_4, SiC and other covalent substances such as AlN, BN, Si, exhibit good bonding i.e. strong compacts are obtained on hot-pressing of fine (about 1μm) powders of these substances even though the degree of densification may be small. It is the peculiar lack of bonding in pure Si_3N_4, i.e., interparticle contact growth, which raises suspicion that additional complications are involved. Such could result for instance from competing mass transport mechanisms, the high Si_3N_4 dissociation pressure, an unfavorable grain boundary to surface energy ratio, oxide coatings of the Si_3N_4 crystallites, or from the presence of a non-wetting liquid.

In order to investigate grain boundary formation in Si_3N_4 experiments under high pressure hot-pressing conditions were performed. It was expected that a temperature could be found at which Si_3N_4 would form a dense hard polycrystalline solid. It is believed that in this type of solid the final stages of the densification process requires internal atomic mass transport and therefore such a temperature would be an indirect measure of the atomic mobility in Si_3N_4 and at the same time a lower limit below which consolidation of Si_3N_4 could not be obtained.

High fractional densities obtained under very high pressure, however, would not be a sufficient evidence for diffusional mass transport. Nadeau (1973) who studied high pressure hot-pressing of SiC, observed that with some powders 95% theoretical density was obtained at 5 GPa at low temperatures predominantly by rearrangement and fragmentation of powder particles. The resulting compacts, although they had a dense ceramic appearance and could be polished, were not well bonded ans showed low hardness and specific microstructures related to particle fragmentation. In the present study, therefore, microhardness was selected as the principal evaluation procedure. Current characteristics of the specimens, density, X-ray diffraction and microstructural features were also obtained.

II. EXPERIMENTAL

The apparatus and procedure for ultra-high pressure consolidation of Si_3N_4 powders has been essentially the same as used for diamond synthesis and has been adequately discussed previously (Hall, 1960). The only necessary change has been the use of a boron nitride liner within the carbon heater to prevent Si_3N_4-carbon interaction. The liner consisted of a thin-walled tube and bottom and top spacers, all of which were machined from commercial BN[+] stock to fit into the carbon heater. Pellets 0.95 cm x 0.64 cm were prepressed from the powders, repressed isostatically at 200 MPa (30,000 psi) and prefired in nitrogen or NH_3 at 1200°C. Two of

---

[+]Boron nitride HBN - Union Carbide

these fired pellets would fit into the cell. Preliminary experiments
showed that 1400°C was the lowest temperature that could be used to
obtain bonding and high densities. Cell components, on the other
hand, limited the upper temperature to about 1600°C. Temperature
was determined from power readings and calibrations and was accurate
within ± 50°C. The specimens were held at temperature for about 20
min. under 5.5 GPa.

The experiments were done with 5 different Si_3N_4 powders whose
characteristics are given in Table I. No. 1 and 2 were synthesized
from SiH_4 and NH_3 (Prochazka, 1977) and differ in silicon and oxygen
content. No. 3 was prepared by ammonolysis of $SiCl_4$ and was the
finest in the present series. No. 4 was procured[†] and characterized
in this laboratory. No. 5 was of the same origin but processed by
vibratory milling in benzene with cemented carbide balls for 6
hours. About 0.9% tungsten was picked up during milling and only
marginal refinement of the crystallite size was achieved, however.
The metal content except for W in No. 5 was very low, <0.02%, and
was actually the criterion for this powder selection. No additional
processing besides homogenization with a mortar and pestle was used.
After hot-pressing the pills were about 0.6x0.3 cm in size and
rarely free of cracks. Frequently a few cracks perpendicular to the
pressing direction would form on pressure relaxation due to differences in compressibility of the cell components. An example of a
section of a typical pressing is shown in Figure 1. If the specimen

Table I. Characteristics of Si_3N_4 Powders Used in Hot-Pressing
Experiments

No.	Origin	X-ray	Surface Area m^2/g	Known Impurities	Color
1	Synthesized in house	Amorphous	26	0.5% O_2 8% Si*	brown
2	"	Amorphous	14	2.0% O_2 1% Si	cream
3	$SiCl_4$ +NH_3	Amorphous	43	n.d.	white
4	GTE Sylvania	α-Si_3N_4 β trace	4.3	1% O_2 0.05 Cl_2	off white
5	Same[+] milled	α-Si_3N_4 β trace	4.5	same +0.9% WC	gray

*Free Si content estimated from weight gain on nitriding at 1450°C.

[+] Powder milled vibratory for 6 hours in benzene with 1/8" cemented carbide
balls.

†GTE Sylvania Silicon Nitride SN-502.

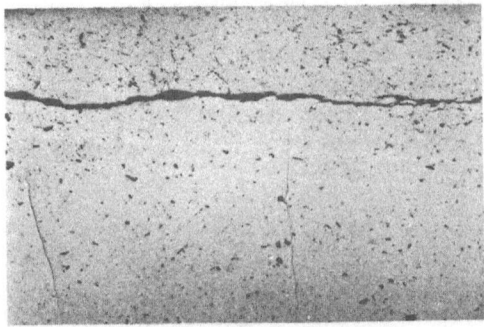

Fig. 1.   Section of a Si_3N_4 specimen hot-pressed at 5.5 GPa.
Specimen No. 9, Table II, magnification 25x.

Fig. 2. Polished and etched section of Specimen No. 2. Double
stage replica magnified 15,000x. White rounded spots
correspond to etched out silicon grains.

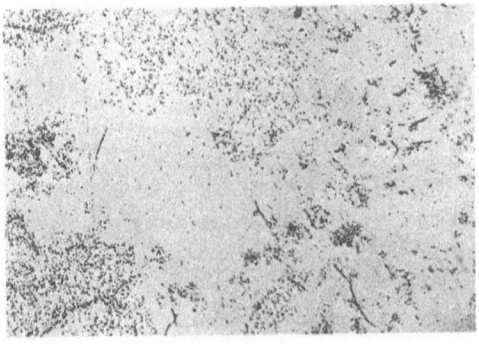

Fig. 3. Residual porosity associated with silicon free areas in
Specimen No. 4. As polished magnified 250x.

was poorly bonded it usually fragmented.

Density was determined by liquid displacement or by floating of specimen fragments in mixtures of CCl_4 and CH_2I_2 of known density (Nadeau, 1973). However neither technique is accurate in this situation because of specimen cracking; errors of +0.05 g/cc are anticipated. The cracks made it also impossible to determine whether or not the specimens had open porosity. For metallography and hardness indentation, specimens were mounted, ground and polished with 3µ diamond grit. Etching with HF-based etchants was essentially unsuccessful. Some specimens developed pits or lines of fine pits, presumably due to segregated impurities along grain boundaries, but in essence, the specimens did not respond. With molten hydroxide (250°C), etching did occur; however, due to the very fine grain structure, the specimens over etched easily. Difficulties with revealing the grain structure by etching made grain size determination somewhat uncertain and the estimates in Table II were obtained by combining results of SEM fractography, optical microscopy and study of double stage replicas. Annealing experiments were done by packing the specimens into Si_3N_4 powder in a BN capsule and heating progressively in nitrogen at 1650°, 1750° and 1850°C for 40 minutes. After each step the specimen was reexamined metallographically. None of the specimens, however, survived 1850°C without serious degradation.

III.   RESULTS

The hot-pressing conditions and the specimens' characteristics have been summarized in Table II, and the specifics are discussed below.

Figure 2 shows a replica of an etched section of specimen No. 2. The rounded white spots were identified as etched out silicon grains by selective etching with $HF-HNO_3-H_2O$ indicating that free silicon segregated and formed separate grains, somewhat larger than the average Si_3N_4 grains estimated to be near 0.2µm in size. The fracture surface was found pore free and revealed predominantly transgranular fracture. The determined microhardness value (3000 kg/mm^2) was higher than values reported for hot-pressed Si_3N_4 previously and confirmed that excellent bonding was achieved.

Specimen No. 3 showed no change in appearance and no grain growth on annealing at 1650° and 1750° but degraded severely at 1850°C as a result of Si_3N_4 dissociation.

The second series (No. 4 and 5) was prepared from a batch of powder which was much closer to stoichiometry, but still contained some free silicon. Its content was estimated to be <2% as no Si diffraction lines could be detected in a powder specimen annealed at 1400°C in Argon. The presence of Si was concluded from observation

Table II.   Results of Hot Pressing of Si_3N_4 at 5 GPa

Spec. No.	Starting Powder[+]	Pretreatment of the Pressings	Consolidation Temperature °C ± 50	Appearance	Density g/cc	Phases Present	Micro-hardness (Knoop) @ .200g load	Grain Size Estim. (μ)
1	1	NH₃, 1200°	1400	Dark gray, fragmented	2.75	n.d.	n.d.	—
2	1	NH₃, 1200°	1500	Dark gray	3.11	β-Si₃N₄ Si trace*	3080	0.2
3	1	N₂, 1200°	1600	Dark gray	3.05	β-Si₃N₄ Si trace*	3000	0.2
4	2	N₂, 1200°	1500	Gray, cracked	3.15	β-Si₃N₄	n.d.	n.d.
5	2	N₂, 1200°	>1550	Gray	3.15	β-Si₃N₄	2600[++] 1800	<0.2
6	3	NH₃, 1200°	1500	White, fragmented	2.57	n.d.	n.d.	n.d.
7	3	NH₃, 1300°	1550	White, Cracked	2.90	β-Si₃N₄	250	bimodal 0.2 and 0.8
8	4	N₂, 1200°	1500	Ivory, Cracked	2.65	n.d.	n.d.	n.d.
9	4	N₂, 1200°	1550	Ivory,	2.89	β-Si₃N₄	1200	0.5
10	5	N₂, 1200°	1500	Gray, Cracked	3.15	α-Si₃N₄ <10% β	2500	≈1
11	5	N₂, 1200°	1550	Gray, Cracked	3.19	n.d.	n.d.	n.d.

+ See Table I for identification

++ Brown and white area resp. (see text)

*About 5%

of a polished section in oblique light which revealed white and
brown areas suggesting that some Si was present and was nonuniformly
distributed. The brown areas were pore free and some showed scattered silicon grains, about $1/2\mu$ in size. The white areas, particularly in specimen 4, prepared at lower temperature were associated
with residual porosity as shown on Figure 3. The microhardness
reading was obtained separately from the brown and white areas,
confirming excellent bonding in the first (2600 kg/mm^2) and less in
the latter (1800 kg/mm^2).

Specimen No. 6 disintegrated on removal from the pressure cell.
The fragmentation could be suppressed by increasing the temperature
of the pretreatment to 1300°C for an hour with Specimen No. 7. A
micrograph of a replica of an <u>as-polished section</u> of this specimen
is shown in Figure 4 and shows a bimodal grain size distribution
with grains 0.1–0.2 μm and grains near 1 μm. Microhardness was low,
250 kg/mm^2, indicating poor bonding, although the specimen had about
90% of theoretical density. It seems that this density was achieved
predominantly by rearrangement and fragmentation, and little grain
boundary formation took place. It is possible that the increased
pretreatment temperature necessary to obtain a compact from this
ultrafine powder, induced coarsening and that the larger grains,
observable in this section, grew already before the pressure consolidation.

Specimens No. 8 and 9 were obtained by compaction of a crystalline α-Si$_3$N$_4$ powder having a mean surface average crystalline size of
0.45μm. No. 9, hot-pressed at about 1550°C, achieved 90% density,
appeared well bonded and showed a moderately high microhardness. The
dark spots shown in Fig. 1 were identified as particles of a second
phase, which from optical properties, etching behavior and chemical
composition, may have been amorphous silica. SEM revealed residual
pores about 0.3 microns, well rounded and almost spherical. Etching
with HF-HNO$_3$-H$_2$O$_2$ (3:1:1) brought about sizable pit formation resulting
from leached out second phase grains; segments of some grain boundaries
were also revealed but had a discontinuous, spotty appearange (Fig. 5).
The observation that prolonged etching did not change the pattern suggested that only a second phase (discontinuously segregated along grain
boundaries) was attacked, and that Si$_3$N$_4$-Si$_3$N$_4$ grain boundaries were
not revealed. Annealing of specimen 9 at 1650°C led to expansion and
delamination as shown in Fig. 6. It resulted from opening and propagation of numerous cracks perpendicular to the pressing direction.
Although the mechanism of the expansion has not been investigated in
detail, in view of the results of other annealing experiments, it seems
related to the presence of the second phase. Its softening at the
annealing temperature could generate the crack opening stress, for
instance by a large wetting angle between the melt and Si$_3$N$_4$, and
consequently by a tendency of the melt phase to redistribute itself.
The grain size was estimated near 0.5 μm from SEM fractographs and
optical microscopy.

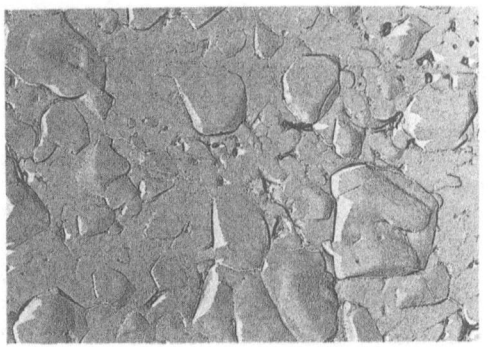

Fig. 4. Double stage replica of as polished section of Specimen
         No. 7. Magnification 15,000x.

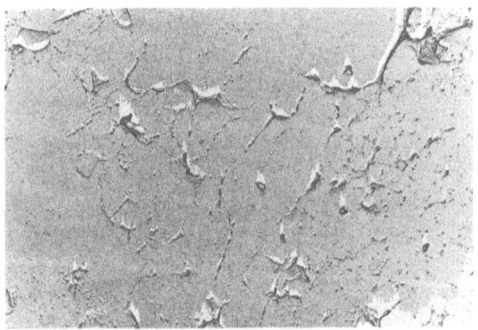

Fig. 5. Etch pattern of Specimen No. 9. Replica, magnified
         10,000x. The revealed segments of grain boundaries con-
         tained probably as silicious second phase.

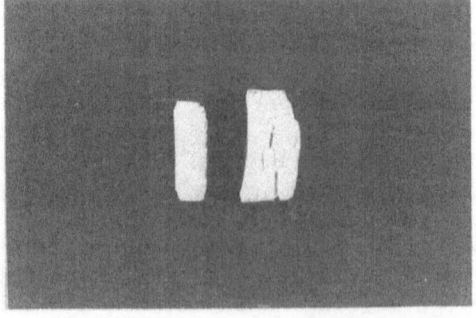

Fig. 6. Segment of Specimen No. 9a, before and b, after annealing
         at 1650°C in nitrogen for 40 minutes magnification 2x.

This would indicate that little or no grain growth took place during the consolidation process which started with an average powder particle of 0.45μm.

Specimens No. 10 and 11 obtained from the same crystalline powder which was vibratory milled for 6 hours, were dark gray, had the highest density of the whole series and a high microhardness value (2600 kg/mm^2). Only very few small pores were observable in polished sections. The etch pattern was similar to that observed in the specimens prepared from the unprocessed powder, i.e., discontinuous segments of grain boundaries were revealed. While all other specimens showed X-ray diffraction lines of only β-Si$_3$N$_4$, the latter two were composed of α-Si$_3$N$_4$, with a minor amount of β. Lines of a specific tungsten containing compound (carbide, silicide) were not identified although some weak extraneous diffractions were observed. These specimens were annealed at 1650° and 1750°C with no apparent deterioration, pore growth or crack formation similar to that observed with Specimen No. 9. However, it appeared from outlines of some grains observable in the etch pattern that appreciable grain growth occurred after annealing at 1750°C. Grains as large as several microns were indicated while the average grain size was estimated about 1μm.

IV. DISCUSSION

The consolidation of Si$_3$N$_4$ under extreme pressure conditions depends critically on powder chemistry. The powders used in the present experiments, although very low in metal content, had variable levels of oxygen and excess silicon. Either or both of these impurities may have contributed to the different response to high pressure compaction, making an unambiguous interpretation of the results difficult. Nevertheless, the present study has firmly established that none of the powders could be consolidated to attain high microhardness at or below 1500°C at 5.5 GPa. Thus about 1550°C seems to be the lowest temperature to obtain good bonding, which we believe is due to development of silicon nitride grain boundaries. Nadeau (1973) observed that temperatures about 1600°C were necessary to obtain bonding in SiC at 5 GPa and a temperature of 2100°C was required to obtain near theoretical densities with pure submicron powders (Prochazka, 1974). Alumina, on the other hand, yielded well bonded, theoretically dense compacts between 900-1200°C at comparable pressures (Yu Ishitobi, 1977). Assuming that the final stages of the structure development in these solids are controlled by diffusional processes, it would be expected that the high temperature behavior of Si$_3$N$_4$ be much closer to that of SiC than Al$_2$O$_3$. A similar conclusion was obtained from creep experiments done with CVD Si$_3$N$_4$ foils (Greskovich, 1975). These results strongly support the expectation that the relatively high creep rates found in present versions of commercial hot-pressed Si$_3$N$_4$ are controlled by extrinsic processes and allow for orders of magnitude improvements.

It was observed that powders rich in silicon achieved high microhardness while those presumably stoichiometric were either poorly bonded or were bonded, at least in part, by a second phase probably containing silica. Similarly the white areas in Specimen 5 showed residual porosity and lower microhardness, while the brown areas were fully dense. It seems therefore that the presence of excess silicon in the starting powders enhanced densification and bonding. Another observation, made in the course of a study of the powders synthesized from silane and ammonia, indicative of a related effect, was the substantial decrease of the crystallization temperature of silicon-rich amorphous Si_3N_4 compared to powders close to stoichiometry (Prochazka, 1977). The former were observed to transform to α-Si_3N_4 (and crystalline Si) between 1250–1350°C while the latter required 1480°. The decrease of transformation temperature due to the presence of silicon was interpreted in terms of an increased atomic mobility resulting from expansion of the structure of the amorphous Si_3N_4 due to the presence of Si-Si bonds substantially longer then Si-N bonds. An alternative mechanism of the effect of silicon could be sought in dissolution and reprecipitation of Si_3N_4 from a silicon melt. However the low temperature and the small volume fraction at which silicon promoted the transformation make this mechanism unlikely.

If the two observations, the enhanced densification under pressure and the decreased temperature of transformation, are indeed related to Si-Si bonds in the amorphous Si_3N_4, then excess silicon probably will not promote densification in crystalline Si_3N_4 as the Si_3N_4 lattice cannot accommodate any appreciable amount of Si beyond stoichiometry. In other words the low temperature limit to obtain bonding in Si_3N_4, determined to be about 1550°C from experiments 1–5, may have been influenced by the presence of silicon, and a temperature higher than this, perhaps by 100–200°C, would be necessary with pure stoichiometric powders. This, then would make pure Si_3N_4 even closer to SiC in intrinsic properties controlled by atomic mobility.

The relatively high microhardness readings for Specimen No. 8, together with the strong indications of the presence of a second phase are consistent with previous experience that such phases may promote densification under hot-pressing conditions. The annealing experiment which led to bloating and deterioration of the specimen, clearly demonstrates the type of degradation which may result when such materials are reheated to the softening temperature of the extra phase. It is noteworthy that the deterioration due to reheating was suppressed by processing of the starting powder with cemented carbide balls.

X-ray analysis showed (with exception of Specimen No. 10 and 11) only diffractions of β-Si$_3$N$_4$ which contrast with results obtained on Si$_3$N$_4$ powder compacts annealed at 1600°C. These usually showed only partial transformation of α to β. Thus the very high pressure either stabilizes the β form of Si$_3$N$_4$ or promotes the rate of the α-to-β transition.

V. CONCLUSIONS

1. The degree of consolidation of Si$_3$N$_4$ under extreme pressure conditions depends critically on powder chemistry. Pure stoichiometric Si$_3$N$_4$ powders could not be hot pressed at 5 GPa at temperatures up to 1600°C to achieve high microhardness. Free silicon and other impurities promote bonding and densification.

2. The results suggest that high creep resistance would result if direct grain-to-grain bonding in high purity Si$_3$N$_4$ is achieved.

3. High pressure tends to stabilize the β-form of Si$_3$N$_4$ or promote the rate of α-to β-transformation.

ACKNOWLEDGMENTS

The work reported in this paper was supported, in part, by a grant from the Advanced Research Project Agency.

REFERENCES

1. Alliegro, R. D., Coffin, L. B., Tinklepough, J. R., 1956, J. Am. Ceram. Soc., 39, 386-389.
2. Deeley, G. G., Herbert, J. M., Moore, N. C., 1961, Powder Met. No. 8, 145-151.
3. Ghoshtagore, R. N. Coble, R. L., 1966, Phys. Rev. 143, 623-626.
4. Greskovich, C., Rosolowski, J. H. Prochazka, S., 1975, "Ceramic Sintering", Final Report No. SRD-75-084, G. E. Corp. R and D, 1975.
5. Hall, H. T., 1960, Rev. Sci. Inst. 3, 125.
6. Kazunori, Kijima, Shin-ichi, Shirasaki, 1976, Jap. J. Chem. Phys., 45, 2669-2671.
7. Morgan, P. E. D., 1973, Final Report, NTIS-AD-778-373.
8. Nadeau, J., 1973, Bulletin Am. Ceram. Soc., 52, 170-174.
9. Prochazka, S., 1974, Ceramics for High Performance Applications, 1974, J. J. Burke, et al., Ed., p. 239-252.
10. Prochazka, S. Greskovich, C., 1977 (to be published), Bull. Am. Ceram. Soc.
11. Terwilliger, G. R., Lange, F. F., 1974, J. Am. Ceram. Soc. 57, 26-29.
12. Yu Ishitobi, Masahika Shimada, MItsue Koizumi, 1977, Bull. Am. Ceram. Soc. 56, 556-558.

THE STRUCTURE OF GRAIN BOUNDARIES IN SILICON NITRIDE BASED ALLOYS

David R. Clarke

Rockwell International Science Center
Thousand Oaks, California 91360

G. Thomas

University of California
Berkeley, California 94720

ABSTRACT

Grain boundaries in silicon-based ceramics have been character-ized by high resolution electron microscopy including the technique of lattice fringe imaging, and this work is illustrated with examples from both hot-pressed silicon nitrides (MgO and Y_2O_3 fluxed) and a magnesium-sialon ($Mg_{1.86}Si_{1.67}Al_{2.47}O_{3.19}N_{3.81}$). Room temperature observations of the glassy phase are consistent with it being only a partially wetting phase, indicating that it cannot form a continuous film. The atomic configuration of the grain boundaries in both materials is presented together with lattice fringe observations of segregation at grain boundaries in the magnesium-sialon.

INTRODUCTION

Grain boundaries and phase boundaries have long held a special significance for the ceramist. In part this is because of their profound influence on both the fabrication of components from powders and on the resulting mechanical and electrical properties. Partly this is also because so little is known about them from existing experimental techniques. However, this situation is changing with the advent of two techniques that promise to revolutionize our under-standing of boundaries in ceramics: high resolution transmission electron microscopy and Auger electron spectroscopy. The importance

of high voltage electron microscopy must not be overlooked, as this
method whilst not yet at the resolution capabilities of 100kV offers
advantages of increased penetration, reduced ionization damage, and
better statistical data. In this present paper the characterization
of boundaries in silicon-based ceramics will be described using
100 kV electron microscopy exclusively.

The application of high resolution electron microscopy to the
investigation of grain boundaries is still very much in its infancy;
the first results were reported at two international conferences
just a year ago.[1,2] Consequently, the correlation between processing
variables and microstructure of the grain boundary regions is very
undeveloped in comparison to those that will be presented in other
papers at this meeting.

After a brief account of the techniques of electron microscopy,
three aspects of the structure of grain boundaries pertinent to
ceramics will be discussed: the location and distribution of intergranular phases, the crystalline structure of the boundary itself,
and the segregation of solute to the grain boundary.

TECHNIQUES AND MATERIALS

Two principal techniques of high resolution electron microscopy
have been used to investigate the nature of boundaries in ceramics:
lattice fringe imaging and high resolution bright and dark field
imaging. As they have been described at length in a recent
review,[3] they will only be mentioned in outline here.

In lattice fringe imaging, the microscope is operated in a phase
contrast mode by recombining two or more diffracted beams, enabling
one or more sets of atomic planes in the sample to be imaged directly.
As a result, microstructural detail down to an atomic spacing can be
examined, for instance, grain boundary ledges in hot-pressed silicon
nitride 3.3Å high can be readily resolved. The other mode, that of
high resolution bright field/dark field imaging, the electron microscope is operated in an amplitude contrast manner by imaging the
transmitted beam alone or in dark field, imaging one of the diffracted
beams. In this case, image detail results from variations in relative scattering of the electron beam produced by adjacent regions and
the resolution is limited by contrast variations. Nevertheless,
intergranular films 10Å in width in hot-pressed silicon nitride have
been detected in this way under favorable conditions.

Results from an investigation of three different types of ceramic
are used here: commercially available silicon nitrides based on additions of MgO (Norton Company HS110, HS130, and NC132), a 10 m/o
Y_2O_3 hot pressed silicon nitride (kindly provided by Dr. F. F. Lange),

and a magnesium-sialon ($Mg_{1.86}Si_{1.67}Al_{2.47}O_{3.19}N_{3.81}$, kindly supplied by Professor K. H. Jack).

INTERGRANULAR FILMS

The properties of a number of materials of contempory interest appear to depend critically on the presence of an intergranular phase. For example, the nonlinear electrical conduction of the ZnO based varistors is attributed to the existence of a Bi_2O_3 rich phase at the ZnO grain boundaries;[4] the lowering of eddy current losses in soft magnetic ferrites can be explained by a thin film layer between the ferrite grains;[5] and the presence of a glassy phase at the grain boundaries has been held responsible for the high temperature strength degradation of MgO fluxed and hot-pressed silicon nitride.[6,7] In each case, good agreement is found between the observed properties and those calculated on the basis that there is a continuous intergranular film separating the grains. However, it is important in either case to know whether or not the inter-granular phase does indeed coat the individual grains, since this may lead to ways to improve the material properties and to further our understanding of the origin of these properties.

Thus, as part of an investigation into the microstructure of silicon-nitride based ceramic alloys, the distribution and location of the intergranular phases in silicon nitrides hot pressed with magnesia and with yttria have been extensively examined. The main findings of those studies[1,2,8,9] have been that (1) the intergranular phase is inhomogeneously distributed throughout the microstructure, being located principally at three- and four-grain junctions,[2] many of the grain boundaries examined are devoid of any phase separating the grains, and (3) intergranular films as narrow as ~8Å are seen at some boundaries and, in many of these cases, the boundary is formed by the low-index plane in one of the two adjacent grains.

Observations illustrating these findings have already been published[1,8,9] so, rather than repeat them with other similar micro-graphs of silicon nitride, the opportunity is taken here to present observations from another silicon nitride alloy that also contains a glassy phase. The alloy, a magnesium substituted silicon-aluminum-nitrogen-oxygen alloy[10] has the composition $Mg_{1.86}Si_{1.67}Al_{2.47}O_{3.19}N_{3.81}$ but for brevity is referred to here simply as a magnesium-sialon. It was originally investigated as a part of a study into compositionally controlled polytypism, described elsewhere in these proceedings,[11] but it is found that the distribution of the glassy phase is the same as that in the hot-pressed silicon nitrides.

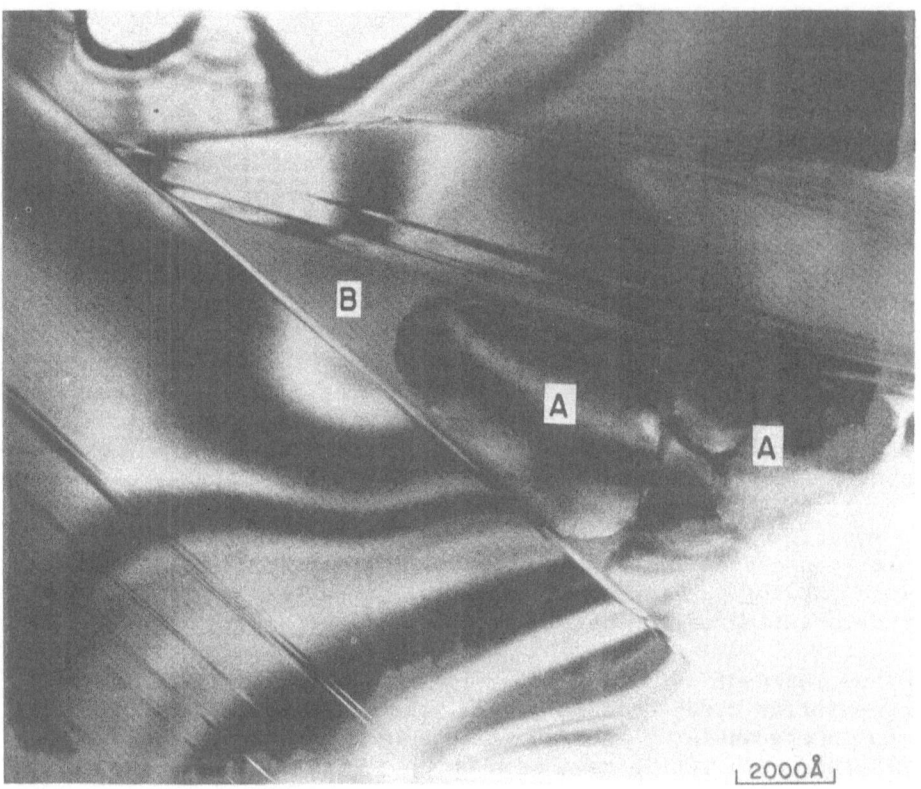

Figure 1. A glassy phase at a three-grain junction in a magnesium-
           sialon. Secondary crystallization has occurred from the
           glassy phase, B, to form the crystallites, A.

     The magnesium-sialon consists of long interlocking plates several
microns in length and about one micron across as may be seen in Fig. 1
of the accompanying paper (Shaw and Clarke, these proceedings). In
addition, at the three- and four-grain junctions, a glassy phase is
frequently seen. Because of the exaggerated plate-like morphology
of the grains, the three-grain junctions are typically wedge shaped
in cross-section, as is shown in Fig. 1. These regions are generally
too small for selected area electron diffraction to be used to determine whether they are glassy or crystalline, but as they remain
featureless on tilting the sample in the microscope, it can be concluded that they are probably glassy.

     In many of these glassy triple-grain junctions, secondary
crystallization has occurred, as in Fig. 1. Unfortunately, again the
regions are too small to obtain electron diffraction patterns from,
so it is impossible to determine what phase they consist of.
Interestingly, however, these secondary crystallites have an ill-
formed morphology in contrast to the highly faceted adjoining grains,

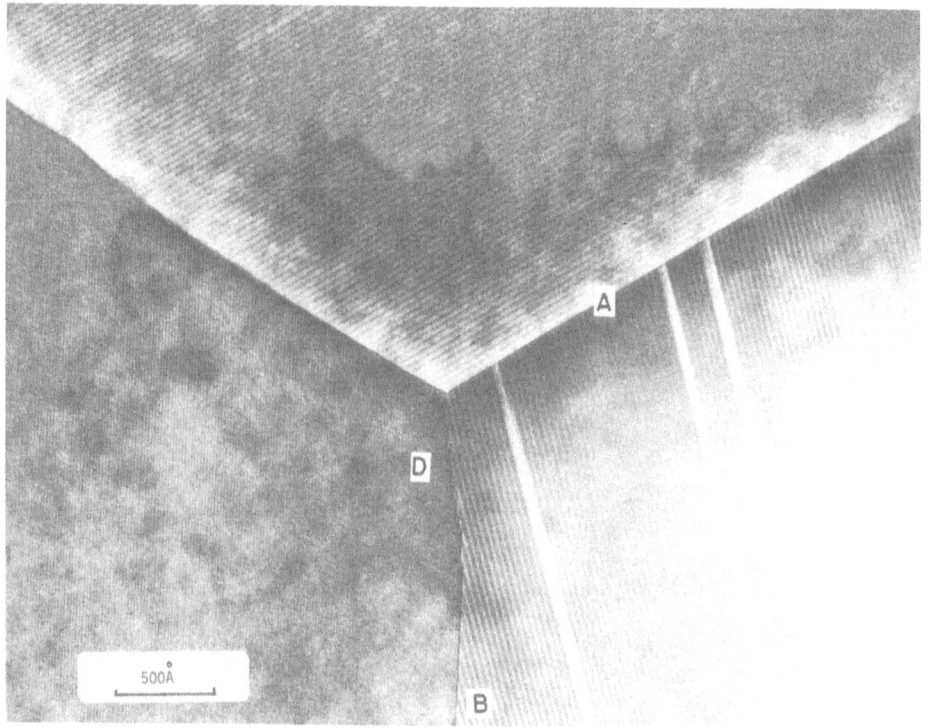

Figure 2. A three-grain junction in magnesium-sialon in which the
           polytype fringes have been imaged. In addition to there
           being no glassy phase in the intergranular regions, the
           crystalline and stepped nature of the boundaries can be
           seen. The fine fringe spacing is 18.8A.

and they lack the long stacking perodicities characteristic also of
the adjacent sialon grains. This secondary crystallization is not
unique to this material; it is also observed in magnesia hot-pressed
silicon nitride where small grains of the silicon oxynitride Si_2N_2O
can be seen surrounded by glass at three-grain Si_3N_4 junctions.

    As with the silicon nitride, many grain boundaries and occasion-
ally even a three-grain junction are free of any glass. This may be
vividly seen in Fig 2 in which the long period polytype fringes have
been imaged. The micrograph also illustrates a number of general
features concerning the structure of grain boundaries and will be
referred to in the following section.

    Another example of a three-grain junction is Fig. 3. Here the
glassy phase can be clearly seen, and also the fact that it has a
nonzero contact angle with the sialon grains, suggesting that it has
been trapped at the junction by the growing plates rather than
wetting them.

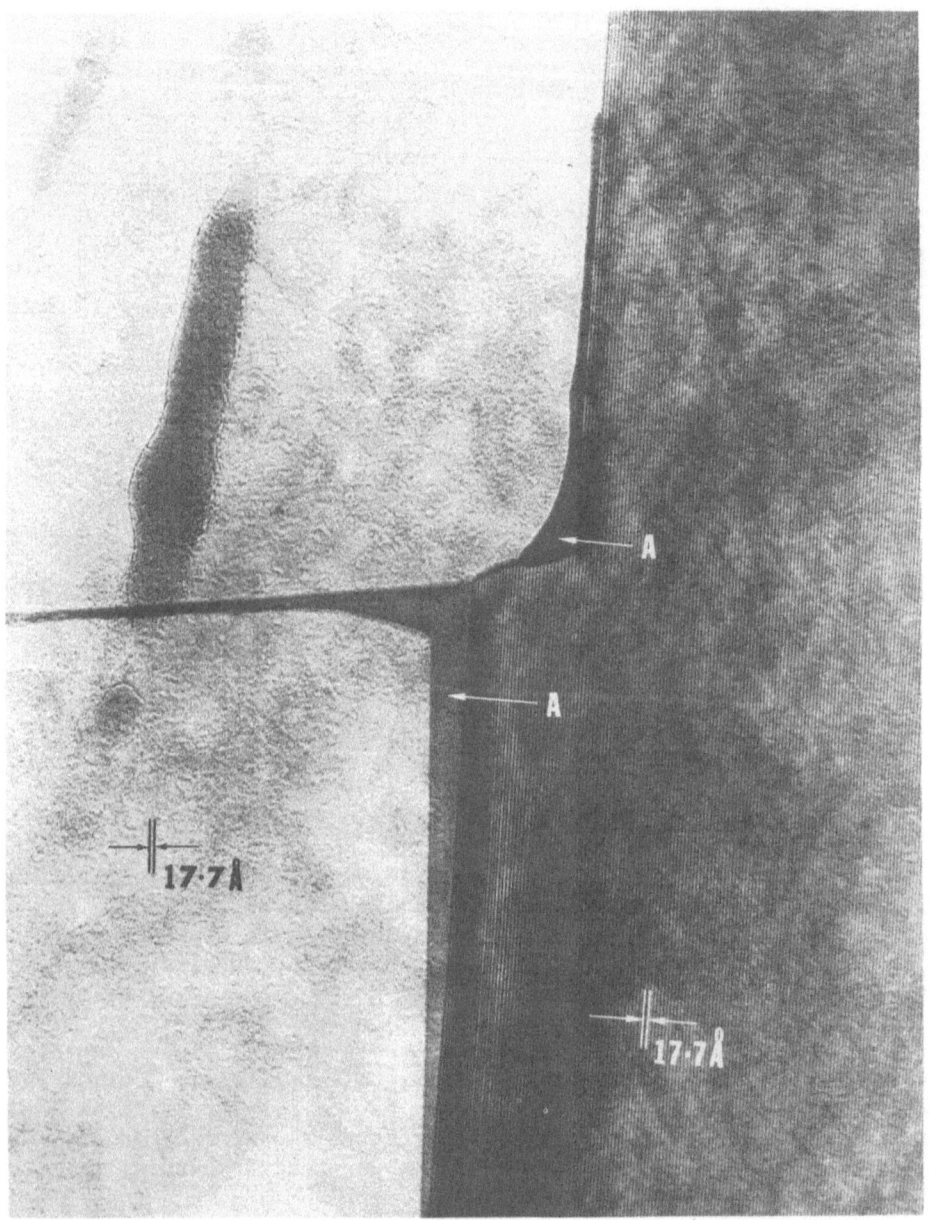

Figure 3.   A three-grain junction in which a glassy phase is present.
The glass has not spread along the boundaries and the
boundaries and the dihedral angles are clearly nonzero.
Magnesium-sialon (countesy of T. M. Shaw).

Inevitably, these observations in both hot-pressed silicon nitride and hot-pressed magnesium-sialon raise doubts as to the generally interpreted morphology and distribution of the glassy phase in these materials, in which it is envisaged that the glass forms a continuous layer surrounding and separating each grain.[7,12] Geometry alone indicates that such an interpretation is a gross simplification except for the case in which the intergranular phase occupies a large volume fraction of the microstructure or when the surface energy of the glassy phase--crystalline phase γ_{SG}--is much less than that between two crystalline grains γ_{GG}.

Firstly, it is necessary to recognize that at equilibrium, the extent to which one phase wets another, will be entirely determined by their relative surface energies. Only when γ_{GS} is less than γ_{GG} will the minor phase completely wet and surround the grains of the major phase. Secondly, there are three distinct types of intergranular regions to be considered: that associated with four grains is a surface. Corresponding to these three regions a minor amount of intergranular phase will appear as isolated particles, open network, or as closed cell foam, respectively.[13]

Since the dihedral angles formed by the glassy phase against the nitride and sialon grains at three-grain junctions are observed to be normally nonzero,[1-3,8,9] these geometric considerations imply that the glassy phase will be distributed between the grains in the same way as any other partially wetting phase will do as was argued by C. S. Smith over a quarter of a century ago.[13] This is schematically illustrated, following Smith, in Fig. 4 and is in agreement with the observations made by electron microscopy.

Local variations, such as complete wetting for certain combinations of grain boundary orientations, occur when the model is refined to include the crystallographic anisotropy of the surface energy. Such an anisotropy exists as observations of $(10\bar{1}0)$[1,9] and $(11\bar{2}0)$[14] habits for silicon nitride crystals indicate and would account for the observations of a continuous grain boundary film in hot-pressed silicon nitride seen at some boundaries, notably those in which a low index plane in one of the adjacent grains forms the boundary. Interestingly, a similar observation[15] is made in the ZnO varistor material, a material in which the intergranular phse is easier to distinguish than in the silicon nitride alloys. Here the Bi_2O_3 rich phase appears only to form an intergranular film when the grain boundary is parallel to the basal plane in one of the adjacent ZnO grains.

The other important factor determining the distribution of the glassy phase in the microstructure is the local microscopic volume fraction of the glass, as this determines the extent to which it encroaches from the three- and four-grain junctions into the two-

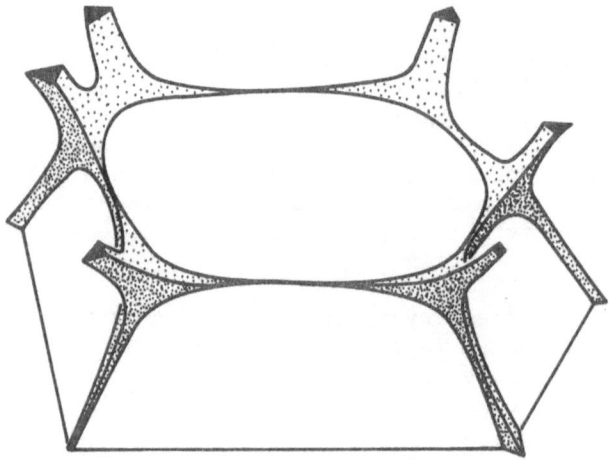

Figure 4.   Schematic illustration of the location of a partially
            wetting phase for which the dihedral angle is 60° based
            on surface energy considerations. It also represents
            the observed distribution of the glassy phase in hot-
            pressed silicon nitride, magnesium-sialon, and the ZnO
            based varistor material (after C. S. Smith).

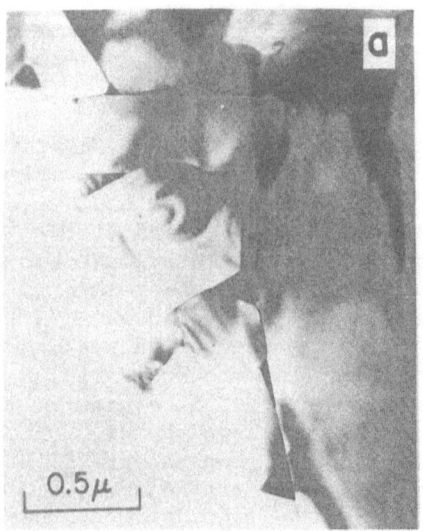
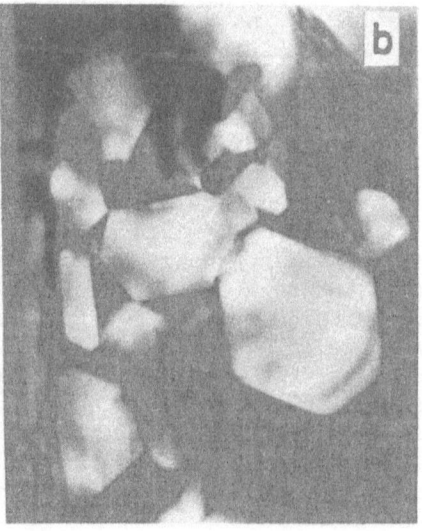

Figure 5.   The two extreme cases for the distribution of the
            yttrium-silicon oxynitride phase (appearing black) in a
            10 m/o Y_2O_3 hot-pressed silicon nitride; a low local volume
            fraction situation (a) and a high local volume fraction (b).

Figure 6.   A lattice fringe micrograph of a low angle grain boundary
            in MgO hot-pressed silicon nitride in which a set of
            (10$\bar{1}$0) planes have been imaged across the boundary. Note
            the presence of a number of terminating fringes. The
            fringe spacing is 6.6Å.

grain junctions. Although in a fully equilibrated structure this
should be a constant, observations indicate that it rarely is, as is
illustrated by the electron micrographs of Figs. 5a and 5b. Both of
these were taken from the same microscope sample cut from the same
billet of a 10 m/o Y_2O_3 hot-pressed silicon nitride. Clearly, the
local volume fraction of the yttrium-rich phase (the black phase) is
different in these two regions despite the contact angle remaining
the same; in one case the yttrium-rich phase is entirely localized
at the three-grain junctions; whereas, in the other case, it
surrounds the silicon nitride grains.

GRAIN BOUNDARY STRUCTURE

     There are two grain boundary structures of interest in ceramics,
the interface between a crystalline and an amorphous grain and the
interface between two crystalline grains.

     Details of the crystal-crystal interface can again be illustrated
with examples drawn from the magnesium sialon system. In those instances where the boundary is straight, it is found that the boundary
plane is formed by a low-index plane in one or the other of the two
grains forming the boundary. This is seen in regions A, and B of
Fig. 2, and has been seen when imaging the fundamental lattice planes
in both MgO and Y_2O_3 hot-pressed silicon nitride, Mg-sialon and in
ZnO. In some cases, the boundary is straight, except for unit cell
high ledges or multiple atomic high ledges.

Those boundaries that are macroscopically curved are seen using lattice fringe imaging to be made up of short segments parallel to a close packed plane in one or other of the grains, as may be seen in region D of Fig. 2. This seems to be a general fact though not unexpected. The curvature of low-angle grain boundaries similarly is made up of short flat segments, but when the lattice planes across the boundary are examined, as in Fig. 6, terminating fringes can be seen with an accompanying strain field contrast. Although image contrast theory has not been fully developed for this particular geometry, there is every reason to believe that these terminating fringes represent grain boundary dislocations.

Only a few observations have been made of the interface between a crystalline and an amorphous region, and until microscopes having atomic resolution are available, it is unlikely that much structural information will be forthcoming. Nevertheless, a common feature of many of the interfaces examined is the presence of ledges one interplanar-spacing high. Examples of these have been presented elsewhere.[3,8,9]

SEGREGATION TO GRAIN BOUNDARIES

Auger electron spectroscopy combined with sputter ion etching has, in less than ten years, proved to be a remarkably versatile tool for investigating segregation of solute to grain boundaries in both metals and alloys.[16] However, it is limited to studying only those boundaries exposed by intergranular fracture. For this reason, we have begun to explore whether lattice fringe images can be used to detect and measure segregation, in the same way that periodic compositional variations due to spinodal transformation have recently been measured from lattice fringe micrographs.[17] By this method, changes in composition would be manifest as changes in local lattice fringe spacings, and measurement of these changes would be related to compositional changes.

In order that the maximum effect be seen, it is necessary that a set of lattice planes parallel to the boundary plane be imaged and the boundary observed edge on. This has been done in producing the lattice fringe image of Fig. 7 taken of a boundary in the magnesium-sialon in which one set of fringes are imaged in the grain to the right. The spacing of the fringe adjacent to the boundary plane is noticeably larger than the spacing between fringes in the interior of the grain. The spacing of the second fringe is also larger but is less perceptably so. Together, these indicate a change in composition adjacent at the boundary. The scale of segregation is quite large, since the fringe periodicity in this case corresponds to the polytype spacing in magnesium-sialon and not to the fundamental close packed layer spacing. Nevertheless, it is still on a very much

Figure 7. A lattice fringe micrograph of a grain boundary in mag-
           nesium-sialon. The fringes (32.4A) have only been
           revealed in the grain to the right, and the grain boundary
           is located at the position of the furtherest left fringe.
           The interfringe spacing is greater at the boundary than
           to the right of the boundary, indicative of segregation
           to the boundary.

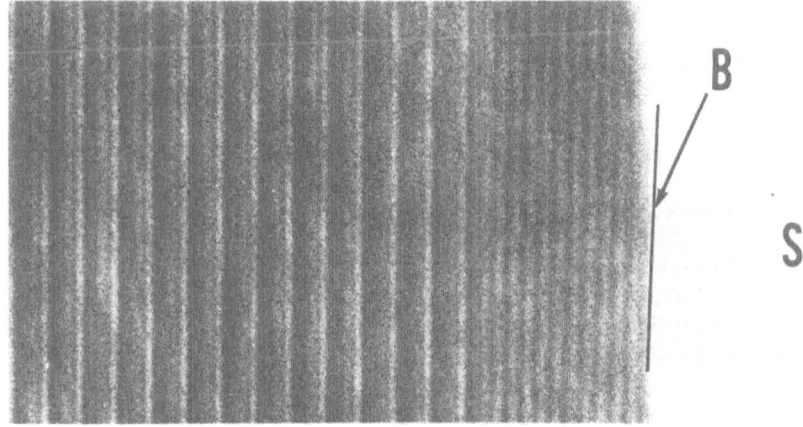

Figure 8. Another grain boundary region is magnesium-sialon with the
           boundary at B. Due to the recording conditions, the second
           grain, S, appears white. There is a band ~100Å wide at
           the grain boundary in which the polytype spacing is smaller
           than that in the rest of the grain. Again, this is indi-
           cative of a segregation effect. The broad fringe spacing
           is 32.4Å.

finer scale than has been possible with electron-optical methods to date, e.g., the electron microprobe. The change in spacing of the first fringe corresponds to a 20% alteration in the lattice spacing. The related change in composition is unclear at the time of writing, since the compositionally controlled polytypism in this alloy is not fully understood.[11]

Another example of a change in fringe spacing in the vicinity of a magnesium-sialon grain boundary is shown in Fig. 8. The spacing in the band adjacent to the boundary is clearly finer than it is in the rest of the grain, indicative again of a local compositional change. This is a frequent observation in this materials and suggests that the compositional change due to segregation effects does not occur smoothly away from the boundary as usually indicated by Auger electron spectroscopy but rather changes discretely.

Although we have deliberately chosen a material in which there is a direct crystallographic relationship between a polytype spacing and composition to demonstrate the applicability of lattice fringe imaging, these and similar observations indicate that the technique will have widespread utility for measuring segregation profiles. One outstanding advantage of the technique is that the segregation can be related directly to the crystallographic orientation of the boundary.

CONCLUSION

Although the techniques of high resolution electron microscopy, and in particular, that of lattice fringe imaging, are only just being brought to bear on ceramics, the information already obtained clearly demonstrates that they will have an important role to play in investigating the effects processing variables have on the microstructure and, hence, properties of ceramics.

Using these techniques, it has been possible to show that the distribution of the glassy phase in both hot-pressed silicon nitride and hot-pressed magnesium-sialon at room temperature is consistent with that obtained for any partially wetting pohase in a polyphase material. The next step is to determine which combinations of grain boundary orientations and grain surface orientations are completely wet by the glassy phase so that a more realistic microstructure model based on an only partially wetting phase can be used for formulating high temperature strength degradation theories.

The crystalline grain boundaries in both silicon nitride and magnesium-sialon are concluded to be faceted, to contain interplanar high ledges, and to be formed by low-index planes in the adjacent grains. Furthermore, segregation effects in hot-pressed magnesium-sialon have been detected, thus opening the way for detailed measurements to be made of segregation profiles in ceramics.

ACKNOWLEDGEMENTS

This work was initiated while both authors were supported by the National Science Foundation (Grant DMR-75-11352). One of us (D.R.C.), also gratefully acknowledges the financial support of the Rockwell International Independent Research and Development Program for this work.

We also wish to thank Professor K. H. Jack of Newcastle University for kindly supplying us with the magnesium-sialon alloy, and it is a pleasure for us to acknowledge useful discussions with our colleagues T. Shaw and Dr. F. F. Lange.

REFERENCES

1. D. R. Clarke, "Nitrogen Ceramics," ed. F. Riley, Proc. NATO
    Advanced Study Institute, Canterbury, UK, 1976, in press.
2. G. Thomas, D. R. Clarke, and O. Van der Biest, in "Ceramic
    Microstructures '76," ed. R. M. Fulrath and J. A. Pask, Proc.
    6th International Materials Symp., Berkeley, 1976, Westview
    Press, Col., 1977.
3. D. R. Clarke, J. Am. Ceram. Soc., Proc. Symp. Electron Microscopy
    Applied to Ceramic Practice, in press.
4. L. M. Levinson and H. R. Philipp, J. App. Phys. $\underline{46}$, 1332, 1975.
5. A. Stuijts, in "Ceramic Microstructures '76," ed. R. F. Fulrath
    and J. A. Pask, Westview Press, Col., 1977.
6. F. F. Lang, J. Am. Ceram. Soc. $\underline{57}$, 84, 1974.
7. R. Kossowsky, D. G. Miller, and E. S. Diaz, J. Mater. Sci. $\underline{10}$,
    983, 1975.
8. D. R. Clarke and G. Thomas, J. Am. Ceram. Soc., $\underline{60}$, 491, 1977.
9. D. R. Clarke and G. Thomas, J. Am. Ceram. Soc., in press.
10. K. H. Jack, J. Mater. Sci., $\underline{11}$, 1135, 1976
11. T. M. Shaw and D. R. Clarke, these proceedings.
12. F. F. Lange, "Deformation of Ceramic Materials," ed. R. C. Bradt,
    and R. E. Tressler, Plenum Press, pp. 361-381, 1975.
13. C. S. Smith, Trans. AIME, $\underline{175}$, 15, 1948.
14. W. P. Clancy, Microscope, $\underline{22}$, 279, 1974.
15. D. R. Clarke, J. App. Phys., 49, 1978.
16. L. A. Harris, J. App. Phys., $\underline{39}$, 1428, 1968.
17. R. Sinclair, R. Gronsky, and G. Thomas, Acta. Met., $\underline{24}$, 789, 1976.

EVOLUTION OF MICROSTRUCTURE IN POLYCRYSTALLINE SILICON CARBIDE

S. Shinozaki

Research Staff, Ford Motor Company

Dearborn, Michigan   48121

K. R. Kinsman

Electric Power Research Institute

Palo Alto, California   94303

This paper describes the results of an organized study of the structural aspects of the development of microstructure in polycrystalline
silicon carbide. While silicon carbide has long seen specialized
application in the polycrystalline form, it is only recently that
technological advances have permitted the high resolution observation of structure by transmission electron microscopy. Most of our
current understanding of the subject is built upon X-ray diffraction analyses of single crystals prepared by precipitation.from
solvents or by growth in the vapor phase. The consolidation processes involved in the preparation of polycrystalline material incorporate these and other phenomena. Much of the single crystal
information already in hand provides a basis for the understanding of
polycrystalline structure, particularly that relating to the fine
structure within grains (crystals). However, the polycrystalline
microstructure depends not only upon the initial fine structure but
also upon the changes that take place as a result of intercrystalline
interaction and processing variables in addition to time and temperature.

The structure of polycrystalline silicon carbide is a result of the
lattice defect processes that are characteristic of different consolidation schemes. This investigation emphasizes these differences,

however, subtle, particularly to the extent that the initial structure and the subsequent solid state alteration of that structure in
the course of normal processing are related and interdependent. At
the same time, important similarities in the course of structural
development are described.

TECHNIQUE

The progress of microstructure development was monitored as a function of time and temperature in polycrystalline silicon carbide(s)
representative of the three basic processes of consolidation: vapor-
solid (chemical vapor deposition), liquid-solid (Ford reaction sintering process), and solid state (G.E. solid state sintering) reaction.
Specimens so chosen were isothermally reacted in a carbon crucible
(unless otherwise noted) by an induction furnace equipped for use in
vacuum or other controlled atmospheres. Both optical and transmission
electron microscopy were employed in complementary fashion. The
techniques of lattice resolution and quantitative electron diffraction analysis were used in concert to identify and to follow the systematic changes in the ordered structures (polytypes) encountered
throughout this study. Samples transparent to electrons at 100KV
were prepared by ion beam thinning and viewed in a Philips EM 300
electron microscope equipped with a goniometer stage. Lattice resolution was arranged by the selection of an objective aperature sized
to include only the ordered lattice spacings equal to or larger than
4H spacings (Ramsdell notation used throughout).

RESULTS AND DISCUSSION

General Mechanics of Microstructure Formation

The microstructure of polycrystalline silicon carbide is considered
on two levels. The grain structure, observable on the scale of optical microscopy, is expected to develop in response to the normal
volumetric and surface diffusional processes integral to the different methods of consolidation. The structure within the grains is
comprised of ordered lamella which cause characteristic and often complex modulations in diffracted intensity and which have formed the
primary basis of structural knowledge to date. There are often
changes within these polytypes in response to the variables of temperature and time, as well as interdependency between the structures
on both levels.

Crystallographically, silicon carbide is a close packed structure and
the cubic (fcc) and hexagonal (or rhombohedral) symmetries are based
upon the stacking sequence of the close packed (molecular) layers.
Beyond the simple ABCABC... cubic sequence and ABAB... hexagonal
stacking are a large number of ordered arrangements (polytypes) having hexagonal or rhombohedral symmetry. Formation of these layers
from vapor or solvent is aided kinetically by surface defects (ledges,

kinks, vacancies) or surface emergent defects (such as screw dislocations) leading to regenerative (spiral) ledge growth. The conditions of growth, particularly the rate as influenced by vapor deposition variables and solvent supersaturation, dictate the relative
planar defect density and arrangement. There is a certain redundancy
among these structures in the sense that all can evolve one from the
other through the judicious rearrangement of "stacking faults" or
glide of partial dislocations in the solid state. This simple crystallographic fact is the most important mechanistic feature contributing to the microstructure. Much of the generation of silicon carbide microstructure thus involves the rearrangement of as-grown defects
(stacking faults, twins) into ordered structures of varying complexity
and the continual evolution to simpler polytypes through dislocation
of close packed planes. Mixtures of these structures are common and
crystallographic coherency is maintained along common close packed
planes.

In considering the basic processes of consolidation, first the initial
structure will be described. Its role in subsequent microstructual
development is then considered in light of the above mechanistic considerations as they apply to both the coarse and fine scale development
As it happens, some structural features are more pronounced in certain
of the processes. This fact will be used to highlight the structural
relationships tutorially in the space available.

Chemical Vapor Deposited SiC (vapor-solid)

Chemical vapor deposited silicon carbide usually forms columnar grains
whose major axes are approximately normal to the macroscopic solid:
vapor interface and within which the close packed planes are oriented
approximately parallel to that interface.[1]   Individually, the
grains develop by layered growth principally of close packed planes
nucleating at surface ledges, kinks or emergent screw dislocations.[2]
The initial crystal structure depends upon the deposition conditions
which can be adjusted to provide dominantly β or α phases or a mixture.[3]

At the typically high deposition rates used and because of the influence of surface defects, the resultant grains are usually highly internally faulted (Fig. 1a). Annealing or rearrangement of these
planar defects, which necessarily must occur through the glide of
Shockley partial dislocations, doesn't take place rapidly at the deposition temperature. As a result, fault densities can be observed that
yield virtually continuous $(10.\ell)$ streaking (Fig. 1b). These structures have been described as "one dimensionally disordered", a terminology that reflects the apparent lack of discrete spots along certain
directions in the diffraction pattern. Quantitative analysis of diffracted intensity and lattice resolution of the diffracting volume
established that the specification of relative "order" is, to be

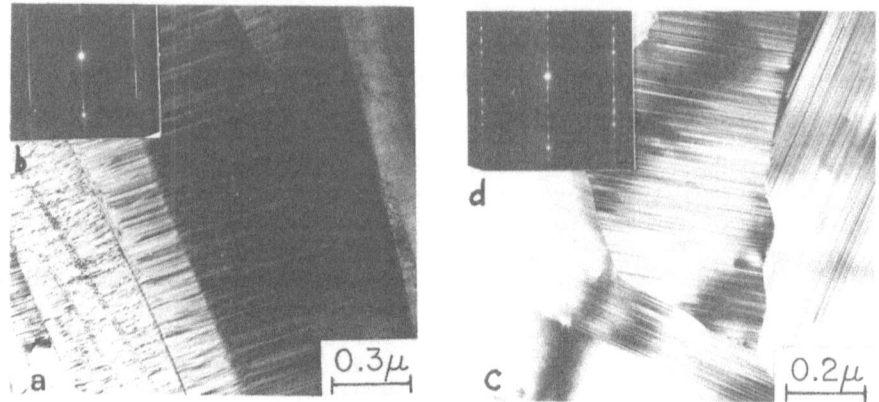

Fig. 1.   CVD processed material: TEM of (a) initial heavily faulted columnar grains and (c) structure annealed 2 hours at 1820°C.   (b) and (d) are 10.l rows in the respective SAD patterns.

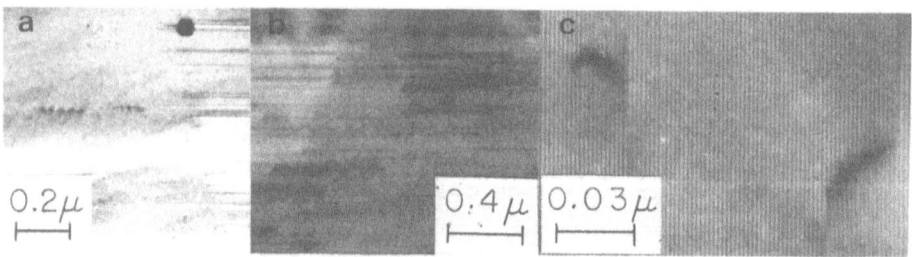

Fig. 2.   Illustration of the role of dislocations in the structural development of SiC (a) reaction sintered (aged ½ hr. at 1600°C), (b) as sintered (G.E.) and (c) reaction sintered (aged 2 hr. at 1600°C).

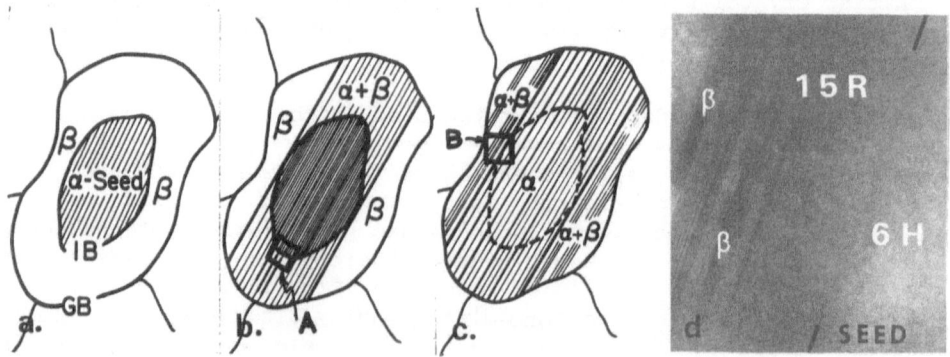

Fig. 3.   Schematic sequential microstructural development within a grain of reaction sintered SiC (a-c).   (d) is a high resolution lattice image of the structure typical of stage (c), area B.

precise, a question both of the scale and sensitivity of the measuring techniques. For example, periodicities in diffracted intensity (however slight) are detected within the most smoothly continuous streaks observed and indicate the existence of, at least, thin lamella of recognizable ordered structure. Lattice resolution confirms that these structures are comprised of stacking faults, thin twins, thin lamella both of a number of the ordered (polytype) α structures and the β phase all arranged (microsyntactically) on a single variant of the close packed lattice. The realization that a phase mixture is required in order to approximate smoothly continuous streaking is important.[1] It is only slightly less significant than the observation that truly phase-pure α or β is exceedingly rare.[4] Given the small volume free energy differences associated with the variation in stacking of these close-packed structures, and the fact that planar defects in each phase are often crystallographically redundant, this perhaps is to be anticipated on considerations of entropy. The volume fraction of the minority phase often is very small but can in special circumstances contribute importantly to the morphology of the developing microstructure.

In CVD silicon carbide, subsequent isothermal reaction causes the α phase to develop within the established β grains, forming through the agency of lateral glide of Shockley partial dislocations. The α phase thus forms in a plate morphology and the extent to which each of the $\{111\}_\beta$ variants participates is dependent on the vagaries of dislocation nucleation and subsequent interaction with other lattice defects. Evidence of dislocation participation in the development of the microstructure is demonstrated in Fig. 2. During the initial stages of chemical vapor deposition the β grain structure tends to be more equiaxed than the subsequently dominant columnar grains. A typical isothermally transformed structure (1820°C for 2 hrs) within such equiaxed grains is shown in Fig. 1c. The corresponding partial SAD pattern in Fig. 1d. depicts a predominant 6H (α) structure combined with some local diffuse intensity, attributable to the existence of lamellar mixtures of the other short range polytypes (4H, 15R, 21R and 8H) and remnant β (3C) phase.

Concomitant with continued isothermal reaction the structure on a fine scale changes from a highly faulted fine lamellar "disorder" through a usually complex mixture of polytypes to a dominantly single α polytype; in the CVD material studied (after 26 hours at 1820°C) this was 6H with a trace of 15R and 3C. The mechanics of this fine scale evolution is generic to the influence of isothermal reaction in all of the materials studied, changing only in detail to reflect local conditions or processing histories.[1,4] Direct lattice resolution combined with quantitative SAD pattern analyses indicates that considerable care must be exercised in recording and interpreting diffraction patterns. Often complex long period polytypes as "identified" by SAD pattern alone are found to be spatially arranged in a

way that only averages to the identified structure. Long period
structures are comprised of ordered arrangements of lamella of the
simple short range polytypes; slight departures from the "expected"
structural sequencing both in lamella thickness and distribution
are often found. On the other hand, very long period polytypes,
virtually free of faults, are also found in which the observed structure correlates exactly with the diffraction pattern. Isothermal
reaction first refines and then simplifies the structures present
through a process of coarsening of lamella of the constitutive (short
period) polytypes. Depending upon the relative complexity of the
initial structures, various intermediate polytypes form and disperse
during evolution to the terminal short period structures. Thus processing history, as it influences initial structures and local conditions is intimately involved with the detail of this evolution as
reviewed in a very general way in Table I. The general trends, however influenced by vagaries of processing, are similar among all the
structures.

Reaction Sintered SiC (liquid-solid)

Reaction sintered silicon carbide as produced by the Ford or REFEL
method is a composite of silicon carbide grains and some (SiC en-
riched) silicon residual from the conversion of carbon to silicon carbide. The grain structure is comprised of seed crystals enclosed in
β silicon carbide which precipitates epitaxially from the supersaturated liquid silicon.[5] As a consequence of this liquid-solid
reaction, the initial defect density in the "new" β silicon carbide is
neither high nor unusual in character, while the detailed structure of
the enclosed (typically) α seed crystal reflects its separate and
more complex processing history. Thus the basic unit grain as por-
trayed schematically in cross section in Fig. 3, contains an interphase boundary delineating the epitaxial association of the seed
crystal with the surrounding β phase, and can be viewed operationally
as a partially (centrally) transformed single crystal.

The sequential morphological development upon continued isothermal
reaction is schematically described in Fig. 3a-c. There are three
zones of interest. The seed single crystal invariably is predominantly a mixture of thin lamellae of α polytypes, containing both
dislocations and planar faults, all of which are a particular result
of the processing (ball milling) leading to the fine particles. The
lamellae, in keeping with the operative crystallographic constraints,
form with a basal plane habit, and the surrounding epitaxial β phase
thus maintains exact alignment thereto of one $\{111\}_\beta$ variant. The
transformation of the surrounding β phase is kinetically favored in
the direction <u>parallel</u> to, and spatially as an extension of, the
lamellar habit planes. In that direction any crystallographic dis-
registry along the interphase boundary is comprised essentially of
partial dislocations of the type that are responsible mechanistically
for conversion of β to α. In that orientation the α/β interphase

TABLE I

Dominant Structures as a Consequence of Reaction Time

Consolidation Process	Initial Structure(s)	Intermediate Structure(s)	Terminal Structure(s)
CVD	β(3C)-faulted	Complex inter-leaved lamellar mixtures of short period polytypes	6H (Tr. 3C, 15R)
Reaction Sintered	α seed - 6H plus complex long period polytypes. β(3C)	"	6H (Tr. 3C, 4H) 4H (molten Si) (Tr. 3C, 6H, 15R)
Sintered	β(3C) α (6H and Tr. 15R, 8H)	"	6H, 15R [Tr. = trace]

boundary thus functions as a readily activated source for the continued growth of the seed α phase structures into the neighboring
β phase. Fig. 2a. illustrates such dislocations as they extend the
existent polytype within a region such as that identified as A in
Fig. 3b. This transformation sequence leads initially to a macroscopic blocky α morphology with an aspect ratio dictated by the width
of the seed crystal, measured normal to the habit plane, and by intersection of the lengthening α phase with the surrounding β grain
boundaries. Development of the α phase in the remaining areas relies
upon nucleation of dislocations principally at the β/β grain boundaries, a kinetically less favored process. While other transformation variant options are available, in practice the tendency at this
stage is for these regions to adopt the same variant as the seed
(whether initially or during subsequent coarsening). In this instance, as in the CVD case, the macroscopic grain size is not altered
perceptibly by continued reaction and the β to α transformation is
confined within the initial β grains. The morphological development
is a natural consequence of the defect mechanisms at play, modified
only by the detail of the role of the seed crystal.

The evolution of polytypes in this material is perturbed sequentially
by the different local "processing" histories within the fundamental
structural unit described in Fig. 3. The polytype distribution initially within the seed crystal nominally extends into the β phase.
However, even here some distinction arises because of local lattice
defect interactions. It is observed that upon annealling (in vacuum

or Argon) at 1800°C the seed and related area evolve through what
appears to be the generic sequence of complex structures and lamellar
coarsening to arrive at predominantly a low fault density 6H structure, very much like that shown in Fig. 2c. The kinetically distinct
bounding areas, since their polytype sequencing is dependent on the
vagaries of another (β grain boundary) nucleation source, progresses
both at a different rate and through a different polytype mix. The
result is that these areas also eventually convert to 6H but generally
at a slower overall rate. Thus, the local polytype and β lamella
distribution, after aging for hours at 1800°C, typically appears as
the lattice resolution micrograph (Fig. 3d) of the area marked B in
Fig. 3c.

However, an important departure occurs if the specimens are annealed
in an atmosphere in which the remnant saturated silicon is held in
equilibrium. In that instance, 4H is the terminal structure. The
generation of this polytype during growth from a silicon-rich melt is
known.[6]   The particular circumstance of our observation in reaction sintered SiC suggests an important influence of local chemical
conditions on the stability of ordered phases.

Sintered SiC (solid state)

While the path of morphological development in solid state sintered
SiC is the most complex of the three processes recounted in this
study, the fine scale structural evolution seems nominally independent of morphology and proceeds in a way similar to that described
for the other processes. The α phase morphology develops continuously but is best considered conceptually in stages. The initial
structure is a fine grained (∿ 5μ dia.) equiaxed polycrystalline
matrix comprised principally of β phase grains that have coarsened
somewhat during sintering.[7]

Comparable to the processes previously described, lamella of α polytypes develop first within the confines of the initial β grain structure (Fig. 4a). At that point in the reaction, a sensitive balance
is struck between the driving force for continued α growth and reduction in total interfacial energy through β grain growth.[4]   The
result is that α (lamellae) plates thence extend beyond the initial
β grain size, a major departure from morphological observations in
the two other systems. In response to the constraint imposed by the
strict α/β crystallographic relationships, the plate edge (coincident
with the high angle β grain boundary) advances accompanied by local
adjustments in β grain boundary position (Fig. 4b,c). The mechanistic result is a cooperative local β grain growth to accommodate α
plate lengthening which effectively "recrystallizes" the β grains
in the growth path, reduces high angle interfacial area and leads to a
morphologically unique envelopment of α by β (Fig. 4a-d). This
structural entity can be an order of magnitude longer than the
neighboring β grain size. Study of Fig. 4 indicates that the role

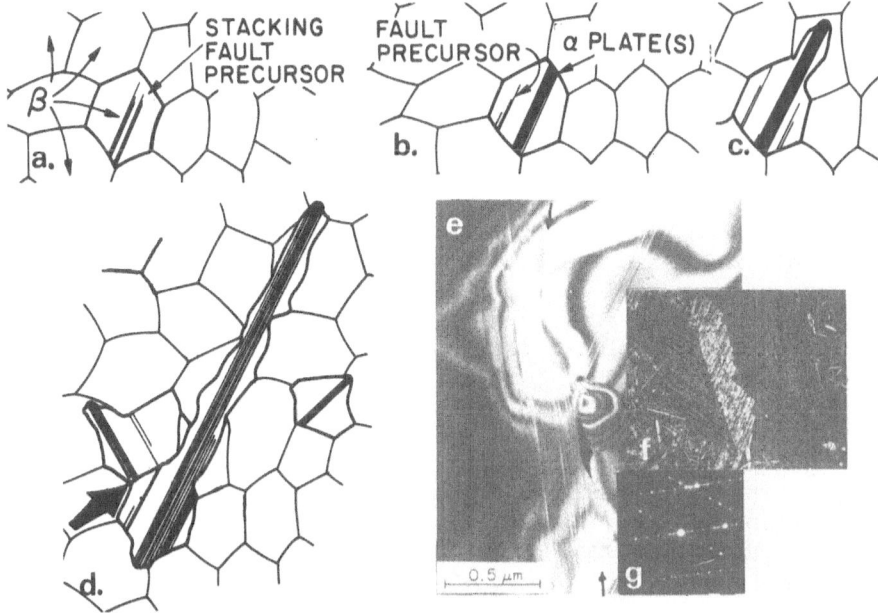

Fig. 4. Schematic sequential microstructual development of fine grained sintered β SiC (a-d). TEM of the "feather" morphology (optical micrograph (f)) is in (e), the corresponding SAD in (g).

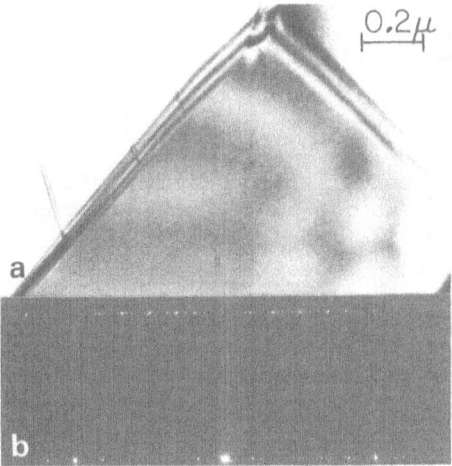

Fig. 5. Long period polytype (99R) produced in sintered SiC. (a) is the lattice image and (b) the corresponding 10.ℓ row of the SAD pattern.

of the β phase in this mechanism is specialized and can function even
when there is only a small amount of β present so long as it is
properly dispersed and the local conditions for grain growth persist.

In the next stage, continued nucleation leads to packets of plates
of the same orientation. Concommitant with this is a tendency for
certain of these packets to develop in association with each other
to form a more or less distinct "feather" morphology (Fig. 4f.).
Quantitative study by TEM of the structure of the feather midrib
(Fig. 4e,g.) indicates that the most common angular misorientation
(~38°) corresponds to a common {111} coincident site lattice while
the only other characteristic angle, though of far less frequency,
is the 70° misorientation between {111}β. Thus it appears that true
feather formation is favored by the specific β grain boundary texture
and probably initiates as indicated (by arrow) in Fig. 4d. Happen-
stance natural impingement of single packets during growth accounts
for the range of "near" feather morphologies also encountered. Thus,
while the "feather" is distinct, it is a secondary and special result
of the tendency to cluster plates into packets. This is favored at
high reaction temperatures while at lower temperatures (1600°C) a
simple more random profusion of single plates is favored and the
packets are not observed. This feature seems very sensitive to
competitive grain boundary nucleation kinetics as influenced by grain
boundary structure. This structure in turn can be affected by slight
adjustments in processing. For example when the same (G.E.) powder
is hot pressed and sintered, the feather structures do not develop[8]
although the unique α plate morphology is sustained.   Apparently
a critical reduction in the frequency of β coincident site orientations occurs as a result of the hot pressing. Further, while the
unique α plate morphology in principal can be maintained with a very
small volume fraction of β phase, arguments thus far preclude
formation of feathers except where β is the dominant phase.

Continued reaction eventually transforms virtually all of the β
phase, and the resulting α structure develops into a coarse equiaxed
microstructure. The kinetics and fine structure of this stage can
be augmented by selected impurities. Reaction of samples in a boron
nitride crucible for example result both in exaggerated grain growth
and in the formation of exceptional long period polytypes (Fig. 5).
The latter is unusual because long period polytypes are not the rule
in the unaided development of microstructure in this material and
because at this stage polytype simplification (not complication) is
anticipated.

The evolution of the fine structure in the sintered material seems
independent of the details of coarse morphological development. The
structure on this scale reflects the relatively benign processing
history which begins with low defect density β powder sintered into
polycrystalline form. The α phase forms initially in thin (10-100 Å)
lamellar mixtures principally of 6H but also 15R, 21R and other of

the simple short range polytypes. Aside from a coarsening of this mix, no organization into long range ordered polytypes occurs normally and eventually the 6H and 15R structures dominate (Table I).

CONCLUDING REMARKS

A physical rationale is provided, tested, and found to account for microstructure development in three differently prepared polycry- stalline silicon carbides. It recognizes the fundamental lattice defect-based mechanism of the β→α transformation and demonstrates how the different local processing-related physical constraints logically influence all levels of microstructure evolution. While the basis for understanding is in place, the results summarized in this paper clarify the need for continued systematic research. For example, while in general it can be said that the fine structural details (polytype kind and distribution) evolve predictably in the sense that thin lamellae coarsen and short range polytypes seem ultimately most stable, the role of environment (e.g., molten silicon or boron nitride) clearly has a strong and unexplained influence upon both kinetics and (ordered) phase stability.

Microstructure has been considered at two levels of resolution. Understanding of factors that control morphological development has the most practical significance since structure on this scale influences the physical properties currently of most interest (e.g., grain size, fracture strength). It is now clear that certain of those factors are subtle (e.g., the distribution of small amounts of β phase). On the other hand, the details of the fine structure, particularly the genesis of the α polytypes, while influenced by some of the same factors that affect morphological development, are not likely to be as practically important. However, the results of lattice resolution and quantitative electron diffraction pattern analyses reported here emphasize the critical need to continue use of those techniques in combination. The fact that other less sensi- tive techniques are revealed to provide a misleading impression of the number and type of ordered structures present is important when considering a credible experimental basis for theories of polytypic behavior.

REFERENCES

1. S. Shinozaki and K. R. Kinsman, <u>Acta Met.</u>, in press.
2. W. F. Knippenberg, <u>Philips Research Reports</u>, <u>18</u>, 161 (1963).
3. R. Holzl, Chemetal Corp., privat communication, (1977).
4. K. R. Kinsman and S. Shinozaki, <u>Acta Met.</u>, in press.
5. H. Sata, S. Shinozaki and M. Yessik, J. Appl. Phys., <u>45</u>, 1630 (1974).
6. Y. Inomata, Z. Inove, M. Mitomo and H. Tanaka, J. Cer. Assoc. Jap., <u>77</u>, 83 (1969.
7. S. Prochazka, C. A. Johnson and R. A. Giddings, <u>G. E. Report No. SRD-75-126</u> (1974).
8. A. H. Heuer, Case-Western Reserve University, private communication, (1977).

DISCUSSION

K. H. Jack (University of Newcastle): The impression should not be given that X-ray diffraction methods are useless. Indeed the two techniques, electron microscopy and X-ray diffraction, are complementary. It is true that the latter generally gives an average structure, but it would be impossible to interpret lattice images without this information.

Although they may be described by the same Ramsdell symbols, the structures of the SiC polytypes and the Sialon polytypes are not the same. Moreover, there exist Sialon polytypes with the same metal:non-metal atom ratio but with different crystal structures. Detailed studies of the Sialon polytypes where structure is dependent on composition might well throw some light on the reasons for the existence of the SiC polytypes.

H. Palmour III (North Carolina State University): As I have watched the truly impressive high resolution TEM images shown in this and in preceding papers, I have been reminded that in an earlier Conference held here in 1964 (the first in this Series, concerned with the Role of Grain Boundaries and Surfaces in Ceramics), the first examples of direct imaging of grain boundaries in ion beam thinned materials to be shown in the U.S. were presented by my good friend and current seatmate, Dr. Max Paulus.* The impact of the then new ion beam thinning technique has now become very evident, and I want to express to Dr. Paulus the appreciation we all share for this significant contribution to our field of science.

---

*M. Paulus, pp. 183-186 in The Role of Grain Boundaries and Surfaces in Ceramics, W. W. Kriegel and H. Palmour III, Eds., Mat. Sci. Research Vol. 3, Plenum Press, New York, 1966.

THERMAL GRADIENT DEPOSITION OF SiC DIFFUSION TRACERS

R. F. Davis, J. D. Hong and M. Hon

North Carolina State University

Raleigh, North Carolina

INTRODUCTION

The periodic desire for high temperature and/or corrosion
resistant semiconductor devices has invariably caused considerable
attention to be focused on α- and β-SiC. However, the favorable
properties of refractoriness and chemical inertness also introduce
difficulties in fabrication. As such, numerous techniques
including sublimation, chemical vapor deposition, sputtering and
the several forms of liquid phase epitaxy (LPE) have been used in
an attempt to form defect-free crystals and devices (see selected
papers in ref. 1-3).

One of the forms of the last technique is the traveling solvent
method (TSM) of deposition. Gillessen and von Muench[4] have applied
the stability criterion of Nelson, et al.[5] to different LPE SiC growth
methods and shown that TSM provides a greater opportunity for stable
growth than the more common technique of crystallization from a
saturated silicon melt in a crucible.

The primary objective of the research reported herein was the
deposition and chemical reaction of ^{14}C-and ^{30}Si-containing SiC
powders with bulk single or polycrystalline pieces of this material.
However, it is now well recognized that even pressed compacts of
powders of covalent compounds such as SiC, Si_3N_4 and BN do not
sinter, i.e., they undergo no extensive macroscopic densification
without the aid of applied pressure and/or special additives[6,7].
Similarly, tracer-containing powders of these substances applied
by common methods such as painting do not chemically react with a
bulk substrate of the same composition even when heated to $\geq$ 2273K.

Interface reaction controlled diffusion is thus a possibility which
may limit the accuracy of the determination of the pertinent mass
transport parameters. This is particularly true for the covalent
compounds, as they invariably contain as oxide surface layer.

In order to achieve the objective noted above, additional
techniques for tracer deposition were explored. However, only TSM
allowed the use of minute amounts of costly tracer material without
the potential for selective evaporation of Si or radioactive
contamination.

PRINCIPLES OF TSM

TSM is based on a variant of the original temperature gradient
technique of Pfann[8] and subsequently used by this author for the
growth of silicon[8,9]. In its application, a temperature gradient
is established across a thin molten solvent zone sandwiched between
the source and the seed material. At the temperature of reaction,
the solute (e.g., SiC) enters the solvent at both interfaces;
however, more is dissolved at the hotter interface. The resultant
concentration gradient provides a driving force for the solute to
diffuse toward the cooler interface which becomes supersaturated
and deposition of the solute occurs. The solvent zone effectively
moves through the source crystal until it emerges at some external
surface or the experiment is terminated.

The requirements of a solvent are that it should readily wet
the solid, that the solubility should increase with increasing
temperature (i.e. the slope of the liquidus in the solvent-solute
system should be positive) and that the solvent be insoluble in the
solute. (This last feature is particularly significant for semiconductor materials.) The solvent must also have a low vapor
pressure at the temperature of growth, particularly if the growth
time is to be prolonged.

Cr^{10-14} has been the most widely used solvent for the growth
of SiC by TSM although Si (in the presence of I to reduce
evaporation[15]) and rare earth elements such as Dy, Gd, Th, Yb[16] and
Y[17] have also been used as the solvent. The major difficulty with
Cr is that it does not wet the surface of SiC (actually a thin
layer of SiO_2[14]) without extensive cleaning and immediate coating
by thermal evaporation or the addition to the Cr of more reactive
metals such as Si or Ta[11]. Thin zones of Cr have a tendency to
break up and thick zones are susceptible to convection currents;
thus, an optimum thickness of 100-200µm is normally used. Most of
the above mentioned rare earths are very expensive and the results
obtained with a majority of them have not been well documented.

By contrast, Perusek[17] has experienced excellent success with
the use of Y. in growing both SiC crystals and p-n junctions of this
latter material containing various dopants. The principal advantages
of using Y were found to be shorter processing time, the entrapment
of impurities and higher solubility for SiC. The considerable
oxidation rate of molten Y also apparently reduces the SiO_2 surface
films, as no precleaning is necessary. Because of the availability
of this material from several sources and in several forms and the
ability of TSM to deposit small amounts of SiC through a Y zone,
this combination of material and method were selected in order to
reach the present research objective noted previously.

EXPERIMENTAL PROCEDURE

The most common experimental arrangement used by previous
authors[10-17] has been a sandwich composed of two single crystals
of SiC and the intermediate solvent metal mounted on either an rf
heated graphite pedestal or a graphite strip situated in a vacuum
evaporator. Because of the necessity for the deposition of powdered
materials, the procedures of the present research were somewhat
altered from these more common schemes.

Powders of -325 mesh natural Si and diluted ^{14}C or ^{30}Si and
natural C were wet-mixed in a Si/C atom ratio of 1.2 in glycerol to
avoid component separation due to the difference in the respective
densities, brushed on a source crystal and heated to 1723K - 14.4 x
10^3s and 1823K - 7.2 x 10^3s under 8.1 x 10^4 Pa* of argon to form
β-SiC. No free Si or C was detected after this reaction. The
source crystal-tracer combination (with tracer side up) and an
optically polished and HF cleaned seed crystal were placed separately
on a flat graphite strip and heated to 1273K under a vacuum of 3 x
10^{-3} Pa to enhance the tenacity of a 10^{-5}m coating (determined by
weight gain) produced by thermal evaporation of Y sponge** from a
W coil at 1873K and located 0.02m above the samples. The coated
samples were subsequently arranged in the form of a sandwich and
transferred to a second graphite boat configuration, Fig. 1, containing a central cavity whose wall height surpassed that of the total
SiC/tracer/SiC assembly. When heated in an evaporator, this arrangement created a thermal gradient which allowed the controlled
deposition of tracer-containing SiC on the seed crystal.

To affect this deposition, the temperature of the Y (measured
at the top surface of the source crystal) was raised to 1798K (T_m

---

*1 atm. = 1.0133 x 10^5 Pascals
**Lunex Inc., Pleasant Valley, Iowa, 99.09% pure, produced by fusion
  of anhydrous yttrium chloride.

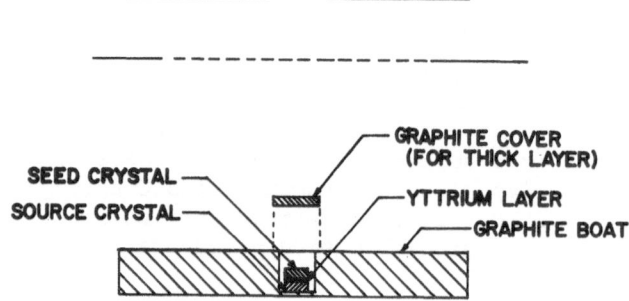

Figure 1.   Schematic of the graphite boat and SiC/tracer/SiC
            assembly used for producing the thermal gradient
            epitaxy on the seed crystal.

of Y = 1782K) for 120s under 8.1×10^4 Pa of Argon to allow the Y
to melt and react with the tracer and the seed crystal to form a
continuous layer. To achieve complete wetting, a flux of SiC down
the thermal gradient and reaction with the top crystal to form a
thin ($0.25 - 0.5 \times 10^{-6}$m) tracer film, the temperature was slowly
raised to 1858K for 6×10^2s and quickly cooled. For the deposition
of a thick (50×10^{-6}m) layer, the graphite holder was covered to
reduce heat loss and the temperature was raised from 1798K to 2073K
and maintained for 1.5×10^3s. The temperature was then slowly
decreased to 1773K over a period of 1.2×10^3s to reduce the Y
solubility at the cool end and thus deposit more tracer onto the
seed crystal. All temperatures in this research were determined by
optical pyrometry.

After cooling to room temperature, each assembly was separated
and freed of residual Y by immersion in a magnetically stirred
40% HCl solution and cleaned in methanol for 6×10^2s.

RESULTS AND DISCUSSION

In all experiments, the deposited SiC crystals were chemically
reacted with and strongly bonded to the seed crystal and to each
other such that they could not be removed without considerable
grinding. An important contribution to the efficacy of this
reaction is strongly believed to be due to the removal by the Y
of the oxide film on the surface of the seed crystal.

The thin film tracer layer deposited at 1858K contained a
mixture of very small ($\sim 10^{-6}$m) α platelets and β needles as shown

in Figs. 2(A) and (B), respectively. By contrast, the thick film deposition at 2073K, Figs. 3(A), (B), resulted in the formation of only the α form having a much larger crystal size (up to ~40 x 10^{-6}m) than found in the lower temperature deposits. The type(s) of crystals which form and their morphology, size and distribution are independent of the polytype of the starting powder on the seed crystal; these parameters appear to be only a function of the temperature of deposition. This is in agreement with the results of other SiC growth methods such as CVD[18], i.e., the presence of β is increased as the temperature of deposition is lowered. The exact reasons for this phenomenon are not known at this time.

In related TSM research using two single crystals separated by a Cr solvent, Griffiths[13] found that below 2073K the deposited crystals were also α and β; whereas, above this temperature only the seed crystal type was propagated. Combining these results with those of x-ray diffraction and structural defect considerations, this author concluded that below 2073K, SiC both dissolved in and precipitated from the Cr in the form of small, crystallographically mismatched aggregates which quickly became interlocked and subsequently limited the crystal size. Higher temperatures, however, appeared to cause solution and precipitation of individual Si and C atoms having notably higher surface mobilities which allowed enhanced growth of the forming crystals.

The above argument concerning temperature dependent aggregates vs. atomic solution and precipitation also appears valid in the case of Yt, although at different temperatures. In Figs. 3(A),(B) of the thick film, one observes small cubes throughout the microstructure which have been identified by x-ray dispersive analysis as pure Si. This is an indication that atomic Si exists in the Yt at high temperatures; this feature does not form in the lower temperature deposition, Figs. 2(A),(B). As in the case above, the resulting SiC crystals are much larger in the higher temperature case and are of one form (α) only.

Another factor in the present research which is believed to contribute to the formation of small crystals is the considerable morphological irregularity and lack of flatness caused by the presence of the tracer powder. This is thought to introduce perturbations in the temperature gradient and the thermal conductivity in the Y solvent zone which causes deposition at numerous seed crystal sites and at different rates. The interface between the layer and substrate retains the flatness of the original crystal as shown in the side view, Fig. 4, of a thick layer deposit. Occasional porosity can also be seen in the as-grown layer.

The presence of Y was also evident in very small amounts (0.3 - 2%) only on top of the tracer deposit as determined by x-ray

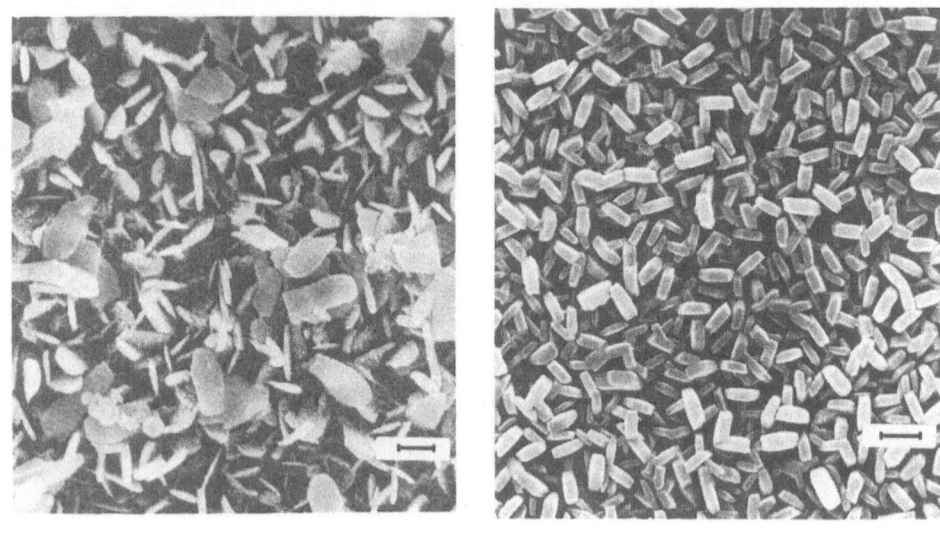

(A)   (B)

Figure 2.   Microstructure of the top surface of a thin film
            of SiC deposited at 1858K showing a) a mixture of
            α platlets and β needles and b) a clump of short
            β-needles. The former microstructure is the most
            typical. Bar = 1 micron.

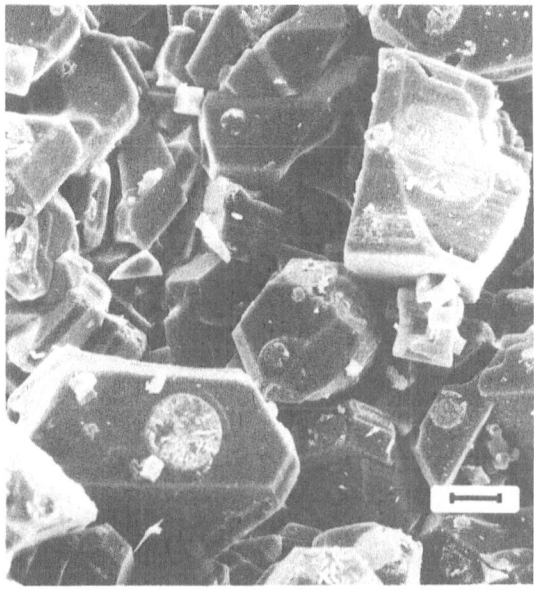

(A)

(B)

Figure 3. Microstructure of the top surface of a thick layer
of SiC deposited at 2073K; only the α-form is present.
The small cubes evident in both pictures are elemental
silicon. The microstructure does not change as a
function of position within the layer. Bar = 10 microns.

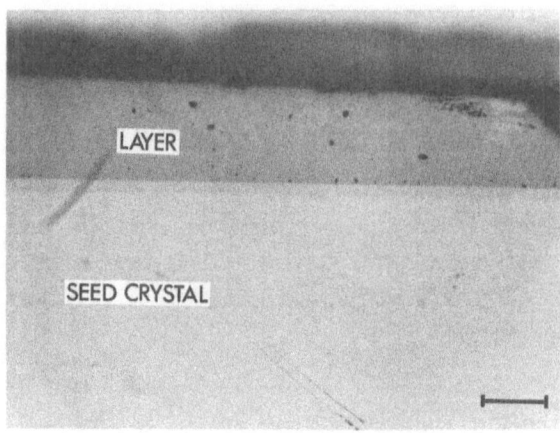

Figure 4.   Side view of a thick layer of SiC deposited at
           2073K.   Some porosity can be seen within the
           layer.   Bar = 100 microns.

dispersive techniques and electron microprobe analysis. However,
following the diffusion anneal, no Y was detected in any portion
of the assembly up to 8 x 10^{-6}m into the crystal by the above
methods as well as thermal neutron activation analysis. Additional
tests involving the direct coating of a SiC substrate with Y
followed by heating at 2150°C – 1 hour (the time of the regular
diffusion anneal) resulted in no detectable Y diffusion into the
SiC piece. Thus the present investigators conclude that Y is not
soluble in solid SiC and thus does not affect the diffusion of
either C or Si.

The wt.% solubility of SiC in Y increased as a linear function
of temperature up to 2073K. The results are described by the
equation:

$$\text{Wt.\% SiC} = 0.065T - 94.30 \tag{1}$$

In summary, the utilization of TSM provides an unique diffusion
tracer deposition technique which is particularly useful for those
covalent solids which show marginal reaction in the pure aggregate
state, contain thin films of strongly bonded species such as an
oxide and readily dissolve in a low vapor pressure molten solvent.

ACKNOWLEDGMENTS

The authors gratefully acknowledge the support of the National
Science Foundation under Grant DMR74–22481 and express their

appreciation to R. Perusek and J. Blank for their valuable ideas concerning this effort, to IBM for the x-ray dispersive analysis, and to Mrs. J. Brown for the typing of the manuscript.

BIBLIOGRAPHY

1. Silicon Carbide, A High Temperature Semiconductor, ed. by J. R. O'Conner and J. Smiltens, Pergamon Press, New York, 1960.
2. Silicon Carbide, 1968, ed. by H. K. Henisch and R. Roy as a special issue of the Mat. Res. Bul., 4, (1969).
3. Silicon Carbide, 1973, ed. by R. C. Marshall, J. W. Faust, Jr. and C. E. Ryan, U. South Carolina Press, Columbia, 1974.
4. K. Gillessen and W. van Muench, J. Cryst. Growth, 19, 263 (1973).
5. W. E. Nelson, et al., AFCRL-66-579 Rept. #3 (1966).
6. S. Prochazka in Special Ceramics 6, ed. by P. Popper, pp. 171-181, The British Ceramic Research Association, 1975.
7. C. Greshovich and J. H. Rosolowski, J. Am. Cer. Soc., 59, 336 (1976).
8. W. G. Pfann, Trans. AIME, 203, 961 (1955).
9. W. G. Pfann, Zone Melting, 2nd ed., John Wiley and Sons, Inc., New York, pp. 254-268 (1966).
10. L. B. Griffiths and A. I. Mlavsky, J. Electrochem. Soc., 111, 805 (1964).
11. M. Λ. Wright, Ibid., 112, 1114 (1965).
12. W. F. Kippenberg and G. Verspui, Phillips. Res. Rpts., 21, 113 (1966).
13. L. B. Griffiths, J. Phys. Chem. Solids., 27, 257 (1966).
14. M. Kumagawa, M. Ozeki and S. Yamada, Jap. J. App. Phys., 9, 1422 (1970).
15. W. von Muench and K. Gillessen, in ref. 3, pp. 51-57.
16. V. I. Pavlochenko, et al., Sov. Phys.-Sol. St., 10, 2205 (1969).
17. R. J. Perusek, U. S. Patent 3,669,763 (1972).
18. J. R. Weiss and R. J. Diefendorf, in ref. 3, pp. 80-91.

CONTRIBUTORS

ADVISORY COMMITTEE

R. M. Fulrath*
University of California, Berkeley, California 94720

R. S. Gordon
University of Utah, Salt Lake City, Utah 84117

R. N. Katz
Army Materials and Mechanics Research Center, Watertown,
Massachusetts 02172

G. C. Kuczynski
Notre Dame University, Notre Dame, Indiana 46556

F. F. Lange
Rockwell International Science Center, Thousand Oaks,
California 91360

M. G. McLaren
Rutgers University, Piscataway, New Jersey 08854

G. Y. Onoda
University of Florida, Gainesville, Florida 32611

J. A. Pask
University of California, Berkeley, California 94720

J. L. Pentecost
Georgia Institute of Technology, Atlanta, Georgia 30332

J. S. Reed
Alfred University, Alfred, New York 14802

R. W. Rice
U. S. Naval Research Laboratory, Washington, D. C. 20375

R. Roy
The Pennsylvania State University, University Park,
Pennsylvania 16802

A. L. Stuijts
N. V. Phillips Research Laboratories, Eindhoven, The
Netherlands

V. J. Tennery
Oak Ridge National Laboratory, Oak Ridge, Tennessee 37830

E. D. Whitney
University of Florida, Gainesville, Florida 32611

O. J. Whittemore, Jr.
University of Washington, Seattle, Washington 98195

*Deceased

663

CONTRIBUTORS

Conference Staff
North Carolina State University

Co-Chairmen

   R. F. Davis, Associate Professor, Materials Engineering and
      Engineering Research Services Division
   H. Palmour III, Research Professor, Engineering Research
      Services Division
   T. M. Hare, Research Associate, Engineering Research Services
      Division

Honorary Chairman

   W. W. Kreigel, Professor Emeritus, Materials Engineering

Arrangements

   Bruce Winston, Continuing Education

Publicity

   Mary N. Yionoulis, Engineering Publications and Information,
      Office of Information Services

Administrative Liaison

   R. F. Stoops, Engineering Research Services Division
   W. W. Austin, Materials Engineering
   R. A. Mabry, Continuing Education

CONFERENCE SESSION CHAIRMEN

R. F. Davis
   North Carolina State University, Raleigh, North Carolina 27650

V. D. Frechette
   Alfred University, Alfred, New York 14802

T. M. Hare
   North Carolina State University, Raleigh, North Carolina 27650

W. J. Huppman
   Max-Planck Institute for Metals Research, Stuttgart, West
   Germany

W. W. Kreigel
  3501 26th Place W., Seattle, Washington 98199

H. Palmour III
  North Carolina State University, Raleigh, North Carolina 27650

J. A. Pask
  University of California, Berkeley, California 94720

J. L. Pentecost
  Georgia Institute of Technology, Atlanta, Georgia 30332

AUTHORS

L. A. Addington
  Eagle-Picher Research Lab, Miami, Oklahoma 74354

E. M. Anderson
  Babcock & Wilcox,Co., Lynchburg, Virginia 24505

H. U. Anderson
  University of Missouri, Rolla, Missouri 65401

P. F. Becher
  Naval Research Laboratory, Washington, D. C. 20375

B. A. Bender
  Lehigh University, Bethlehem, Pennsylvania

J. V. Biggers
  Pennsylvania State University, University Park,
  Pennsylvania 16802

W. R. Bitler
  Pennsylvania State University, University Park,
  Pennsylvania 16802

H. K. Bowen
  Massachusetts Institute of Technology, Cambridge,
  Massachusetts 02139

R. C. Bradt
  Pennsylvania State University, University Park,
  Pennsylvania 16802

R. M. Cannon
  Massachusetts Institute of Technology, Cambridge,
  Massachusetts 02139

W. R. Cannon
    Massachusetts Institute of Technology, Cambridge,
    Massachusetts 02139

T. Carbone
    Alfred University, Alfred, New York 14802

F. M. A. Carpay
    Philips Research Laboratory, Eindhoven, The Netherlands

A. Choudry
    University of Rhode Island, Kingston, Rhode Island 02881

U. Chowdhry
    CRD, E. I. duPont de Nemours & Co., Wilmington, Delaware

N. O. Clark
    Rutgers University, New Brunswick, New Jersey 08903

D. R. Clarke
    Rockwell International Science Center, Thousand Oaks,
    California 91360

R. L. Coble
    Massachusetts Institute of Technology, Cambridge,
    Massachusetts 02139

T. J. Curci
    Pennsylvania State University, University Park,
    Pennsylvania 16802

I. B. Cutler
    University of Utah, Salt Lake City, Utah 84112

H. H. Davis
    Babcock & Wilcox Co., Lynchburg, Virginia 24505

R. F. Davis
    North Carolina State University, Raleigh,
    North Carolina 27650

L. C. De Jonghe
    Cornell University, Ithaca, New York 14853

R. C. Garvie
    CSIRO, Division of Tribophysics, Melbourne, Australia

A. Gatti
    General Electric Company, Space Sciences Laboratory,
    Philadelphia, Pensylvania 19101

G. E. Gazza
    Army Materials and Mechanics Research Center, Watertown,
    Massachusetts 02172

B. B. Ghate
    Bell Telephone Laboratories, Inc., Allentown,
    Pennsylvania 18103

P. J. Gielisse
    University of Rhode Island, Kingston, Rhode Island 02881

E. Goo
    Cornell University, Ithaca, New York 14853

R. S. Gordon
    University of Utah, Salt Lake City, Utah 84112

T. J. Gray
    Olin Corp., New Haven, Connecticut 06511

S. M. Han
    Hanyang University, Research Institute of Industrial Sciences,
    Seoul, South Korea

D. L. Hankey
    Pennsylvania State University, University Park,
    Pennsylvania 16802

T. M. Hare
    North Carolina State University, Raleigh, North Carolina 27650

R. J. Hart
    Bell Telephone Laboratories, Inc., Allentown,
    Pennsylvania 18103

L. L. Hench
    University of Florida, Gainesville, Florida 32611

M. Hoch
    University of Cincinnati, Cincinnati, Ohio 45221

R. L. Holman
    Xerox Corporation, Rochester, New York 14644

M. Hon
    North Carolina State University, Raleigh, North Carolina 27650

J. D. Hong
    North Carolina State University, Raleigh, North Carolina 27650

M. L. Huckabee
    North Carolina State University, Raleigh, North Carolina 27650

R. R. Hughan
    CSIRO, Division of Tribophysics, Melbourne, Australia

W. J. Huppmann
    Max-Planck Institute for Metals Research, Stuttgart, W. Germany

K. H. Jack
    University of Newcastle-upon-Tyne, Newcastle-upon-Tyne, England

A. D. Jatkar
    University of Utah, Salt Lake City, Utah 84112

D. L. Johnson
    Northwestern University, Evanston, Illinois 60201

D. W. Johnson, Jr.
    Bell Laboratories, Murray Hill, New Jersey 07974

O. W. Johnson
    University of Utah, Salt Lake City, Utah 84112

P. F. Johnson
    University of Florida, Gainesville, Florida 32611

R. N. Katz
    Army Materials and Mechanics Research Center, Watertown,
    Massachusetts 02172

T. Kim
    University of Rhode Island, Kingston, Rhode Island 02881

Y. S. Kim
    Bell Telephone Laboratories, Inc., Allentown,
    Pennsylvania 18103

K. R. Kinsman
    Electric Power Research Institute, Palo Alto,
    California 94303

S. K. Kurtz
    Philips Laboratory, Briarcliff Manor, New York 10510

F. F. Lange
    Rockwell International Science Center, Thousand Oaks,
    California 91360

S. Lukasiewicz
    Alfred University, Alfred, New York 14802

L. G. McCoy
    Battelle Columbus Laboratories, Columbus, Ohio 43201

R. W. McCulloch
    Oak Ridge National Laboratory, Oak Ridge, Tennessee 37830

B. J. McEntire
    University of Utah, Salt Lake City, Utah 84112

B. A. McFarland
    Naval Research Laboratory, Washington, D. C. 20375

M. G. McLaren
    Rutgers University, Piscataway, New Jersey 08854

W. S. Machin
    Babcock & Wilcox Co., Lynchburg, Virginia 24505

G. L. Messing
    University of Florida, Gainesville, Florida 32611

G. R. Miller
    University of Utah, Salt Lake City, Utah 84112

M. L. Miller
    University of Utah, Salt Lake City, Utah 84112

C. S. Morgan
    Oak Ridge National Laboratory, Oak Ridge,
    Tennessee 37830

P. E. D. Morgan
    University of Pittsburgh, Pittsburgh, Pennsylvania 15261

H. Mörtel
    University of Erlangen-Nurnberg, Erlangen, Germany

K. M. Nair
    University of Cincinnati, Cincinnati, Ohio 45221

D. E. Niesz
    Battelle Columbus Laboratories, Columbus, Ohio 43201

M. J. Noone
    General Electric Company, Philadelphia, Pennsylvania 19101

G. Y. Onoda
    University of Florida, Gainesville, Florida 32611

W. R. Ott
    Rutgers University, New Brunswick, New Jersey 08903

H. Palmour III
    North Carolina State University, Raleigh, North Carolina 27650

R. T. Pascoe*
    CSIRO, Division of Tribophysics, Melbourne, Australia

J. A. Pask
    University of California, Berkeley, California 94720

A. E. Pasto
    Oak Ridge National Laboratory, Oak Ridge, Tennessee 37830

M. Paulus
    Laboratoire d'Etude et de Synthese des Microstructures, CNRS,
    Paris, France

S. Pejovnik
    University of Ljubljana, Ljubljana, Yugoslavia

G. W. Phelps
    Rutgers University, New Brunswick, New Jersey 08903

R. L. Pober
    Massachusetts Institute of Technology, Cambridge,
    Massachusetts 02139

R. A. Potter
    Babcock & Wilcox Co., Lynchburg, Virginia 24505

S. Prochazka
    General Electric Co., Corporate Research and Development
    Center, Schenectady, New York 12301

J. S. Reed
    Alfred University, Alfred, New York 14802

R. W. Rice
    Naval Research Laboratory, Washington, D. C. 20375

W. A. Rocco
    General Electric Company, Schenectady, New York 12301

*Deceased

M. D. Sacks
   University of California, Berkeley, California 94720

C. Scott
   Alfred University, Alfred, New York 14802

T. M. Shaw
   University of California, Berkeley, California 94720

S. Shinozaki
   Ford Motor Company, Dearborn, Michigan 48121

J. H. Sommers
   Columbia University, New York, New York

J. J. Stiglich
   Eagle-Picher, Miami, Oklahoma 74354

S. Strijbos
   Philips Research Laboratories, Eindhoven, The Netherlands

L. Tarhay
   Pennsylvania State University, University Park,
   Pennsylvania 16802

V. J. Tennery
   Oak Ridge National Laboratory, Oak Ridge, Tennessee 37830

G. Thomas
   University of California, Berkeley, California 94720

P. A. Vermeer
   Delft University of Technology, Delft, The Netherlands

R. W. Vest
   Purdue University, West Lafayette, Indiana 47907

A. V. Virkar
   University of Utah, Salt Lake City, Utah 84112

E. D. Whitney
   University of Florida, Gainesville, Florida 32611

O. J. Whittemore
   University of Washington, Seattle, Washington 98195

R. R. Wills
   Battelle Columbus Laboratories, Columbus, Ohio 43201

M. F. Yan
     Bell Laboratories, Murray Hill, New Jersey 07974

C. E. Zimmer
     Babcock & Wilcox Co., Lynchburg, Virginia 24505